WILLIAM JOHN MACQUORN RANKINE

MISCELLANEOUS SCIENTIFIC PAPERS

FROM THE TRANSACTIONS AND PROCEEDINGS OF THE ROYAL AND OTHER SCIENTIFIC AND PHILOSOPHICAL SOCIETIES AND THE SCIENTIFIC JOURNALS

WITH
A MEMOIR OF THE AUTHOR
BY P. G. TAIT

Elibron Classics
www.elibron.com

Elibron Classics series.

ISBN 1-4021-7192-7 (paperback)
ISBN 1-4021-0745-5 (hardcover)

This Elibron Classics Replica Edition is an unabridged facsimile of the edition published in 1881 by Charles Griffin and Company, London.

MISCELLANEOUS

SCIENTIFIC PAPERS.

H. ADLARD, Sc.

CHARLES GRIFFIN & COMPANY, LONDON.

MISCELLANEOUS

SCIENTIFIC PAPERS:

BY

W. J. MACQUORN RANKINE, C.E., LL.D., F.R.S.,

LATE REGIUS PROFESSOR OF CIVIL ENGINEERING AND MECHANICS IN THE UNIVERSITY OF GLASGOW.

FROM THE TRANSACTIONS AND PROCEEDINGS OF THE ROYAL AND OTHER SCIENTIFIC AND PHILOSOPHICAL SOCIETIES, AND THE SCIENTIFIC JOURNALS.

WITH

A MEMOIR OF THE AUTHOR

BY P. G. TAIT, M.A.,

PROFESSOR OF NATURAL PHILOSOPHY IN THE UNIVERSITY OF EDINBURGH.

EDITED BY

W. J. MILLAR, C.E.,

SECRETARY TO THE INST. OF ENGINEERS AND SHIPBUILDERS IN SCOTLAND.

With Portrait, Plates, and Diagrams.

LONDON:

CHARLES GRIFFIN AND COMPANY,

STATIONERS' HALL COURT.

MDCCCLXXXI.

GLASGOW
PRINTED BY BELL AND BAIN, 41 MITCHELL STREET

PREFACE.

THE republication of a Selection of the Papers of the late Professor Macquorn Rankine was originally projected by several of his personal friends shortly after his decease, the object being to combine, in a suitable volume-form, papers which were to be found only in the Records of Scientific Societies, and in the Scientific and Engineering Journals, and thus to present to the many admirers of the talented author a memorial of his great worth and ability.

Introductory to the Selection of Papers now published is an exhaustive Memoir by Professor Tait, who kindly consented to embody in this form the main features of Rankine's life, together with his recollections of one with whom he had been intimately associated as a fellow-worker. There remains, therefore, for the Editor merely to point out, briefly, the principle that has guided him in making the selection from the papers placed in his hands for the purpose of republication.

The object, then, kept in view has been the preservation of such papers as are most characteristic of their author in his capacity of a scientific and mathematical inquirer. Professor Rankine was not, nor did he claim to be, a popular writer; his command of mathematical expression was such, that he naturally embodied his reasoning and conclusions in symbols. All his writings, however, are marked by a power of statement so clear and logical, that the reader, even should he fail entirely to follow the demonstrations, cannot but be benefited in the attempt to master them.

Besides the papers in the present Collection, Professor Rankine contributed many others to Scientific and Engineering Societies and Journals. A number of these, from their nature, (descriptions of works and of machines,) were but of passing interest. The papers now published are of permanent value, and deal rather with general scientific principles and their applications to practice.

The leading characteristics of Professor Rankine's writings are too well-known to require comment here. One special feature, however, in his method of treatment, may be pointed out—viz., the carefully arranged division of the subject into sections. Starting with a general statement of the object of the paper, he advances by degrees in the argument, giving full reference to the subordinate parts of the paper, which are marked by numerals or letters to distinguish them, thus progressing with logical precision until the conclusion is reached. Another noteworthy feature may also be referred to—viz., that of the introduction of new and suitable words or phrases, proposed by the author, to convey more clearly his meaning. Lastly, we observe throughout all Professor Rankine's writings the most minute accuracy of statement, and the most scrupulously honourable care to give to all fellow-workers in the same field with himself their just due, whilst pointing out what he considers to be original on his own part.

The principal papers in the volume are those relating to Thermodynamics and to Hydrodynamics, where such subjects as the Action of Heat in the Steam-Engine, and the Forms of Waves and Water-Lines of Ships, are discussed at length—the scientific and mathematical investigation of these questions being perhaps most eminently characteristic of Professor Rankine.

The papers have been grouped into three Divisions, so as to bring as nearly as possible kindred subjects together; and in every case the name of the Society or Journal through which the paper was originally brought forward, together with the date of its publication, has been given.

It need hardly be added, that the papers appear without change of any kind. In their present form they stand precisely as they finally left their author's pen; and no pains have been spared to ensure perfect accuracy in the reproduction.

The Editor, like many others, has a grateful remembrance of Professor Rankine, having enjoyed the great privilege of being one of his students in the old College of Glasgow, and having afterwards had the advantage of his friendship in many ways. Rankine's lectures, although simpler than his text-books, were marked by the same clearness of arrangement, and were enforced by his distinct and vigorous enunciation, and admirably illustrated by carefully prepared diagrams. As chairman of the meetings of the societies of which he was president, his methodical habits and business qualifications were of marked service. These qualities were also evidenced in the drawing up of reports in committee; one of the last services which he rendered to engineering science being in connection with an experimental inquiry on safety valves on behalf of the Institution of Engineers and Shipbuilders in Scotland—his decease, unfortunately, occurring before the completion of the experiments.

The Papers classed under Part I. relate more or less to Temperature and Elasticity. The First, originally published in 1849, gives an approximate equation for the elasticity of vapour in contact with its liquid, (this equation appears to have been one of the results obtained by Professor Rankine whilst investigating the molecular constitution of matter,) and concludes with the statement of a proposition borne out by experiment, which "may be safely and usefully applied in practice."

In the Second Paper, published in the same year, is given a formula for calculating the expansion of liquids by heat. This formula, the author states, he had found useful whilst considering the comparative volumes of liquids at various temperatures.

The Third Paper, on "The Centrifugal Theory of Elasticity, as applied to Gases and Vapours," published in 1851, shows how the laws of the pressure and expansion of gases may be deduced from the hypothesis of molecular vortices. The investigation was begun in 1842, but was laid aside on account of the want of experimental data; it was again resumed after the publication of the results of M. Regnault's experiments on gases and vapours, and laid before the Royal Society of Edinburgh in February, 1850. The hypothesis is defined to be "that which assumes *that each atom of matter consists of a nucleus or central point enveloped by an elastic atmosphere, which is retained in its position by attractive forces; and that the elasticity due to heat arises from the centrifugal force of these atmospheres revolving or oscillating about their nuclei or central points.*" After showing that somewhat similar ideas had been entertained by philosophers at different times, there follows a supposition which, he says, "so far as I am aware, is peculiar to my own researches. It is this, *that the vibration which, according to the undulatory hypothesis, constitutes radiant light and heat, is a motion of the atomic nuclei or centres, and is propagated by means of their mutual attractions and repulsions.*" Tables are given showing the closeness of agreement between the formulæ made use of, and the experimental results.

In the Fourth Paper we have an extension of the preceding one; by means of a fresh investigation the complete applicability of the hypothesis of molecular vortices to all substances in all conditions is shown, and this demonstration is followed by the deduction from that hypothesis of the law of the equivalence of heat and power.

The Fifth Paper deals with the Laws of Elasticity in reference to the strength of structures—the relations between pressures and strains; it shows how the Laws of Elasticity are simplified by adopting the supposition of atomic centres of force, but also points out that this supposition requires modification, and by means of the hypothesis of molecular vortices simplifies the investigation of these Laws.

In the Sixth Paper the distinction between *Strain* and *Stress* is made clear, and a nomenclature adopted (from Greek equivalents) descriptive of their relations. Crystalline forms are considered, together with their action on light.

The Seventh Paper treats of "The Vibrations of Plane-Polarised Light." The principles laid down in the paper are shown to be incompatible with the "idea of a luminiferous ether enveloping ponderable particles;" inasmuch as that "the luminiferous medium is a system of atomic nuclei or centres of force, whose office is to give *form* to matter; while the atmospheres by which they are surrounded give of themselves merely *extension*."

In the Eighth Paper an attempt is made to diminish the difficulties attending the undulatory theory of light, by proposing a theory of oscillations round axes, instead of the theory of vibrations.

In the Ninth Paper, we have a mathematical investigation into the relations existing between the velocity of sound in elastic substances and the elasticity of the materials.

Part II. relates principally to Energy and the Mechanical Action of Heat. The first Paper of this division (No. X.) is entitled "The Reconcentration of the Mechanical Energy of the Universe"—a remarkable speculation, laid before the British Association at Belfast, and published in 1852, alluded to by Professor Jevons in his *Principles of Science*, and acknowledged by Sir Wm. Thomson to be very characteristic of Rankine. This paper, after referring to all experimental evidence as being in favour of the doctrine of the mutual convertibility of the different kinds of the physical energy of the universe, and of the tendency towards a uniform temperature of matter, goes on to point out how it is conceivable that ultimately the diffused energy may be gathered into foci, and renewed stores of chemical power produced from the now inert compounds when passing through the intense heat of these foci.

The next Paper deals with the Law of the Transformation of Energy, and the ratio of Work done to Energy expended in various forms of engines; and is followed by a comprehensive paper entitled "Outlines of the Science of Energetics," treating of the laws of physical phenomena—the Science of Mechanics ("the only example yet existing of a complete physical theory")—the use of hypotheses—definitions of Energy, Work, &c., and Efficiency of Engines.

The Thirteenth Paper is mainly descriptive of the use of the term *Energy*, and gives Professor Rankine's reason for his introduction of the phrase "*Potential Energy*."

No. XIV., with its supplements XV.–XIX., on "The Mechanical Action of Heat," is an important contribution of considerable length to the Royal Society of Edinburgh, extending over the years 1850–1853. It is based upon the hypothesis of molecular vortices, and relates chiefly to the "mutual conversion of heat and mechanical power, by means of the expansion and contraction of gases and vapours."

In the Twentieth Paper the various conditions existing in a heated substance are shown geometrically by curves, a method adopted first, it is stated, by James Watt in his Steam-engine Indicator. The efficiency of thermodynamic engines is considered, and Stirling and Ericsson's hot-air engines described.

The next Paper is on "Formulæ for the Maximum Pressure and Latent Heat of Vapours," followed by one on "The Density of Steam," in which the general equation of thermodynamics, stated in paper No. XIV., is again given, to show the connection existing between it and the law of the density of steam.

In Paper No. XXIII. the Two Laws of Thermodynamics are stated, and it is shown that the derivation of the Second Law from steady molecular motion (*e.g.*, in circular streams or in circulating streams of any figure), as given in previous papers, may be more simply effected than by the methods adopted in these papers.

The following Paper, published in *The Engineer* in 1867, refers to the want of popular illustrations of The Second Law of Thermodynamics, and explains the nature of the two laws, with the particular questions to which they are respectively applicable.

"*The First Law* informs us that when mechanical work is done by means of heat, a quantity of heat disappears. . . . To calculate this disappearance of heat, the work done must be sensibly *external*, and subject to direct measurement."

"*The Second Law* informs us how to deduce the whole amount of work done, *internal* and *external*, from the knowledge which we have of the *external* work."

An illustration is given by the expansion of a perfect or sensibly perfect gas; but it is pointed out that it is different when we have to deal with fluids in the act of evaporating, instanced by a case showing where the *second law* is applicable.

Paper No. XXVI. is on "The Working of Steam in Compound Engines," defines such engines, and states their advantages, with rules for the construction of indicator-diagrams.

This is followed by Papers on "The Theory of Explosive Gas-Engines," and on "The Explosive Energy of Heated Liquids." In the first of these it is shown that, in calculations respecting the practical use of heat-engines, it is convenient to use pressures and volumes rather than temperatures. The mixtures of gas and air most suitable for gas-engines are also given.

The Papers in Part III. relate to Wave-Forms, Propulsion of Vessels, Stability of Structures, &c.

The first of the series, No. XXIX., which is entitled "On the Exact Form of Waves near the Surface of Deep Water," shows that the form of such waves is trochoidal, and states that this form was first pointed out by Mr. Scott Russell.

The next Paper, "On Plane Water-Lines," investigates the curves suitable for the water-lines of a ship. Water-line curves are designated Neoids, Cyclogenous Neoids (or water-line curves generated by circles), Oögenous Neoids (or those generated by oval bodies), and Lissoneoids, or water-lines of smoothest gliding. It is noticed that although, from lengthened practice in the art of shipbuilding, the forms of water-lines have attained a high degree of excellence, yet that this is due rather to empirical means, than to a knowledge of general principles. The system of Chapman is shown to be wholly empirical, consisting of parabolic forms; and Mr. Scott Russell's is instanced as the first useful theory of ships' water-lines, being based on wave figures. The various forms of water-lines are then considered in reference to their fitness for different classes of vessels.

Other two Papers follow, the first of which is intended to assist those who are not familiar with the higher mathematics in understanding the subject of Stream-Lines. A stream-line is defined as the line traced by a particle in a current of fluid, and an elementary method is given for determining circular stream-lines, a subject mathematically investigated in the preceding paper. The other Paper is an investigation "to determine the relations which must exist between the laws of the elasticity of any substance, whether gaseous, liquid, or solid, and those of the wave-like propagation of a finite longitudinal disturbance in that substance." A Paper on "The Theoretical Limit of the Efficiency of Propellers" follows, showing the theoretical limit of efficiency which improvements in propellers may attain; states the formulæ for reaction and effective power, shows at what relative velocities the propeller is most effective, and compares the advantages of various forms of propeller, with numerical examples.

Paper No. XXXIV., "On the Design and Construction of Masonry Dams," originally consisted of a report to the municipality of Bombay, made in 1870, in reference to proposed extensions of the Water-Works there, and was afterwards published in *The Engineer*.

This Paper enters into the question of stability of structures, showing the best and most economical form which a high masonry reservoir wall should have, and is of value to the civil engineer when proposing to adopt a masonry or concrete wall, instead of an ordinary embankment, for reservoir purposes.

Papers Nos. XXXV. and XXXVI. are extensions of methods adopted by Professor Rankine in his *Manual of Applied Mechanics*, in connection with the stability of structures of various figures, such as Frames and Arches; and the last paper of the series, No. XXXVII., is a mathematical demonstration of a property of certain curves, bearing on the forms of the slopes of wave-crests.

In conclusion, the Editor desires to acknowledge the courtesy shown by the executive officers of the various Societies and Journals to which the papers, thus brought together, were originally contributed; and to express his thanks—not only for the permission, readily accorded, to republish—but also for the kindness which supplied, in many instances, copies of the Papers selected.

W. J. MILLAR.

GLASGOW,
October, 1880.

CONTENTS.

PART I.

PAPERS RELATING TO TEMPERATURE, ELASTICITY, AND EXPANSION OF VAPOURS, LIQUIDS, AND SOLIDS.

PART II.

PAPERS RELATING TO ENERGY AND ITS TRANSFORMATIONS, THERMODYNAMICS, MECHANICAL ACTION OF HEAT IN THE STEAM-ENGINE, &c.

PART III.

PAPERS RELATING TO WAVE FORMS, PROPULSION OF VESSELS, STABILITY OF STRUCTURES, &c.

MEMOIR.

THE life of a genuine scientific man is, from the common point of view, almost always uneventful. Engrossed with the paramount claims of inquiries raised high above the domain of mere human passions, he is with difficulty tempted to come forward in political discussions, even when they are of national importance; and he regards with surprise, if not with contempt, the petty municipal squabbles in which local notoriety is so eagerly sought. To him the discovery of a new law of nature, or even of a new experimental fact, or the invention of a novel mathematical method, no matter who has been the first to reach it, is an event of an order altogether different from, and higher than, those which are so profusely chronicled in the newspapers. It is something true and good for ever, not a mere temporary outcome of craft or expediency. With few exceptions, such men pass through life unnoticed by, almost unknown to, the mass of even their educated countrymen. Yet it is they who, far more than any autocrats or statesmen, are really moulding the history of the times to come. Man has been left entirely to himself in the struggle for creature comforts, as well as for the higher appliances which advance civilisation; and it is to science, and not to so-called statecraft, that he must look for such things. Science can and does provide the means; statecraft can but more or less judiciously promote, regulate, or forbid their use or abuse. One is the lavish and utterly unselfish furnisher of material good, the other the too often churlish and ignorant dispenser of it. In the moral world their analogues are charity and the relieving officer! So much it is necessary to say for the sake of the general reader; to the world of science no apology need be made. In *it* Rankine's was and is a well-known name.

It is high eulogy, but strictly correct, to say that Rankine holds a prominent place among the chief scientific men of the last half century. He was one of the little group of thinkers to whom, after the wondrous Sadi Carnot, the world is indebted for the pure science of modern thermodynamics. Were this all, it would be undoubtedly much. But his services to applied science were relatively even greater. By his admirable teaching, his excellent text-books, and his original memoirs, he has done more than any other man of recent times for the advancement of British Scientific Engineering. He did not, indeed, himself design or construct gigantic structures, but he taught, or was the means of teaching, that invaluable class of men to whom the projectors of such works entrust the calculations on which their safety as well as their efficiency mainly depend. For behind the great architect or engineer, and concealed by his portentous form, there is the real worker, without whom failure would be certain. The public knows but little of such men. Not every von Moltke has his services publicly acknowledged and rewarded by his imperial employer! But he who makes possible the existence of such men confers lasting benefit on his country. And it is quite certain that Rankine achieved the task.

William John Macquorn Rankine was born in Edinburgh on the 5th July, 1820. He was a Scot of Scots. His father was descended from the Rankines of Carrick and the Cochranes of Dundonald. His maternal grandfather was Grahame of Drumquhassle, a descendant of the Grahams of Dougalston. In Rankine's MSS. there is to be found a tracing of the various steps of his pedigree from Robert the Bruce. His father, David Rankine, was in youth Lieutenant in the 21st Regiment (Rifle Brigade); but, as will be seen by what follows, was also a man of great general information, especially in practical matters. He was employed in later life in constructing railways, and afterwards became Secretary to the Caledonian Railway Company. Rankine repeatedly notes in his journal the hints and instruction he had received from his father. He was profoundly attached to his parents; and one of the most touching notes in his journal is the brief record of his lasting obligations to them for early instruction in the fundamental principles of the Christian religion and the character of its Founder.

It will be convenient to give, in the first place, a brief sketch of Rankine's career, and to reserve for a time such comments upon his more important investigations and treatises as would materially interfere with the continuity of the sketch.

From Rankine's private journal it appears that his first introduction to arithmetic and elementary mathematics, mechanics, and physics was obtained mainly from his father. He attended Ayr Academy in 1828-9, and Glasgow High School in 1830. After this he seems to have been for some years privately instructed in Edinburgh, his state of health preventing his being sent to a public school. In December, 1834, his uncle, Archibald

Grahame, presented him with a copy of Newton's *Principia*, which he read carefully. He remarks—"This was the foundation of my knowledge of the higher mathematics, dynamics, and physics. My knowledge of the higher mathematics was obtained chiefly by private study." About this period he paid much attention to the theory of music. In 1836 he studied Practical and Theoretical Chemistry under David Boswell Reid; and in November of the same year entered Edinburgh University. He there attended the Natural Philosophy course under Professor Forbes, and gained (before completing his seventeenth year) the Gold Medal for an essay on the "Undulatory Theory of Light." In the summer of 1837, he studied Natural History under Professor Jameson, and Botany under Professor Graham. He attended the Natural Philosophy Class a second time in 1837-8, and obtained an extra prize for an essay on "Methods in Physical Investigation." He records in his journal, that in 1836-8, during leisure, he read much metaphysics, chiefly Aristotle, Locke, Hume, Stewart, and Degerando. I have learned from himself that about this period he "wasted" a great deal of time in the fascinating but too often delusive pursuit of "Theory of Numbers."

In 1837-8 he made his first acquaintance with the practice of engineering, having assisted his father in superintending the works of the Leith branch of the Edinburgh and Dalkeith Railway. In the latter year he became a pupil of the late Sir John Macneill, C.E., having among his fellow-pupils many who have since risen to eminence. His journal records the names of G. W. Hemans, J. W. Bazalgette, W. R. Le Fanu, Matthew Blackiston, John Moffat, and Jonas S. Stawell.

During the succeeding four years he was employed by Macneill on various surveys, and schemes for river improvements, water-works, and harbours for Ireland. Also, for some time, on the Dublin and Drogheda Railway. While engaged on this railway in 1841, he contrived the method of setting out curves "by chaining and angles at circumference combined," which has since been known as "Rankine's method."

In 1842 appeared his first published work, a pamphlet entitled *An Experimental Inquiry into the advantage of Cylindrical Wheels on Railways.* This was based upon experiments suggested to him by his father, and carried out by them together.

In the same year Queen Victoria visited Edinburgh for the first time, and Rankine was charged with the superintendence of the erection of the huge bonfire which blazed on the top of Arthur's Seat. He constructed it with radiating air passages under the fuel, and succeeded, as he complacently records, in partially vitrifying the rock!

During this and the succeeding year he sent several papers to the Institute of Civil Engineers, for some of which prizes were awarded to him. He records that most of them were based on suggestions by his father, especially that on the "Fracture of Axles." He showed that such fractures arose through gradual deterioration or fatigue, involving the gradual extension inwards of a crack originating at a square-cut shoulder. In this paper the importance of continuity of form and fibre was first shown, and the hypothesis of spontaneous crystallisation was disproved.

In 1844-5 he was employed under Locke and Errington on the Clydesdale Junction Railway project; and afterwards, till 1848, on various schemes promoted by the Caledonian Railway Company. In 1845-6 he engineered a project for Edinburgh and Leith Water-works, which was defeated by the opposition of the Edinburgh Water Company.

About 1848 he seems to have commenced that extensive series of researches on molecular physics which occupied him at intervals during the rest of his life, and which constitutes his chief claim to distinction in the domain of pure science. The first paper he published on the subject, with the title "Elasticity of Steam," appeared in the *Edinburgh New Philosophical Journal* in July, 1849; and at the end of that year he sent to the Royal Society of Edinburgh his great paper on the "Mechanical Action of Heat." It was not, however, read to the Society till February, 1850. On the contents of this, and his subsequent papers dealing with similar subjects, some remarks will be made below. In July, 1850, he read to the British Association at Edinburgh another paper on a closely connected subject, "Elastic Solids."

In 1852 the Loch Katrine Water-works scheme for the supply of Glasgow was revived by Rankine and John Thomson. This scheme, now successfully carried out, was first proposed by Lewis Gordon and Laurence Hill, Junior.

In 1853, one of Rankine's most characteristic papers in pure science, "On the General Law of Transformation of Energy," was read by him to the Glasgow Philosophical Society. In the same year, along with the late J. R. Napier, he projected and patented a new form of air-engine. The patent was afterwards abandoned.

He was now elected to the Fellowship of the Royal Society, and sent to that body his next great paper on Thermodynamics—viz., "On the Geometrical Representation of the Expansive Action of Heat," which was printed in the *Philosophical Transactions.*

From January to April, 1855, he acted in Glasgow University as substitute for Professor Lewis Gordon, on whose resignation he was appointed to the Chair of Engineering, which he held till his death. His inaugural discourse, delivered on Dec. 10 of the last-mentioned year, bore the title "*De concordiâ inter Scientiarum Machinalium Contemplationem et Usum.*" In this year he wrote, among several contributions to Nichol's *Cyclopædia*, an article on "Heat, Mechanical Action of," the earliest formal treatise on Thermodynamics in the English language. In 1856, the preparation of his course of lectures led him to the invention of some remarkable methods connected with Transformation of Structures. These are based on the discovery of "reciprocal diagrams" of frames and forces, since greatly extended and simplified by Clerk-Maxwell. The remarkable storm which occurred on February 7 of this year, directed his inquiries to the "Stability of Chimneys," on which he has published a valuable article.

In 1857, he resigned the associateship of the Institute of Civil Engineers; and shortly afterwards, on the establishment of the Institute of Engineers in Scotland, he delivered the opening address as first President. At this time he was busily engaged on a *Treatise on Shipbuilding*, his *Manual of Applied Mechanics*, an article on the same subject for the *Encyclopædia Britannica*, and an investigation (based on J. R. Napier's experiments) of the theory of skin-resistance of ships. He also sent to the French Academy of Sciences a memoir,—"*De l'Equilibre intérieure d'un corps solide, élastique, et homogène.*"

In July, 1859, an offer of service was sent to the Lieutenancy by the "Glasgow University Rifle Volunteers." It was accepted in October, and Rankine received his commission as Captain. He spent the greater part of November at the Hythe School of Musketry, and, on his return, instructed the officers and sergeants of his corps. In the same year appeared his valuable *Manual of the Steam-Engine and other Prime Movers.* In 1860, he was made Senior Major, and commanded the second battalion of his regiment at the memorable Volunteer Review, held by the Queen in the Queen's Park, Edinburgh.

In 1861, he finished his *Manual of Civil Engineering*, which was published early in the following year. At the International Exhibition in 1862, at London, he was a Juror in Class VIII., "Machinery in General."

In 1864, he resigned his commission in the Volunteers, "finding it impossible to attend at once to duties as field-officer and as professor, to engineering business, and to literary work, especially *Treatise on Shipbuilding.*"

In 1865, he was appointed Consulting Engineer to the Highland and Agricultural Society of Scotland; and became a regular contributor to the *Engineer*, in which many excellent articles of his appeared.

In 1866, was published his *Treatise on Shipbuilding, Theoretical and Practical.* Though four names were announced on the title page as joint authors, by far the greater part of the work was written by Rankine; but the proofs were revised by all four.

1869 produced *Machinery and Millwork*, the fourth of Rankine's great engineering treatises. The other three had then reached their eighth, sixth, and fifth editions respectively. In his journal for this year occurs the following note:—

"*Sept.* 16. Thomas Graham, Master of the Mint, died [son of a cousin of my mother's father]. I applied for vacancy to Chancellor of Exchequer (Lowe). Application well supported by friends, and civilly received; but the office was virtually abolished, being conjoined with the Chancellorship of the Exchequer."

Rankine lost his father in 1870, and his mother in the following year. Both had passed the age of seventy. The loss of his parents, to whom he was so fondly attached, seems to have accelerated the development of the illness which had for some years been growing upon him. He was well enough, in 1871, to contribute most valuable matter to the proceedings of the "Committee on Designs for Ships of War," which was appointed shortly after the loss of the "Captain." He investigated for that Committee the "Stability of Unmasted Ships of Low Freeboard," and the "Stability of Ships under Canvas."

In February, 1872, Rankine completed his memoir of his friend, John Elder, and in July reported on the cause of the explosion of the Tradeston Flour Mill. In May the increase of the endowment of his chair, which he had in vain sought from Government, was given by Mr. Elder's widow; and the income of Rankine's post was at last made sufficient to maintain its occupant. But by this time his energy was fast failing, the simplest work fatigued him; and he died on December 24, of a general decline rather than of any special disease. He had been for some years liable to violent headaches, and towards the close of his life these affected his sight. They were probably symptoms of heart disease, which ultimately developed paralysis. The gradual decay of his physical powers is painfully evident in the last pages of his journal, where, though the substance is correct and to the point, the handwriting becomes more and more irregular at each succeeding entry.

Such are the more prominent events in the life of this great and good man. Even now, after the lapse of eight years, it is difficult to realise the fact of his death. He was so many-sided, and yet so complete in himself, that the mental image of him formed by each of his friends remains almost as clear and distinct as if it had been formed but a few days ago.

Of the man himself it is not easy to speak in terms which, to a stranger, would appear unexaggerated. His appearance was striking and prepossessing in the extreme, and his courtesy resembled almost that of a gentleman of the old school. His musical taste had been highly cultivated, and it was always exceedingly pleasant to see him take his seat at the piano to accompany himself as he sang some humorous or grotesquely plaintive song—words and music alike being generally of his own composition. Some of the best of these songs have been collected in a small volume, *Songs and Fables* (Second Edition; Glasgow, Maclehose, 1874). We extract one which gives, in a very telling form, one point of view of a much-debated semi-scientific question :—

THE THREE-FOOT RULE.

WHEN I was bound apprentice, and learned to use my hands,
Folk never talked of measures that came from foreign lands :
Now I'm a British workman, too old to go to school;
So whether the chisel or file I hold, I'll stick to my three-foot rule.

Some talk of millimètres, and some of kilogrammes,
And some of décilitres, to measure beer and drams;
But I'm a British workman, too old to go to school:
So by pounds I'll eat, and by quarts I'll drink, and I'll work by my three-foot rule.

A party of astronomers went measuring of the earth,
And forty million mètres they took to be its girth;
Five hundred million inches, though, go through from pole to pole;
So let's stick to inches, feet, and yards, and the good old three-foot rule.

The great Egyptian pyramid's a thousand yards about;
And when the masons finished it, they raised a joyful shout;
The chap that planned that building, I'm bound he was no fool;
And now 'tis proved beyond a doubt, he used a three-foot rule.

Here's a health to every learned man that goes by common sense,
And would not plague the workman on any vain pretence;
But as for those philanthropists, who'd send us back to school,
Oh, *bless* their eyes, if ever they tries to put down the three-foot rule.

When the "Red Lions" met during the British Association week of 1871, in Edinburgh, Rankine was hailed with universal acclaim as the Lion-King. His versatility in that singular post was very much akin to that of Professor Edward Forbes; though their paths in science were widely different. His conversation was always interesting, and embraced with equal seeming ease all topics, however various. He had the still rarer qualification of being a good listener also. The evident interest which he took in all that was said to him had a most reassuring effect on the speaker; and he could turn without apparent mental effort from the prattle of young children to the most formidable statement of new results in mathematical or physical science. Then his note-book was at once produced, and in a few lines he jotted down the essence of the statement, to be pondered over at leisure, provided it did not at once appear to him how it was to be verified. The questions which he asked on such occasions were always almost startlingly to the point, and showed a rapidity of thought not often met with in minds of such calibre as his, where the mental inertia which enables them to overcome obstacles often prevents their being quickly set in motion. His kindness, shown in the readiness with which he undertook to read proof sheets for a friend, or even to contribute a portion of a chapter (where the subject was one to which he had paid special attention), was, for a man so constantly at work, absolutely astonishing. The writer of this brief notice has several times availed himself of such assistance. It was given almost as soon as asked, and it was invariably of sterling value. Nothing is more precious to a writer on scientific subjects (especially when questions of priority are involved) than the assistance of a friendly—though, if necessary, severe—critic, such as was Rankine.

We must not refrain from pointing out, in connection with his scientific merits, how very good and how exemplary for scientific writers and investigators his character was. He was ambitious; that is obvious from the number and variety of his books and papers, and the quite unnecessary display of symbols in several of his less popular writings. But he was the very soul of honour in respect to giving all credit to others, and in never attempting in anything, small or great, to go a hairbreadth beyond the line of right as to his own claims. He showed a particularly good and generous temper in cases of difference on scientific questions—a temper which proved the true metal, unalloyed by any mean quality.

Rankine was, in many subjects, an almost self-taught man, and the direction of his earlier scientific work seems not to have been a very profitable one. But, once on the right tack, his progress became very rapid. Every mathematician worthy of the name has made himself; some, as Rowan Hamilton, by attacking at an early age the grander works of Lagrange and Laplace; others by attempting original flights without the assistance of books. Rankine published only one or two papers on subjects of pure mathematics; and even these, though not containing any direct allusion to physics, were connected somewhat closely with kinematical or physical investigations, such as the deformation of an elastic solid.

The number of Rankine's scientific papers seems absolutely enormous, when we consider the minute and scrupulous care with which he attended to every point of detail in the writing and printing of them. How he managed, in addition to these, to find time for the composition of his many massive (not heavy) and elaborate volumes—all marked with the most striking stamp of originality—for his memoirs, and his almost weekly communications to *The Engineer* and other professional papers, must always remain matter for conjecture. In the Royal Society's splendid *Catalogue of Scientific Papers* we find that from 1843 to 1872 (both inclusive) he published, in recognised scientific journals alone, upwards of a hundred and fifty papers—many of these being exhaustive essays on mathematical or physical questions, and all, save one or two, containing genuine contributions to the advance of science. Leaving out of account the more strictly professional of these papers, we find among the titles of the rest such heads as the following:—*Molecular Vortices, Elasticity of Solids, Isorrhopic Axes, Compressibility of Water at Different Temperatures, Centrifugal Theory of Elasticity, Oscillatory Theory of Light, General Law of Transformation of Energy, Plane Water-Lines, Oogenous Neoids.* To indicate even briefly the nature and importance of the varied contents of these papers alone, would require vastly more time and space than are at present at our disposal. The more important of them are included in the present volume; others of less importance, or of less characterised originality, may be consulted by the reader in the scientific publications where they originally appeared.

Unquestionably the greatest pure scientific work of Rankine's is contained in his numerous papers bearing on the Dynamical Theory of Heat, and on Energy generally. As Sir William Thomson has remarked, even the mere title of his earliest paper on this subject, "Molecular Vortices," is an important contribution to physical science. The mode in which Rankine, in 1849, attacked the true theory of heat, which had just been recalled to the attention of scientific men by the admirable experiments and numerical determinations of Joule, was quite different from that adopted by any one of his concurrents; and though objections may fairly be raised to certain parts, even his first paper constituted a remarkable contribution to our physical knowledge. The essential characteristic of his method is the introduction of a hypothesis as to the *nature* of the motions and displacements (of the ultimate parts of bodies) upon which temperature depends, and in which heat, whether latent or sensible, consists. He thought it necessary to defend this mode of investigation, and did so in a remarkable address to the Philosophical Society of Glasgow, from which we extract the following passages, which are valuable not alone from their intention, but also from the insight they give us into the character of the man:—

"In order to establish that degree of probability which warrants the reception of a hypothesis into science, it is not sufficient that there should be a mere loose and general agreement between its results and those of experiment. Any ingenious and imaginative person can frame such hypotheses by the dozen. The agreement should be mathematically exact to that degree of precision which the uncertainty of experimental data renders possible, and should be tested in particular cases by numerical calculation. The highest degree of probability is attained when a hypothesis leads to the prediction of laws, phenomena, and numerical results which are afterwards verified by experiment; as when the wave-theory of light led to the prediction of the true velocity of light in refracting media, of the circular polarisation of light by reflexion, and of the previously unknown phenomena of conical and cylindrical refraction; and as when the hypothesis of atoms in chemistry led to the prediction of the exact proportions of the constituents of innumerable compounds. . . . I think I am justified in claiming for the hypothesis of molecular vortices, as a means of advancing the theory of the mechanical action of heat, the merit of having fulfilled the proper purposes of a mechanical hypothesis in physical science, which are to connect the laws of molecular phenomena by analogy with the laws of motion, and to suggest principles such as the second law of thermodynamics and the laws of the elasticity of imperfect gases, whose conformity to fact may afterwards be tested by direct experiment. And I make that claim the more confidently, that I conceive the hypothesis in question to be in a great measure the development, and the reduction to a precise form, of ideas concerning the molecular condition which constitutes heat, that have been entertained from a remote period by the leading minds in physical science. . . . I wish it, however, to be clearly understood, that although I attach great value and importance to sound mechanical hypotheses as means of advancing physical science, I firmly hold that they can never attain the certainty of observed facts; and accordingly, I have laboured assiduously to show that the two laws of thermodynamics are demonstrable as facts, independently of any hypothesis; and in treating of the practical application of those laws, I have avoided all reference to hypothesis whatsoever."

The application of the doctrine, that *heat and work are convertible*, to the discovery of new relations among the properties of bodies, was made about the same time by three scientific men—W. Thomson, Rankine, and Clausius.

Of these, Thomson cleared the way for the new theory by his account of the almost forgotten work of Carnot on the "Motive Power of Heat." This excessively important investigation was published in 1824, when the world of science was not prepared for its reception, and had been allowed to drop out of notice. Thomson gave a very full abstract of its contents in the *Transactions of the Royal Society of Edinburgh*, 1849, and pointed out that they would require modification if the new theory were adopted, as Carnot had throughout assumed that heat is a substance, and therefore indestructible. He showed, besides, that Carnot's method was capable of giving an *absolute* definition of temperature; independent, that is, of the properties of any particular substance. He also experimentally verified a deduction made by his brother, James Thomson (from Carnot's theory), as to the alteration of the freezing point of water by pressure.

Rankine (late in 1849) and Clausius (early in 1850) took the first step towards the formation of a true theory of the action of heat on bodies, by showing (by perfectly different modes of attacking the question) the nature of the modifications which Carnot's theory required. The recent publication of Carnot's MSS. proves that that very remarkable man had himself recognised the necessity for such modifications (and had all but succeeded in making them) before his premature death. Thomson, in 1851, put the foundations of the theory in the form they have since retained.

In Rankine's paper of 1849, he applied the theory to the determination of the relation between the latent heat of steam and its density, and made a very noteworthy prediction of the true value of the specific heat of air, at a time when the experimental results which were considered the best were far from the truth. [Rankine's results were soon after verified by the experimental researches of Joule and Regnault.] He also showed that saturated steam, pressing out a piston in a vessel impervious to heat, must cool so as to keep constantly at the temperature of saturation; and that, besides, a portion of it liquefies.

A very excellent statement of the claims of Rankine in thermodynamics is given in the following quotation from an article by Clerk-Maxwell (*Nature*, 1878, Vol. XVII., p. 257):—

"Of the three founders of theoretical thermodynamics, Rankine availed himself to the greatest extent of the scientific use of the imagination. His imagination, however, though amply luxuriant, was strictly scientific. Whatever he imagined about molecular vortices, with their nuclei and atmospheres, was so clearly imaged in his mind's eye, that he, as a practical engineer, could see how it would work.

"However intricate, therefore, the machinery might be which he imagined to exist in the minute parts of bodies, there was no danger of his going on to explain natural phenomena by any mode of action of this machinery which was not consistent with the general laws of mechanism. Hence, though the construction and distribution of his vortices may seem to us as complicated and arbitrary as the Cartesian system, his final deductions are simple, necessary, and consistent with facts.

"Certain phenomena were to be explained. Rankine set himself to imagine the mechanism by which they might be produced. Being an accomplished engineer, he succeeded in specifying a particular arrangement of mechanism competent to do the work, and also in predicting other properties of the mechanism which were afterwards found to be consistent with observed facts.

"As long as the training of the naturalist enables him to trace the action only of particular material systems, without giving him the power of dealing with the general properties of all such systems, he must proceed by the method so often described in histories of science—he must imagine model after model of hypothetical apparatus, till he finds one which will do the required work. If this apparatus should afterwards be found capable of accounting for many of the known phenomena, and not demonstrably inconsistent with any of them, he is strongly tempted to conclude that his hypothesis is a fact, at least until an equally good rival hypothesis has been invented. Thus Rankine,* long after an explanation of the properties of gases had been founded on the theory of the collisions of molecules, published what he supposed to be a proof that the phenomena of heat were invariably due to steady closed streams of continuous fluid matter.

"The scientific career of Rankine was marked by the gradual development of a singular power of bringing the most difficult investigations within the range of elementary methods. In his earlier papers, indeed, he appears as if battling with chaos, as he swims, or sinks, or wades, or creeps, or flies,

> 'And through the palpable obscure finds out
> His uncouth way;'

but he soon begins to pave a broad and beaten way over the dark abyss; and his latest writings show such a power of bridging over the difficulties of science, that his premature death must have been almost as great a loss to the diffusion of science as it was to its advancement.

"The chapter on thermodynamics in his book on the steam-engine was the first published treatise on the subject, and is the only expression of his views addressed directly to students.

"In this book he has disencumbered himself to a great extent of the hypothesis of molecular vortices, and builds principally on observed facts, though he, in common with Clausius, makes several assumptions, some expressed as axioms, others implied in definitions, which seem to us anything but self-evident. As an example of Rankine's best style we may take the following definition:—

"'A Perfect Gas is a substance in such a condition that the total pressure exerted by any number of portions of it, at a given temperature, against the sides of a vessel in which they are enclosed, is the sum of the pressures which each portion would exert if enclosed in the vessel separately at the same temperature.'

"Here we can form a distinct conception of every clause of the definition; but when we come to Rankine's Second Law of Thermodynamics we find

* "On the Second Law of Thermodynamics," *Phil. Mag.*, Oct., 1865, § 12, p. 244; but in his paper "On the Thermal Energy of Molecular Vortices," *Trans. R.S. Edin.*, XXV., p. 557 (1869), he admits that the explanation of gaseous pressure by the impacts of molecules has been proved to be possible.

that though, as to literary form, it seems cast in the same mould, its actual meaning is inscrutable.

"'THE SECOND LAW OF THERMODYNAMICS.—If the total actual heat of a homogeneous and uniformly hot substance be conceived to be divided into any number of equal parts, the effects of those parts in causing work to be performed are equal.'

"We find it difficult enough, even in 1878, to attach any distinct meaning to the total actual heat of a body, and still more to conceive this heat divided into equal parts, and to study the action of each of these parts; but as if our powers of deglutition were not yet sufficiently strained, Rankine follows this up with another statement of the same law, in which we have to assert our intuitive belief that

"'If the absolute temperature of any uniformly hot substance be divided into any number of equal parts, the effects of those parts in causing work to be performed are equal.'

"The student who thinks that he can form any idea of the meaning of this sentence is quite capable of explaining, on thermodynamical principles, what Mr. Tennyson says of the great Duke—

> 'Whose eighty winters freeze with one rebuke
> All great self-seekers trampling on the right.'"

Rankine's researches on heat were for the most part connected, as we have already said, with a theory of the constitution of bodies, and a speculation as to the physical nature of a hot body, to which he gave the name of Theory of Molecular Vortices. In this theory, the invisibly small parts of bodies apparently at rest are supposed to be in a state of motion, the rapidity of which may be compared with that of a cannon ball. It was distinguished from other theories, which attribute motion to bodies apparently at rest, by the further assumption that this motion is like that of very small vortices, each whirling about its own axis, and that the centrifugal force of this motion contributes to the elasticity of the body. A theory of a similar kind has since been applied by Clerk-Maxwell to the explanation of magnetic phenomena; and Sir W. Thomson has made the rigorous investigation of vortices possible by his paper on "Vortex Motion," and has also contributed to the philosophy of speculation by his theory of "Vortex Atoms."

Rankine's researches on the general theory of elastic bodies are characterised by the fact that while, in laying the foundation of the theory, he confines himself to the use of rigorous methods, and does not shrink from any mechanical difficulty in their application, he always prepares the way for the application of the results to practice, by making the definitions so clear, the methods so simple, the results so definite, that they can be mastered by the exercise of a little thought, without special mathematical training. This quality is prominent also in his researches on fluid motion, three of which are of special importance.

I. The theory of the propagation of waves, such as those of sound in elastic media, is generally supposed to belong to the most abstruse departments of mathematical science. Even Newton made some oversights in his investigation, and it required more than a century of hard mathematical development before the theory reached its present state—which is still very imperfect. Rankine, by the introduction of a few new conceptions in the elementary part of the investigations, has rendered it possible for any one acquainted with elementary dynamics to follow the theory up to the point at which it was left by Laplace, and almost as much further as it has yet been carried.

II. The theory of waves on the surface of water, when their height is not regarded as infinitely small, is still more difficult than that of sound waves. Stokes has, indeed, in a masterly series of investigations, arrived at a second, and for some purposes a third, approximation. An *exact* solution, however, but of a particular case only, was arrived at by Rankine. He was not aware that it had been given by Gerstner in 1802, having been deduced from an assumption, generally erroneous, but true for this special case. Unfortunately, as this theory essentially involves rotation of fluid elements, it is not a solution of the usual problem of waves at the surface of a perfect liquid, for it implies a kind of motion which could not be produced in such a substance if originally in a state of rest.

III. Rankine's third investigation is that of lines of motion of water flowing past a ship. He begins with the mathematical theory of such lines, but soon applies his results to the determination of good forms of the lines of a ship, and the investigation of the principal causes of the resistance to the motion of the ship, and the means of diminishing that resistance.

No other person has done so much as either Rankine or William Froude to promote naval dynamics, and the application of science to the shaping of ships, and to the estimation of their performances.

To Rankine, the Scientific Sub-Committee of the late Admiralty Committee on Designs owed most of its reports, and a very large proportion of their effectiveness. Even those most disposed to disparage that Committee and its work, have made exception as regards the Reports of the Scientific Sub-Committee, and, in particular, Rankine's contributions. (It seems to us that only ignorance or unfavourable bias could attempt to disparage the committee at all; for it undoubtedly did, though in an unostentatious manner, very good service indeed.)

Rankine's works on *Applied Mechanics*, on the *Steam-Engine*, and on *Engineering*, contain many valuable and original methods; and while the publication of any one of them would have established the fame of one of our average scientific men, that on the steam-engine could not have been produced by any but an original discoverer of a high order. Some of the investigations contained in this series of volumes are as remarkable for the material aid they afford to the man of practice as for the light they throw upon his work.

The following gives a striking instance of Rankine's tact under a novel and somewhat puzzling combination of circumstances. In August, 1858, he wrote to the *Philosophical Magazine* the annexed short letter, which was printed in the September number of that journal:—

"In the course of last year there were communicated to me, in confidence, the results of a great body of experiments on the engine power required to propel steamships of various sizes and figures, at various speeds. From those results I deduced a general formula for the resistance of ships, having such figures as usually occur in steamers, which, on the 23rd of December, 1857, I communicated to the owner of the experimental data, and he has since applied it to practice with complete success.

"As the experimental data were given to me in confidence, I am for the present bound in honour not to disclose the formula which I deduced from them; but as I am desirous not to delay longer the placing it upon record, I have recourse to the old fashion of sending it to you in the form of an anagram, in which the letters that occur in its verbal statement are arranged in alphabetical order, and the number of times that each letter occurs is expressed by figures:—20 A. 4 B. 6 C. 9 D. 33 E. 8 F. 4 G. 16 H. 10 I. 5 L. 3 M. 15 N. 14 O. 4 P. 3 Q. 14 R. 13 S. 25 T. 4 U. 2 V. 2 W. 1 X. 4 Y. (219 letters in all). I hope I may soon be released from my present obligation to secrecy."

There could be no doubt that this refers to a remarkable investigation which Rankine carried out for his friend James R. Napier, who had asked him to estimate the horse-power necessary to propel at a given rate a vessel which he was about to construct. Guided by this consideration, I found, in 1872, the following sentence of Rankine's in *The Civil Engineer and Architect's Journal* (October 1, 1861), but without any reference whatever to the anagram or to the *Philosophical Magazine*:—

"The resistance of a sharp-ended ship exceeds the resistance of a current of water of the same velocity in a channel of the same length and mean girth, by a quantity proportional to the square of the greatest breadth, divided by the square of the length of the bow and stern."

Curiously enough, Rankine seems to have made an arithmetical mistake, or a mis-spelling, because this sentence exactly fits all of the above numerical data, with the exception that it contains just one E too much, and has, therefore, 220 letters in all. Rankine's private MSS., to which I have recently had access, show that my guess was correct, but do not enable me to find how the numerical error just noticed arose.

Mr. Napier informed me that in all his business relations with Rankine, nothing had so much impressed him as the rapid and keen insight with which he seemed at once to seize upon the most essential points in the solution of a practical question, though stated to him for the first time; how he first shook himself free from the petty complications, and gave almost immediately an approximate estimate embracing all the larger bearings of the question; and then, much more formally and deliberately, and with the minute accuracy and system of a man of business, proceeded to work the question with the desired exactness. Mr. Napier said that on the occasion of his first consultation with Rankine on the matter referred to in the anagram above, Rankine's very first words pointed out to him what a large proportion of the resistance to a vessel's motion is due to friction, and how ill-considered was the then growing demand for long and narrow ships.

Rankine's text-books on engineering subjects are in many respects the most satisfactory that have been published in any country. At the time of their publication they have always been in advance of the professional knowledge of the day, but they possess much greater merits than that of mere novelty. Rankine was peculiarly happy in discriminating between those branches of engineering knowledge which grow from daily experience, and those which depend on unchangeable scientific principles. In his books he dealt almost exclusively with the latter, which may, and certainly will, be greatly extended, but so far as they have been established can never change. Hence his books are a mine which smaller men may work for many years, rendering his knowledge more generally available by giving it a popular setting of their own. By the bulk of the engineering profession the books are considered hard reading, but as engineering education improves they will more and more be recognised as both wonderfully complete and essentially simple. Rankine, by his education as a practical engineer, was eminently qualified to recognise the problems of which the solution is required in practice; but the large scope of his mind would not allow him to be content with giving merely the solution of those particular cases which most frequently occur in engineering as we now know it. His method invariably is to state the problem in a very general form, find the solution, and then apply this solution to special cases. This method does not make the books easy reading for students, nor does it give the most convenient work of reference for the practical man; but it has produced writings the value of which is permanent, instead of being ephemeral.

In his *Applied Mechanics* we have the best existing work on the application of the doctrines of pure mechanics to general engineering problems.

No specious reasoning has been detected in this great work, a fact which should for ever dispel the old and false antithesis between theory and practice—a contrast drawn by practical men who never understood fully any theory, and assented to by scientific men who were not candid enough to point out where their theories were incomplete. In the work named *Civil Engineering*, Rankine applied the general doctrines of applied mechanics to the special problems which the civil engineer of to-day meets with in his practice, and his volume contains much valuable statistical information. In his work on *Prime Movers* we have a most thoroughly original statement of the thermodynamic theory, so far as it bears on the design and use of the steam-engine. This work especially shows Rankine's clear discrimination of what is permanent and can be taught, from that which must vary from day to day, and can only be acquired by personal experience; the distinction between the *science* and the *art* of the engineer.

His treatise on *Machinery and Millwork* gives the mechanical engineer instruction of a kind analogous to that which the civil engineer may derive from the book called *Civil Engineering*. The problems stated generally in the *Applied Mechanics* are in it applied to the special cases which arise in the design of machinery. Several of these Manuals have been recently translated; *Prime Movers* into French, *Civil Engineering* into German, &c.; and *Machinery and Millwork* will soon appear in Italian.

Most of the common treatises on engineering subjects are mere rechauffés or compilations; and no library becomes sooner worthless than that of the engineer, the practice of this year being wholly different from that of five years since. Really original papers and monographs rapidly lose their interest and importance, except as historical landmarks, but Rankine's works will retain their value after this generation has passed away.

In concluding the scientific part of this brief notice of a true man, we need scarcely point out to the reader how much of Rankine's usefulness was due to steady and honest work. The unscientific are prone to imagine that talent (especially when, as in Rankine's case, it rises to the level of genius) is necessarily rapid and off-hand in producing its fruits. No greater mistake could be made. The most powerful intellects work slowly and patiently at a new subject. Such was the case with Newton, and so it is still. Rapid they may be, and in general are, in new applications of processes long since mastered; but it is only your pseudo-scientific man who forms his opinion at once on a new subject. This truth was prominently realised in Rankine, who was prompt in reply when his knowledge was sufficient, but patient and reticent when he felt that more knowledge was necessary. With him thought was never divorced from work—both were good of their kind—the thought profound and thorough, the work a workman-like expression of the thought. Few, if any, practical engineers have contributed so much to abstract science, and in no case has scientific study been applied with more effect to practical engineering. Rankine's name will ever hold a high place in the history of science, and will worthily be associated with those of the great men we have recently lost. And, when we think who these were, how strangely does such a list—including the names of Babbage, Boole, Brewster, Leslie Ellis, Faraday, Forbes, Herschel, Rowan Hamilton, Clerk-Maxwell, Rankine, and others, though confined to physical or mathematical science alone—contrast with the astonishing utterance of the Prime Minister of Great Britain and Ireland, to the effect that the present is by no means an age abounding in minds of the first order! Ten such men lost by this little country within the last dozen years or so—any one of whom would have made himself an enduring name had he lived in any preceding age, be it that of Hooke and Newton, or that of Cavendish and Watt! Nay more, even such losses as these have not extinguished the hopes of science amongst us. Every one of these great men has, by some mysterious influence of his genius, kindled the sacred thirst for new knowledge in younger but kindred spirits, many of whom will certainly rival, some even may excel, their teachers!

For the dates and statements of fact in this Memoir, I am indebted mainly to Rankine's private MSS., access to which has been given me by his relatives. Some special details I have had from his own lips. I have also to acknowledge my obligations to Sir William Thomson, to Professor Jenkin, and specially to the late Mr. J. R. Napier, who was one of Rankine's most enthusiastic admirers. He furnished me with much of the more technical part of the materials for a notice of Rankine's scientific work, which I wrote immediately after his death for the *Glasgow Herald*, December 28, 1872, and of which I have made considerable use in what precedes.

Of Rankine's purely scientific work I have spoken from actual acquaintance with his writings; but I have found it necessary to apply for assistance while attempting to discuss the merits of his more practical investigations.

P. G. TAIT.

College, Edinburgh,
October, 1880.

PART I.

PAPERS RELATING TO TEMPERATURE, ELASTICITY, AND EXPANSION OF VAPOURS, LIQUIDS, AND SOLIDS.

PART I.

PAPERS RELATING TO TEMPERATURE, ELASTICITY, AND EXPANSION OF VAPOURS, LIQUIDS, AND SOLIDS.

I.—ON AN EQUATION BETWEEN THE TEMPERATURE AND THE MAXIMUM ELASTICITY OF STEAM AND OTHER VAPOURS. (See PLATE I.)*

IN the course of a series of investigations founded on a peculiar hypothesis respecting the molecular constitution of matter, I have obtained, among other results, an equation giving a very close approximation to the maximum elasticity of vapour in contact with its liquid at all temperatures that usually occur.

As this equation is easy and expeditious in calculation, gives accurate numerical results, and is likely to be practically useful, I proceed at once to make it known, without waiting until I have reduced the theoretical researches, of which it is a consequence, to a form fit for publication.

The equation is as follows:—

$$\text{Log. P} = \alpha - \frac{\beta}{t} - \frac{\gamma}{t^2}, \quad . \quad . \quad . \quad (1.)$$

where P represents the maximum pressure of a vapour in contact with its liquid;

t, the temperature, measured on the air-thermometer, from a point which may be called the ABSOLUTE ZERO, and which is—

274°·6 of the *Centigrade scale* below the freezing point of water;

462°·28 of *Fahrenheit's scale* below the *ordinary* zero of that scale,

* Originally published in the *Edinburgh New Philosophical Journal* for July, 1849.

supposing the boiling point to have been adjusted under a pressure of 29·922 inches of mercury, so that 180° of Fahrenheit may be exactly equal to 100 Centigrade degrees;

461°·93 below the ordinary zero of Fahrenheit's scale, when the boiling point has been adjusted under a pressure of 30 inches of mercury, 180° of Fahrenheit being then equal to 100°·0735 of the Centigrade scale.

The *form* of the equation has been given by theory; but three constants, represented by α, β, and γ, have to be determined for each fluid by experiment.

The inverse formula, for finding the temperature from the pressure, is of course

$$\frac{1}{t}=\sqrt{\frac{\alpha-\log. \mathrm{P}}{\gamma}+\frac{\beta^2}{4\gamma^2}}-\frac{\beta}{2\gamma}. \quad . \quad . \quad . \quad (2.)$$

It is obvious that for the determination of the three constants, it is sufficient to know accurately the pressures corresponding to three temperatures; and that the calculation will be facilitated if the reciprocals of those temperatures, as measured from the absolute zero, are in arithmetical progression.

In order to calculate the values of the three constants for the vapour of water, the following data have been taken from M. Regnault's experiments:—

Temperatures in Centigrade Degrees.		Common Logarithms of the Pressure in Millimètres of Mercury.	Remarks.
Above the Freezing Point.	Above the Absolute Zero.		
° 220·	° 494·6	° 4·2403	Measured by M. Regnault on his curve, representing the mean results of his experiments.
100·	374·6	2·8808136	Logarithm of 760 millimètres.
26·86	301·46	1·4198	Calculated by interpolation from M. Regnault's general table.

These data give the following results for the vapour of water, the pressures being expressed in millimètres of mercury, and the temperatures in Centigrade degrees of the air-thermometer:—

$$\text{Log. } \gamma=5{\cdot}0827176 \qquad \text{Log. } \beta=3{\cdot}1851091$$

$$\alpha=7{\cdot}831247$$

TABLE I.—VAPOUR OF WATER.

Temperatures in Centigrade Degrees of the Air-Thermometer from		Pressures in Millimètres of Mercury, according to		Difference between Calculation and Experiment in Millimètres.	Corresponding Differences of Temperature.	Common Logarithms of the Pressures in Millimètres according to		Differences between Calculation and Experiment in Logarithms.	Corresponding Differences of Temperature.	Temperatures above the Freezing Point
the Freezing Point.	the Absolute Zero.	the Formula.	M. Regnault's Experiments.			the Formula.	M. Regnault's Experiments.			
°	°				°					
−30	244·6	0·35	0·34	+0·01	−0·42					
−20	254·6	0·89	0·91	−0·02	+0·25					
−10	264·6	2·07	2·08	−0·01	+0·06					
0	274·6	4·47	4·60	−0·13	+0·38					
+10	284·6	9·05	9·16	−0·11	+0·18					
20	294·6	17·33	17·39	0·06	+0·06					
*26·86	301·46	26·29	26·29	0·00	0·00					
30	304·6	31·57	31·55	+0·02	−0·01					
40	314·6	55·05	54·91	+0·14	−0·05					
50	324·6	92·26	91·98	+0·28	−0·06					
60	334·6	149·15	148·79	+0·36	−0·05					
70	344·6	233·48	233·09	+0·39	−0·04					
80	354·6	355·04	354·64	+0·40	−0·03					
90	364·6	525·70	525·45	+0·25	−0·01				°	°
*100	374·6	760·00	760·00	0·00	0·00	2·880813	2·8808	0·0000	0·00	100
110	384·6	1074·82	1073·70	+1·12	−0·03	3·031362	3·0307	+0·0007	0·05	110
120	394·6	1490·1	1489·0	+1·1	−0·02	3·173204	3·1734	0·0002	+0·01	120
130	404·6	2028·0	2029·0	−1·0	+0·02	3·307061	3·3076	−0·0005	+0·04	130
140	414·6	2713·8	2713·0	+0·8	−0·02	3·433576	3·4332	+0·0004	0·03	140
150	424·6	3575·5	3572·0	+3·5	−0·04	3·553334	3·5537	−0·0004	+0·03	150
160	434·6	4643·6	4647·0	−3·4	+0·03	3·666853	3·6676	0·0007	+0·06	160
170	444·6	5951·2	5960·0	−8·8	+0·06	3·774603	3·7750	0·0004	+0·04	170
180	454·6	7533·7	7545·0	11·3	+0·07	3·877005	3·8772	−0·0002	+0·02	180
190	464·6	9428·5	9428·0	+ 0·5	0·00	3·974443	3·9743	+0·0001	0·01	190
200	474·6	11675	11660	+15	−0·06	4·067268	4·0674	0·0001	+0·01	200
210	484·6	14315	14308	+ 7	−0·02	4·155796	4·1561	0·0003	+0·03	210
*220	494·6	17390	17390	0	0·00	4·240300	4·2403	0·0000	0·00	220
230	504·6	20945	20915	+30	−0·08	4·321083	4·3207	+0·0004	−0·05	230
(1.)	(2.)	(3.)	(4.)	(5.)	(6.)	(7.)	(8.)	(9.)	(10.)	(11.)

Table I. exhibits a comparison between the results of the formula and those of M. Regnault's experiments, for every tenth degree of the Centigrade air-thermometer, from 30° below the freezing point to 230° above it, being within one or two degrees of the whole range of the experiments.

M. Regnault's values are given, as measured by himself, on the curves representing the mean results of his experiments, with the exception of the pressures at 26°·86, one of the data already mentioned, and that at −30°, which I have calculated by interpolation from his table, series *h*.

Each of the three data used in determining the constants is marked with an asterisk *.

In the columns of differences between the results of the formula and those of experiment, the sign + indicates that the former exceed the latter, and the sign − the reverse.

Beside each such column of differences is placed a column of the corresponding differences of temperature, which would result in calculating the temperature from the pressure by the inverse formula. These are found by multiplying each number in the preceding columns by $-\frac{dt}{dP}$ or by $\frac{-dt}{d \log. P}$ as the case may require.

In comparing the results of the formula with those of experiment, as exhibited in Table I., the following circumstances are to be taken into consideration:—

First, That the uncertainty of barometric observations amounts in general to at least one-tenth of a millimètre.

Secondly, That the uncertainty of thermometric observations is from one-twentieth to one-tenth of a degree under ordinary circumstances, and at high temperatures amounts to more.

Thirdly, That, in experiments of the kind referred to in the table, those two sorts of uncertainty are combined.

The fifth column of the table shows that, from 30° below the freezing point to 20° above it, where the minuteness of the pressures makes the barometric errors of most importance, the greatest difference between experiment and calculation is $\frac{13}{100}$ of a millimètre, or $\frac{1}{200}$ of an inch of mercury, a very small quantity in itself, although, from the slowness with which the pressure varies at low temperatures, the corresponding difference of temperature amounts to $\frac{38}{100}$ of a degree.

The sixth and tenth columns show that, from 20° to 230° above the freezing point, the greatest of the discrepancies between experiment and observation corresponds to a difference of temperature of only $\frac{8}{100}$ of a degree, and that very few of those discrepancies exceed the amount corresponding to $\frac{1}{20}$ of a degree.

A comparison between the sixth and tenth columns shows that, for four of the temperatures given—viz., 120°, 150°, 200°, and 210°—the pressures deduced from M. Regnault's curve of actual elasticities, and from his logarithmic curve respectively, differ from the pressures given by the formula in opposite directions.

If the curves represented by the formula were laid down on M. Regnault's diagram, they would be almost undistinguishable from those which he has himself drawn, except near the freezing point, where the scale of pressures is very large, the heights of the mercurial column being magnified eight-fold on the plate. In the case of the curves of logarithms of pressures above one atmosphere, the coincidence would be almost perfect.

The formula may, therefore, be considered as accurately representing the results of all M. Regnault's experiments throughout a range of temperatures from 30° of the Centigrade scale below the freezing point to 230° above it, and of pressures from $\frac{1}{2200}$ of an atmosphere up to 28 atmospheres.

It will be observed that equation (1.) bears some resemblance to the formula proposed by Professor Roche in 1828, viz.:—

$$\text{Log. } P = A - \frac{B}{T + C}.$$

where T represents the temperature measured from the ordinary zero point, and A, B, and C, constants, which have to be determined from three experimental data. It has been shown, however, by M. Regnault, as well as by others, that though this formula agrees very nearly with observation throughout a limited range of temperature, it errs widely when the range is extensive. I have been unable to find Professor Roche's memoir, and I do not know the reasoning from which he has deduced his formula.

The use in computation of the equations I have given, whether to calculate the pressure from the temperature, or the temperature from the pressure, is rapid and easy. In Table II. they are recapitulated, and the values of the constants for different measures of pressure and temperature are stated.

In calculating the values of a, the specific gravity of mercury has been taken as 13·596.

Temperatures measured by mercurial thermometers are in all cases to be reduced to the corresponding temperatures on the air-thermometer, which may be done by means of the table given by M. Regnault in his memoir on that subject.

TABLE II.—VAPOUR OF WATER.

Formula for calculating the Maximum Elasticity of Steam (P), from the Temperature on the Air-Thermometer, measured from the Absolute Zero (t) :—

$$\text{Log. P} = a - \frac{\beta}{t} - \frac{\gamma}{t^2}.$$

Inverse Formula for calculating the Temperature from the Maximum Elasticity of Steam :—

$$\frac{1}{t} = \sqrt{\frac{a - \log. \text{P}}{\gamma} + \frac{\beta^2}{4\gamma^2}} - \frac{\beta}{2\gamma}.$$

Values of the Constants depending on the Thermometric Scale :

For the Centigrade scale :—

Absolute zero 274°·6 below the freezing point of water.

$\text{Log. } \beta = 3{\cdot}1851091$ $\text{Log. } \gamma = 5{\cdot}0827176$

$\frac{\beta}{2\gamma} = 0{\cdot}0063294$ $\frac{\beta^2}{4\gamma^2} = 0{\cdot}00004006$

For Fahrenheit's scale; boiling point adjusted at 29·922 inches:—

Absolute zero 462°·28 below ordinary zero.

$\text{Log. } \beta = 3{\cdot}4403816$ $\text{Log. } \gamma = 5{\cdot}5932626$

$\frac{\beta}{2\gamma} = 0{\cdot}0035163$ $\frac{\beta^2}{4\gamma^2} = 0{\cdot}000012364$

For Fahrenheit's scale; boiling point adjusted at 30 inches :—

Absolute zero 461°·93 below ordinary zero.

$\text{Log. } \beta = 3{\cdot}4400625$ $\text{Log. } \gamma = 5{\cdot}5926244$

$\frac{\beta}{2\gamma} = 0{\cdot}0035189$ $\frac{\beta^2}{4\gamma^2} = 0{\cdot}000012383$

Values of the Constant a depending on the Measure of Elasticity :

For millimètres of mercury, ... $a = 7{\cdot}831247$

English inches of mercury, ... 6·426421

Atmospheres of 760 mil. = 29·922 inches = 14·7 lbs. on the square inch = 1·0333 kil. on the centimètre^2, ...	4·950433
Atmospheres of 30 inches = 761 mil. ·99 = 14·74 lbs. on the square inch = 1·036 kil. on the centimètre^2, ...	4·949300
Kilogrammes on the square centimètre, ...	4·964658
Pounds avoirdupois on the square inch, ...	6·117817

N.B.—All the Constants are for *common logarithms.*

I have applied similar formulæ to the vapours of *alcohol* and *ether*, making use of the experiments of Dr. Ure.

In order to calculate the constants, the following experimental data have been taken, assuming that, on Dr. Ure's thermometers, 180° were equal to 100 Centigrade degrees.

	Temperatures on Fahrenheit's Scale from the Ordinary Zero.	Temperatures on Fahrenheit's Scale from the Absolute Zero.	Pressures in Inches of Mercury.	REMARKS.
	°	°		
For Alcohol of the specific gravity 0·813,	250	712·3	132·30	From Dr. Ure's Table.
	173	635·3	30·00	Do.
	111·02	573·32	6·30	Interpolated in the same Table.
For Ether, boiling at 105° F., under 30 inches of pressure,	200	662·3	142·8	From Dr. Ure's Table.
	148·8	611·1	66·24	Interpolated.
	105	567·3	30·00	From the Table.
For Ether, boiling at 104° F., under 30 inches of pressure,	104	566·3	30·00	From Dr. Ure's Table.
	66·7	529·0	13·76	Interpolated.
	34	496·3	6·20	From the Table.

The values of the constants in equation (1.), calculated from these data, are as follows, for inches of mercury and Fahrenheit's scale:—

	α	Log. β.	Log. γ.
Alcohol, specific gravity 0·813, . . .	6·16620	3·3165220	5·7602709
Ether, boiling point 105°F., ...	5·33590	3·2084573	5·5119893
Ether, boiling point 104°F., ...	5·44580	3·2571312	5·3962460

Absolute zero 462°·3 below ordinary zero.

The curves represented by the formulæ for those three fluids are laid down on the diagram which accompanies this memoir (Plate I.), and which has been reduced to one-fourth of the original scale. The horizontal divisions represent the scale of Fahrenheit's thermometer, numbered from the ordinary zero; the vertical divisions, pressures of vapour, according to the scales specified on the respective curves. The points corresponding to the experimental data are surrounded by small circles.

The curve for alcohol extends from 32° to 264° of Fahrenheit. It is divided into two portions, having different vertical scales, suitable to high and low pressures respectively.

The curve for the less volatile ether extends from 105° to 210°; that for the more volatile ether, from 34° to 104°.

The results of Dr. Ure's experiments are marked by small crosses.

The irregular and sinuous manner in which those crosses are distributed, indicates that the errors of observation, especially at high temperatures, must have been considerable. This does not appear surprising, when we recollect how many causes of uncertainty affect all the measurements required in such experiments, especially the thermometric observations, and how little those causes have been understood until very recently. The data from which the constants have been calculated are, of course, affected by the general uncertainty of the experiments.

When those circumstances are taken into account, it is obvious, from inspection of the diagram, that the curves representing the formulæ agree with the points representing the experiments as nearly as the irregularity of the latter and the uncertainty of the data permit; and that there is good reason to anticipate that, when experiments shall have been made on the vapours of alcohol and ether with a degree of precision equal to that attained by M. Regnault in the case of the vapour of water, the equation will be found to give the elasticities of those two vapours as accurately as it does that of steam.

Although the diagram affords the best means of judging of the agreement between calculation and experiment, three tables (III., IV., and V.) are annexed, in order to show the numerical amount of the discrepancies at certain temperatures. The data, as before, are marked with asterisks.

It is worthy of remark, in the case of alcohol, that although the lowest of the experimental data is at the temperature of 111°·02, the formula agrees extremely well with the experiments throughout the entire range of 79 degrees *below* that point.

Plate 1.

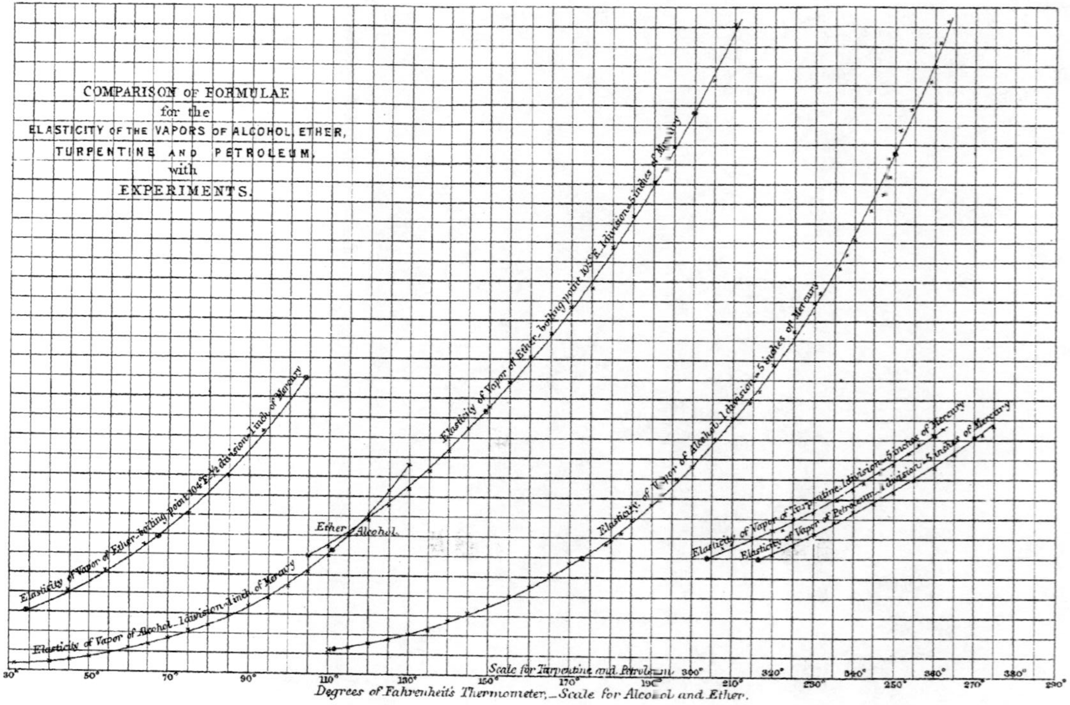

COMPARISON OF FORMULAE
for the
ELASTICITY OF THE VAPORS OF ALCOHOL, ETHER,
TURPENTINE AND PETROLEUM,
with
EXPERIMENTS.
Elasticity of Vapor of Ether-boiling point 104°F. 1 division-1 inch of Mercury
Elasticity of Vapor of Alcohol-1 division-1 inch of Mercury
Elasticity of Vapor of Ether-boiling point 105°F. 1 division-5 inches of Mercury
Elasticity of Vapor of Alcohol-1 division-5 inches of Mercury
Ether
Alcohol.
Elasticity of Vapor of Turpentine-1 division-5 inches of Mercury
Elasticity of Vapor of Petroleum-1 division-5 inches of Mercury
30°
50°
70°
90°
110°
130°
150°
170°
190°
210°
230°
250°
270°
290°
300°
320°
340°
360°
380°
Scale for Turpentine and Petroleum.
Degrees of Fahrenheit's Thermometer,-Scale for Alcohol and Ether.

TABLE III.—VAPOUR OF ALCOHOL, OF THE SPECIFIC GRAVITY 0·813.

Temperatures in Degrees of Fahrenheit from the Ordinary Zero.	Pressures in Inches of Mercury according to		Differences between Calculation and Experiment in Inches.	Corresponding Differences of Temperature.
	the Formula.	Dr. Ure's Experiments.		
°				°
32	0·41	0·40	+0·01	−0·5
40	0·57	0·56	+0·01	−0·4
50	0·84	0·86	−0·02	+0·7
60	1·22	1·23	−0·01	+0·2
70	1·74	1·76	−0·02	+0·3
80	2·43	2·45	−0·02	+0·2
90	3·36	3·40	−0·04	+0·4
100	4·56	4·50	+0·06	−0·5
110	6·12	6·00	+0·12	−0·7
*111·02	6·30	6·30	0·00	0·0
120	8·10	8·10	0·00	0·0
130	10·61	10·60	+0·01	−0·0
140	13·73	13·90	−0·17	+0·5
150	17·60	18·00	−0·40	+0·9
160	22·32	22·60	−0·28	+0·5
170	28·06	28·30	−0·24	+0·4
*173	30·00	30·00	0·00	0·0
100	34·06	34·73	+0·23	−0·3
190	43·21	43·20	+0·01	−0·0
200	52·96	53·00	−0·04	+0·0
210	64·47	65·00	−0·53	+0·5
220	77·92	78·50	−0·58	+0·4
230	93·54	94·10	−0·56	+0·4
240	111·58	111·24	+0·34	−0·2
*250	132·30	132·30	0·00	0·0
260	155·98	155·20	+0·78	−0·3
264	165·58	166·10	−0·52	+0·2
(1.)	(2.)	(3.)	(4.)	(5.)

TABLE IV.—VAPOUR OF ETHER; BOILING POINT, 105°F.

Temperatures in Degrees of Fahrenheit above the Ordinary Zero.	Pressures in Inches of Mercury according to		Differences between Calculation and Experiment in Inches of Mercury.	Corresponding Differences of Temperature.
	the Formula.	Dr. Ure's Experiments.		
°				°
*105	30·00	30·00	0·00	0·0
110	33·08	32·54	+0·54	−0·9
120	39·98	39·47	+0·51	−0·7
125	43·83	43·24	+0·59	−0·8
130	47·95	47·14	+0·81	−1·0
140	57·10	56·90	+0·20	−0·2
*148·8	66·24	66·24	0·00	0·0
150	67·53	67·60	−0·07	+0·1
160	79·35	80·30	−0·95	+0·9
170	92·68	92·80	−0·12	+0·1
175	99·94	99·10	+0·84	−0·6
180	107·62	108·30	−0·68	+0·4
190	124·29	124·80	−0·51	+0·3
*200	142·80	142·80	0·00	0·0
205	152·78	151·30	+1·48	−0·7
210	163·27	166·00	−2·73	+1·1

TABLE V.—VAPOUR OF ETHER; BOILING POINT, 104°F.

Temperatures in Degrees of Fahrenheit above the Ordinary Zero.	Pressures in Inches of Mercury according to		Differences between Calculation and Experiment in Inches of Mercury.	Corresponding Differences of Temperature.
	the Formula.	Dr. Ure's Experiments.		
°				°
*34	6·20	6·20	0·00	0·0
44	8·02	8·10	−0·08	+0·4
54	10·24	10·30	−0·06	+0·2
64	12·94	13·00	−0·06	+0·2
*66·7	13·76	13·76	0·00	0·0
74	16·19	16·10	+0·09	−0·2
84	20·06	20·00	+0·06	−0·1
94	24·64	24·70	−0·06	+0·1
*104	30·00	30·00	0·00	0·0
(1.)	(2.)	(3.)	(4.)	(5.)

The results of Dr. Ure's experiments on the vapours of *turpentine* and *petroleum* are so irregular (as the diagram shows), and the range of temperatures and pressures through which they extend so limited, that the value of the constant γ cannot be determined from them with precision. I have, therefore, endeavoured to represent the elasticities of those two vapours approximately by the *first two terms* of the formula only, calculating the constants from two experimental data for each fluid. The equation thus obtained

$$\text{Log. P} = \alpha - \frac{\beta}{t},$$

is similar in form to that of Professor Roche.

The data and the values of the constants are as follows:—

Temperatures on Fahrenheit's Scale from		Pressures in Inches of Mercury.	Values of the Constants for Fahrenheit's Scale and Inches of Mercury.
the Ordinary Zero.	the Absolute Zero.		
°	°	*Turpentine.*	
360	822·3	60·80	$\alpha=5·98187$
304	766·3	30·00	Log. $\beta=3·5380701$
		Petroleum.	
370	832·3	60·70	$\alpha=6·19451$
316	778·3	30·00	Log. $\beta=3·5648490$

Although the temperatures are much higher than the boiling point of water, I have not endeavoured to reduce them to the scale of the air-thermometer, as it is impossible to do so correctly without knowing the nature of the glass of which the mercurial thermometer was made.

The diagram shows that the formula agrees with the experiments as well as their irregularity entitles us to expect.

The following tables give some of the numerical results :—

TABLE VI.—VAPOUR OF TURPENTINE.

Temperatures in Degrees of Fahrenheit from the Ordinary Zero.	Pressures in Inches of Mercury according to the Formula (of two terms).	Pressures in Inches of Mercury according to Dr. Ure's Experiments.	Differences between Calculation and Experiment in Inches of Mercury.	Corresponding Differences of Temperature.
°				°
*304	30·00	30·00	0·00	0·0
310	32·52	33·50	−0·98	+2·3
320	37·09	37·06	+0·03	−0·0
330	42·16	42·10	+0·06	−0·1
340	47·78	47·30	+0·48	−0·9
350	53·98	53·80	+0·18	−0·3
*360	60·80	60·80	0·00	0·0
362	62·24	62·40	−0·16	+0·0

TABLE VII.—VAPOUR OF PETROLEUM.

Temperatures in Degrees of Fahrenheit, from the Ordinary Zero.	Pressures in Inches of Mercury according to the Formula (of two terms).	Pressures in Inches of Mercury according to Dr. Ure's Experiments.	Differences between Calculation and Experiment in Inches of Mercury.	Corresponding Differences of Temperature.
°				°
*316	30·00	30·00	0·00	0·0
320	31·71	31·70	+0·01	−0·0
330	36·35	36·40	−0·05	+0·1
340	41·52	41·60	−0·08	+0·2
350	47·27	46·86	+0·41	−0·7
360	53·65	53·30	+0·35	−0·5
*370	60·70	60·70	0·00	0·0
375	64·50	64·00	+0·50	−0·7
(1.)	(2.)	(3.)	(4.)	(5.)

I have also endeavoured, by means of the first two terms of the formula, to approximate to the elasticity of the vapour of *mercury*, as given by the experiments of M. Regnault. The data and the constants are as follows :—

Temperatures in Centigrade Degrees from the Freezing Point.	Temperatures in Centigrade Degrees from the Absolute Zero.	Pressures in Millimètres of Mercury.	Values of the Constants in the Formula Log. P $= \alpha - \frac{\beta}{t}$.	
°	°			
358	632·6	760·00	α for millimètres	=7·5305
			,, for English inches	6·1259
177·9	452·5	10·72	Log. β Centigrade scale	3·4685511
			,, Fahrenheit's scale, boiling point adjusted at 29·922 inches, ...	3·7238263

The following table exhibits the comparative results of observation and experiment :—

TABLE VIII.—VAPOUR OF MERCURY.

Temperatures in Centigrade Degrees from the Freezing Point.	Pressures in Millimètres of Mercury according to		Differences between Calculation and Experiment in Millimètres.
	the Formula (of two terms).	M. Regnault's Experiments.	
° 72·74	0·115	0·183	−0·068
100·11	0·480	0·407	+0·073
100·6	0·49	0·56	−0·07
146·3	3·49	3·46	+0·03
*177·9	10·72	10·72	0·00
200·5	21·85	22·01	−0·16
*358·0	760·00	760·00	0·00

The discrepancies are obviously of the order of errors of observation, and the formula may be considered correct for all temperatures below 200°C., and for a short range above that point. From its wanting the third term, however, it will probably be found to deviate slightly from the truth between 200° and 358°; while above the latter point it must not be relied on.

I have not carried the comparison below 72°, because in that part of the scale the whole pressure becomes of the order of errors of observation.

In conclusion, it appears to me that the following proposition, to which I have been led by the theoretical researches referred to at the commencement of this paper, is borne out by all the experiments I have quoted, especially by those of greatest accuracy, and may be safely and usefully applied to practice.

If the maximum elasticity of any vapour in contact with its liquid be ascertained for three points on the scale of the air-thermometer, then the constants of an equation of the form

$$\text{Log. P} = \alpha - \frac{\beta}{t} - \frac{\gamma}{t^2}$$

may be determined, which equation will give, for that vapour, with an accuracy limited only by the errors of observation, the relation between the temperature (t), *measured from the absolute zero* (274·6 *Centigrade degrees below the freezing point of water*), *and the maximum elasticity* (P), *at all temperatures between those three points, and for a considerable range beyond them.*

II.—ON A FORMULA FOR CALCULATING THE EXPANSION OF LIQUIDS BY HEAT.*

HAVING been lately much engaged in researches involving the comparative volumes of liquids at various temperatures, I have found the following formula very useful:

$$\text{Log. } V = Bt + \frac{C}{t} - A.$$

Log. V represents the common logarithm of the volume of a given mass of liquid, as compared with its volume at a certain standard temperature, which, for water, is the temperature of its maximum density, or 4°·1 Centigrade, and for other liquids 0° Centigrade.

t is the temperature measured from the absolute zero mentioned in my paper on the Elasticity of Vapours, in the *Edinburgh New Philosophical Journal* for July, 1849 (*see preceding Paper*), and is found by adding 274°·6 to the temperature according to the Centigrade scale.

A, B, and C, are three constants, depending on the nature of the liquid, whose values for the Centigrade scale, corresponding to water, mercury, alcohol, and sulphuret of carbon, are given below.

	A.	Log. B.	Log. C.
Water, ...	0·4414907	$\bar{4}$·8987546	1·7890286
Mercury, ...	0·0229130	$\bar{5}$·9048766	$\bar{1}$·3703897
Alcohol, ...	0·2615033	$\bar{4}$·8414452	1·2893056
Sulphuret of Carbon, ...	0·2540074	$\bar{4}$·8483872	1·2192054

The data from which the constants have been computed have been taken from the following authorities:—for water, from the experiments of Hallström; for mercury, from those of Regnault; and for alcohol and sulphuret of carbon, from those of Gay-Lussac. As the experiments of M. Gay-Lussac give only the apparent expansion of the liquids in glass, I have assumed, in order to calculate the true expansion, that the dilatation

* Originally published in the *Edinburgh New Philosophical Journal* for October, 1849.

of the glass used by him was ·0000258 of its volume for each Centigrade degree. This is very nearly the mean dilatation of the different kinds of glass. M. Regnault has shown that, according to the composition and treatment of glass, the coefficient varies between the limits ·000022 and ·000028.

Annexed are given tables of comparison between the results of the formula and those of experiment. The data from which the constants were calculated are marked with asterisks.

The table for water shows, that between 0° and 30° Centigrade, the formula agrees closely with the experiments of Hallström, and that from 30° to 100° its results lie between those of the experiments of Gay-Lussac and Deluc.

The experiments of Gay-Lussac originally gave the apparent volume of water in glass as compared with that at 100°. They have been reduced to the unit of minimum volume by means of Hallström's value of the expansion between 4°·1 and 30°, and the coefficient of expansion of glass already mentioned.

In the fifth column of the table of comparison for mercury, it is stated which of the experimental results were taken from M. Regnault's own measurements on the curve, representing the mean results of his experiments, and which from his tables of actual experiments, distinguishing the series.

In the experimental results for alcohol and sulphuret of carbon, the respective units of volume are the volumes of those liquids at their boiling points, and the volumes given by the formula have been reduced to the same units.

Expansion of Water.

Temperature on the Centigrade Scale.	Volume as compared with that at 4°·1 C. according to the Formula.	the Experiments.	Difference between Calculation and Experiment.	Authorities for the Experiments.
° 0	1·0001120	1·0001082	+ ·0000038	Hallström.
*4·1	1·0000000	1·0000000	·0000000	Do.
10	1·0002234	1·0002200	+ ·0000034	Do.
20	1·0015668	1·0015490	+ ·0000178	Do.
*30	1·0040245	1·0040245	·0000000	Do.
...	...	1·0041489	− ·0001244	Deluc.
40	1·00750	1·00748	+ ·00002	Gay-Lussac.
...	...	1·00774	− ·00024	Deluc.
60	1·01718	1·01670	+ ·00048	Gay-Lussac.
...	...	1·01773	− ·00055	Deluc.
80	1·03007	1·02865	+ ·00142	Gay-Lussac.
...	...	1·03092	− ·00085	Deluc.
100	1·04579	1·04290	+ ·00289	Gay-Lussac.
...	...	1·04664	− ·00085	Deluc.

Expansion of Mercury.

Temperature on the Centigrade Scale.	Volume as compared with that at 0° C. according to		Difference between Calculation and Experiment.	Remarks.
	the Formula.	M. Regnault's Experiments.		
*0°	1·000000	1·000000	·000000	Curve.
90·22	1·016333	1·016361	– ·000028	Series I.
100·00	1·018134	1·018153	– ·000019	Curve.
100·52	1·018230	1·018267	– ·000037	Series I.
*150·00	1·027419	1·027419	·000000	Curve.
198·79	1·036597	1·036468	+ ·000129	Series II.
205·07	1·037786	1·037805	– ·000019	Series IV.
205·57	1·037905	1·037910	– ·000005	Series III.
*300·00	1·055973	1·055973	·000000	Curve.

Expansion of Alcohol.

Temperature on the Centigrade Scale.	Volume as compared with that at 78°·41 C. according to		Difference between Calculation and Experiment.
	the Formula.	M. Gay-Lussac's Experiments.	
3·41°	·91795	·91796	– ·00001
*18·41	·93269	·93269	·00000
33·41	·94803	·94799	+ ·00004
*48·41	·96449	·96449	·00000
63·41	·98183	·98210	– ·00027
*78·41	1·00000	1·00000	·00000

Expansion of Sulphuret of Carbon.

Temperature on the Centigrade Scale.	Volume as compared with that at 46°·60 C. according to		Difference between Calculation and Experiment.
	the Formula.	M. Gay-Lussac's Experiments.	
* – 13°·40	·93224	·93224	·00000
+ 1·60	·94768	·94776	– ·00008
*16·60	·96417	·96417	·00000
31·60	·98163	·98163	·00000
*46·60	1·00000	1·00000	·00000

III.—ON THE CENTRIFUGAL THEORY OF ELASTICITY, AS APPLIED TO GASES AND VAPOURS.*

1. The following paper is an attempt to show how the laws of the pressure and expansion of gaseous substances may be deduced from that which may be called the *hypothesis of molecular vortices*, being a peculiar mode of conceiving that theory which ascribes the elasticity connected with heat to the centrifugal force of small revolutions of the particles of bodies.

The fundamental equations of this theory were obtained in the year 1842. After having been laid aside for nearly seven years, from the want of experimental data, its investigation was resumed in consequence of the publication of the experiments of M. Regnault on gases and vapours. Its results having been explained to the Royal Society of Edinburgh in February, 1850, a summary of them was printed as an introduction to a paper on the Mechanical Action of Heat in the twentieth volume of the *Transactions* of that body. I now publish the investigation in detail in its original form, with the exception of some intermediate steps of the analysis in the second and third sections, which have been modified in order to meet the objections of Professor William Thomson, of Glasgow, to whom the paper was submitted after it had been read, and to whom I feel much indebted for his friendly criticism.

This paper treats exclusively of the relations between the density, heat, temperature, and pressure of gaseous bodies in a statical condition, or when those quantities are constant. The laws of their variation belong to the theory of the mechanical action of heat, and are investigated in the other paper already referred to.

The present paper consists of six sections.

The first section explains the hypothesis.

The second contains the algebraical investigation of the statical relations between the heat and the elasticity of a gas.

The third relates to temperature and real specific heat.

The fourth treats of the coefficients of elasticity and dilatation of gases, and compares the results of the theory with those of M. Regnault's experiments.

* Read before the Royal Society of Edinburgh, February 4, 1850, and published in the *Philosophical Magazine* for December, 1851.

The fifth treats of the laws of the pressure of vapours at saturation.

The sixth relates to the properties of mixtures of gases of different kinds.

I have endeavoured throughout this paper to proceed as directly as possible to results capable of being compared with experiment, and to carry theoretical researches no further than is necessary in order to obtain such results with a degree of approximation sufficient for the purpose of that comparison.

SECTION I.—ON THE HYPOTHESIS OF MOLECULAR VORTICES.

2. The hypothesis of molecular vortices may be defined to be that which assumes—*that each atom of matter consists of a nucleus or central point enveloped by an elastic atmosphere, which is retained in its position by attractive forces, and that the elasticity due to heat arises from the centrifugal force of those atmospheres, revolving or oscillating about their nuclei or central points.*

According to this hypothesis, *quantity of heat* is the *vis viva* of the molecular revolutions or oscillations.

Ideas resembling this have been entertained by many natural philosophers from a very remote period; but, so far as I know, Sir Humphry Davy was the first to state the hypothesis I have described in an intelligible form. It appears since then to have attracted little attention, until Mr. Joule, in one of his valuable papers on the Production of Heat by Friction, published in the *London and Edinburgh Philosophical Magazine* for May, 1845, stated it in more distinct terms than Sir Humphry Davy had done. I am not aware, however, that any one has hitherto applied mathematical analysis to its development.

3. In the present stage of my researches, there are certain questions connected with the hypothesis as to which I have not found it necessary to make any definite supposition, and which I have therefore left indeterminate. Those questions are the following:—

First, Whether the elastic molecular atmospheres are continuous, or consist of discrete particles. This may be considered as including the question, whether elasticity is to a certain extent a primary quality of matter, or is wholly the result of the repulsions of discrete particles.

Secondly, Whether at the centre of each molecule there is a real nucleus having a nature distinct from that of the atmosphere, or a portion of the atmosphere in a highly condensed state, or merely a centre of condensation of the atmosphere, and of resultant attractive and repulsive forces. Therefore, although the word *nucleus* properly signifies a small central body, I shall use it in this paper, for want of a better term, to signify an *atomic centre*, whether a real nucleus or a centre of condensation and force. I assume, however, that the volume of the nucleus, if any, is inappreciably small as compared with that of the atmosphere.

4. I have now to state a supposition, which, so far as I am aware, is peculiar to my own researches. It is this:—*that the vibration which, according to the undulatory hypothesis, constitutes radiant light and heat, is a motion of the atomic nuclei or centres, and is propagated by means of their mutual attractions and repulsions.*

It will be perceived at once, that from the combination of this supposition with the hypothesis of molecular vortices, it follows that the absorption of light and of radiant heat consists in the transference of motion from the nuclei to their atmospheres; and conversely, that the emission of light and of radiant heat is the transference of motion from the atmospheres to the nuclei.

It appears to me that the supposition I have stated possesses great advantages over the ordinary hypothesis of a luminiferous ether pervading the spaces between ponderable particles, especially in the following respects:—

First, The propagation of transverse vibrations requires the operation of forces, which, if not altogether attractive, are of a very different nature from those capable of producing gaseous elasticity, and which it is difficult to ascribe to such a substance as the ether is supposed to be; while attractive forces between the atomic centres are perfectly consistent with their being kept asunder by the elasticity of their atmospheres.

Second, The immense velocity of light and radiant heat is a natural consequence of this supposition, according to which the vibrating masses must be extremely small as compared with the forces exerted by them.

Third, According to the most probable view of the theory of dispersion, the unequal refrangibility of undulations of different lengths is a consequence of the distances between the particles of the vibrating medium having an appreciable magnitude as compared with the lengths of the undulations. This is scarcely conceivable of the ether, but easily conceivable of the atomic nuclei.

Fourth, The manner in which the propagation of light and of radiant heat is affected by the molecular arrangement of crystalline bodies, is rendered much more intelligible if the vibrations are supposed to be those of the atomic nuclei, on whose mutual forces and positions the form of crystallisation must depend.

[*Note.*—The consequences of this supposition, in the theory of double refraction and polarisation, are pointed out and shown to be corroborated by Professor Stokes' experiments on diffraction, in a paper read to the Royal Society of Edinburgh on the 2nd of December, 1850, and published in the *Philosophical Magazine* for June, 1851.]

SECTION II.—INVESTIGATION OF THE GENERAL EQUATIONS BETWEEN THE HEAT AND THE ELASTICITY OF A GAS.

5. I now proceed to investigate the *statical* relations between the heat and the elasticity of a gaseous body; that is to say, their relations when both are invariable. The *dynamical* relations between those phenomena which involve the principles of the mutual conversion of heat and mechanical power by means of elastic fluids, and of the latent heat of expansion and evaporation, form the subject of another paper.

6. It is obvious that, in the condition of perfect fluidity, the forces resulting from attractions and repulsions of the atomic centres or nuclei upon their atmospheres and upon each other, must be considered as being sensibly functions merely of the general density of the body, and as being either wholly independent of the relative positions of the particles, or equal for so many different positions as to be sensibly independent of them; for otherwise a certain degree of viscosity would arise, and constitute an approach to the solid state. For the same reason, in the state of perfect fluidity each atomic atmosphere must be considered as being sensibly of uniform density in each spherical layer described round the nucleus with a given radius, and the total attractive or repulsive force on each indefinitely small portion of an atmosphere must be considered as acting in a line passing through its nucleus; that force, as well as the density, being either independent of the direction of that line, or equal for so many different and symmetrical directions as to be sensibly independent of the direction.

7. An indefinite number of equal and similar atoms, under such conditions, will arrange themselves so that the form of their bounding surfaces will be the rhombic dodecahedron, that being the nearest to a sphere of all figures which can be built together in indefinite numbers.

8. I may here explain that by the term *bounding surfaces of the atoms*, I understand a series of imaginary surfaces lying between and enveloping the atomic centres, and so placed that at every point in these surfaces the resultant of the joint actions of all the atomic centres is null. To secure the permanent existence of each atom, it must be supposed that the force acting on each particle of atomic atmosphere is centripetal towards the nearest nucleus or centre.

The variation of that force in the state of perfect fluidity must be so extremely small in the neighbourhood of those surfaces, that no appreciable error can arise, if, for the purpose of facilitating the calculation of the elasticity of the atmosphere of an atom at its bounding surface, the form of that surface is treated as if it were a sphere, of a capacity equal to that of the rhombic dodecahedron.

9. If the several atoms exercised no mutual attractions nor repulsions,

the total elasticity of a body would be equal to the elasticity of the atomic atmospheres at their bounding surfaces. Supposing such attractions and repulsions to exist, they will produce an effect, which, in the state of perfect fluidity, will be a function of the mean density of the body; and which, for the gaseous state, will be very small as compared with the total elasticity. Therefore, if p be taken to represent the superficial elasticity of the atomic atmospheres, P the actual or total elasticity of the fluid, and D its general density,

$$P = p + f(D), \quad . \quad . \quad . \quad . \quad . \quad (1.)$$

where $f(D)$ is a function of the density, which may be positive or negative according to the nature of the forces operating between distinct atoms.

10. The following relations must subsist between the masses of the atmosphere and nucleus, and the density and volume of each atom:

Let R represent the radius of the sphere already mentioned, whose capacity is equal to the volume of an atom, that volume being equal to $\frac{4\pi}{3}R^3$.

Let μ denote the mass of the atmosphere of an atom, m that of the nucleus, and $M = \mu + m$ the whole mass of the atom (so that if there is no real nucleus, but merely a centre of condensation, $m = 0$, and $M = \mu$).

Then D being the general density of the body, $\frac{\mu}{M}D$ is the mean density of the atomic atmosphere, and $M = \frac{4\pi}{3}R^3 D$.

If uR be taken to denote the distance of any spherical layer of the atmosphere from the nucleus, the density of the layer may be represented by

$$\frac{\mu}{D} D\psi(u),$$

and the function ψu will be subject to this equation of condition,

$$\mu = \int_{u=0}^{u=1} du \left(4\pi R^3 \frac{\mu}{M} D u^2 \psi u \right),$$

which is equivalent to

$$1 = 3\int_0^1 du(u^2 \psi u). \quad . \quad . \quad . \quad . \quad . \quad (2.)$$

11. So far as our experimental knowledge goes, the more substances are rarefied—that is to say, the more the forces which interfere with the operation of the elasticity of the atomic atmospheres are weakened—the more nearly do they approach to a condition called that of *perfect gas*, in

which the elasticity is simply proportional to the density. I therefore assume the elasticity of the atomic atmosphere at any given point to be represented by multiplying its density at that point by a constant coefficient b, which may vary for different substances, but, as I have already stated, without deciding whether that elasticity is a primary quality or the result of the repulsion of particles. Consequently, the superficial atomic elasticity

$$p=b\frac{\mu}{M}D\psi(1), \quad . \quad . \quad . \quad . \quad (3.)$$

$\psi(1)$ being the value of ψu, which corresponds to the bounding surface of the atom, where $u=1$.

12. Let an oscillatory movement have been propagated from the nuclei to every part of their atmospheres, the size of the orbits of oscillation being everywhere very small as compared with the radii of the atoms, and let this movement have attained a permanent state, which will be the case when every part of each atmosphere, as well as each nucleus, moves with the same mean velocity, v—*mean velocity* signifying that part of the velocity which is independent of periodic changes. It is necessary to suppose that the propagation of this movement to all parts of a molecular atmosphere is so rapid as to be practically instantaneous.

We shall conceive all the masses and densities referred to, to be measured *by weight*. Then taking g to represent the velocity generated by the force of gravity at the earth's surface in unit of time, the whole mechanical power to which the oscillatory movement in question is equivalent in *one atom* will be represented in terms of gravity by

$$\frac{Mv^2}{2g}=q; \quad . \quad . \quad . \quad . \quad . \quad (4.)$$

that is to say, the weight of the atom, M, falling through the height $\frac{v^2}{2g}$ due to the velocity v; and this is the mechanical measure of the quantity of heat in one atom in terms of gravity.

13. Any such motion of the particles of a portion of matter confined in a limited space, will in general give rise to a centrifugal tendency with respect to that space. In order to obtain definite results with respect to that centrifugal tendency in the case now under consideration, it is necessary to define, to a certain extent, the general character of the supposed movement.

In the first place, it is periodical; secondly, it is similar with respect to so large a number of radii drawn in symmetrical directions from the atomic centre, as to be sensibly similar in its effects with respect to all directions round that centre. This symmetry exists in the densities of the different particles of the atomic atmosphere in a gas, and in the forces which act upon them; and we are therefore justified in assuming it to exist in their motions.

Two kinds of motion possess these characteristics:

First, Radial oscillation, by which a portion of a spherical stratum of atmosphere surrounding an atomic centre, being *in equilibrio* at a certain distance from that centre, oscillates periodically to a greater and a less distance. This forms part of the *vis viva* of the molecular movements; but it can only affect the superficial atomic elasticity by periodic small variations, having no perceptible effect on the external elasticity.

Second, Small rotations and revolutions of particles of the atomic atmosphere round axes in the direction of radii from the atomic centre, by which each spherical layer is made to contain a great number of equal and similar vortices, or equal and similar groups of vortices having their axes at right angles to the layer, and similarly situated with respect to a great many symmetrical directions round the atomic centre.

Let us now consider the condition, as to elasticity, of a small vortex of an atmosphere whose elasticity is proportional to its density, inclosed within a cylindrical space of finite length, and not affected by any force at right angles to the axis except its own elasticity. Let Z denote the external radius of the cylinder, ρ_1 its external density, ρ its mean density, ρ' the density at any distance z from the axis (all the densities being measured by weight), w the uniform velocity of motion of its parts. The condition of equilibrium of any cylindrical layer is, that the difference of the pressures on its two sides shall balance the centrifugal force; consequently (b being the coefficient of elasticity)

$$0 = \frac{w^2 \rho'}{gz} - b \frac{d\rho'}{dz}.$$

The integral of this equation is

$$\rho' = a z^{\frac{w^2}{bg}}.$$

The coefficient a is determined by the following relation, analogous to that of equation (2), between the densities:

$$\rho \frac{Z^2}{2} = \int_0^Z dz(\rho' z) = \frac{a}{\frac{w^2}{bg} + 2} Z^{\frac{w^2}{bg}+2};$$

whence

$$a = \rho\left(\frac{w^2}{2bg} + 1\right) Z^{-\frac{w^2}{bg}}$$

And the general value of the density is

$$\rho' = \rho\left(\frac{w^2}{2bg} + 1\right)\left(\frac{z}{Z}\right)^{\frac{w^2}{bg}}. \quad \quad . \quad . \quad . \quad . \quad (5.)$$

Making $z=Z$, and multiplying by the coefficient of elasticity b, we obtain for the elasticity of the atmosphere, at the cylindrical surface of the vortex,

$$b\rho_1 = b\rho + \frac{w^2\rho}{2g}; \quad . \quad . \quad . \quad (5a.)$$

which exceeds the mean elasticity $b\rho$ by a quantity equivalent to the weight of a column of the mean density ρ, and of the height due to the velocity w, and independent of the radius of the vortex.

Supposing a spherical layer, therefore, to contain any number of vortices of any diameter, in which the mean density is equal, it is necessary to a permanent condition of that layer that the velocities in all these vortices should be equal, in order that their lateral elasticities may be equal.

Although the mean elasticity at the plane end, or any plane section at right angles to the axis of a vortex, is simply $=b\rho$, being the same as if there were no motion, yet the elasticity is variable from point to point, and the law of variation depends on the velocity. Therefore, if two vortices are placed end to end, it is necessary to a stable condition of the fluid, not only that their terminal planes should coincide, and that their mean elasticities should be *in equilibrio*, but also that their velocities should be equal, or subject only to periodical deviations from a state of equality.

Therefore, the mean velocity of vortical motion, independent of small periodic variations, is the same throughout the whole atomic atmosphere; and the mean total velocity, independent of small periodic variations, being uniformly distributed also, the *vis viva* of the former may be expressed as a constant fraction of that of the latter, so that

$$w^2 = \frac{v^2}{k}, \quad . \quad . \quad . \quad . \quad (5b.)$$

$\frac{1}{k}$ being the mean value of a coefficient which is subject to small periodical variations only.*

* As it has been represented to me that I have, without stating sufficient grounds, assumed the velocity of revolution w to be constant throughout each individual vortex, I add this note to assign reasons for that supposition.

First, Unless w, the velocity of revolution of a particle, is independent of z, its radius vector, the atomic atmosphere cannot be in a permanent condition.

For if w is a function of z, the external elasticity of a vortex will be a function of its diameter. If the whole atmosphere is in motion, vortices of different diameters must exist in the same spherical layer; and if their external elasticities are different, their condition cannot be permanent.

Second, Whatsoever may be the nature of the forces by which velocity is communicated throughout the atmosphere, the tendency of those forces must be to equalise that velocity, and thus to bring about a permanent condition.

This coefficient, being the ratio of the *vis viva* of motion of a peculiar kind to the whole *vis viva* impressed on the atomic atmospheres by the action of their nuclei, may be conjectured to have a specific value for each substance, depending, in a manner as yet unknown, on some circumstance in the constitution of its atoms. It will afterwards be seen that this circumstance is the chemical constitution.

Let the entire atmosphere of an atom be conceived to be divided into a great number of very acute pyramids meeting at the centre, and having even numbers of faces, equal and opposite in pairs; and let one of these pyramids, intersecting a spherical layer whose distance from the nucleus is $\mathrm{R}u$ and thickness $\mathrm{R}du$, cut out a frustum, containing and surrounded by vortices. Consider one pair of the faces of that frustum; their length being $\mathrm{R}du$, let their breadth be h, and their distance asunder f. Then they make with each other the angle at the apex of the pyramid

$$2 \sin^{-1} \cdot \frac{f}{2\mathrm{R}u};$$

their common area is $h\mathrm{R}du$; and the sum of the volumes of the two triangular frusta of the spherical layer, included by diagonal planes drawn between their radial edges, is

$$\frac{fh\mathrm{R}du}{2},$$

the sum of all such triangular frusta being the whole volume of the spherical layer.

The additional pressure due to the centrifugal force of vortices, viz.—

$$\frac{v^2\rho}{2gk}$$

acts on the two lateral faces, its total amount for each being

$$\frac{v^2\rho}{2gk}h\mathrm{R}du.$$

The transverse components of this pair of forces balance each other. Their radial components, amounting to

$$\frac{f}{\mathrm{R}u} \times \frac{v^2\rho}{2gk}h\mathrm{R}du = \frac{v^2\rho fhdu}{2gku},$$

constitute a centrifugal force relatively to the atomic centre, acting on the pair of triangular frusta whose mass is

$$\frac{\rho fh\mathrm{R}du}{2}.$$

The condition of permanent, or periodical, equilibrium of this pair of frusta, requires that this centrifugal force shall be balanced by the variation of the mean elasticity of the atmosphere at the two surfaces of the spherical layer, combined with the attraction of the nucleus. The action of the former of these forces is represented by

$$-b\frac{d\rho}{du}du \times \frac{fh}{2}.$$

Let the accelerating force of attraction towards the nucleus be represented by

$$-\frac{\phi(\mathrm{R}u)}{\mathrm{R}},$$

ϕ being a function, which, by the definition of an atomic bounding surface in article 8, is null at that surface, or when $u=1$. Then the attraction on the pair of frusta is

$$-\frac{\rho fhdu\phi(\mathrm{R}u)}{2}.$$

Add these three forces together; let the sum be divided by

$$\frac{1}{2}\rho fhdu,$$

and let the density ρ be denoted, as in article 10, by

$$\frac{\mu}{\mathrm{M}}\mathrm{D}\psi(u);$$

then the following differential equation is obtained as the condition of a permanent state of the atomic atmosphere:

$$\frac{v^2}{gku}-\frac{b}{\psi(u)}\cdot\frac{d\psi(u)}{du}-\phi=0. \quad . \quad . \quad . \quad (5c.)$$

This equation will be realised for each layer at its mean position, on each side of which its radial oscillations are performed.

The variation of this expression being of opposite sign to the variation of $\frac{d\psi(u)}{du}$, shows that any small disturbance of the density produces a force tending to restore that distribution to the state corresponding to the position of equilibrium of the layers, and therefore that the state indicated by equation (5*c*) is *stable*.

14. The integral of equation (5*c*) is

$$\psi u = u^{\frac{v^2}{gkb}} e^{a-\frac{1}{b}\int_1^u du\cdot\phi}. \quad . \quad . \quad . \quad (6.)$$

The arbitrary constant a is determined from the equation of condition (2) in the following manner:—

Substituting for ψu in equation (2) its value as given above, we obtain

$$1 = 3\int_0^1 du \left(u^{\frac{v^2}{gkb}+2} e^{a - \frac{1}{b}\int_1^u du \cdot \phi} \right),$$

or

$$e^{-a} = 3\int_0^1 du \left(u^{\frac{v^2}{gkb}+2} e^{-\frac{1}{b}\int_1^u du\, \phi} \right); \quad . \quad . \quad . \quad (7.)$$

which integration having been effected, we shall obtain for the value of the superficial elasticity of the atomic atmospheres,

$$p = b\frac{\mu}{\mathrm{M}}\mathrm{D}\psi(1) = b\frac{\mu}{\mathrm{M}}\mathrm{D}e^a. \quad . \quad . \quad . \quad (8.)$$

To obtain an infinite series for approximating to the value of the integral in equation (7), let the following substitutions be made:—

$$\left.\begin{array}{l} \mathrm{Log.}_e u = \lambda \\ \frac{v^2}{gkb} + 3 = 3\theta \\ e^{-\frac{1}{b}\int_1^u du \cdot \phi} = \omega, \end{array}\right\} \quad . \quad . \quad (9.)$$

and let the values of the successive differential coefficients of ω with respect to λ, when $\lambda = 0$, $\phi = 0$, and $\omega = 1$, be denoted by

$$(\omega'), (\omega''), (\omega'''), \&c.$$

Then

$$e^{-a} = 3\int_{-\infty}^0 d\lambda . e^{3\theta\lambda}\omega.$$

The value of which (when the function ϕ is such as to admit of its having a finite value) is

$$e^{-a} = \frac{1}{\theta}\left(1 - \frac{(\omega')}{3\theta} + \frac{(\omega'')}{9\theta^2} - \frac{(\omega''')}{27\theta^3} + \&c.\right),$$

whence

$$\psi(1) = e^a = \theta\left(1 + \frac{(\omega')}{3\theta} + \frac{(\omega')^2 - (\omega'')}{9\theta^2} + \frac{(\omega')^3 - 2(\omega')(\omega'') + (\omega''')}{27\,\theta^3} + \&c.\right) \quad . \quad (10.)$$

Now, because $(\omega') = -\frac{1}{b}\phi_{(u=1)} = 0$,

$$\psi(1) = e^a = \theta\left(1 - \frac{(\omega'')}{9\theta^2} + \&c.\right),$$

which may be represented by

$$\left(\frac{v^2}{3\,gkb} + 1\right)\left(1 - \mathrm{F}(\mathrm{D}, \theta)\right); \quad . \quad . \quad (10a.)$$

$F(D,\theta)$ being a quantity which becomes continually less as the density becomes less and the heat greater. The complete expression for the elasticity of a gas is therefore, according to equations (1), (8), and (10*a*),

$$P = p + f(D) = \frac{\mu}{M} D \left(\frac{v^2}{3gk} + b \right) \left(1 - F(D,\theta) \right) + f(D); \quad (11.)$$

when each atom contains a quantity of heat measured by the mechanical power corresponding to the velocity v in the weight M, or

$$q = \frac{Mv^2}{2g},$$

according to equation (4).

SECTION III.—OF TEMPERATURE, AND OF REAL SPECIFIC HEAT.

15. The definition of temperature consists of two parts:—First, the definition of that condition of two portions of matter when they are said to be *at the same temperature;* and, second, the definition of the measure of *differences of temperature.*

Two bodies are said to be *at the same temperature* when there is no tendency for one to become hotter by abstracting heat from the other; that is to say (calling the two bodies A and B), when there is either no tendency to transmission of heat between them, or when A transmits as much heat to B as B does to A. Now it is known by experiment, that any surface or other thing which affects the transmission of heat being placed between B and A, has exactly the same influence upon the same quantity of heat passing in either direction; therefore, to produce equilibrium of temperature between A and B, the powers of their atoms to communicate heat must be equal.

15*a*. If we apply to vortices at the surface of contact of the atmospheres of two atoms of the same or different kinds, the conditions of permanency laid down in article 13 for vortices in the same atmosphere, these conditions take the following form:—

First, The superficial atomic mean elasticities must be the same; in other words, the superficial atomic mean densities must be inversely as the coefficients of elasticity of the atmospheres. This is the condition of *equilibrium of pressure.*

Second, The law of variation of the elasticity from the centre to the circumference of a vortex, as expressed in equation (5), must be the same for both atoms; and this law depends on the quantity $\frac{w^2}{b} = \frac{v^2}{bk}$; therefore the condition of *equilibrium of heat* is, that the square of the velocity of vortical motion, divided by the coefficient of atmospheric elasticity, shall be the same for each atom. Of this quantity, therefore, and of constants common to all substances, *temperature* must be a function.

Taking the characteristics (A) and (B) to distinguish the quantities proper to the two atoms, we have the following equation:—

$$\left.\begin{aligned} b\frac{\mu}{\mathrm{M}}\mathrm{D}\psi(1)(\mathrm{A}) &= b\frac{\mu}{\mathrm{M}}\mathrm{D}\psi(1)(\mathrm{B}) \\ \frac{v^2}{kb}(\mathrm{A}) &= \frac{v^2}{kb}(\mathrm{B}), \\ \text{temperature} &= \phi\left(\frac{v^2}{kb}, \text{universal constants}\right). \end{aligned}\right\} \quad . \quad . \quad (12.)$$

16. In a *perfect gas*, equation (11) is reduced to

$$\mathrm{P} = p = b\frac{\mu}{\mathrm{M}}\mathrm{D}\left(\frac{v^2}{3gkb} + 1\right), \quad . \quad . \quad . \quad (12a.)$$

the pressure being simply proportional to the mean elasticity of the atmospheric part of the gas multiplied by a function of the heat, which, as equation (12) shows, is a function of the *temperature*, from its involving only $\frac{v^2}{kb}$ and universal constants.

Therefore, in two perfect gases at the same pressure and temperature, the mean elasticities of the atmospheric parts are the same, and consequently—

The mean specific gravities of the atmospheric parts of all perfect gases are inversely proportional to the coefficients of atmospheric elasticity.

Let n therefore represent the number of atoms of a perfect gas which fill unity of volume under unity of pressure at the temperature of melting ice, so that $n\mathrm{M}$ is the total specific gravity of the gas, and $n\mu$ that of its atmospheric part; then

$$bn\mu = \text{constant for all gases}, \quad . \quad . \quad (12b.)$$

and consequently

$$\frac{v^2}{kb} \propto \frac{n\mu v^2}{k}. \quad . \quad . \quad . \quad . \quad . \quad (12c.)$$

Therefore,

Temperature is a function of universal constants, and of the vortical vis viva *of the atomic atmospheres of so much of the substance as would, in the condition of perfect gas, fill unity of volume under unity of pressure at some standard temperature.*

The equation (12a) further shows, that *in any two perfect gases the respective values of the quotient of the pressure by the density corresponding to the same temperature, bear to each other a constant ratio for all temperatures,* being that of the values of the coefficient $b\frac{\mu}{\mathrm{M}}$.

Therefore the pressure of a perfect gas at a given density, or its volume under a given pressure, is the most convenient *measure* of temperature.

Let P_0 represent the elasticity of a perfect gas of the density D at the temperature of melting ice, P that of the same gas at the same density, at a temperature distant T degrees of the thermometric scale from that of melting ice, and C a constant coefficient depending on the scale employed; then the value of T is given by the equation

$$\left.\begin{aligned} T &= C\frac{P-P_0}{P_0} \\ \text{or} \quad T+C &= C\frac{P}{P_0} \end{aligned}\right\} \quad . \quad . \quad . \quad . \quad . \quad (13.)$$

The value of the constant C is found experimentally as follows:—Let P_1 represent the elasticity of the gas at the temperature of water boiling under the mean atmospheric pressure, T_1 the number of degrees, on the scale adopted, between the freezing and boiling points of water; then

$$\left.\begin{aligned} T_1 &= C\frac{P_1-P_0}{P_0} \\ \text{and} \quad C &= T_1\frac{P_0}{P_1-P_0} \end{aligned}\right\} \quad . \quad . \quad . \quad . \quad (14.)$$

C is in fact the reciprocal of the coefficient of increase of elasticity with temperature, or the reciprocal of the coefficient of dilatation, of a perfect gas at the temperature of melting ice.

17. As it is impossible in practice to obtain gases in the theoretical condition referred to, the value of C can only be obtained by approximation. From a comparison of all M. Regnault's best experiments, I have arrived at the following values, which apply to all gaseous bodies.

For the Centigrade scale, C = 274°·6, being the reciprocal of 0·00364166.

For Fahrenheit's scale, if adjusted so that 180° are equal to 100° Centigrade,—

C for temperatures measured from the freezing-point of water

$$=494°{\cdot}28.$$

C for temperatures measured from the ordinary zero

$$=494°{\cdot}28-32°=462°{\cdot}28.$$

The point C degrees below the ordinary zero of thermometric scales may be called the *absolute zero of temperature;* for temperatures measured from that point are proportional to the elasticities of a theoretically perfect gas of constant density.

Temperatures so measured may be called *absolute temperatures*. Throughout this paper I shall represent them by the Greek letter τ, so that

$$\tau = \mathrm{T} + \mathrm{C}. \quad . \quad . \quad . \quad . \quad (15.)$$

It is to be observed, that the *absolute zero of temperature* is not the *absolute zero of heat*.

18. If we now substitute for P in equation (13) its value according to equation (12*a*), we obtain the following result:—

$$\tau = \mathrm{T} + \mathrm{C} = \mathrm{C}\frac{\mu}{\mathrm{M}} \cdot \frac{\mathrm{D}}{\mathrm{P}_0}\left(\frac{v^2}{3gk} + b\right).$$

Let n represent, as before, the theoretical number of atoms in unity of volume under unity of pressure, at the temperature of melting ice, of the gas in question, supposing the disturbing forces, represented by $-\mathrm{F}(\mathrm{D},\theta)$ and $f(\mathrm{D})$, to be inappreciable; then $n\mathrm{M}$ is the weight of unity of volume under those circumstances, and it is evident that

$$\frac{\mathrm{D}}{\mathrm{P}_0} = n\mathrm{M}.$$

Consequently

$$\tau = \mathrm{T} + \mathrm{C} = \mathrm{C}n\mu\left(\frac{v^2}{3gk} + b\right), \quad . \quad . \quad . \quad (16.)$$

being the complete expression for that function of heat called *temperature*.

It follows that the function θ, which enters into the expressions for the elasticity of gases, is given in terms of temperature by the equation

$$\theta = \frac{v^2}{3gkb} + 1 = \frac{\tau}{\mathrm{C}n\mu b}. \quad . \quad . \quad . \quad (16a.)$$

If, according to the expression 4, for the quantity of heat in one atom we substitute $\frac{2gq}{\mathrm{M}}$ for v^2 in equation (16), we obtain the following equations:

$$\left.\begin{aligned} \tau &= \mathrm{C}n\mu\left(\frac{2q}{3k\mathrm{M}} + b\right), \\ q &= \frac{\mathrm{M}v^2}{2g} = \frac{3k\mathrm{M}}{2}\left(\frac{\tau}{\mathrm{C}n\mu} - b\right); \end{aligned}\right\}$$

and if Q represent the quantity of heat in *unity of weight*,

$$\mathrm{Q} = \frac{q}{\mathrm{M}} = \frac{v^2}{2g} = \frac{3k}{2}\left(\frac{\tau}{\mathrm{C}n\mu} - b\right). \quad . \quad . \quad (17.)$$

19. The real specific heat of a given substance is found by taking the differential coefficient of the quantity of heat with respect to the tempera-

ture. Hence it is expressed in various forms by the following equations, in which the coefficient $\frac{1}{k}$ is supposed not to vary sensibly with the temperature.

Real specific heat of one atom,

$$\frac{dq}{d\tau}=\frac{3kM}{2Cn\mu};$$

real specific heat of unity of weight,

$$\frac{dQ}{d\tau}=\frac{3k}{2Cn\mu}; \qquad (18.)$$

real specific heat of so much of a perfect gas as occupies unity of volume under unity of pressure at the temperature of melting ice,

$$n\frac{dq}{d\tau}=\frac{3kM}{2C\mu}.$$

The coefficient $\frac{kM}{\mu}$, representing the ratio of the total *vis viva* of the motions of the molecular atmospheres to the portion of *vis viva* which produces elasticity, multiplied by the ratio of the total mass of the atom to that of its atmospheric part, is the specific factor in the capacity of an atom for heat. The view which I have stated as probable in article 13—that the first factor of this coefficient is, like the second, a function of some permanent peculiarity in the nature of the atom—is confirmed by the laws discovered by Dulong: that the specific heats of all simple atoms bear to each other very simple ratios, and generally that of equality; that the same property is possessed by the specific heats of certain groups of similarly constituted compound atoms; and that the specific heats of equal volumes of all simple gases, at the same temperature and pressure, are equal.

The coefficient $\frac{kM}{\mu}$ varies in many instances to a great extent for the same substance in the solid, liquid, and gaseous states. So far as experiment has as yet shown, it appears not to vary, or not sensibly to vary, with the temperature; and this I consider probable *à priori*, except at or near the points of fusion of solid substances.

Apparent specific heat differs from real in consequence of the consumption and production of certain quantities of heat by change of volume and of molecular arrangement, which accompany changes of temperature.

This subject belongs to the theory of the mechanical action of heat.

SECTION IV.—OF THE COEFFICIENTS OF ELASTICITY AND DILATATION OF GASES.

20. If in equation (11) we substitute for $\frac{v^2}{3gk}+b$ its value $\frac{\tau}{Cn\mu}$ we obtain the following value for the elasticity of a gas,

$$P=\frac{D}{nM}\cdot\frac{\tau}{C}\Big(1-F(D,\theta)\Big)+f(D); \quad . \quad . \quad (19.)$$

in which $\frac{D}{nM}$ denotes the ratio of the actual weight of unity of volume to the weight of unity of volume under unity of pressure, at the absolute temperature C, in the theoretical state of perfect gas;

τ is the absolute temperature;

$-F(D,\theta)$ is a function of the temperature and density, representing the effect of the attraction of the atomic nucleus or centre in diminishing the superficial elasticity of its atmosphere;

And $f(D)$ is a function of the density only, representing the effect of the mutual attractions and repulsions of the atoms upon the whole elasticity of the body.

From this equation are now to be determined, so far as the experiments of M. Regnault furnish the requisite data, the laws of the deviation of gases from that theoretical state in which the elasticity is proportional to the density multiplied by the absolute temperature.

21. The value of $-F(D,\theta)$ is given by the infinite series of equations (10), (10a), substituting in which for θ its value $\frac{\tau}{Cn\mu b}$ we obtain the following result:—

$$-\frac{\tau}{C}F(D,\theta)=-\frac{A_1}{\tau}-\frac{A_2}{\tau^2}-\frac{A_3}{\tau^3}-\&c.;$$

A_1, A_2, A_3, &c., being a series of functions, the value of which is given by the following equation:—

$$-A_m=\frac{C^m n^{1+m} M \mu^m b^{1+m}}{3^{1+m}}\mathfrak{T}_{1+m}; \quad . \quad . \quad (20.)$$

$\mathfrak{T}_{1+m}$ being the coefficient of x^{1+m} in the development of the reciprocal of the series

$$1-(\omega')x+(\omega'')x^2-(\omega''')x^3+\&c.,$$

when (ω') &c. have the values given in equation (9).

Equation (19) is thus transformed into

$$P=\frac{D}{nM}\Big(\frac{\tau}{C}-\frac{A_1}{\tau}-\frac{A_2}{\tau^2}-\&c.\Big)+f(D). \quad . \quad . \quad (21.)$$

The series in terms of the negative powers of the absolute temperature converges so rapidly, that I have found it sufficient, in all the calculations I have hitherto made respecting the elasticity of gaseous bodies, to use the first term only, $-\frac{A}{\tau}$.

22. Instead of making any assumption respecting the laws of the attractions and repulsions which determine the functions A and $f(D)$, I have endeavoured to represent those functions by empirical formulæ, deduced respectively from the experiments of M. Regnault on what he terms the *coefficient of dilatation of gases at constant volume*, which ought rather to be called the *coefficient of increase of elasticity with temperature*, and from his experiments on the *compressibility of elastic fluids at constant temperature*.

From the data thus obtained I have calculated, by means of the theory, *the coefficients of dilatation of gases under constant pressure*, which, as a test of the accuracy of the theory, I have compared with those deduced by M. Regnault from experiment.

23. The mean coefficient of increase of elasticity with temperature at constant volume between 0° and 100° of the Centigrade thermometer is found by dividing the difference of the elasticities at those two temperatures by the elasticity at 0°, and by 100°, the difference of temperature. It is therefore represented by

$$E=\frac{P_1-P_0}{100^\circ P_0}; \quad . \quad . \quad . \quad . \quad . \quad (22.)$$

where E represents the coefficient in question, and P_0 and P_1 the elasticities at 0° and 100° Centigrade respectively.

Now by equation (21), neglecting powers of $\frac{1}{\tau}$ higher than the first, we have

$$P_0=\frac{D}{nM}\left(1-\frac{A}{C}\right)+f(D)$$

$$P_1=\frac{D}{nM}\left(\frac{100^\circ+C}{C}-\frac{A}{100^\circ+C}\right)+f(D),$$

whence

$$E=\frac{D}{nM\,P_0}\left(\frac{1}{C}+\frac{A}{C\,(C+100^\circ)}\right). \quad . \quad . \quad . \quad . \quad (23.)$$

Supposing the value of $\frac{D}{nMP_0}$ to be known, this equation affords the means of calculating the values of the function A corresponding to various densities, from those of the coefficient E as given by experiment.

As a gas is rarefied $\frac{D}{nMP_0}$ approximates to unity, A diminishes without limit, and the value of E consequently approximates to $\frac{1}{C}$, the reciprocal of the absolute temperature at 0° Centigrade. This conclusion is verified by experiment; and by means of it I have determined the values already given—viz., $C = 274°{\cdot}6$ Centigrade, and $\frac{1}{C} = {\cdot}00364166$ for the Centigrade scale.

24. In order to calculate the values of $\frac{D}{nMP_0}$, I have made use of empirical formulæ, deduced from those given by M. Regnault in his memoir on the Compressibility of Elastic Fluids. In M. Regnault's formulæ, the unit of pressure is one metre of mercury, and the unit of density the actual density corresponding to that pressure. In the formulæ which I am about to state, the unit of pressure is an atmosphere of 760 millimetres of mercury, or 29·922 inches; and the unity of density, the theoretical density in the perfectly gaseous state at 0° Centigrade, under a pressure of one atmosphere, which has been found from M. Regnault's formulæ by making the pressure = 0 in the value of $\frac{MnP_0}{D}$. M. Regnault's experiments were made at temperatures slightly above the freezing point, but not sufficiently so to render the formulæ inaccurate for the purpose of calculating the ratio in question, $\frac{D}{nMP_0}$.

The formulæ are as follows:—

Supposing $\frac{D}{Mn}$ given,

$$\frac{CnMP}{\tau D} = 1 + a\frac{D}{Mn} + \beta\left(\frac{D}{M}\right)^2;$$

which, when T is small, or τ nearly $= C$, gives an approximate value of $\frac{nMP_0}{D}$.

Supposing P_0 given,

$$\frac{\tau D}{CnMP} = 1 + \gamma P_0 + \varepsilon P_0^2;$$

which, when T is small, gives an approximate value of $\frac{D}{nMP_0}$. . . . (24.)

The values of the constants a, β, γ, ε, and of their logarithms, are given, together with the mean temperatures above the freezing point at which

M. Regnault's experiments were made, for atmospheric air, carbonic acid gas, and hydrogen.

ATMOSPHERIC AIR. T=4°·75.

	Constants.	Logarithms.
$\alpha = -\gamma =$	$-\ \cdot000860978$	$\bar{4}\cdot9349920$
$\beta =$	$+\ \cdot000011182$	$\bar{5}\cdot0485140$
$\varepsilon =$	$-\ \cdot000009700$	$\bar{6}\cdot9867717$

CARBONIC ACID GAS. T = 3°·27.

$\alpha = -\gamma =$	$-\ \cdot00641836$	$\bar{3}\cdot8074242$
$\beta =$	$-\ \cdot0000041727$	$\bar{6}\cdot6204126$
$\varepsilon =$	$+\ \cdot0000865535$	$\bar{5}\cdot9372846$

HYDROGEN. T = 4°·75.

$\alpha = -\gamma =$	$+\ \cdot000403324$	$\bar{4}\cdot6056546$
$\beta =$	$+\ \cdot0000048634$	$\bar{6}\cdot6869401$
$\varepsilon =$	$-\ \cdot0000044981$	$\bar{6}\cdot6530291$

The three substances above-mentioned are the only gases on which experiments have yet been made, under circumstances sufficiently varied to enable me to put the theory to the test I have described in article 22.

25. M. Regnault has determined the values of the coefficient of elasticity E for carbonic acid at four different densities, and for atmospheric air at ten. By applying equations (23) and (24) to those data, I have ascertained that the function A for these two gases may be represented empirically, for densities not exceeding that corresponding to five atmospheres, by the formulæ given below, which lead to formulæ for the coefficient E.

For Carbonic Acid,

$$A = a\frac{D}{nM};$$

where log. a = 0·3344538, and consequently

$$E = \frac{D}{nMP_0} \cdot \frac{1}{C}\left(1 + \frac{a}{C + 100^\circ} \cdot \frac{D}{nM}\right)$$

$$\log. \frac{a}{C + 100^\circ} = \bar{3}\cdot7608860. \qquad (25.)$$

For Atmospheric Air,

$$A = a\left(\frac{D}{nM}\right)^{\frac{6}{10}};$$

where log. a=0·3176168, and consequently

$$E = \frac{D}{nMP_0} \cdot \frac{1}{C}\left(1 + \frac{a}{C+100°} \cdot \left(\frac{D}{nM}\right)^{\frac{6}{10}}\right) \quad . \quad . \quad (26.)$$

$$\log. \frac{a}{C+100°} = \bar{3}·7440490$$

The value of log. $\frac{1}{C}$ is $\bar{3}$·5612995.

The following table shows that those empirical formulæ accurately represent the experiments, the greatest differences being less than one-half of ·0000136, which M. Regnault, in the seventy-first page of his memoir, assigns as the limit of the errors of observation due to barometric measurements alone.

As the coefficient E for hydrogen has been determined for one density only, it is impossible to obtain an empirical formula for that gas. The single ascertained value of E is nevertheless inserted in the table.

Table of Coefficients of Increase of Elasticity with Temperature at Constant Volume.

	Pressure at 0° Cent. in Atmospheres $= P_0$.	Density $\frac{D}{nM}$.	Coefficient E according to the Formulæ.	Coefficient E according to Experiment.	Difference between the Formulæ and Experiment.
			CARBONIC ACID.		
I.	0·9980	1·00448	·0036865	·0036856	+ ·0000009
II.	1·1857	1·19487	·0036951	·0036943	+ ·0000008
III.	2·2931	2·32788	·0037465	·0037523	− ·0000058
IV.	4·7225	4·87475	·0038647	·0038598	+ ·0000049
			ATMOSPHERIC AIR.		
I.	0·1444	0·1444	·0036484	·0036482	+ ·0000002
II.	0·2294	0·2294	·0036507	·0036513	− ·0000006
III.	0·3501	0·3502	·0036535	·0036542	− ·0000007
IV.	0·4930	0·4932	·0036564	·0036587	− ·0000023
V.	0·4937	0·4939	·0036564	·0036572	− ·0000008
VI.	1·0000	1·00085	·0036652	·0036650	+ ·0000002
VII.	2·2084	2·2125	·0036810	·0036760	+ ·0000050
VIII.	2·2270	2·2312	·0036812	·0036800	+ ·0000012
IX.	2·8213	2·8279	·0036880	·0036894	− ·0000014
X.	4·8100	4·8289	·0037081	·0037091	− ·0000010
			HYDROGEN.		
	1·0000	0·9996	No formula.	·0036678	

26. The empirical formulæ (24), representing the experiments of M. Regnault on the compressibility of carbonic acid gas, atmospheric air, and hydrogen at certain temperatures, give for these temperatures the values of a function which is theoretically expressed by

$$\frac{CnMP}{\tau D} = 1 - \frac{CA}{\tau^2} + \frac{CnMf(D)}{\tau D}. \quad . \quad . \quad (27.)$$

It is evident, that supposing the value of $\frac{CnMP}{\tau D}$ for any given density to be known by experiment, and that of A to be calculated from the value of the coefficient E, or from the empirical formulæ (25) and (26), the corresponding value of the function $\frac{nMf(D)}{D}$ may be determined by means of equation (27).

By this method I have obtained the following empirical formulæ for calculating the values of that function:—

For Carbonic Acid,

$$\frac{nMf(D)}{D} = h \frac{D}{nM},$$

where log. $h = \bar{3}{\cdot}1083932$.

For Atmospheric Air,

$$\frac{nMf(D)}{D} = h \left(\frac{D}{nM}\right)^{\frac{1}{2}},$$

where log. $h = \bar{3}{\cdot}8181545$.

(28.) [applies to both formulæ above]

As only one value of $\frac{nMf(D)}{D}$ for hydrogen can at present be ascertained, it is impossible to determine a formula for that gas. The single value in question is—

$$\text{For } P_0 = 1 \text{ atmosphere, } \frac{nMf(D)}{D} = {\cdot}01059. \quad . \quad (29.)$$

27. I now proceed to determine theoretically, from the data which have already been obtained, the *mean coefficients of dilatation at constant pressure*, between 0° and 100° of the Centigrade scale, for the three gases under consideration, at various pressures.

Let E′ represent the coefficient required; S_0 and S_1 the respective values of $\frac{nM}{D}$ for 0° and 100° under the pressure P, that is to say, the volumes occupied by the weight nM at those temperatures; A_0 and A_1,

f_0 and f_1, the corresponding values of A and $f(\mathrm{D})$. Then from equation (21) we deduce the following results:—

$$\mathrm{S}_0 = \frac{1}{\mathrm{P}}\left(1 - \frac{\mathrm{A}_0}{\mathrm{C}} + \mathrm{S}_0 f_0\right)$$

$$\mathrm{S}_1 = \frac{1}{\mathrm{P}}\left(1 + \frac{100^\circ}{\mathrm{C}} - \frac{\mathrm{A}_1}{\mathrm{C}+100^\circ} + \mathrm{S}_1 f_1\right);$$

and consequently

$$\mathrm{E}' = \frac{\mathrm{S}_1 - \mathrm{S}_0}{100^\circ \mathrm{S}_0}$$

$$= \frac{1}{\mathrm{S}_0 \mathrm{P}}\left(\frac{1}{\mathrm{C}} + \frac{\mathrm{A}_0}{100\mathrm{C}} - \frac{\mathrm{A}_1}{100(\mathrm{C}+100)} - \frac{\mathrm{S}_0 f_0 - \mathrm{S}_1 f_1}{100}\right). \quad . \quad (30.)$$

In applying the empirical formulæ (25), (26), and (28), to determine the values of A_1 and $\mathrm{S}_1 f_1$ in the above equation, it will produce no appreciable error to use $\frac{\mathrm{C}}{\mathrm{C}+100}\mathrm{D}_0$ as an approximate value of D_1 for that purpose only. By making the necessary substitutions, the following formulæ are obtained:—

For Carbonic Acid,

$$\mathrm{E}' = \frac{\mathrm{D}_0}{n\mathrm{MP}}\left(\frac{1}{\mathrm{C}} + a \cdot \frac{\mathrm{D}_0}{n\mathrm{M}}\right),$$

where log. $a = \bar{5}\cdot5189349$.

For Atmospheric Air,

$$\mathrm{E}' = \frac{\mathrm{D}_0}{n\mathrm{MP}}\left(\frac{1}{\mathrm{C}} + a\left(\frac{\mathrm{D}_0}{n\mathrm{M}}\right)^{\frac{6}{10}} - \beta \cdot \left(\frac{\mathrm{D}_0}{n\mathrm{M}}\right)^{\frac{1}{2}}\right),$$

where log. $a = \bar{5}\cdot4717265$
log. $\beta = \bar{6}\cdot9759738$ . . (31.)

28. The following table exhibits a comparison between the results of the formulæ and those of M. Regnault's experiments. It is not, like the preceding table (article 25), the verification of empirical formulæ, but is a test of the soundness of the theoretical reasoning from which equations (30) and (31) have been deduced.

It is impossible, from the want of a sufficient number of experimental data, to give a formula similar to (31) for hydrogen. I have calculated, however, the value of the coefficient E′ for that gas, corresponding to the pressure of one atmosphere, on the assumption that at that pressure a formula similar to that for carbonic acid gas is applicable without sensible error.

The table shows only one instance in which the difference between the result of the theory and that of experiment exceeds ·0000136; the limit, according to M. Regnault, of the errors of observation capable of arising from one cause alone,—the uncertainty of barometric measurements. That discrepancy takes place in one of the determinations of the coefficient E′ for carbonic acid gas under the pressure of one atmosphere. In the other determination the discrepancy is less than the limit.

The agreement between theory and experiment is most close for the highest pressures; and M. Regnault has shown that the higher the pressure the less is the effect of a given error of observation in producing an error in the value of the coefficient.

The theory is therefore successful in calculating the coefficients of dilatation of gases, so far as the means at present exist of putting it to the test.

TABLE OF COEFFICIENTS OF DILATATION UNDER CONSTANT PRESSURE, SHOWING A COMPARISON BETWEEN THEORY AND EXPERIMENT.

Pressure in Atmospheres.	Coefficient E′ according to the Theory.	Coefficient E′ according to M. Regnault's Experiments.	Difference between Theory and Experiment.
		CARBONIC ACID GAS.	
1·000	·0036988	First Memoir. ·0037099	− ·0000111
		Second Memoir. ·003719	− ·0000202
3·316	·0038430	First Memoir. ·0038450	− ·0000020
		ATMOSPHERIC AIR.	
1·0000	·0036650	First Memoir. ·0036706	− ·0000056
		Second Memoir. ·003663	+ ·0000020
		·003667	− ·0000020
3·3224	·0036955	First Memoir. ·0036944	+ ·0000011
3·4474	·0036969	·0036965	+ ·0000004
		HYDROGEN.	
1·0000	·0036598	·0036613	− ·0000015

SECTION V.—OF THE ELASTICITY OF VAPOUR IN CONTACT WITH THE SAME SUBSTANCE IN THE LIQUID OR SOLID STATE.

29. As the most important phenomena of evaporation take place from the liquid state, I shall generally use the word *liquid* alone throughout this section in speaking of the condition opposed to the gaseous state; but all the reasonings are equally applicable to those cases in which a substance evaporates from the solid state.

30. In considering the state of a limited space entirely occupied by a portion of a substance in the liquid form, and by another portion of the same substance in the form of vapour, both being at rest, the most obvious condition of equilibrium is, that the *total elasticity* of the substance in each of the two states must be the same; that is to say,

$$\mathrm{P} = p_0 + f(\mathrm{D}_0) = p_1 + f(\mathrm{D}_1), \quad . \quad . \quad . \quad (32.)$$

where p_0 represents the superficial atomic elasticity in the liquid state, p_1 that in the gaseous state, and $f(\mathrm{D}_0)$, $f(\mathrm{D}_1)$, the corresponding values of the pressures, positive or negative, due to mutual actions of distinct atoms.

31. A second condition of equilibrium is, that the superficial elasticities of two contiguous atoms must be equal at their surface of contact. Hence, although there may be an abrupt change of *density* at the bounding surface between the liquid and the vapour, there must be no change of superficial atomic elasticity except by inappreciable degrees; and at that bounding surface, if there is an abrupt change of density (as the reflexion of light renders probable), there must be two densities corresponding to the same superficial atomic elasticity.

32. A third condition of equilibrium is to be deduced from the mutual attractions and repulsions of the atoms of liquid and of vapour. In a gas of uniform density, those forces, acting on each individual particle at an appreciable distance from the bounding surface, balance each other, and have accordingly been treated as merely affecting the total elasticity of the body by an amount denoted by $f(\mathrm{D})$; but near the bounding surface of a liquid and its vapour, it is obvious that the action of the liquid upon any atom must be greater than that of the vapour. A force is thus produced which acts on each particle in a line perpendicular to that bounding surface, and which is probably attractive towards the liquid, very intense close to the bounding surface, but inappreciable at all distances from it perceptible to our senses. Such a force can be balanced only by a gradual increase of superficial atomic elasticity in a direction towards the liquid. Hence, although at perceptible distances from the surface of the liquid the density of vapour is sensibly uniform, the layers close to that surface are probably in a state of condensation by attraction, analogous to that of the earth's atmosphere under the influence of gravity.

Professor Faraday has expressed an opinion, founded on his own experiments and those of MM. Dulong and Thenard, that a state of condensation exactly resembling that which I have described is produced in gases by the superficial attraction of various substances, especially platinum, and gives rise to chemical actions which have been called *catalytic*.

To express this third condition algebraically, let the boundary between the liquid and the vapour be conceived to be a plane of indefinite extent, perpendicular to the axis of x; and let positive distances be measured in a direction from the liquid towards the vapour.

Let x, $x+dx$ represent the positions of two planes, perpendicular to the axis of x, bounding a layer whose thickness dx is very great as compared with the distance between two atomic centres, but very small as compared with any perceptible distance, and let a portion of the layer be considered whose transverse area is unity.

Let ρ represent the mean density of the layer. Then it is acted upon by a force

$$-\rho X dx,$$

the resultant of the actions of all the neighbouring atoms, which has the negative sign, because it is attractive towards the liquid, X being a function of the position of the layer in question, and of the densities and positions of all the neighbouring layers.

The superficial atomic elasticity behind the layer being p, and in front of it $p+\frac{dp}{dx}dx$, it is also acted on by the force

$$-\frac{dp}{dx}dx;$$

hence its condition of equilibrium is

$$\frac{dp}{dx}+\rho X=0. \qquad (33.)$$

In order to integrate this equation, so as to give a relation applicable at perceptible distances from the surface, let x_0, x_1 represent the positions of two planes perpendicular to the axis of x, the former situated in the liquid, the latter in the vapour, and so far asunder that the densities beyond them are sensibly uniform, and equal respectively to D_0 for the liquid and D_1 for the vapour, the corresponding superficial atomic elasticities being p_0 and p_1. Then dividing equation (33) by ρ, and integrating between the limits x_0 and x_1, the result obtained is

$$\int_{p_0}^{p_1}\frac{dp}{\rho}=-\int_{x_0}^{x_1}dx\,.\,X. \qquad (34.)$$

Had we a complete knowledge of the laws of molecular forces in the solid, liquid, and gaseous states, this equation, taken in conjunction with the two conditions previously stated, would be sufficient to determine formulæ for calculating the total elasticity and the respective densities of a liquid and its vapour when in contact in a limited space, at all temperatures.

33. In the absence of that knowledge, I have used equation (34), so as to indicate the *form* of an approximate equation suitable for calculating the elasticity of vapour in contact with its liquid at all ordinary temperatures, the coefficients of which I have determined empirically,—for water and mercury, from the experiments of M. Regnault, and for alcohol, ether, turpentine, and petroleum, from those of Dr. Ure.

It has been shown (equation 19) that the superficial atomic elasticity is expressible approximately in terms of the density and temperature for gases by

$$p=\rho\cdot\frac{\tau}{CnM}\left(1-F\left(\rho,\frac{\tau}{Cn\mu b}\right)\right),$$

where the function F is a very rapidly converging series, in terms of the negative powers of the absolute temperature, the coefficients being functions of the density. It is probable that a similar formula is applicable to liquids, the series being less convergent.

It follows that the density is expressible approximately in terms of the superficial atomic elasticity by

$$\rho=p\frac{CnM}{\tau}\left(1+\Phi\left(p,\frac{\tau}{Cn\mu b}\right)\right),$$

the function Φ being also a converging series in terms of the negative powers of the absolute temperature, and the coefficients being functions of p.

Making this substitution in the first side of equation (34), and abbreviating $\Phi\left(p,\frac{\tau}{Cn\mu b}\right)$ into Φ, we obtain the following result:—

$$\int_{p_0}^{p_1}dp\cdot\frac{1}{\rho}=\frac{\tau}{CnM}\int_{p_0}^{p_1}dp\cdot\frac{1}{p(1+\Phi)}$$

$$=\frac{\tau}{CnM}\left(\log._e p_1-\log._e p_0-\int_{p_0}^{p_1}dp\cdot\frac{\Phi}{p(1+\Phi)}\right)$$

$$=-\int_{x_0}^{x_1}dx\,.\,X; \qquad (35.)$$

from which, making

$$\log._e p_0+\int_{p_0}^{p_1}dp\cdot\frac{\Phi}{p(1+\Phi)}=\Psi,\text{ and } CnM\int_{x_0}^{x_1}dx\,.\,X=\Omega,$$

the following value results for the hyperbolic logarithm of the superficial atomic elasticity of the vapour at sensible distances from the surface of the liquid:—

$$\log._e p_1=\Psi-\frac{\Omega}{\tau}. \qquad (36.)$$

In the cases which occur in practice, the density of the vapour is very small as compared with that of the liquid. Hence it follows that in such cases the value of Ψ depends chiefly on the superficial atomic elasticity of the liquid, and that of Ω on its density. The density is known to diminish with the temperature, but slowly. The superficial atomic elasticity, according to equation (32), is expressed by

$$p_0 = p_1 + f(D_1) - f(D_0),$$

where p_1 and $f(D_1)$ are obviously small as compared with $f(D_0)$, a function of the density of the liquid, so that the variations of p_0 and of Ψ with the temperature are comparatively slow also.

Therefore, when the density of the vapour is small as compared with that of the liquid, the principal variable part of the logarithm of its superficial atomic elasticity, and consequently of its whole pressure, is negative, and inversely proportional to the absolute temperature; and

$$\alpha - \frac{\beta}{\tau}$$

(α and β being constants) may be regarded as the first two terms of an approximate formula for the logarithm of the pressure.

A formula of two terms, similar to this, was proposed about 1828 by Professor Roche. I have not been able to find his memoir, and do not know the nature of the reasoning from which he deduced his formula. It has since been shown, by M. Regnault and others, to be accurate for a limited range of temperature only. The quantity corresponding in it to τ is reckoned from a point determined empirically, and very different from the absolute zero.

Thus far the investigation has been theoretical. The next step is to determine empirically what other terms are requisite in order to approximate to the effect of the function $f(D)$, and of the variation of the functions Ψ and Ω.

The analogy of the formulæ for the dilatation of gases, the obvious convenience in calculation, and the fact that the deviations of the results of the first two terms from those of experiment are greatest at low temperatures, naturally induced me to try, in the first place, the effect of a third term inversely proportional to the *square* of the absolute temperature, making the entire formula for the logarithm of the pressure of vapour in contact with its liquid

$$\log. P = \alpha - \frac{\beta}{\tau} - \frac{\gamma}{\tau^2},$$

and the inverse formula, for calculating the absolute temperature from the pressure, . . . (37.)

$$\frac{1}{\tau} = \sqrt{\frac{\alpha - \log. P}{\gamma} + \frac{\beta^2}{4\gamma^2}} - \frac{\beta}{2\gamma}.$$

the values of the constants a, β, γ, being determined by the ordinary methods from three experimental data for each substance.

34. The agreement of those formulæ with the results of experiment proved so remarkable, that, as they are calculated to be practically useful, I thought it my duty not to delay their publication until I should have an opportunity of submitting my theoretical researches to the Royal Society of Edinburgh. I therefore communicated the formulæ to the *Edinburgh New Philosophical Journal* for July, 1849, together with the full details of their comparison, graphic and tabular, with the experiments of M. Regnault upon water and mercury, and with those of Dr. Ure upon alcohol, ether, turpentine, and petroleum, but without giving any account of the reasoning by which I had been led to them.

Without repeating those details here, I may state that the agreement between the results of the formulæ and those of observation is in every case as close as the precision of the experiments renders possible. This is remarkable, especially with respect to the experiments of M. Regnault on the elasticity of steam, which extend throughout a range of temperatures from 30° below zero of the Centigrade scale to 230° above it, and of pressures from $\frac{1}{2200}$ of an atmosphere to 28 atmospheres, and which, from the methods of observation adopted, especially those of measuring temperature, necessarily surpass by far in precision all other experiments of the same kind. From 20° to 230° Cent. the greatest discrepancy between calculation and experiment corresponds to a difference of $\frac{8}{100}$ of a Centigrade degree, and very few of the other differences amount to so much as $\frac{1}{20}$ of a degree. Below 20°, where the pressure varies so slowly with the temperature that its actual value is the proper test of the formula, the greatest discrepancy is $\frac{13}{100}$ of a millimetre of mercury, or $\frac{1}{200}$ of an inch. If the curves representing the formulæ were laid down on M. Regnault's diagram, they would be scarcely distinguishable from those which he has himself drawn to exhibit the mean results of his experiments.

Annexed is a table of the values of the constants a, log. β, log. γ, $\frac{\beta}{2\gamma}$, $\frac{\beta^2}{4\gamma^2}$, for the fluids for which they have been calculated. As the existing experiments on mercury, turpentine, and petroleum, are not sufficiently extensive to indicate any precise value for the coefficient γ (which requires a great range of temperatures to evince its effect), I have used for these fluids, as an approximation, the first two terms of the formula only, $a - \frac{\beta}{\tau}$.

For different measures of pressure, the contact a evidently varies equally with the complement of the logarithm of the unit of pressure.

For different thermometric scales β varies inversely as the length of a

TABLE OF THE CONSTANTS IN THE FORMULÆ FOR THE ELASTICITIES OF VAPOURS IN CONTACT WITH THEIR LIQUIDS.

(1.) Names of the Fluids.	(2.) Scale of Pressures.	(3.) Scale of Temperatures.	(4.) a.	(5.) Log. β.	(6.) Log. γ.	(7.) $\frac{\beta}{2\gamma}$	(8.) $\frac{\beta^2}{4\gamma^2}$	(9.) Range of Temperatures.	(10.) Range of Pressures.
Water, ...	Millims. of mercury.	Centigrade.	7·831247	3·1851091	5·0827176	·0063294	·00004006	Centigrade. −30° to + 230°	Millimetres. 0·35 to 20945
Water, ...	Inches of mercury.	Fahrenheit.	6·426421	3·4403816	5·5932626	·0035163	·000012364	Fahrenheit. −22° to +446°	Inches. 0·014 to 824·63
Alcohol, spec. gr. 0·813.	Inches of mercury.	Fahrenheit.	6·16620	3·3165220	5·7602709	·0017998	·000003239	+32° to 264°	0·41 to 165·58
Ether, boiling at 105° F.	Inches of mercury.	Fahrenheit.	5·33590	3·2084573	5·5119893	·0024856	·000006178	105° to 210°	30·00 to 163·27
Ether, boiling at 104° F.	Inches of mercury.	Fahrenheit.	5·44580	3·2571312	5·3962460	·0036296	·000013174	34° to 104°	6·20 to 30·00
Turpentine, . .	Inches of mercury.	Fahrenheit.	5·98187	3·5380701	...	...	...	304° to 362°	30·00 to 62·24
Petroleum, . . .	Inches of mercury.	Fahrenheit.	6·19451	3·5648490	...	...	...	316° to 375°	30·00 to 64·50
Mercury, . . .	Millims. of mercury.	Centigrade.	7·5305	3·4685511	...	...	...	Centigrade. 72°·74 to 358°	Millimetres. 0·115 to 760
Mercury, . . .	Inches of mercury.	Fahrenheit.	6·1259	3·7238236	...	...	...	Fahrenheit. 162°·93 to 676°·4	Inches. ·0046 to 29·92

degree, γ inversely as the square of that length, $\frac{\beta}{2\gamma}$ directly as the length of a degree, and $\frac{\beta^2}{4\gamma^2}$ directly as the square of that length.

For all the fluids except water, it will probably be found necessary to correct more or less the values of the constants, when more precise and extensive experiments have been made, especially those for the more volatile ether, and for turpentine, petroleum, and mercury, which have all been determined from data embracing but a small range of pressures.

In reducing the constants for the Centigrade scale to those for Fahrenheit's scale, 180° of the latter have been assumed to be equal to 100° of the former. In order that this may be the case, the boiling point of Fahrenheit's scale must be adjusted under a barometric pressure of 760 millimetres, or 29·922 inches, of mercury, whose temperature is 0° Centigrade.

In the ninth and tenth columns of the table are given the limits on the scales of temperature and pressure between which the formulæ have been compared with experiment. It is almost certain that the formula for the pressure of steam may be employed without material error for a considerable range beyond, and probably also that for the pressure of vapour of alcohol; but none of the formulæ are to be regarded as more than approximations to the exact physical law of the elasticity of vapours, for the determination of which many data are still wanting, that can only be supplied by extensive series of experiments.

The following are some additional values of the constant a for steam, corresponding to various units of pressure used in practice :—

Units of Pressure.	Values of a.
ATMOSPHERES of 760 millimetres of mercury—	
= 29·922 inches of mercury	
= 14·7 lbs. on the square inch	
= 1·0333 kilog. on the square centim.,	4·950433
ATMOSPHERES of 30 inches of mercury—	
= 761·99 millimetres	
= 14·74 lbs. on the square inch	
= 1·036 kilog. on the square centim.,	4·949300
Kilogrammes on the square centimetre, . .	4·964658
Kilogrammes on the circular centimetre, . .	4·859748
Pounds avoirdupois on the square inch, . .	6·117662
Pounds avoirdupois on the circular inch, . .	6·012752
Pounds avoirdupois on the square foot, . .	8·276025

All the numerical values of the constants are for common logarithms.

35. According to the principles which form the basis of calculation in this section, every substance, in the solid or liquid state, is surrounded by an atmosphere of vapour, adhering to its surface by molecular attraction; and even when the presence of vapour is imperceptible at all visible distances from the body's surface, the elasticity of the strata close to that surface may be considerable, and sufficient to oppose that resistance to being brought into absolute contact, which is well known to be very great in solid bodies, and perceptible even in drops of liquid. It is possible that this may be the only cause which prevents all solid bodies from cohering when brought together.

The action of an atmosphere of vapour, so highly dense and elastic as to operate at visible distances, may assist in producing the *spheroidal state* of liquids.

If the particles of clouds are small vesicles or bubbles (which is doubtful), the vapour within them may, according to these principles, be considerably more dense than that which pervades the external air, and may thus enable them to preserve their shape.

Section VI.—Of Mixtures of Gases and Vapours of Different Kinds.

36. The principle stated in Section II. article 11, that *the elasticity of the atomic atmosphere is proportional to its density*, might be otherwise expressed by saying, that *the elasticity of any number of portions of atomic atmosphere, compressed into a given space, is equal to the sum of the elasticities which such portions would respectively have, if they occupied the same space separately.*

If the same principle here laid down for portions of atomic atmosphere of any one kind of substance, be considered as true also of portions of atomic atmosphere of substances of different kinds mixed, and if it be supposed that when two or more gases are mixed there is no mutual force exerted between atoms of different kinds, except the elastic pressure of the atomic atmospheres, it will then evidently follow,—

First, that the mixed gases will only be *in equilibrio* when the particles of each of them are diffused throughout the whole space which contains them.

Secondly, that the particles of each gas taken separately will be in the same condition as to density, elasticity, arrangement, and mutual action, and also as to gravitation, or any other action of an external body, as if that gas occupied the space alone.

Thirdly, that the joint elasticity of the mixed atomic atmospheres at any given point, will be the sum of the elasticities which they would respectively have had at that point if each gas had occupied the space alone.

Fourthly, that the value of the elasticity, positive or negative, resulting from the attractions and repulsions of separate atoms, will be the sum of the values it would have had if each gas had occupied the space alone; and,

Fifthly, that the total elasticity of the mixed gases will be the sum of the elasticities which each would have had separately in the same space.

If there are any mutual actions between the particles of different gases except the elasticity of the molecular atmospheres, these conclusions will no longer be rigidly true; but they will still be approximately true if the forces so operating are very small. This is probably the actual condition of mixed gases.

37. On applying the same principle to the case of a gas mixed with a vapour in contact with its liquid, it is obvious that if the attractions and repulsions of the particles of the gas upon those of the vapour are null or inappreciable, the direct effect of the presence of the gas upon the elasticity assumed by the vapour at a given temperature will also be null or inappreciable.

The gas, however, may have a slight indirect influence, by compressing the liquid, and consequently increasing its superficial atomic elasticity and its attractive power, on which the functions Ψ and Ω in equation (36) depend. The probable effect of this will be to make the elasticity of the vapour somewhat less than if no gas were present. There appear to be some indications of such an effect; but they are not sufficient to form a basis for calculation.

Supposing the gas, on the contrary, to exercise an appreciable attraction on the particles of vapour, the elasticity of the latter will be increased. Traces of an effect of this kind are perceptible in M. Regnault's experiments on the vapour of mercury in which air was present.

38. I have already referred to the property ascribed by Professor Faraday to various substances, of attracting, and retaining at their surfaces, layers of gas and vapour in a high state of condensation. Supposing a solid body to acquire, in this manner, a mixed atmosphere, consisting partly of its own vapour and partly of foreign substances, the total elasticity of that atmosphere at any point will be equal, or nearly equal, to the sum of the elasticities which each ingredient would have had separately; and thus solid metals, glass, charcoal, earthy matters, and other substances, may acquire a great power of resisting cohesion, although producing no perceptible vapours of their own at ordinary temperatures.

IV.—ON THE CENTRIFUGAL THEORY OF ELASTICITY AND ITS CONNECTION WITH THE THEORY OF HEAT.*

SECTION I.—RELATIONS BETWEEN HEAT AND EXPANSIVE PRESSURE.

1. IN February, 1850, I laid before the Royal Society of Edinburgh a paper in which the laws of the pressure and expansion of gases and vapours were deduced from the supposition that that part of the elasticity of bodies which depends upon heat arises from the centrifugal force of the revolutions of the particles of elastic atmospheres surrounding nuclei or atomic centres. A summary of the results of this supposition, which I called the Hypothesis of Molecular Vortices, was printed in the *Transactions* of this Society, Vol. XX., as an introduction to a series of papers on the Mechanical Action of Heat; and the original paper has since appeared in detail in the *Philosophical Magazine*.

In that paper the *bounding surfaces of atoms* were defined to be imaginary surfaces, situated between and enveloping the atomic nuclei, and symmetrically placed with respect to them, and having this property—that at these surfaces the attractive and repulsive actions of the atomic nuclei and atmospheres upon each particle of atomic atmosphere balance each other. The pressure of the atomic atmospheres at those imaginary boundaries is the part of the total expansive pressure of the body which varies with heat, the effect of the centrifugal force of molecular vortices being to increase it.

In the subsequent investigation it was assumed that, owing to the symmetrical action of the particles of gases in all directions, and the small amount of those attractive and repulsive forces which interfere with the elasticity of their atmospheres, no appreciable error would arise from treating the boundary of the atmosphere of a single atom, in calculation, as if it were spherical, an assumption which very much simplified the analysis.

An effect, however, of this assumption was to make it doubtful whether the conclusions deduced from the hypothesis were applicable to any substances except those nearly in the state of perfect gas. I have, therefore,

* Read before the Royal Society of Edinburgh on December 15, 1851, and published in Vol. XX., Part iii., of the *Transactions* of that Society.

in the present paper, investigated the subject anew, without making any assumption as to the arrangement of the atomic centres, or the form of the boundaries of their atmospheres. The equations deduced from the hypothesis between expansive pressure and heat, are therefore applicable to all substances in all conditions; and it will be seen that they are identical with those in the original paper, showing that the assumption that the atomic atmospheres might be treated, in calculation, as if spherical, did not give rise to any error.

By the aid of certain transformations in those equations I have been enabled, in investigating the principles of the mutual transformation of heat and expansive power, to deduce *Joule's law* of the equivalence of heat and mechanical power directly from them, instead of taking it (as I did in my previous papers) as a consequence of the principle of *vis viva*. *Carnot's law* is also deduced directly from the hypothesis, as in one of the previous papers.

2. *Classification of Elastic Pressures.*—The pressures considered in the present paper are those only which depend on the volume occupied by a given weight of the substance, not those which resist change of figure in solids and viscous liquids. Certain mathematical relations exist between those two classes of pressures; but they do not affect the present investigation.

To illustrate this symbolically, let V represent the volume occupied by unity of weight of the substance, so that $\frac{1}{V}$ is the mean density; Q, the quantity of heat in unity of weight, that is to say, the *vis viva* of the molecular revolutions, which, according to the hypothesis, give rise to the expansive pressure depending on heat; and let P denote the total expansive pressure. Then,

$$P = F(V, Q) + f(V), \quad . \quad . \quad . \quad . \quad (1.)$$

In this equation, F(V, Q) is the pressure of the atomic atmospheres at the surfaces called their boundaries, which varies with the centrifugal force of the molecular vortices as well as with the mean density; and $f(V)$ is a portion of pressure due to the mutual attractions and repulsions of distinct atoms, and varying with the number of atoms in a given volume only. If the above equation be differentiated with respect to the hyperbolic logarithm of the density, we obtain the coefficient of elasticity of volume

$$\frac{1}{\mathfrak{d}} = -\frac{dP}{\frac{dV}{V}} = -\frac{d}{\frac{dV}{V}} F(V, Q) - \frac{d}{\frac{dV}{V}} f(V). \quad . \quad . \quad (1A.)$$

where $\mathfrak{d}$ denotes the cubic compressibility.

The latter portion of this coefficient, $-\frac{\frac{d}{dV}}{V}f(V)$, consists of two parts, one of which is capable of being resolved into forces, acting along the lines joining the atomic centres, and gives rise to rigidity or elasticity of figure as well as to elasticity of volume, while the other, which is not capable of being so resolved, gives rise to elasticity of volume only. The ratio of each of those parts to their sum must be a function of the heat, the former part being greater and the latter less, as the atomic atmosphere is more concentrated round the nucleus; that is to say, as the heat is less; but their sum, so far as elasticity of volume is concerned, is a function of the density only.

That is to say, as in equation (12) of my paper on the Laws of the Elasticity of Solids (*Cambridge and Dublin Mathematical Journal*, February, 1851), let the total coefficient of elasticity of volume be denoted thus—

$$\frac{1}{\mathfrak{d}} - J + \psi(C_1, C_2, C_3), \qquad \text{(1B.)}$$

C_1, C_2, C_3, being the coefficients of rigidity round the three axes of elasticity, and J a coefficient of fluid elasticity; then

$$\left.\begin{aligned} J &= -\frac{\frac{d}{dV}}{V}F(V, Q) - \psi(V, Q)\cdot\frac{\frac{d}{dV}}{V}f(V) \\ \phi(C_1, C_2, C_3) &= -\Big(1 - \psi(V, Q)\Big)\cdot\frac{\frac{d}{dV}}{V}f(V) \end{aligned}\right\} \qquad \text{(1C.)}$$

For the present, we have to take into consideration that portion only of the expansive pressure which depends on density and heat jointly, and is the means of mutually converting heat and expansive power; that is to say, the pressure at the boundaries of the atomic atmospheres, which I shall denote by

$$p = F(V, Q).$$

Pressures throughout this paper are supposed to be measured by units of weight upon unity of area; densities by the weight of unity of volume.

3. *Determination of the External Pressure of an Atomic Atmosphere.*— Let a body be composed of equal and similar atomic nuclei, arranged in any symmetrical manner, and enveloped by an atmosphere, the parts of which are subject to attractive and repulsive forces exercised by each other and by the nuclei. Let it further be supposed that this atmosphere, at each point, has an elastic pressure proportional to the density at that point, multiplied by a specific coefficient depending on the nature of the

substance, which I shall denote by h. (This coefficient was denoted by b in previous papers.)

Let ρ and p' denote the density and pressure of the atomic atmosphere at any point; then

$$p' = h\rho$$

$$\text{Let } -g\frac{d\Phi}{dx}, -g\frac{d\Phi}{dy}, -g\frac{d\Phi}{dz},$$

be the accelerative forces operating on a particle of atomic atmosphere in virtue of the molecular attractions and repulsions which I have made explicitly negative, attractions being supposed to predominate. The property of the surfaces called the boundaries of the atoms is this

$$\left(\frac{d\Phi}{dx}\right)_1 = 0, \left(\frac{d\Phi}{dy}\right)_1 = 0, \left(\frac{d\Phi}{dz}\right)_1 = 0,$$

the suffix $_1$ being used to distinguish the value of quantities at those surfaces. Hence Φ_1 is a maximum or minimum. Those surfaces are symmetrical in form round each nucleus, and equidistant between pairs of adjacent nuclei. Their equation is

$$\Phi - \Phi_1 = 0.$$

Let M denote the total weight of an atom; μ that of its atmospheric part, and $M - \mu$ that of its nucleus; then

MV is the volume of the atom,—

$\frac{\mu}{MV}$ the mean density of the atmospheric part, measured by weight, the nucleus being supposed to be of insensible magnitude;—

and we have the following equations

$$\left.\begin{aligned} MV &= \iiint_{(1)} dx\,dy\,dz, \\ \mu &= \frac{\mu}{MV}\iiint_{(1)} dx\,dy\,dz = \iiint_{(1)} \rho\,dx\,dy\,dz. \end{aligned}\right\} \quad . \quad (2.)$$

The suffix $(_1)$ denoting that the integration is to be extended to all points within the surface

$$(\Phi - \Phi_1 = 0).$$

According to the hypothesis now under consideration, *Heat* consists in a revolving motion of the particles of the atomic atmosphere, communicated to them by the nuclei. Let v be the common mean velocity possessed by the nucleus of an atom and the atmospheric particles, when the distribution of this motion has been equalised. I use the term *mean* velocity to denote that the velocity of each particle may undergo small periodic changes, which it is unnecessary to consider in this investigation.

Then the quantity of heat in unity of weight is

$$Q = \frac{v^2}{2g}$$

being equal to the mechanical power of unity of weight falling through the height $\frac{v^2}{2g}$. The quantity of heat in one atom is of course MQ, and in the atmospheric part of an atom, μQ.

I shall leave the form of the paths described by the atmospheric particles indeterminate, except that they must be closed curves of permanent figure, and included within the surface $(\Phi - \Phi_1 = 0)$. Let the nucleus be taken as the origin of co-ordinates, and let α, β, γ, be the direction-cosines of the motion of the particles at any point (x, y, z). Then the equations of a permanent condition of motion at that point, are

$$\left.\begin{aligned} -\frac{1}{\rho}\cdot\frac{dp'}{dx} - \frac{d\Phi}{dx} - 2\,Q\left(\alpha\frac{d}{dx} + \beta\frac{d}{dy} + \gamma\frac{d}{dz}\right)\alpha = 0 \\ -\frac{1}{\rho}\cdot\frac{dp'}{dy} - \frac{d\Phi}{dy} - 2\,Q\left(\alpha\frac{d}{dx} + \beta\frac{d}{dy} + \gamma\frac{d}{dz}\right)\beta = 0 \\ -\frac{1}{\rho}\cdot\frac{dp'}{dz} - \frac{d\Phi}{dz} - 2\,Q\left(\alpha\frac{d}{dx} + \beta\frac{d}{dy} + \gamma\frac{d}{dz}\right)\gamma = 0 \end{aligned}\right\} \quad . \quad (3.)$$

Let r be the radius of curvature of the path of the particles through (x, y, z), and α' β' γ', its direction-cosines; then the above equations obviously become

$$\left.\begin{aligned} -\frac{1}{\rho}\frac{dp'}{dx} - \frac{d\Phi}{dx} - 2Q\frac{\alpha'}{r} = 0 \\ -\frac{1}{\rho}\frac{dp'}{dy} - \frac{d\Phi}{dy} - 2Q\frac{\beta'}{r} = 0 \\ -\frac{1}{\rho}\frac{dp'}{dz} - \frac{d\Phi}{dz} - 2Q\frac{\gamma'}{r} = 0 \end{aligned}\right\} \quad . \quad . \quad (3A.)$$

If these equations are integrable,

$$\frac{\alpha'}{r}dx + \frac{\beta'}{r}dy + \frac{\gamma'}{r}dz$$

must be an exact differential. Let $-\phi$ be its primitive function; the negative sign being used, because α', β', γ' must be generally negative. Then the integral of the equations (3) is

$$\log._e \rho = \frac{1}{h}\int\frac{dp'}{\rho} = \frac{1}{h}(2\,Q\phi - \Phi) + \text{constant};$$

or taking ρ_1 to denote the pressure at the bounding surface of the atom :—

$$\rho = \rho_1 e^{\frac{2Q}{h}(\phi - \phi_1) - \frac{1}{h}(\Phi - \Phi_1)} \quad . \quad . \quad . \quad (4.)$$

Our present object is to determine the superficial-atomic density, ρ_1, and thence the pressure $p = h\rho_1$, in terms of the mean density $\frac{1}{V}$ and heat Q. For this purpose we must introduce the above value of ρ into equation (2), giving

$$\mu = \rho_1 \cdot \iiint_{(1)} e^{\frac{2Q}{h}(\phi - \phi_1) - \frac{1}{h}(\Phi - \Phi_1)} dx\,dy\,dz$$

whence

$$p = h\rho_1 = h\mu \div \iiint_{(1)} e^{\frac{2Q}{h}(\phi - \phi_1) - \frac{1}{h}(\Phi - \Phi_1)} dx\,dy\,dz. \quad (5.)$$

Let the volume of the atom be conceived to be divided into layers, in each of which ϕ has a constant value. Then we may make the following transformations.

$$\left. \begin{aligned} \iiint dx\,dy\,dz &= k\mathrm{MV} \int e^{k(\phi - \phi_1)} \frac{\psi}{\psi_1} d\phi \\ \iiint e^{-\frac{1}{h}(\Phi - \Phi_1)} dx\,dy\,dz &= k\,\mathrm{MV} \int e^{k(\phi - \phi_1)} \frac{\omega}{\omega_1} d\phi \end{aligned} \right\} \quad (6.)$$

k being a specific constant, and ψ and ω functions of ϕ, and of the nature and density of the substance.

The lower limit of integration of ϕ must be made $-\infty$, that it may include orbits of indefinitely small magnitude described round the atomic centre.

The nature of the function ψ is limited by the following condition,

$$1 = k \int_{-\infty}^{\phi_1} e^{k(\phi - \phi_1)} \frac{\psi}{\psi_1} d\phi \quad . \quad . \quad . \quad (7.)$$

$$\text{Let } \frac{2Q}{hk} + 1 = \theta.$$

Then these transformations give the following result for the pressure at the bounding surface of an atom :—

$$\left. \begin{aligned} p = h\rho_1 &= \frac{h\mu}{\mathrm{MV}} \div \int_{-\infty}^{k\phi_1} e^{\theta k(\phi - \phi_1)} \frac{\omega}{\omega_1} k\,d\phi \\ &= \frac{h\mu}{\mathrm{MV}} \cdot \frac{\theta\omega_1}{\omega_1 - \frac{\omega'_1}{\theta} + \frac{\omega''_1}{\theta^2} - \frac{\omega'''_1}{\theta^3} + \&c.} \end{aligned} \right\} \quad . \quad (8.)$$

ω'_1, &c., being the successive differential coefficients of ω with respect to $k\phi$, when $\phi = \phi_1$.

4. The following transformation will be found useful in the sequel.

Let λ be the indefinite value of $\log._e V$, and λ_1 its actual value in the case under consideration. Let G be the same function of λ which ω is of $k\phi$, and let G′, G″, &c., be its successive differential coefficients with respect to λ.

Let

$$H_1 = \int_{-\infty}^{\lambda_1} e^{\theta(\lambda - \lambda_1)} G \, d\lambda = \frac{G_1}{\theta} - \frac{G_1'}{\theta^2} + \frac{G_1''}{\theta^3} - \&c.$$

Then

$$p = \frac{h\mu G_1}{MVH_1}. \quad . \quad . \quad . \quad . \quad . \quad (9.)$$

The function H has the following properties, which will be afterwards referred to:—

$$\left.\begin{aligned} &\frac{dH_1}{d\lambda_1} + \theta H_1 - G_1 = 0 \\ &\int_{-\infty}^{\lambda_1} H d\lambda = -\frac{dH_1}{d\theta} \end{aligned}\right\} \quad . \quad . \quad . \quad . \quad (10.)$$

5. *Case of a Perfect Gas.*—As a substance is rarefied, it gradually approaches a condition in which the pressure, under like circumstances as to heat, varies proportionally to the density. This is because the effect of the molecular attractions and repulsions on the pressure diminishes with the density, so that Φ, ω, and G approximate to constant quantities. In the limiting or perfectly gaseous condition, therefore,

$$H_1 = \frac{G_1}{\theta}$$

and

$$p = \frac{h\mu\theta}{MV} = \frac{h\mu}{MV}\left(\frac{2Q}{hk} + 1\right). \quad . \quad . \quad . \quad . \quad (11.)$$

6. *Equilibrium of Heat: Nature of Temperature and Real Specific Heat.*—When the atmospheres of atoms of two different substances are in contact at their common bounding surface, it is necessary to a permanent condition that the pressure in passing that surface should vary continuously.

Let (*a*) and (*b*) be taken as characteristics, to distinguish the specific quantities peculiar to the two media respectively. Let dm denote the volume of an indefinitely thin layer, close to the bounding surface. Then the following equations must be fulfilled to ensure a permanent condition:—

$$p\,(a) = p\,(b); \; \frac{dp'}{dm}(a) = \frac{dp'}{dm}(b) \text{ when } p' = p. \quad . \quad . \quad (12.)$$

By making the proper substitutions in equation (4), it appears that

$$p' = pe^{k\theta(\phi - \phi_1)} \frac{\psi_1}{\omega_1} \cdot \frac{\omega}{\psi}.$$

Hence

$$\frac{dp'}{dm}(p' = p) = p\left(\theta \frac{d(k\phi)}{dm} + \frac{d \cdot \frac{\omega}{\psi}}{dm}\right).$$

Now, p is the same for both media:

$$\frac{\psi_1}{\omega_1} \cdot \frac{\omega}{\psi} = e^{-\frac{1}{h}(\Phi - \Phi_1)}$$

is either a maximum or a minimum, so that its differential is null; and dm is a continuous function of $k\phi$, so that

$$\frac{d(k\phi)}{dm}(a) = \frac{d(k\phi)}{dm}(b).$$

There remains only the function of heat:

$$\theta = \frac{2Q}{hk} + 1.$$

Therefore the condition of a permanent state of molecular motion, that is to say, the condition of *equilibrium of heat*, is that this function shall be the same for the two substances; or that

$$\frac{2Q_a}{h_a k_a} = \frac{2Q_b}{h_b k_b}. \quad . \quad . \quad . \quad . \quad . \quad (13.)$$

Hence, *temperature* depends on the above function only; for the definition of temperature is, that bodies at the same temperature are in a permanent condition as to heat, so far as their mutual action is concerned.

The ratio of the *real specific heat* of (a) to that of (b) is obviously

$$h_a k_a : h_b k_b. \quad . \quad . \quad . \quad . \quad . \quad (14.)$$

7. *Measure of Temperature and Specific Heat.*—The function θ is proportional to the pressure of a perfect gas at a constant density. That pressure, therefore, is the most convenient *measure* of temperature.

Let τ denote *absolute* temperature, as measured by the pressure of a perfect gas at constant density, and reckoned from a certain *absolute zero*, 274°·6 Centigrade, or 494°·28 Fahrenheit below the temperature of melting ice. Let κ be a constant which depends on the length of a degree on the thermometric scale, and is the same for all substances in nature. Then

$$\left.\begin{aligned} \tau &= \kappa\theta = \frac{2\kappa Q}{hk} + \kappa \\ Q &= (\tau - \kappa)\frac{hk}{2\kappa} \end{aligned}\right\} \quad . \quad . \quad . \quad (15.)$$

and the real specific heat of the substance, that is to say, the depth of fall, under the influence of gravity, which is equivalent to a rise of one degree of temperature in the body, is represented by

$$\mathfrak{k} = \frac{h k}{2 \kappa} \quad . \quad . \quad . \quad . \quad . \quad (16.)$$

The pressure of a perfect gas is represented in terms of temperature by

$$p = \frac{h \mu \tau}{\mathrm{M V} \kappa} \quad . \quad . \quad . \quad . \quad . \quad (17.)$$

It may also be expressed thus: let τ_0 denote the absolute temperature of melting ice in degrees of the scale employed, and V_0 the volume of unity of weight of the substance in the theoretical state of perfect gas, at the temperature of melting ice and pressure unity:—then

$$p = \frac{\mathrm{V}_0}{\mathrm{V}} \frac{\tau}{\tau_0}; \quad . \quad . \quad . \quad . \quad . \quad (18.)$$

On comparing this with equation (17) we see that

$$\left.\begin{array}{l} \dfrac{h \mu}{\kappa \mathrm{M}} = \dfrac{\mathrm{V}_0}{\tau_0}; \\ \kappa = \dfrac{h \mu \tau_0}{\mathrm{M V}_0}; \ \dfrac{h \mu}{\mathrm{M V}_0} = \dfrac{\kappa}{\tau_0} \end{array}\right\} \quad . \quad . \quad . \quad (19.)$$

Now h is the specific elasticity of the atomic atmosphere of the substance; $\frac{\mu}{\mathrm{M V}_0}$ is the mean specific gravity of that atmosphere, when the body is in the theoretical state of perfect gas; and κ and τ_0 are the same for all substances in nature. Therefore, *for every substance in nature, the mean specific gravity of the atomic atmosphere in the theoretical state of perfect gas is inversely proportional to the specific elasticity of that atmosphere.*

Real specific heat may also be thus expressed:—

$$\mathfrak{k} = \frac{\mathrm{V}_0}{\tau_0} \cdot \frac{k \mathrm{M}}{2 \mu} \quad . \quad . \quad . \quad . \quad . \quad (20.)$$

in which $\frac{\mathrm{V}_0}{\tau_0}$ corresponds to $\frac{1}{\mathrm{C} n \mathrm{M}}$ in my former papers, and $\frac{k \mathrm{M}}{2 \mu}$ to $\frac{3 k \mathrm{M}}{2 \mu}$ or $\frac{1}{\mathrm{N}}$.

The latter factor appears to depend on the chemical constitution of the substance, being the same for all simple gases.

8. *Total Pressure of Substances in general, expressed in terms of Temperature.* In equation (9) let $\frac{\tau}{\kappa}$ be put for θ: then

$$P = p + f(V) = f(V) + \frac{h\mu}{\kappa M}\frac{\tau}{V}G_1 \div \left\{G_1 - \frac{\kappa G_1'}{\tau} + \frac{\kappa^2 G_1''}{\tau^2} - \&c.\right\}$$

$$= f(V) + \frac{V_0}{\tau_0}\frac{\tau}{V}\left\{1 - \frac{A_1}{\tau} - \frac{A_2}{\tau^2} - \frac{A_3}{\tau^3} - \&c.\right\} \quad . \quad (21.)$$

where

$$A_1 = -\frac{\kappa G'}{G_1};\; A_2 = -\frac{\kappa^2}{G_1^2}(G_1'^2 - G_1'')$$

$$A_3 = -\frac{\kappa^3}{G_1^3}(G_1'^3 - 2G_1'G_1'' + G_1'''); \; \&c.$$

This formula is identical with that which I employed in my former paper to represent the pressure of an imperfect gas, and which I found to agree with M. Regnault's experiments, when the coefficients A and the function $f(V)$ had been calculated empirically.

SECTION II.—RELATIONS BETWEEN HEAT AND EXPANSIVE POWER.

9. *Variations of Sensible and Latent Heat: Fundamental Equation of the Theory.*—If the forms, positions, and magnitudes of the paths described by the revolving particles of the atomic atmospheres be changed, whether by a variation of mean density, or by a variation of temperature, an increase or diminution of the *vis viva* of their motion, that is to say, of the heat of the body, will take place in virtue of that change of the paths of motion—an increase when they are contracted, and a diminution when they are dilated.

Let $\delta.Q$ represent, when positive, the indefinitely small quantity of heat which must be communicated to unity of weight of a substance, and when negative, that which must be abstracted from it, in order to produce the indefinitely small variation of temperature $\delta\tau$ simultaneously with the indefinitely small variation of volume δV. Let $\delta.Q$ be divided into two parts

$$\delta Q + \delta Q' = \delta . Q,$$

of which δQ, being directly employed in varying the *velocity* of the particles, is the variation of the *actual* or *sensible* heat possessed by the body; while $\delta Q'$, being employed in varying their *orbits*, represents the amount of the mutual transformation of heat with expansive power and molecular action, or the variation of what is called the *latent* heat; that is to say, of a molecular condition constituting a source of power, out of which heat may be developed. ($\delta Q'$ in this paper corresponds to $-\delta Q'$ in my former papers.)

The variation of sensible heat has evidently this value,

$$\delta Q = k\,\delta\tau. \quad . \quad . \quad . \quad . \quad (22.)$$

Let δx, δy, δz, be the displacements of the orbit of the particles of atomic atmosphere at the point (x, y, z). A molecule $\rho\, dx\, dy\, dz$ is acted upon by the accelerative forces (see equation 3A).

$$-2Q\frac{d\phi}{dx};\ -2Q\frac{d\phi}{dy};\ -2Q\frac{d\phi}{dz};$$

parallel to the three axes respectively.

The sum of the actions of those forces on the molecule $\rho\, dx\, dy\, dz$ during the change of temperature and volume, is

$$-2Q\left(\frac{d\phi}{dx}\delta x + \frac{d\phi}{dy}\delta y + \frac{d\phi}{dz}\delta z\right)\rho\, dx\, dy\, dz,$$
$$= -2Q\delta\phi\rho\, dx\, dy\, dz.$$

The sum of such actions upon all the particles in unity of weight is equal in amount and opposite in sign to the variation of latent heat; that is to say,

$$\delta Q' = \frac{2Q}{M}\iiint_{(1)}\rho\,\delta\phi\, dx\, dy\, dz. \quad . \quad . \quad (23.)$$

To determine the value of the variation $\delta\phi$, let it be divided into two parts, thus:—

$$\delta\phi = \delta\phi_1 + \delta\Delta\phi,$$

where $\Delta\phi = \phi - \phi_1$.

First, With respect to $\delta\phi_1$, it is obvious that because, according to equations (6, 7),

$$MV = kMV\int_{-\infty}^{\phi 1} e^{k\Delta\phi}\frac{\psi}{\psi_1}d\phi,$$

we must have

$$\delta V = kV\delta\phi_1 \text{ and } \delta\phi_1 = \frac{\delta V}{kV}.$$

Hence the first part of the integral (23) is

$$\frac{2Q}{M}\delta\phi_1\iiint_{(1)}\rho\, dx\, dy\, dz = \frac{2\mu Q}{MkV}\cdot\delta V$$
$$= \frac{h\mu}{\kappa M}(\tau - \kappa)\frac{\delta V}{V}. \quad . \quad . \quad . \quad . \quad (23A.)$$

To determine the second part of the integral we have the condition that the quantity of atomic atmosphere inclosed within each surface at which $\Delta\phi$ has some given value is invariable; that is to say,

$$\left(\delta\Delta\phi\frac{d}{d\Delta\phi} + \delta V\frac{d}{dV} + \delta\tau\frac{d}{d\tau}\right)\left(k\rho_1 MV\int_{-\infty}^{\phi} e^{k\theta\Delta\phi}\frac{\omega}{\omega_1}d\phi\right) = 0.$$

Hence

$$\delta\Delta\phi = \frac{-\left(\delta V\frac{d}{dV}+\delta\tau\frac{d}{d\tau}\right)\left(k\rho_1 M V\int_{-\infty}^{\phi} e^{k\theta\Delta\phi}\frac{\omega}{\omega_1}d\phi\right)}{k\rho_1 M V e^{k\theta\Delta\phi}\frac{\omega}{\omega_1}}$$

The value of the second part of the integral (23) is now found to be:—

$$\frac{2Q}{M}\int\int\int_{(1)}\rho\delta\Delta\phi dxdydz = \frac{2Q}{M}k\rho_1 M V\int_{-\infty}^{\phi_1} e^{k\theta\Delta\phi}\frac{\omega}{\omega_1}\delta\Delta\phi d\phi$$

$$= -\frac{2Q}{kM}\left(\delta V\frac{d}{dV}+\delta\tau\frac{d}{d\tau}\right)\left\{\rho_1 M V\int_{-\infty}^{k\phi_1}\int_{-\infty}^{k\phi} e^{k\theta\Delta\phi}\frac{\omega}{\omega_1}k^2 d\phi^2\right\}.$$

In the double integral, let $\lambda = \log._e V$ be put for $k\phi$, G for ω, and H for the single integral, as in equation (9). Then the double integral becomes

$$\frac{1}{G_1}\int_{-\infty}^{\lambda_1} H d\lambda = -\frac{1}{G_1}\frac{dH_1}{d\theta} \text{ by Eq. (10).}$$

$$= -\frac{\kappa}{G_1}\frac{dH_1}{d\tau}.$$

Also, because $\rho_1 MV = \frac{\mu G_1}{H_1}$ by Eq. (9), and $\frac{2Q}{kM} = \frac{h}{\kappa}(\tau - \kappa)$, the second part of the integral (23) is found to be

$$\frac{h\mu}{M}(\tau-\kappa)\left(\delta\tau\frac{d}{d\tau}+\delta V\frac{d}{dV}\right)\frac{dH_1}{H_1 d\tau}. \quad \quad (23B.)$$

Hence, adding together (23A) and (23B) we find for the total variation of latent heat

$$\delta Q' = \frac{h\mu}{M}(\tau-\kappa)\left\{\delta\tau . \frac{d^2\log._e H_1}{d\tau^2} + \delta V \cdot \left(\frac{1}{\kappa V} + \frac{d^2\log._e H_1}{d\tau dV}\right)\right\}. \quad (24.)$$

To express this in terms of quantities which may be known directly by experiment, we have, by equations (10) and (9)—

$$\frac{dH_1}{H_1 d\lambda} + \theta - \frac{G_1}{H_1} = 0, \text{ that is to say,}$$

$$\frac{d\log._e H_1}{dV} = \frac{G_1}{H_1 V} - \frac{\tau}{\kappa V} = \frac{M}{h\mu}p - \frac{\tau}{\kappa V};$$

and, therefore,

$$\log._e H_1 = \frac{M}{h\mu}\int p dV - \frac{\tau}{\kappa}\log._e V + f(\tau) + \text{constant.}$$

$f(\tau)$ is easily found to be $= -\log._e \tau$ for a perfect gas, and, being independent of the density, is the same for all substances in all conditions; hence we find (the integrals being so taken that for a perfect gas they shall $= 0$)

$$* \quad \frac{d \log._e \mathrm{H}_1}{d\tau} = \int \left(\frac{\mathrm{M}}{h\mu} \frac{dp}{d\tau} - \frac{1}{\kappa \mathrm{V}} \right) d\mathrm{V} - \frac{1}{\tau},$$

$$\frac{d^2 \log._e \mathrm{H}_1}{d\tau^2} = \frac{\mathrm{M}}{h\mu} \int \frac{d^2 p}{d\tau^2} d\mathrm{V} + \frac{1}{\tau^2},$$

$$\frac{d^2 \log._e \mathrm{H}_1}{d\tau d\mathrm{V}} = \frac{\mathrm{M}}{h\mu} \frac{dp}{d\tau} - \frac{1}{\kappa \mathrm{V}},$$

and, therefore,

$$\delta \mathrm{Q}' = (\tau - \kappa) \left\{ \delta\tau \cdot \left(\frac{h\mu}{\mathrm{M}\tau^2} + \int \frac{d^2 p}{d\tau^2} d\mathrm{V} \right) + \delta\mathrm{V} \cdot \frac{dp}{d\tau} \right\} \quad . \quad (25.)$$

is the variation of latent heat, expressed in terms of the pressure, volume, and temperature; to which, if the variation of sensible heat, $\delta \mathrm{Q} = \mathfrak{k}\delta\tau$, be added, the complete variation of heat, $\delta \mathrm{Q} + \delta \mathrm{Q}' = \delta . \mathrm{Q}$, in unity of weight of the substance, corresponding to the variations $\delta \mathrm{V}$ and $\delta\tau$ of volume and temperature, will be ascertained.

It is obvious that equation (25), with its consequences, is applicable to any mixture of atoms of different substances *in equilibrio* of pressure and temperature; for in that case τ, $\frac{dp}{d\tau}$, and $\frac{d^2p}{d\tau^2}$ are the same for each substance. We have only to substitute for $\frac{h\mu}{\mathrm{M}}$ the following expression:—

$$n_1 \frac{h_1 \mu_1}{\mathrm{M}_1} + n_2 \frac{h_2 \mu_2}{\mathrm{M}_2} + \&c.,$$

where n_1, n_2, &c., are the proportions of the different ingredients in unity of weight of the mixture, so that $n_1 + n_2 + \&c. = 1$.

Equation (25) agrees exactly with equation (6) in the first section of my original paper on the Theory of the Mechanical Action of Heat. It is the fundamental equation of that theory; and I shall now proceed to deduce the more important consequences from it.

10. *Equivalence of Heat and Expansive Power: Joule's Law.*—From the variation of the heat communicated to the body, let us subtract the variation of the expansive power given out by it, or

$$\mathrm{P}\delta\mathrm{V} = \{p + f(\mathrm{V})\} \delta\mathrm{V}.$$

The result is the variation of the total power exercised upon or communicated to unity of weight of the substance, supposing that there is no chemical, electrical, magnetic, or other action except heat and pressure; and its value is—

* This coefficient corresponds to $-\frac{\mathrm{U}}{\kappa}$ in the notation of my previous paper on the Mechanical Action of Heat.

$$\delta\Psi = \delta Q + \delta Q' - P\delta V = \delta\tau . \left\{ \mathfrak{k} + \frac{h\mu}{M}\left(\frac{1}{\tau} - \frac{\kappa}{\tau^2}\right) + (\tau - \kappa)\int \frac{d^2 p}{d\tau^2} d V \right\}$$

$$+ \delta V . \left\{ (\tau - \kappa)\frac{dp}{d\tau} - p - f(V) \right\} . \qquad (26.)$$

This expression is obviously *an exact differential*, and its integral is the following function of the volume and temperature :—

$$\Psi = \mathfrak{k}(\tau - \kappa) + \frac{h\mu}{M}\left(\log._e \tau + \frac{\kappa}{\tau}\right) + \int \left\{ (\tau - \kappa)\frac{dp}{d\tau} - p \right\} d V - \int f(V) d V. \quad (27.)$$

Accordingly, the total amount of power which must be exercised upon unity of weight of a substance, to make it pass from the absolute temperature τ_0 and volume V_0 to the absolute temperature τ_1 and volume V_1, is

$$\Psi(V_1, \tau_1) - \Psi(V_0, \tau_0).$$

This quantity consists partly of expansive or compressive power, and partly of heat, in proportions depending on the mode in which the intermediate changes of temperature and volume take place; but the total amount is independent of these changes.

Hence, *if a body be made to pass through a variety of changes of temperature and volume, and at length be brought back to its primitive volume and temperature, the algebraical sum of the portions of power applied to and evolved from the body, whether in the form of expansion and compression, or in that of heat, is equal to zero.*

This is one form of the law, proved experimentally by Mr. Joule, of the equivalence of heat and mechanical power. In my original paper on the Mechanical Action of Heat, I used this law as an axiom, to assist in the investigation of the equation of latent heat. I have now deduced it from the hypothesis on which my researches are based—not in order to prove the law, but to verify the correctness of the mode of investigation which I have followed.

Equations (26) and (27), like equation (23), are made applicable to unity of weight of a mixture, by putting $\Sigma n \mathfrak{k}$ for $\mathfrak{k}$, and $\Sigma n \frac{h\mu}{M}$ for $\frac{h\mu}{M}$.

The train of reasoning in this article is the converse of that followed by Professor William Thomson of Glasgow, in article 20 of his paper on the Dynamical Theory of Heat, where he proves from Joule's law that the quantity corresponding to $\delta\Psi$ is an exact differential.

11. *Mutual Conversion of Heat and Expansive Power: Carnot's Law of the Action of Expansive Machines.*—If a body be made to pass from the volume V_0 and absolute temperature τ_0 to the volume V_1 and absolute temperature τ_1, and be then brought back to the original volume and

temperature, the total power exerted (Ψ) will have, in those two operations, equal arithmetical values, of opposite signs. Each of the quantities Ψ consists partly of heat and partly of expansive power, the proportion depending on the mode of intermediate variation of the volume and temperature, which is arbitrary. If the mode of variation be different in the two operations, the effect of the double operation will be to transform a portion of heat into expansive power, or *vice versâ*.

Let (*a*) denote the first operation, (*b*) the *reverse* of the second. Then

$$\Psi_b - \Psi_a.$$

The terms of Ψ which involve functions of τ only, or of V only, are not affected by the mode of intermediate variation of those quantities. The term on which the mutual conversion of heat and expansive power depends, is therefore

$$\int \left\{ (\tau-\kappa)\frac{dp}{d\tau} - p \right\} d\mathrm{V}\,(b) = \int \left\{ (\tau-\kappa)\frac{dp}{d\tau} - p \right\} d\mathrm{V}\,(a),$$

or,

$$\int \left(\frac{d\,\mathrm{Q}'}{d\mathrm{V}} - p \right) d\,\mathrm{V}\,(b) = \int \left(\frac{d\,\mathrm{Q}'}{d\mathrm{V}} - p \right) d\,\mathrm{V}\,(a).$$

Hence,

$$\int \frac{d\,\mathrm{Q}'}{d\,\mathrm{V}} d\,\mathrm{V}\,(a) - \int \frac{d\,\mathrm{Q}'}{d\,\mathrm{V}} d\,\mathrm{V}\,(b) = \int p\,d\,\mathrm{V}(a) - \int p\,d\,\mathrm{V}\,(b),$$

which last quantity is the amount of the *heat transformed into expansive power*, or the total latent heat of expansion in the *double operation*.

Let

$$\int \frac{dp}{d\tau} d\,\mathrm{V} = \int \frac{1}{\tau-\kappa} \cdot \frac{d\,\mathrm{Q}'}{d\,\mathrm{V}} d\,\mathrm{V} = \mathrm{F}.$$

Then because

$$\frac{d\,\mathrm{Q}'}{d\,\mathrm{V}} d\,\mathrm{V} = (\tau-\kappa)\,d\,\mathrm{F},$$

we have

$$\int_{\mathrm{V}_0}^{\mathrm{V}_1} p\,d\mathrm{V}(a) - \int_{\mathrm{V}_0}^{\mathrm{V}_1} p\,d\mathrm{V}(b) = \int_{\mathrm{F}_0}^{\mathrm{F}_1} (\tau-\kappa)\,d\,\mathrm{F}\,(a) - \int_{\mathrm{F}_0}^{\mathrm{F}_1} (\tau-\kappa)\,d\mathrm{F}\,(b)$$

$$= \int_{\mathrm{F}_0}^{\mathrm{F}_1} (\tau_a - \tau_b)\,d\mathrm{F} = \int_{\mathrm{V}_0}^{\mathrm{V}_1} \frac{\tau_a - \tau_b}{\tau_a - \kappa} \frac{d\,\mathrm{Q}'_a}{d\,\mathrm{V}} d\,\mathrm{V}. \quad . \quad (28.)$$

In which τ_a and τ_b are the pair of absolute temperatures, in the two operations respectively, corresponding to *equal values of* F.

This equation gives a relation between the heat transformed into expansive power by a given pair of operations on a body, the latent heat of expansion in the first operation, and the mode of variation of temperature in the two operations. It shows that the proportion of the original latent heat of expansion finally transformed into expansive power, is a function of the temperatures alone, and is therefore independent of the nature of the body employed.

Equation (28) includes Carnot's law as a particular case. Let the limits of variation of temperature and volume be made indefinitely small. Then

$$dp\,d\mathrm{V} = \frac{d\tau}{\tau - \kappa} \cdot \frac{d\mathrm{Q}'}{d\mathrm{V}}\,d\mathrm{V},$$

and dividing by $d\tau d\mathrm{V}$

$$\frac{dp}{d\tau} = \frac{1}{\tau - \kappa} \cdot \frac{d\mathrm{Q}'}{d\mathrm{V}}.$$

This differential equation is also an immediate consequence of equation (25).

If $\frac{\mu}{\mathrm{J}}$ be put for $\frac{1}{\tau - \kappa}$, and J M for $\frac{d\mathrm{Q}'}{d\mathrm{V}}$, it becomes identical with the equation by which Professor William Thomson expresses Carnot's law, as deduced by him and by Mr. Clausius from the principle, that *it is impossible to transfer heat from a colder to a hotter body without expenditure of mechanical power.*

The investigation which I have now given is identical in principle with that in the fifth section of my paper on the Mechanical Action of Heat; but the result is expressed in a more comprehensive form.

Equation (28), like (25), (26), and (27), is applicable to a mixture, composed of any number of different substances, in any proportions, provided the temperature, the pressure, and the coefficients $\frac{dp}{d\tau}, \frac{d^2p}{d\tau^2}$, are the same throughout the mass.

12. *Apparent Specific Heat.*—The general value of apparent specific heat of unity of weight is

$$\left. \begin{aligned} \mathrm{K} = \frac{d\mathrm{Q}}{d\tau} + \frac{d\mathrm{Q}'}{d\tau} + \frac{d\mathrm{Q}'}{d\mathrm{V}} \cdot \frac{d\mathrm{V}}{d\tau} = \mathrm{k} + (\tau - \kappa) \left\{ \frac{h\mu}{\mathrm{M}\tau^2} + \int \frac{d^2p}{d\tau^2} d\mathrm{V} \right. \\ \left. + \frac{d\mathrm{V}}{d\tau} \cdot \frac{dp}{d\tau} \right\} \end{aligned} \right\} \quad (29.)$$

agreeing with equation (13) of my previous paper.

The value in each particular case depends on the mode of variation of volume with temperature. Specific heat at constant volume is

$$\mathrm{K}_{\mathrm{V}} = \mathrm{k} + (\tau - \kappa) \left(\frac{h\mu}{\mathrm{M}\tau^2} + \int \frac{d^2p}{d\tau^2} d\mathrm{V} \right). \quad . \quad . \quad (30.)$$

When the pressure is constant, we must have

$$\frac{d\mathrm{P}}{d\mathrm{V}} d\mathrm{V} + \frac{dp}{d\tau} d\tau = 0,$$

and, consequently,

$$\frac{d\mathrm{V}}{d\tau} = -\frac{\frac{dp}{d\tau}}{\frac{d\mathrm{P}}{d\mathrm{V}}},$$

therefore specific heat at constant pressure is

$$\mathrm{K}_p = \mathrm{K}_v + (\tau - \kappa)\frac{\left(\frac{dp}{d\tau}\right)^2}{-\frac{d\mathrm{P}}{d\mathrm{V}}}. \quad . \quad . \quad (31.)$$

This agrees with equation (16) of Professor Thomson's paper, if $\frac{\mathrm{J}}{\mu}$ in his notation $= \tau - \kappa$.

If the body be a perfect gas, then

$$\left.\begin{aligned}
\mathrm{K} &= \frac{\mathrm{V}_0}{\tau_0}\left(\frac{k\mathrm{M}}{2\mu} + \frac{\kappa}{\tau} - \frac{\kappa^2}{\tau^2} + \frac{\tau - \kappa}{\mathrm{V}} \cdot \frac{d\mathrm{V}}{d\tau}\right) \\
\mathrm{K}_v &= \frac{\mathrm{V}_0}{\tau_0}\left(\frac{k\mathrm{M}}{2\mu} + \frac{\kappa}{\tau} - \frac{\kappa^2}{\tau^2}\right) \\
\mathrm{K}_p &= \mathrm{K}_v + \frac{\mathrm{V}_0}{\tau_0}\left(1 - \frac{\kappa}{\tau}\right) = \frac{\mathrm{V}_0}{\tau_0}\left(\frac{k\mathrm{M}}{2\mu} + 1 - \frac{\kappa^2}{\tau^2}\right)
\end{aligned}\right\} \quad (32.)$$

The fact that the specific heats of all simple gases for unity of weight are inversely proportional to their specific gravities, shows that $\frac{k\mathrm{M}}{2\mu}$ is the same for them all.

13. *Velocity of Sound in Fluids.*—Let a denote the velocity of sound in a fluid, and $d\,.\,\mathrm{P}$ the total differential of the pressure. Then

$$a = \sqrt{\left(g \cdot \frac{d\,.\,\mathrm{P}}{d\,.\,\frac{1}{\mathrm{V}}}\right)} = \sqrt{\left\{g\mathrm{V}^2\left(-\frac{d\mathrm{P}}{d\mathrm{V}} - \frac{d\mathrm{P}}{d\tau}\cdot\frac{d\tau}{d\mathrm{V}}\right)\right\}}. \quad (33.)$$

If it were possible to maintain the temperature of each particle of the fluid invariable during the passage of sound, this velocity would be simply

$$\sqrt{\left(g \cdot \frac{d\mathrm{P}}{d\,.\,\frac{1}{\mathrm{V}}}\right)}.$$

But we have reason to believe that there is not time, during the passage of sound, for an appreciable transfer of heat from atom to atom, so that for each particle

$$d\,Q + d\,Q' = 0; \text{ or, } K = 0 \text{ in equation (29).}$$

To fulfil this condition, we must have

$$\frac{d\,\tau}{d\,V} = -\frac{\tau - \kappa}{K_v} \cdot \frac{d\,p}{d\,\tau}.$$

Consequently,

$$a = \sqrt{\left\{ g V^2 \left(-\frac{d\,P}{d\,V} + \frac{\tau - \kappa}{K_v} \cdot \left(\frac{d\,p}{d\,\tau}\right)^2 \right) \right\}},$$

or, by equation (31),

$$a = \sqrt{\left(g \frac{d\,P}{d \cdot \frac{1}{V}} \cdot \frac{K_P}{K_V} \right)}. \quad . \quad . \quad . \quad (34.)$$

That is to say, *the action of heat increases the velocity of sound in a fluid beyond what it would be, if heat did not act, in the ratio of the square root of the specific heat at constant pressure, to the square root of the specific heat at constant volume.*

This is Laplace's law of the propagation of sound, which is here shown to be applicable, not only to perfect gases, but to all fluids whatsoever.

V.—LAWS OF THE ELASTICITY OF SOLID BODIES.*

INTRODUCTION.

1. THE science of the elasticity of solid bodies, considered with reference to its most important application, the determination of the strength of structures, consists of three parts:

First. The investigation of what may be specially termed the *Laws of Elasticity;* that is to say, the mutual relations which must exist between the elasticities of different kinds possessed by a given solid, and between the different values of those elasticities in different directions.

Secondly. The integration of the equations of equilibrium and motion of the particles of an elastic solid. The results of this process enable us to determine the relative displacements of the particles from their natural positions in a solid body of a given material and figure, subjected to a given combination of forces.

Thirdly. The application of the results derived from the first two branches of the theory to our experimental knowledge of the pressures and relative displacements to which the particles of known materials may safely be subjected in practice. This enables us to compute the strength of actual structures.

2. Notwithstanding the great amount of attention which has been paid to the strength of materials, and the numerous and elaborate experiments which have been made respecting it, few examples exist of the sound application of physical and mathematical principles to practice in connection with this subject. This has arisen chiefly from the fact, that the first and second branches of the inquiry have to a great extent been carried on without reference to their application to the third, and the third conducted without regard to the principles of the first and second. The results of investigation, on correct principles, into the theory of elasticity have been limited in their applications, with a few exceptions, to the laws of the propagation of vibratory movements; and those few exceptions relate almost exclusively to bodies of equal elasticity in all directions—a class which excludes many of the most useful materials of

* Read before the British Association at Edinburgh, on August 1, 1850, and published in the *Cambridge and Dublin Mathematical Journal*, May, 1851.

construction. On the other hand, when it has been found necessary to adopt theoretical principles, for the purpose of reducing the results of experiments on the strength and elasticity of materials to a system, assumptions have often been made, with a view chiefly to simplicity in calculation, of a kind inconsistent with the real nature of elastic bodies.

3. The present inquiry relates to the first part of the theory of elasticity—viz., the laws of the relations which must exist between the elasticities of different kinds possessed by a given substance, and between their various values in different directions.

SECTION I.—COMPOSITION AND RESOLUTION OF STRAINS AND MOLECULAR PRESSURES.

4. At the outset of the inquiry two preliminary problems present themselves: the composition and resolution of relative molecular displacements, and the composition and resolution of pressures such as the parts of elastic bodies exert upon each other. The former is a question of pure geometry; the latter, of pure statics. They are usually considered simultaneously, on account of the analogy which exists between their solutions. This is not the result of the physical connection between the two classes of phenomena, and it would still exist although there were no such physical connection; it is merely a consequence of the analogy between forces in statics and straight lines in geometry.

Those two problems have been so fully investigated by MM. Cauchy, Lamé, and Clapeyron, as to leave nothing further to be done. The theorems and formulæ which they have obtained are many and important. In the present paper I shall state those principles and results only to which there will be occasion to refer in the sequel.

5. It is desirable that some single word should be assigned to denote the state of the particles of a body when displaced from their natural relative positions. Although the word *strain* is used in ordinary language indiscriminately to denote relative molecular displacement, and the force by which it is produced, yet it appears to me that it is well calculated to supply this want. I shall therefore use it, throughout this paper, in the restricted sense of *relative displacement of particles*, whether consisting in dilatation, condensation, or distortion; while under the term *pressure* I shall include every kind of force which acts between elastic bodies, or the parts of an elastic body, as the cause or effect of a state of strain, whether that force is tensile, compressive, or distorting.

The nature and magnitude of a simple and uniform strain are defined by three things:

First. The direction of the lines along which the particles of the body are displaced from their natural position.

Secondly. The direction along which the rate of variation of the displacement from point to point is a maximum. This direction is normal to a series of planes of equal displacement, and may be called the *strain-normal.*

Thirdly. The amount of that rate of variation, being the differential coefficient of the displacement with respect to distance along the strain-normal.

6. A strain may be resolved into three components, in which the directions of displacement shall be respectively parallel to three rectangular axes, while the strain-normal remains unchanged, by multiplying its amount by the direction-cosines of the total displacement.

Each of these three components may itself be resolved into three components, in which, the direction of displacement remaining unchanged, the strain-normals are respectively parallel to the three axes, by multiplying its amount by the direction-cosines of the original strain-normal.

Thus every strain is reducible to nine components.

These nine components, however, are equivalent to but six distinct strains. If we consider the strains as thus reduced to three rectangular axes, we shall find that they are of two kinds: *longitudinal,* that is to say, strains of *linear extension or condensation,* where the displacements are parallel to the strain-normals; and *transverse,* or strains of *distortion,* when these directions are at right angles. Thus, if x, y, z, denote the three rectangular axes, and ξ, η, ζ, small molecular displacements respectively parallel to them, then

$$\frac{d\xi}{dx}, \frac{d\eta}{dy}, \frac{d\zeta}{dz},$$

are longitudinal strains, which are dilatations when positive, and condensations when negative. I shall denote them respectively by N_1, N_2, N_3; their sum, when positive, is the *cubic* dilatation of the particles, and when negative, the cubic condensation.

Transverse strains, or distortions, are represented by the six differential coefficients of the displacements with respect to axes at right angles to them—viz.,

$$\frac{d\eta}{dz}, \frac{d\zeta}{dy}; \frac{d\zeta}{dx}, \frac{d\xi}{dz}; \frac{d\xi}{dy}, \frac{d\eta}{dx}.$$

Let the axis of x be perpendicular to the plane of the paper. Let ABCD be the section, by the plane yz, of a prism which in its natural state is square, and has its faces normal to the axes of y and z. A distortion in the plane yz, relatively to these axes, is measured by the deviation from rectangularity of this originally square section, that deviation being considered positive which makes the angles B and D acute. Now, so far as the positions of the particles in this prism relatively to each other are concerned, it is immaterial whether that deviation from rectangularity is produced by keeping the sides A D and B C parallel to their original positions, and giving angular motion to A B and D C—a change represented by $\frac{d\eta}{dz}$; or by keeping A B and D C parallel to their original positions, and giving angular motion to A D and B C—a change represented by $\frac{d\zeta}{dy}$; or by combining those two operations: so that the total transverse strain in the plane yz is represented by the sum of these two coefficients,

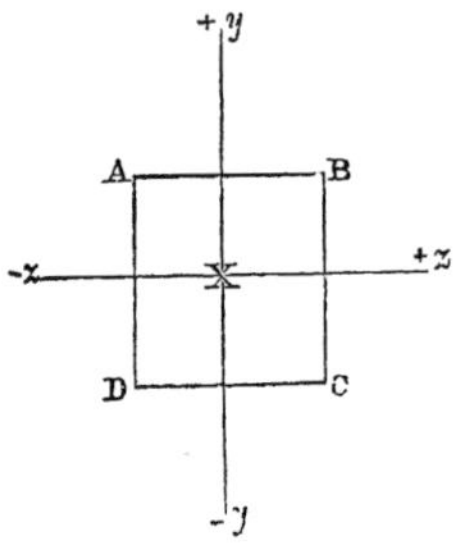

$$\frac{d\eta}{dz} + \frac{d\zeta}{dy} = 2\,T_1.$$

Similar reasoning gives, for the total distortion in the plane zx,

$$\frac{d\zeta}{dx} + \frac{d\xi}{dz} = 2\,T_2,$$

and in the plane xy,

$$\frac{d\xi}{dy} + \frac{d\eta}{dx} = 2\,T_3,$$

The factor 2 is used in these expressions for the sake of convenience in the employment of certain formulæ, to be afterwards quoted.

The halved-differences of the pairs of differential coefficients,

$$\frac{1}{2}\left(\frac{d\eta}{dz} - \frac{d\zeta}{dy}\right);\ \frac{1}{2}\left(\frac{d\zeta}{dx} - \frac{d\xi}{dz}\right);\ \frac{1}{2}\left(\frac{d\xi}{dy} - \frac{d\eta}{dx}\right),$$

represent rotations of the prism as a whole round the axes of x, y, z, respectively, which have no connection with the positions of its particles relatively to each other.

The component strains into which all others can be resolved with respect to a given set of axes are thus reduced to six, three longitudinal and three transverse.

7. A pressure, like a strain, is defined by three things:

1st. The direction of the pressure.

2nd. The position of the surface at which the pressure is exerted.

3rd. The amount of the pressure as expressed in units of force per unit of area of the surface of action.

A pressure on a plane, in whatsoever direction it may act, may be resolved into three rectangular components, one normal to the plane and two tangential. The normal pressure may be compressive or tensile: when compressive, it is considered as positive; when tensile, negative.

In an elastic solid which is *in equilibrio*, let a cube be conceived to exist with its faces normal to the axes of co-ordinates, and let the pressures throughout its extent be uniform. This cube exerts on the matter round it, and is reacted on by three pairs of normal pressures, at the faces respectively normal to the axis of x, y, z, which may be denoted by P_1, P_2, P_3, the pressures at opposite faces being equal.

Let A B C D represent the section of this cube by the plane yz. On the faces A B and C D, parallel to xz, let a pair of tangential forces act in the directions denoted by the order of the letters, tending to produce distortion by making the angles B and D acute and A and C obtuse. Let a pair of forces of similar tendency act on the faces C B and A D, parallel to zy. These two pairs of forces are equal and opposite to those which the cube, in consequence of the transverse displacements of its particles, exerts on the surrounding portion of the solid. No displacement of the relative situations of a system of particles can give the system a tendency to revolve as a whole round an axis. Such a tendency must exist in the cube unless the tangential forces on the faces A B, C D are equal to those on the faces C B, A D.

Therefore, *the tangential pressure parallel to* z, *on a plane normal to* y, *is equal to the tangential pressure parallel to* y, *on a plane normal to* z: a theorem first proved by Cauchy.

The common value of those forces may be denoted by Q_1, as they are both perpendicular to x.

Similar reasoning shows that the two pairs of tangential forces perpendicular to y have one common value, Q_2.

In like manner, those perpendicular to z may be denoted by Q_3.

Thus the pressures exerted by and on the cube are reduced to six, three normal and three tangential.

8. The composition of pressures applied to different planes, and their reduction to new axes, depends on the following principle :—

Conceive a small triangular pyramid, with its apex at the origin of rectangular co-ordinates, its sides being formed by the three co-ordinate planes, and its base by a plane in any given direction intersecting them. Let pressures in one given direction act on the three sides, and be balanced by a pressure in the same direction on the base. Each of the three sides is equal to the base multiplied by the cosine of the angle between the normal to the base and the normal to the side in question. Therefore, the total pressure on the base is equal to the sum of the pressures on the sides, each multiplied by the cosine of the angle between the normal to the side in question and the normal to the base. If the normal to this base is one of three new axes of rectangular co-ordinates, the total pressure thus found may be reduced to normal and tangential pressures, by multiplying it by its direction-cosines with respect to the new axes.

9. I annex, for convenience of reference, the general formulæ which have been deduced from this principle.

Let x, y, z, be rectangular axes of co-ordinates, and P_1, P_2, P_3, Q_1, Q_2, Q_3, normal and tangential pressures which act as shown in the following table:

Normals.	Planes.	Pressures parallel to x	y	z
x	yz	P_1	Q_3	Q_2
y	zx	Q_3	P_2	Q_1
z	xy	Q_2	Q_1	P_3

Let R_1, R_2, R_3, be the rectangular components of the total pressure at a plane, the direction-cosines of whose normal are a_1, a_2, a_3.

Then

$$\left.\begin{aligned} R_1 &= a_1 P_1 + a_2 Q_3 + a_3 Q_2, \\ R_2 &= a_1 Q_3 + a_2 P_2 + a_3 Q_1, \\ R_3 &= a_1 Q_2 + a_2 Q_1 + a_3 P_3, \end{aligned}\right\} \quad . \quad . \quad . \quad . \quad (1.)$$

Let this normal be taken as the axis of x' in a new set of rectangular axes x', y', z', which make with the original axes the angles whose cosines are given in the following table:—

Original Axes.	New Axes. x'	y'	z'	
x	a_1	b_1	c_1	Direction-cosines.
y	a_2	b_2	c_2	
z	a_3	b_3	c_3	

Let P_1', P_2', P_3', Q_1', Q_2', Q_3', be the normal and tangential pressures, as reduced to the new axes: then

$$\left.\begin{aligned} P_1' &= P_1 a_1^2 + P_2 a_2^2 + P_3 a_3^2 \\ &\quad + 2Q_1 a_2 a_3 + 2Q_2 a_3 a_1 + 2Q_3 a_1 a_2, \\ P_2' &= P_1 b_1^2 + P_2 b_2^2 + P_3 b_3^2 \\ &\quad + 2Q_1 b_2 b_3 + 2Q_2 b_3 b_1 + 2Q_3 b_1 b_2, \\ P_3' &= P_1 c_1^2 + P_2 c_2^2 + P_3 c_3^2 \\ &\quad + 2Q_1 c_2 c_3 + 2Q_2 c_3 c_1 + 2Q_3 c_1 c_2, \\ Q_1' &= P_1 b_1 c_1 + P_2 b_2 c_2 + P_3 b_3 c_3 \\ &\quad + Q_1(b_2 c_3 + b_3 c_2) + Q_2(b_3 c_1 + b_1 c_3) + Q_3(b_1 c_2 + b_2 c_1), \\ Q_2' &= P_1 c_1 a_1 + P_2 c_2 a_2 + P_3 c_3 a_3 \\ &\quad + Q_1(c_2 a_3 + c_3 a_2) + Q_2(c_3 a_1 + c_1 a_3) + Q_3(c_1 a_2 + c_2 a_1), \\ Q_3' &= P_1 a_1 b_1 + P_2 a_2 b_2 + P_3 a_3 b_3 \\ &\quad + Q_1(a_2 b_3 + a_3 b_2) + Q_2(a_3 b_1 + a_1 b_3) + Q_3(a_1 b_2 + a_2 b_1), \end{aligned}\right\} \quad (2.)$$

By the substitution of N for P and T for Q, the formulæ given above are made applicable to the reduction of *strains* to new axes of co-ordinates.

I shall not here recapitulate the many elegant and important theorems which MM. Cauchy and Lamé and Clapeyron have deduced from those equations, as they do not relate to the branch of the theory of elasticity of which this paper treats.

I may mention that in their memoir in the seventh volume of Crelle's *Journal*, MM. Lamé and Clapeyron have used N and T to denote *pressures*, and have expressed *strains* simply by the differential coefficients $\frac{d\,\xi}{d\,x}$, &c.

SECTION II.—PHYSICAL RELATIONS BETWEEN PRESSURES AND STRAINS, SO FAR AS THEY ARE INDEPENDENT OF HYPOTHESES RESPECTING THE MOLECULAR CONSTITUTION OF MATTER.

10. In almost all investigations which have hitherto been made respecting the elasticity of bodies which have different degrees of elasticity in different directions, it has been the practice to take some hypothesis as to the molecular constitution of solid bodies as the basis of calculation from the outset of the inquiry. It appears to me, however, that the more philosophical course is, to ascertain, in the first place, what conclusions can be attained as to the laws of elasticity without the aid of any such hypothesis, and afterwards to inquire how far the theory can be simplified, and what additional results can be gained by introducing suppositions respecting the ultimate constitution of matter.

For the present, therefore, I shall make no assumption as to the questions whether bodies are systems of physical points, or of atoms of definite bulk and figure, or are continuous, or have a constitution intermediate between those three; and I shall use the word *particle* in its literal sense of a *small part.*

11. I shall restrict the present inquiry to homogeneous bodies possessing a certain degree of symmetry in their molecular actions, which consists in this: that the actions upon any given particle of the body of any two equal particles situated at equal distances from it within the sphere of molecular action, in opposite directions, shall be equal and opposite.

Substances may possess higher degrees of molecular symmetry, but this is the lowest.

The statement that a body is *homogeneous* means, when applied to molecular action, that the mutual action of a pair of particles, situated at a given distance from each other in a given direction, shall be equal to that of any other pair of particles equal to the first, situated at an equal distance from each other in a parallel direction.

12. It is known by observation, that strains and pressures are physically

connected. It is also known by observation, that the pressure with which a strain is connected consists in a tendency of the body to recover its natural state, and is opposite or nearly opposite in direction to the strain; thus longitudinal condensation is accompanied with positive normal, or nearly-normal pressure; longitudinal dilatation, with negative normal, or nearly-normal pressure; and distortion in a given plane, with tangential pressure in the same plane, of opposite sign.

It is known by experiment, that when a pressure and the strain with which it is connected are given in direction, and when the strain does not exceed a certain limit, being in most cases the utmost limit to which a structure can be strained without danger to its permanency, the pressure and the strain are sensibly proportional to each other. The quantity by which a strain is to be multiplied to give the corresponding pressure is a *coefficient of elasticity*, and is expressed, like a pressure, by a certain number of units of force per unit of surface.

I have said that a strain and the corresponding pressure referred to the same plane are opposite or *nearly* opposite in direction; for they are not of necessity exactly opposite for all directions of strain, except in substances which are possessed of the highest degree of molecular symmetry; that is to say, which are equally elastic in all directions. For those having lower degrees of symmetry, the following proposition is true:—

THEOREM I. *In an elastic substance which is homogeneous and symmetrical with respect to molecular action, there are three directions at right angles to each other in which a longitudinal strain produces an exactly normal pressure on a plane at right angles to the direction of the strain.*

Those three directions are called *Axes of Elasticity*. The proposition is equivalent to an assertion, that *the lowest degree of symmetry of molecular action necessarily involves symmetry with respect to three rectangular co-ordinate planes.*

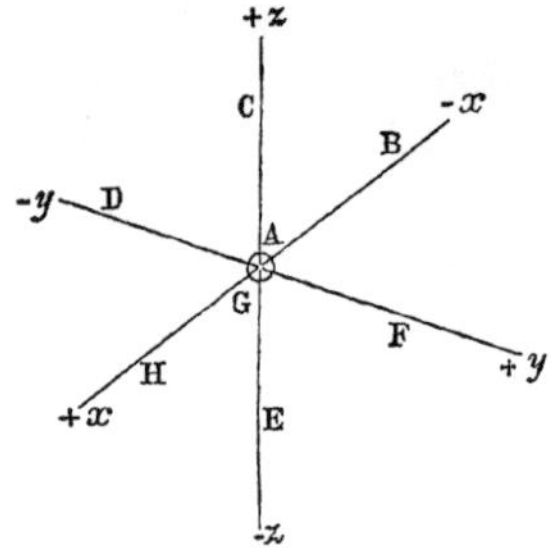

This theorem has been often demonstrated for systems of atoms. But it is easily seen that the truth of these demonstrations depends, not on the special hypotheses which they involve, but on the fundamental condition of symmetry.

The following demonstration involves no hypothesis.

Let a point O in the interior of a body be assumed as the origin of rectangular co-ordinates, the axes being considered as fixed, and the body as movable angularly in all directions about the origin. Space round O is divided by the co-ordinate planes into eight similar indefinitely-extended rectangular three-sided pyramids. Let those pyramids be designated as follows, according to the signs of the co-ordinates comprised in them.

Signs of x	Signs of y	Signs of z	Designation of Pyramid.
+	+	+	A
−	+	+	B
−	−	+	C
+	−	+	D
+	+	−	E
−	+	−	F
−	−	−	G
+	−	−	H

To express the relative situations of these pyramids, as taken in pairs, let the following terms be used:

Diametrically opposite—when the pyramids touch at the apex only; comprising the following pairs,

A, G; B, H; C, E; D, F.

Diagonally opposite—when they touch at an edge; comprising the pairs

A, H; D, E; B, G; C, F;
A, F; B, E; D, G; C, H;
A, C; B, D; E, G; F, H.

Contiguous—when they touch in a face; comprising the pairs

A, B; D, C; H, G; E, F;
A, D; B, C; F, G; E, H;
A, E; B, F; C, G; D, H.

Each pair of contiguous pyramids forms a rectangular wedge, which has an *opposite wedge* touching it along the edge, and a *contiguous wedge* touching it at each of its two faces.

The pairs of opposite wedges are

AB, GH; CD, EF;
AD, FG; BC, EH;
AE, CG; BF, DH.

The pairs of contiguous wedges are

AB, CD; CD, GH; GH, EF; EF, AB;
AD, BC; BC, FG; FG, EH; EH, AD;
AE, BF; BF, CG; CG, DH; DH, AE.

According to the condition of symmetry already stated, the portions of matter comprised in any pair of diametrically opposite pyramids must be symmetrical in their actions on a particle placed at O, or on any pair of equal particles symmetrically placed with respect to O, whatsoever may be the angular position of the body with respect to the axes.

Suppose the body to receive a longitudinal strain in the direction of the axis of z. Let a small circular area ω be conceived to exist in the plane of xy, with its centre at O; and let this area be the base of a cylinder extending indefinitely in a negative direction along the axis of z, and denoted by ωz. The pressure on the plane xy is proportional and parallel to the resultant of the actions of the four pyramids A, B, C, D, on the cylinder ωz, divided by the area ω. The action of each of those pyramids consists of a normal component parallel to z, and a tangential component parallel to the plane xy. In order that the total pressure may be normal, those tangential actions must balance each other, which can only be the case when the tangential action of the wedge AB, parallel to the axis of y, is equal and opposite to that of the contiguous wedge CD, and the tangential action of the wedge BC, parallel to the axis of x, is equal and opposite to that of the contiguous wedge AD.

The pair of contiguous wedges AB, CD, touch in the plane of xz, having the axis of x for their common edge. If the actions of this pair of wedges on ωz, when longitudinally strained along z, are unsymmetrical, this cannot arise from the form, position, or strain of these wedges, which are exactly symmetrical with respect to each layer of particles in ωz, but from the nature of the particles occupying the wedges. Now by rotating the body through a right angle about the axis of x, we can bring the particles which formerly occupied CD into AB, and the particles which formerly occupied GH (which consists of two pyramids diametrically opposite, and therefore molecularly symmetrical to A and B) into CD. In this new situation of the body with respect to the axes of co-ordinates, the resultant of the tangential actions, parallel to y, of the wedges AB and CD, on ωz, though not necessarily *equal*, will be *opposite in direction* to the original resultant; and this change will have been produced, not abruptly, but continuously, so that the value of the resultant must have passed through zero. Therefore, whatsoever may be the situation of the axis of x amongst the particles of the body, it is possible, by rotating the body about that axis, to find a position in which the tangential actions of the wedges AB and CD, parallel to y, on the cylinder ωz, shall balance each other. And by similar reasoning it may be proved, that whatsoever may be the situation of the axis of y amongst the particles of the body, it is possible, by rotating the body about that axis, to find a position in which the tangential actions of the wedges BC and AD, parallel to x, on the cylinder ωz, shall balance each other.

Therefore, by combining rotations about the axes of x and y, it is possible to find a position of the solid with respect to the axes of co-ordinates such that the tangential actions of the four pyramids A, B, C, D, on the cylinder ωz, arising from a longitudinal strain along z, shall be in equilibrio, and that the total pressure on $x\,y$ shall be normal.

The direction, with respect to the solid, which fulfils this condition, is called an *axis of elasticity.*

Let $-z\,O+z$, being now an axis of elasticity, be considered as fixed in the solid.

From the manner in which the two pairs of wedges AB, CD. and BC, AD, are composed of the four pyramids A, B, C, D, it is clear that the actions of the pair of diagonally opposite pyramids A, C, are symmetrical, and also those of the diagonally opposite pyramids B, D. From this and the symmetry of the actions of diametrically opposite pyramids it follows, that the actions of the four pairs of contiguous pyramids, A, E; D, H; B, F; C, G, are symmetrical, and also those of the two pairs of diagonally opposite pyramids, E, G; F, H. This symmetry of action (subject to the condition of symmetry of strain) is not disturbed by rotation about the axis of z.

Let the small circular area ω be now conceived to exist in the plane yz, and let the cylinder of which it is the base extend in a negative direction along the axis of x, and be called the cylinder ωx. Let the solid receive a longitudinal strain along the axis of x. The action of A on ωx is symmetrical to that of E, and the action of D to that of H; therefore, the tangential actions of the wedges A D, E H, parallel to z, balance each other. It remains only to make the tangential actions of the wedges A E, D H, parallel to y, on ωx, balance each other, which is to be done by rotation about the axis of z.

The solid is now in such a position that x, as well as z, is an axis of elasticity.

The pairs of contiguous pyramids are now all molecularly symmetrical about their common faces. Therefore the pairs of contiguous wedges A B, E F; A E, B F, are symmetrical in their actions on a cylinder $\omega\,y$, when longitudinally strained along y.

Therefore y also is an axis of elasticity, and the theorem is proved.

It is not necessary to the existence of rectangular axes of elasticity that the body should be homogeneous (in the sense in which I have used the word) throughout its whole extent, but only round each point throughout a space which is large as compared with the sphere of appreciable molecular action of each particle. Hence the rectangular axes of elasticity may vary in direction at different points of the same body; and some, or all of them, may follow the course of a system of curves, as they do in a rope, a piece of bent timber, or a curved bar of fibrous metal.

13. The axes of elasticity are evidently those which ought to be selected as axes of co-ordinates, for the resolution of all pressures and strains, in researches on the laws of elasticity. The strains and pressures being so resolved, we shall have the expression $-A_1 N_1$ for *part* of the normal pressure on the plane yz: A_1 being the *coefficient of longitudinal elasticity* for the axis of x. But this is not the whole of that pressure; for it is known by observation, that the normal pressure on a given plane is augmented by condensation, and diminished by dilatation of the particles, in a direction parallel to the given plane as well as normal to it. The normal pressure P_1 on yz, therefore, depends not only on the longitudinal strain N_1 along x, but also on the longitudinal strains N_2 and N_3 along y and z. Applying similar reasoning to the other normal pressures, they are found to be represented as follows:

Axis.	Plane.	
x	yz	$P_1 = -A_1N_1 - B_3N_2 - B_2'N_3$
y	zx	$P_2 = -B_3'N_1 - A_2N_2 - B_1N_3$
z	xy	$P_3 = -B_2N_1 - B_1'N_2 - A_3N_3$

$$\left.\begin{aligned} P_1 &= -A_1N_1 - B_3N_2 - B_2'N_3 \\ P_2 &= -B_3'N_1 - A_2N_2 - B_1N_3 \\ P_3 &= -B_2N_1 - B_1'N_2 - A_3N_3 \end{aligned}\right\} \quad (3.)$$

The tangential pressures are represented, in terms of the distortions, in the following manner:

Plane.	
yz	$Q_1 = -2C_1T_1$
zx	$Q_2 = -2C_2T_2$
xy	$Q_3 = -2C_3T_3$

$$\left.\begin{aligned} Q_1 &= -2C_1T_1 \\ Q_2 &= -2C_2T_2 \\ Q_3 &= -2C_3T_3 \end{aligned}\right\} \quad (4.)$$

These six equations are merely the representation of observed facts, framed with regard to the principle of axes of elasticity.

They contain *twelve* coefficients of elasticity, which may be thus classified:

A_1, A_2, A_3, are the coefficients of *longitudinal elasticity* for the axes of x, y, z, respectively;

B_1, B_1', are the coefficients of *lateral* elasticity in the plane of yz: the former expressing the effect of a strain along z in producing normal pressure parallel to y; the latter, the effect of a strain along y in producing normal pressure parallel to z.

B_2, B_2', are the coefficients of lateral elasticity in the plane of zx; and

B_3, B_3', in the plane of xy.

C_1, C_2, C_3, are the coefficients of *transverse or tangential elasticity*, or of *rigidity*, in the planes of yz, zx, and xy, respectively. The possession of this species of elasticity is the property which distinguishes solids from fluids, and is that upon which the strength and stability of solid structures entirely depend. When a beam, or any other portion of a solid structure, *takes a set*, as it is called (or undergoes permanent alteration of figure), it is the *rigidity* which has been overstrained and has given way. So far as I am aware, however, it has not hitherto been *directly* referred to in researches on the strength of materials, except in those relative to torsion.

The principal object of the present inquiry is to determine what mutual relations must necessarily exist amongst those twelve coefficients of elasticity in each substance.

14. The three coefficients of rigidity, so far as we have as yet seen, represent the elasticity called into play by three kinds of distortion, measured respectively by the alteration of the angles of the three rectangular sections of a cube whose faces are normal to the three axes of elasticity. I shall now, however, prove that the tangential pressures produced by equal distortions are equal, so long as the plane in which the distortion takes place is unchanged, and are not altered by any change of the direction, in that plane, of the sides of the figure on which the distortion is measured; that is to say—

THEOREM II. *The coefficient of rigidity is the same for all directions of distortion in a given plane.*

Let ABCD be the section at right angles to the edges of a rhombic prism having any angles; and GE and FH two lines normal respectively to the faces of the prism. Let this prism undergo a small alteration in the angles of its section ABCD.

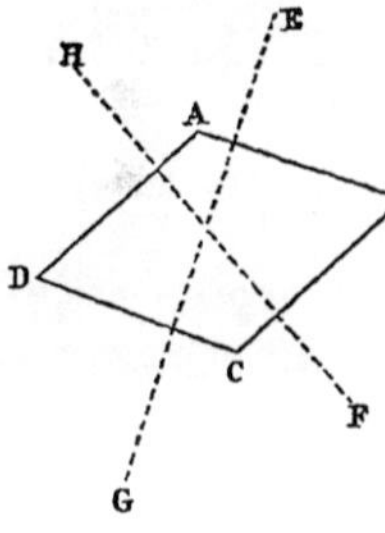

Whether we estimate the *distortion* so produced as a transverse displacement of the particles in lines parallel to AB, and varying along the strain-normal GE, or as a transverse displacement of the particles in lines parallel to AD, and varying along the strain-normal FH, the result, so far as the relative *transverse* displacements of the particles are concerned, will be the same.

Also, the tangential pressures are the same at the pair of faces AB and CD, and at the pair of faces BC and AD; for otherwise a relative displacement among the particles of a body would produce a force tending to make it revolve as a whole round an axis, which is impossible.

Therefore, the tangential forces produced by equal transverse displacements relatively to two strain-normals which make any angle with each other are equal, provided the displacements are in the same plane with the normals; therefore the coefficient of rigidity is the same for all directions in a given plane.

15. This theorem leads to another, which expresses the relations between the twelve coefficients of elasticity, as far as it is possible to

determine them independently of hypotheses respecting the constitution of matter.

THEOREM III. *In each of the co-ordinate planes of elasticity the coefficient of rigidity is equal to one-fourth part of the sum of the two coefficients of longitudinal elasticity for the axes which lie in that plane, diminished by one-fourth part of the sum of the two coefficients of lateral elasticity in the same plane.*

For example, let the plane be that of yz, in which the coefficient of rigidity is C_1, those of longitudinal elasticity A_2 and A_3, and those of lateral elasticity B_1 and B_1'.

Let $2\,T_1'$ represent a distortion in the plane yz, relative to two new axes y', z', in the same plane, and let the angle $\widehat{yy'} = \theta$. Let this distortion be resolved, with respect to the original axes, according to equation (2). Then

$$N_1 = 0;\ N_2 = -\,2\,T_1' \cos\theta \sin\theta;\ N_3 = 2\,T_1' \cos\theta \sin\theta;$$
$$2\,T_1 = 2\,T_1' (\cos^2\theta - \sin^2\theta);\ T_2 = 0;\ T_3 = 0.$$

The corresponding pressures referred to the original axes are

$$P_1 = -\,2\,T_1' (B_3 - B_2') \cos\theta \sin\theta,$$
$$P_2 = -\,2\,T_1' (-\,A_2 + B_1) \cos\theta \sin\theta,$$
$$P_3 = -\,2\,T_1' (-\,B_1' + A_3) \cos\theta \sin\theta,$$
$$Q_1 = -\,2\,C_1\,T_1' (\cos^2\theta - \sin^2\theta),\ Q_2 = 0,\ Q_3 = 0.$$

Let us now determine, according to equation (2), from the above pressures, the *tangential* pressure Q_1' as referred to the new axes. Then

$$Q_1' = -\,2\,T_1' \{C_1 + \cos^2\theta \sin^2\theta\,(A_2 + A_3 - B_1 - B_1' - 4\,C_1)\}.$$

But by the preceding theorem we have also

$$Q_1' = -\,2\,T_1' C_1,$$

for all values of θ, which cannot be true unless the coefficient of $\cos^2\theta \sin^2\theta$ in the first value of Q_1' is $= 0$. Consequently

$$C_1 = \frac{1}{4}(A_2 + A_3 - B_1 - B_1').$$

By applying similar reasoning to the planes of zx and xy it is also proved that

$$C_2 = \frac{1}{4}(A_3 + A_1 - B_2 - B_2'),$$

$$C_3 = \frac{1}{4}(A_1 + A_2 - B_3 - B_3') \qquad (5.)$$

being the algebraical statement of the theorem enunciated.

Thus the number of independent coefficients of elasticity is reduced to *nine*, of which the other three are functions; and this is the utmost reduction of their number which can be made without the aid of suppositions as to the constitution of matter.

The determination by experiment of nine constants for each substance is an undertaking almost hopeless; it is therefore desirable to ascertain whether, by the introduction of some probable hypothesis, their number can be further reduced.

SECTION III.—RESULTS OF THE HYPOTHESIS OF ATOMIC CENTRES.

16. Almost all the investigations of the laws of elasticity which have hitherto appeared, are founded on the hypothesis of Boscovich: that matter consists of physical points or centres of force, or of atoms acting as if their masses were concentrated at their centres; which physical points or atoms occupy space, and produce the phenomena of elasticity, because the forces which act between them, and which depend on their relative distances and positions, tend to make them remain in certain relative positions, and at certain distances apart.

Although the results of this supposition are not verified by all solid substances, still it seems probable that its errors are to be corrected, not by rejecting it, but by combining it with another, to which I shall afterwards refer.

I shall now, therefore, show to what extent the laws of elasticity are simplified by adopting Boscovich's supposition of atomic centres of force, acting on each other by attractive and repulsive forces along the lines joining them. It will be seen, that in consequence of the course adopted, of determining, in the first place, the necessary relations between the coefficients of elasticity which must exist independently of all special hypotheses, this investigation is almost entirely freed from the algebraical intricacy in which it would otherwise be involved.

17. All the consequences peculiar to this hypothesis flow from the following single theorem, in which the term *perfect solid* is used to denote a body whose elasticity is due entirely to the mutual attractions and repulsions of atomic centres of force.

THEOREM IV. *In each of the co-ordinate planes of elasticity of a perfect solid, the two coefficients of lateral elasticity, and the coefficient of rigidity, are all equal to each other.*

Take, for example, the plane of $y z$. The proposition enunciated is equivalent to the assertion, that the tangential pressure parallel to y at the plane of xy, produced by a given transverse strain $2\,T_1 = \frac{d\eta}{dz}$, which consists in a displacement of the atomic centres parallel to y and varying with z, is equal to the normal pressure parallel to z at the same plane xy, produced by a longitudinal strain $N_2 = \frac{d\eta}{dy}$, which consists in condensing or dilating the atomic centres in a direction parallel to y, provided that longitudinal strain is equal in amount to the transverse strain.

The pressure on a given area of the plane xy, is the effect of the joint actions of the atomic centres on the negative side of that plane upon the atomic centres on the positive side.

In the natural or unstrained condition of the body, this pressure is null, showing that those forces neutralise each other. When the body is strained, therefore, the pressure is the resultant of the *variations* of all those forces, arising from the displacements of the atomic centres from their natural relative positions.*

Let m and μ denote a pair of atomic centres, m being situated on the positive side of the plane xy, and μ on the negative side. The force acting between m and μ is supposed to act along the line joining them, and to be a function of its length. When the relative displacement of the atoms is very small as compared with their distance, the variation of this force will be sensibly proportional to the variation of distance, multiplied by some function of the distance. It may therefore be denoted by $\phi r \,.\, dr$, where r denotes the distance (μm). Let this line make with the axes the angles α, β, γ.

Let the strain to be considered, in the first place, be transverse, the displacements being parallel to y and varying with z, the rate of variation being

$$\frac{d\eta}{dz} = 2\,T_1,$$

and the force to be estimated being in the direction y. Then the displacement of m relatively to μ is

$$\Delta\eta = 2\,T_1\, r \cos\gamma.$$

The variation of their distance apart is

$$\delta r = \cos\beta\,\Delta\eta = 2\,T_1\, r \cos\beta \cos\gamma.$$

The variation of the force acting between them is

$$\phi r \,.\, \delta r = 2\,T_1\, r\, \phi r \,.\, \cos\beta \cos\gamma.$$

* Small quantities of the second order relatively to the strains T_1, &c., are here neglected.

And the component of that variation parallel to y, which forms the part of the tangential pressure due to the action of μ on m, is

$$\cos\beta\,\phi r\,.\,\delta r = 2\,T_1\,r\,\phi r\,.\cos^2\beta\cos\gamma. \quad . \quad . \quad \text{(a.)}$$

Next, let the strain be longitudinal, parallel to y, and denoted by

$$N_2 = \frac{d\,\eta}{d\,y}.$$

Then the displacement of m relatively to μ is

$$\Delta\,\eta = N_2\,r\cos\beta.$$

The variation of their distance apart is

$$\delta r = \cos\beta\,\Delta\,\eta = N_2\,r\cos^2\beta.$$

The variation of the force acting between them is

$$\phi r\,.\,\delta r = N_2\,r\,\phi r\cos^2\beta.$$

And the component of that variation parallel to z, which forms the part of the normal pressure on the plane xy due to the action of μ on m, is

$$\cos\gamma\,\phi r\,.\,\delta r = N_2\,r\,\phi r\cos^2\beta\cos\gamma. \quad . \quad . \quad \text{(b.)}$$

On comparing the expressions (a) and (b) it will be seen that the quantities by which $2T_1$ and N_2 are multiplied are identical. Therefore, the tangential force in the direction y on the plane $x\,y$ produced by a distortion in the plane yz, and the normal force in the direction z produced by a longitudinal strain along y, are equal when the strains are equal, *for each pair of atomic centres.* They are therefore equal for a perfect solid, because its elasticity is wholly due to the mutual actions of atomic centres; and the theorem is proved for the plane yz, and may in the same manner be proved for the other co-ordinate planes of elasticity. It is expressed algebraically as follows:

Plane.		
$y\,z$	$B_1 = B_1' = C_1$	
$z\,x$	$B_2 = B_2' = C_2$	. (6.)
$x\,y$	$B_3 = B_3' = C_3$	

18. The combination of these equations with the equations (5) of Theorem III. leads immediately to the following results:

$$\left.\begin{aligned} C_1 &= \frac{A_2 + A_3}{6} \\ C_2 &= \frac{A_3 + A_1}{6} \\ C_3 &= \frac{A_1 + A_2}{6} \end{aligned}\right\} \quad . \quad . \quad . \quad . \quad (7.)$$

$$\left.\begin{aligned} A_1 &= 3\,(C_2 + C_3 - C_1) \\ A_2 &= 3\,(C_3 + C_1 - C_2) \\ A_3 &= 3\,(C_1 + C_2 - C_3) \end{aligned}\right\} \quad . \quad . \quad . \quad (8.)$$

that is to say,

THEOREM V. *In each of the three co-ordinate planes of elasticity of a perfect solid, the coefficient of rigidity is equal to one-sixth part of the sum of the two coefficients of longitudinal elasticity;*
and consequently,

For each axis of elasticity of a perfect solid, the coefficient of longitudinal elasticity is equal to three times the sum of the two coefficients of rigidity for the co-ordinate planes which pass through that axis, diminished by three times the coefficient of rigidity for the plane normal to that axis.

We have now arrived at the conclusion, that in a body whose elasticity arises wholly from the mutual actions of atomic centres, all the coefficients of elasticity are functions of the three coefficients of rigidity. Rigidity being the distinctive property of solids, a body so constituted is properly termed a *perfect solid.*

When the three coefficients of rigidity are equal, the body is a perfect solid, equally elastic in all directions. The equations 6 and 8 become $A = 3\,C$; $B = C$, agreeing with the results deduced by various mathematicians from the hypothesis of Boscovich.

SECTION IV.—RESULTS OF THE HYPOTHESIS OF MOLECULAR VORTICES.

19. The great and obvious deviations from the laws of elasticity, as deduced from the hypothesis of atomic centres, which many substances present, render some modification of it essential.

Supposing a body to consist of a continuous fluid, diffused through space with perfect uniformity as to density and all other properties, such a body must be totally destitute of rigidity or elasticity of figure, its parts having no tendency to assume one position as to *direction* rather than another. It may, indeed, possess elasticity of *volume* to any extent, and display the phenomena of cohesion at its surface and between its parts. Its longitudinal and lateral elasticities will be equal in every direction; and they must be equal to each other by equation (5), which becomes

$$0 = A - B; \ C = 0.$$

If we now suppose this fluid to be partially condensed round a system of centres, there will be forces acting between those centres greater than those between other points of the body. The body will now possess a certain amount of rigidity; but less, in proportion to its longitudinal and lateral elasticities, than the amount proper to the condition of perfect solidity. Its elasticity will, in fact, consist of two parts, one of which, arising from the mutual actions of the centres of condensation, will follow the laws of perfect solidity; while the other will be a mere elasticity of volume, resisting change of bulk equally in all directions.

In a paper on the Mechanical Action of Heat in connection with the Elasticity of Gases and Vapours (*Trans. of the Royal Society of Edinburgh*, Vol. XX., Part I.), I have attempted to develop some of the consequences of a supposition of this kind, called the hypothesis of molecular vortices.* It assumes that each atom of matter consists of a nucleus or central physical point, enveloped by an elastic atmosphere, which is retained in its position by forces attractive towards the atomic centre, and which, in the absence of heat, would be so much condensed round that centre as to produce the condition of perfect solidity in all substances; that the changes of condition and elasticity due to heat arise from the centrifugal force of revolutions among the particles of the atmospheres, diffusing them to a greater distance from their centres, and thus increasing the elasticity which resists change of volume alone, at the expense of that which resists change of figure also; and that the medium which transmits light and radiant heat consists of the nuclei of the atoms, of small mass, but exerting intense forces, vibrating independently, or almost independently of their atmospheres; *absorption* being the communication of that motion to the atmosphere, so that it is lost by the nuclei.

20. A body so constituted, in which the rigidity is considerable, may be called, in general, an *imperfect solid;* and it is obvious that its various elasticities may be represented in the following manner:

THEOREM VI. *In an imperfect solid, according to the hypothesis of molecular vortices, each of the coefficients of longitudinal and lateral elasticity is equal to the same function of the coefficients of rigidity which would be its value in a perfect solid, added to a coefficient of fluid elasticity which is the same in all directions.*

* An abstract of that paper is published in Poggendorff's *Annalen* for 1850, No IX.

Denoting this fourth coefficient by J, we have the following equations, giving the values of the coefficients of longitudinal and lateral elasticity in terms of the coefficient of fluid elasticity, and of the three coefficients of rigidity.

$$\left.\begin{aligned} A_1 &= 3\,(C_2 + C_3 - C_1) + J \\ A_2 &= 3\,(C_3 + C_1 - C_2) + J \\ A_3 &= 3\,(C_1 + C_2 - C_3) + J \\ B_1 &= C_1 + J \\ B_2 &= C_2 + J \\ B_3 &= C_3 + J \end{aligned}\right\} \quad . \quad . \quad . \quad (9.)$$

The utmost number of independent coefficients is thus increased to *four.*

If the coefficients of rigidity be progressively diminished without limit, as compared with the coefficient of fluid elasticity, the body will pass through every stage of the gelatinous state; and when the coefficients of rigidity vanish, its condition will be that of a perfect fluid, in which the longitudinal and lateral elasticities are all equal, and represented by the single coefficient J.

It is to be observed, that in this condition the independent actions of the nuclei or physical points at the atomic centres upon each other, which are the means of radiation, may be very great; their sensible effect on the elasticity of the body being neutralised by other forces, exerted by the parts of the atmospheres.

If two of the coefficients of rigidity are equal (as $C_2 = C_3$), the body is equally elastic in all directions round an axis, which in this case is that of x; and equations (9) become

$$\left.\begin{aligned} A_1 &= 6\,C_2 - 3\,C_1 + J \\ A_2 &= A_3 = 3\,C_1 + J \\ B_1 &= C_1 + J \\ B_2 &= B_3 = C_2 + J \end{aligned}\right\} \quad . \quad . \quad . \quad (9A.)$$

When the three coefficients of rigidity are all equal, the body is an imperfect solid equally elastic in all directions. The results of this condition have been investigated by Professor Stokes, M. Wertheim, and Mr. Clerk Maxwell.

Equations (9) in this case become

$$A = 3\,C + J; \; B = C + J. \quad . \quad . \quad . \quad (9B.)$$

Note respecting Previous Investigations.

(18 *a*.) The investigations of Poisson (*Mem. de l'Acad. des Sciences*, XVIII.), of M. Cauchy (*Exercices des Mathématiques*, passim), and of Mr. Haughton (*Trans. Roy. Irish Acad.*, XXI.), respecting the elasticity of substances unequally elastic in different directions, are all founded on the hypothesis of atomic centres. So far as they relate to substances possessed of rectangular axes of elasticity, they agree in expressing the elasticity of such bodies by means of *six* coefficients, corresponding respectively to those which I have denoted by

$$A_1, A_2, A_3, C_1, C_2, C_3.$$

None of those investigations indicate any relations amongst these six coefficients.

The researches of Mr. Green on the propagation of vibratory movement (*Camb. Trans.*, VII.) differ materially from those which preceded them, inasmuch as they are applicable, not merely to systems of atomic centres or physical points, but to solid substances constituted in any manner whatsoever.* So far as they are applicable to bodies possessed of axes of elasticity, they involve *nine* coefficients: three of longitudinal, three of lateral, and three of transverse elasticity. The following table exhibits a comparison between Mr. Green's notation and that of this paper:

Coefficients of Elasticity.

In the Notation of	Longitudinal.	Lateral.	Transverse.
Mr. Green, . .	G H I	P Q R	L M N
This Paper, .	A_1 A_2 A_3	B_1 B_2 B_3	C_1 C_2 C_3

There is nothing, however, in the researches of Mr. Green to indicate any mutual relations amongst those nine coefficients; and to establish such relations, indeed, it appears to me that the subject must be investigated, not dynamically, but statically.

It may here be observed, that Mr. Green's equations contain three additional coefficients, to represent the effect of a strained condition of the medium on the propagation of vibratory movement; but those three

* A second paper by Mr. Haughton (*Trans. Roy. Irish Acad.*, XXII.) is equally comprehensive.

quantities, being foreign to the subject of this paper, have no expressions corresponding to them in its notation.

In the equations of the propagation of light, Mr. Green effects an apparent reduction in the number of coefficients by introducing the supposition that the vibrations are of necessity wholly tangential to each wave-front. But this supposition is quite at variance with the nature of elastic solids, and is obviously intended by the author as merely an assumption for the purpose of facilitating calculation, and obtaining approximately true results, in the case of luminiferous undulations.

Mr. M'Cullagh's researches on the propagation of light (*Trans. Roy. Irish Acad.*, XXI.) involve a similar assumption.

The result peculiar to the investigations contained in the present paper is the establishment of certain mutual relations amongst the different elasticities of a given substance, whereby the *six* coefficients of Poisson and Cauchy are reduced to functions of *three*, and the *nine* coefficients of Mr. Green to functions of *four*; the former representing the condition of a medium whose elasticity is wholly due to the mutual actions of atomic centres, the latter that of a substance whose condition is intermediate between those of a system of centres of force, and of a continuous and uniformly diffused fluid.

General equations of vibratory movement, in the particular case of uncrystallised media, agreeing with those of Mr. Green, are given by Professor Stokes in his memoir on Diffraction (*Camb. Trans.* IX.) His two coefficients of elasticity have the following values in the notation of this paper :—

Prof. Stokes.	This Paper.
a^2	$= \frac{Ag}{D} = (3\,C + J)\frac{g}{D},$
b^2	$= \frac{Cg}{D}.$

g denotes the accelerating force of gravity; and D, the weight of unity of volume of the vibrating medium.

a and b, in Professor Stokes's paper, are the velocities of propagation of normal and tangential vibrations respectively.

In the researches of Poisson, Navier, Cauchy, Lamé, and others, on the elasticity of bodies equally elastic in all directions, the coefficients are often expressed in terms of two quantities, denoted by k and K, in the following expression for a normal pressure on the plane yz:

$$P_1 = -k N_1 - K(N_1 + N_2 + N_3):$$

k represents a species of longitudinal elasticity, under the condition that the volume remains unchanged; and K, an elasticity resisting change of volume. Their values in the notation of this paper are as follows:—

$$k = A - B = 2\,C; \;\; K = B = C + J.$$

It is evidently impracticable to apply an analogous notation to bodies unequally elastic in different directions.

M. Wertheim has recently made a most elaborate and valuable series of experiments on the elasticity of brass, glass, and caoutchouc, according to a method suggested by M. Regnault, for the purpose of determining the laws of elasticity of uncrystallised substances (*Ann. de Chim. et de Phys.*, Sér. III., tom. XXIII.) He concludes that for brass and glass, and for caoutchouc moderately strained, the following equation is nearly if not exactly true, in the notation to which I have just referred, $k = K$, which, in the notation of this paper, is equivalent to the following:

$$J = C; \;\; B = 2\,C; \;\; A = 4\,C.$$

M. Wertheim has investigated the consequences which must follow in the solution of several problems connected with elasticity, if this law be universally true for solid bodies.

This supposition must be regarded as doubtful; and it is not, indeed, advanced by M. Wertheim as more than a conjecture. So far as our present knowledge goes, it seems more probable that the relations between C and J may be infinitely varied. If the effect of heat is to diminish C and increase J, there may be some temperature for each substance at which M. Wertheim's equation is verified. In the sequel I shall consider more fully the consequences to be deduced from M. Wertheim's experiments on this subject.

SECTION V.—COEFFICIENTS OF PLIABILITY, AND OF EXTENSIBILITY AND COMPRESSIBILITY, LONGITUDINAL, LATERAL, AND CUBIC.

Examples of their Experimental Determination.

21. Coefficients of elasticity serve to determine pressures from the corresponding strains. We have now to consider the determination of strains from pressures.

To determine a distortion from the corresponding tangential pressure, it is sufficient to multiply, using the negative sign, by the reciprocal of the proper coefficient of rigidity. This reciprocal may be called a coefficient of *pliability*.

A similar process, however, cannot be applied to the calculation of longitudinal strains from normal pressures; because, as each normal pressure is a function of all the three longitudinal strains, so each longitudinal strain is a function of all the three normal pressures.

Let the longitudinal strains be represented, in terms of the normal pressures, by the following equations :

$$\left.\begin{aligned} N_1 &= -\mathfrak{a}_1 P_1 + \mathfrak{b}_3 P_2 + \mathfrak{b}_2 P_3 \\ N_2 &= \mathfrak{b}_3 P_1 - \mathfrak{a}_2 P_2 + \mathfrak{b}_1 P_3 \\ N_3 &= \mathfrak{b}_2 P_1 + \mathfrak{b}_1 P_2 - \mathfrak{a}_3 P_3 \end{aligned}\right\} \quad . \quad . \quad (10.)$$

Then the coefficients in these equations are found, by a process of elimination, to have the following values in terms of the coefficients of elasticity.

Let

$$K = 24(C_2^2C_3 + C_2C_3^2 + C_3^2C_1 + C_3C_1^2 + C_1^2C_2 + C_1C_2^2 - C_1^3 - C_2^3 - C_3^3) \\ - 52C_1C_2C_3 + J\{8(C_2C_3 + C_3C_1 + C_1C_2) - 4(C_1^2 + C_2^2 + C_3^2)\}.$$

Then

$$\left.\begin{aligned} \mathfrak{a}_1 &= \frac{1}{K}\{8C_1^2 - 9(C_2 - C_3)^2 + 4C_1J\} \\ \mathfrak{a}_2 &= \frac{1}{K}\{8C_2^2 - 9(C_3 - C_1)^2 + 4C_2J\} \\ \mathfrak{a}_3 &= \frac{1}{K}\{8C_3^2 - 9(C_1 - C_2)^2 + 4C_3J\} \\ \mathfrak{b}_1 &= \frac{1}{K}\{3(C_2 + C_3 - C_1)C_1 - C_2C_3 + 2(C_2 + C_3 - C_1)J\} \\ \mathfrak{b}_2 &= \frac{1}{K}\{3(C_3 + C_1 - C_2)C_2 - C_3C_1 + 2(C_3 + C_1 - C_2)J\} \\ \mathfrak{b}_3 &= \frac{1}{K}\{3(C_1 + C_2 - C_3)C_3 - C_1C_2 + 2(C_1 + C_2 - C_3)J\} \end{aligned}\right\} \quad (11.)$$

The above coefficients may be thus classified :

$\mathfrak{a}_1$, $\mathfrak{a}_2$, $\mathfrak{a}_3$, are the coefficients of *longitudinal extensibility and compressibility* parallel respectively to the three axes of elasticity.

$\mathfrak{b}_1$, $\mathfrak{b}_2$, $\mathfrak{b}_3$, are the coefficients of *lateral extensibility and compressibility* for the three co-ordinate planes of elasticity, serving to determine the effect of a normal pressure on those dimensions of a body which lie at right angles to its direction.

From the manner in which the coefficient J enters into the common denominator K, it is obvious that when the coefficients of rigidity diminish without limit as compared with that of fluid elasticity, the six coefficients of linear extensibility and compressibility increase *ad infinitum*.

In a body whose three coefficients of rigidity are different, the coefficient of cubic compressibility, that is to say, the quotient of the sum of the three longitudinal strains by the mean of the three normal pressures, with the sign changed, has no fixed value unless some arbitrary relation be fixed between those pressures. Let them be supposed, then, to be all equal; let their common value be P, and let the coefficient of cubic compressibility in this case be denoted by $\mathfrak{d}$: then

$$\left.\begin{aligned}
\mathfrak{d} &= -\frac{N_1+N_2+N_3}{P} = \mathfrak{a}_1+\mathfrak{a}_2+\mathfrak{a}_3-2(\mathfrak{b}_1+\mathfrak{b}_2+\mathfrak{b}_3) \\
&= \frac{1}{K}\{8(C_2C_3+C_3C_1+C_1C_2)-4(C_1^2+C_2^2+C_3^2)\} \\
\text{Hence } \frac{1}{\mathfrak{d}} &= J+6(C_1+C_2+C_3) \\
&\quad -\frac{196C_1C_2C_3}{8(C_2C_3+C_3C_1+C_1C_2)-4(C_1^2+C_2^2+C_3^2)}
\end{aligned}\right\} \quad (12.)$$

So that this coefficient is the sum of the three longitudinal coefficients of compressibility, diminished by twice the sum of the three lateral coefficients. It does not, like them, increase *ad infinitum* when the rigidity vanishes; its ultimate value in that case being

$$\frac{1}{J},$$

the reciprocal of the coefficient of fluid elasticity, as might have been expected.

If $C_2=C_3$, so that the body is equally elastic in all directions round the axis of x, equations (11) and (12) take the following forms:

$$\left.\begin{aligned}
K &= 4C_1\{12C_1C_2-6C_1^2-C_2^2+J(4C_2-C_1)\} \\
\mathfrak{a}_1 &= \frac{1}{K}(8C_1^2+4C_1J) = \frac{2C_1+J}{12C_1C_2-6C_1^2-C_2^2+J(4C_2-C_1)} \\
\mathfrak{a}_2 = \mathfrak{a}_3 &= \frac{1}{K}\{8C_2^2-9(C_1-C_2)^2+4C_2J\} \\
\mathfrak{b}_1 &= \frac{1}{K}\{6C_1C_2-3C_1^2-C_2^2+J(4C_2-2C_1)\} \\
\mathfrak{b}_2 = \mathfrak{b}_3 &= \frac{1}{K}(2C_1C_2+2C_1J) = \frac{C_2+J}{24C_1C_2-12C_1^2-2C_2^2+J(8C_2-2C_1)} \\
\mathfrak{d} &= \frac{1}{K}(16C_1C_2-4C_1^2) = \frac{4C_2-C_1}{12C_1C_2-6C_1^2-C_2^2+J(4C_2-C_1)} \\
\frac{1}{\mathfrak{d}} &= J+12C_2+6C_1-\frac{49C_2^2}{4C_2-C_1}
\end{aligned}\right\} \quad (12A)$$

For bodies equally elastic in all directions the coefficients of compressibility and extensibility take the following values:

$$\left.\begin{array}{l} \mathfrak{x} = \dfrac{2C + J}{5C^2 + 3CJ}; \quad \mathfrak{h} = \dfrac{C + J}{10C^2 + 6CJ} \\ \mathfrak{d} = 3\mathfrak{x} - 6\mathfrak{h} = \dfrac{3}{5C + 3J}; \quad \therefore \dfrac{1}{\mathfrak{d}} = J + \frac{5}{3}C. \end{array}\right\} \quad . \quad (12B.)$$

In substances of this kind the coefficient of cubic compressibility is the same, whether the three normal pressures are equal or unequal, being equal to the sum of the three longitudinal strains divided by the mean of the three normal pressures with the sign changed: that is to say,

$$\mathfrak{d} = -3\frac{N_1 + N_2 + N_3}{P_1 + P_2 + P_3}.$$

One of the most frequent errors in investigations respecting the elasticity and strength of materials, and the propagation of sound, has been to confound the coefficients of longitudinal elasticity with the reciprocals of the coefficients of longitudinal compressibility. The equations of this section show clearly how widely these two classes of quantities may differ.

The reciprocal of the longitudinal extensibility, $\frac{1}{\mathfrak{x}}$, is what is commonly termed the *Weight of the modulus of elasticity.*

22. The following formula may be found useful in the determination of the coefficient J of fluid elasticity from experimental data.

Let us suppose that the three coefficients of rigidity of a substance, C_1, C_2, C_3, have been determined by experiments on torsion, and that some one of the coefficients of compressibility and extensibility in equation (11), or those derived from it, has also been determined by experiment. Let the actual value of this coefficient be called $\mathfrak{r}$, and the value which it would have had, had J been $= 0$, $\mathfrak{r}_0$. Also let K_0 denote the value which the denominator K would have had, had J been $= 0$, and let n be the factor by which J is multiplied in the numerator of $\mathfrak{r}$, and m, in the denominator.

Then

$$J = K_0 \cdot \frac{\mathfrak{r}_0 - \mathfrak{r}}{m\,\mathfrak{r} - n}. \quad . \quad . \quad . \quad . \quad (13.)$$

When applied to coefficients of longitudinal extensibility, this formula labours under the disadvantage that a comparatively slight error in the experimental data may cause a serious error in the determination of J. Let us take, for example, an uncrystallised substance, and make successively the two following suppositions,

$$J = 0, \; J = C:$$

it will be found that the results are respectively,

$$\mathfrak{a} = \frac{1}{C} \times 0{\cdot}4, \ \mathfrak{a} = \frac{1}{C} \times 0{\cdot}375,$$

being in the ratio of 16 : 15; so that any uncertainty in the experiments is in this case increased fifteenfold in computing the value of $\frac{J}{C}$. Hence it appears, that without very great precision in the experiments, the coefficient of fluid elasticity cannot be satisfactorily determined by a comparison of the effects of longitudinal tension with those of torsion. It is especially desirable that the two sets of experiments should be made on the same piece of the material.

The best data for calculations of this kind would be experiments on cubic compressibility, in conjunction with experiments on torsion; for, as equations (12), (12A), and (12B) show, in order to determine J we have simply to subtract a certain symmetrical function of the rigidities from the reciprocal of the cubic compressibility. In the process of calculation, the errors in the experiments on rigidity are multiplied, on an average, by $\frac{5}{3}$ only, while those of the experiments on compressibility sustain no augmentation whatsoever.

Next after data of this kind may be ranked experiments on longitudinal extensibility, as compared with the cubic extensibility or compressibility of the same piece of material. Of this method, suggested by M. Regnault, and carried into effect by M. Wertheim, I shall presently speak more fully.

Were it possible to ascertain the velocity of sound in an unlimited mass of an elastic material along each of the axes of elasticity, the coefficients of longitudinal elasticity could be determined with great precision by the formula

$$A = \frac{v^2 D}{g},$$

where v is the velocity of sound, D the weight of unity of volume of the substance, and g the accelerating force of gravity. But it is only practicable to determine the velocity of sound along prismatic or cylindrical rods; and, as I shall show in a subsequent paper, it is impossible, in the present state of our knowledge of the molecular condition of the superficial particles of solid bodies, to assign theoretically the ratio in which the velocity of sound along a rod is less than its velocity in an indefinitely extended mass. That ratio is only known empirically in a few cases, having various values lying between 1 and $\sqrt{\frac{2}{3}}$.

23. The experiments of M. Wertheim, on longitudinal and cubic extensibility (*Ann. de Chim. et de Phys.*, Sér. III., Tom. XXIII.) were made

upon brass and crystal, the results being calculated on the supposition that those substances are homogeneous and equally elastic in all directions. There can be no doubt of the correctness of this supposition with respect to well annealed crystal; and with respect to brass, it is probably very near the truth.

In those experiments, a cylindrical tube of the substance to be examined was strained by longitudinal tension. The increase of length was observed directly. The increase of bulk was found by observing the depression in a capillary tube connected with the summit of the strained tube, of a column of liquid with which they were filled. Let R denote the tensile force reduced to unity of surface; let L be the original length of a given portion of the tube, U its original volume, Δ L and Δ U the increase of those quantities by the tension R; and let the axis of x be that of the tube. Then we have

$$N_1 = \frac{\Delta L}{L}; \quad N_1 + N_2 + N_3 = \frac{\Delta U}{U};$$

$$P_1 = -R; \quad P_2 = 0; \quad P_3 = 0:$$

and consequently, for uncrystallised substances,

$$\mathfrak{a} = \frac{1}{R} \cdot \frac{\Delta L}{L}; \quad \mathfrak{b} = \frac{3}{R} \cdot \frac{\Delta U}{U} \quad \ldots \quad (14.)$$

To determine the coefficients of rigidity and fluid elasticity from these data, we have the following formulæ:

$$\left. \begin{aligned} C &= \frac{1}{3\mathfrak{a} - \frac{1}{3}\mathfrak{b}} = \frac{R}{3\frac{\Delta L}{L} - \frac{\Delta U}{U}} \\ J &= C\left(\frac{3\mathfrak{a}}{\mathfrak{b}} - 2\right) = C\left(\frac{\Delta L}{L} \cdot \frac{U}{\Delta U} - 2\right) \end{aligned} \right\} \quad (14A.)$$

The experiments of M. Wertheim were made on three tubes of brass and five of crystal. In the following table those tubes are designated as M. Wertheim has numbered them. The coefficients $\mathfrak{a}$ and $\mathfrak{b}$, transcribed or calculated from his statement of the mean results of numerous experiments on each tube, express the fraction by which the material is elongated, or increased in bulk, by tension at the rate of *one kilogramme on the square millimetre;* that is to say, 1422·34 pounds avoirdupois on the square inch. (The common logarithm of this number is 3·153004.) The reciprocals of those coefficients, and the coefficients of elasticity, as calculated by equation (14A), are given in kilogrammes on the square millimetre.

TABLE OF COEFFICIENTS CALCULATED FROM M. WERTHEIM'S EXPERIMENTS ON THE EXTENSIBILITY OF BRASS AND CRYSTAL.

	Extensibilities per Kilogramme on the Square Millimètre.		Reciprocals of the Extensibilities in Kilogrammes on the Square Millimètre.		Coefficients of Elasticity in Kilogrammes on the Square Millimètre.			
	Longitudinal. $\mathfrak{a}$	Cubic. $\mathfrak{v}$	$\frac{1}{\mathfrak{a}}$	$\frac{1}{\mathfrak{v}}$	Rigidity. C	Fluid. J	Longitud. A	Lateral. B
BRASS, . . Tube I.	0·0000939	0·0000904	10645·2	11058	3973	4436	16355	8409
,, . . ,, II.	0·0001015	0·0000942	9855·2	10620	3663	4515	15504	8178
,, . . ,, III.	0·0001035	0·0000979	9664·9	10216	3600	4216	15016	7816
CRYSTAL,.........Tube I.	0·0002596	0·0002803	3852·5	3569	1459·2	1136·6	5514·2	2595·8
,, ... ,, II.	0·0002324	...*	4302·6	...	...	...	...	...
,, ... ,, III.	0·0002873	0·0002568	3481·1	3894	1288·0	1747·0	5611·0	3035·0
,, ... ,, IV.	0·0002258	0·0002543	4429·0	3933	1687·4	1120·9	6183·1	2803·3
,, ... ,, V.	0·0002284	0·0002234	4379·1	4476	1637·7	1746·8	6659·9	3384·5

* Tube II. of Crystal was accidentally broken.

The various degrees of elasticity of the brass tubes are ascribed by M. Wertheim to the relative frequency with which they were subjected to wire-drawing, to reduce the thickness of metal. It may be observed, that this operation seems to increase the rigidity more than the fluid elasticity, a fact which might naturally have been expected.

The means of the three sets of results for brass are given in the following table:—

Coefficients of		Kilogrammes on the Sq. Millim.	Lbs. Avoird. on the Square Inch.
Rigidity, ...	C	3745·3	5327100
Fluid Elasticity, ...	J	4389·0	6242700
Longitudinal Elasticity, .	A	15625·0	22224000
Lateral Elasticity, . .	B	8134·3	11579000
Reciprocals of Extensibilities.			
Longitudinal (or Weight of the Modulus of Elasticity), .	$\frac{1}{\mathfrak{a}}$	10054·4	14301000
Cubic, ...	$\frac{1}{\mathfrak{b}}$	10631·0	15121000

Coefficients of Extensibility and Compressibility.		Per Kilog. on the Square Millim.	Per lb. on the Square Inch.
Longitudinal, . . .	$\mathfrak{a}$	0·00009946	0·0000000699
Cubic, ...	$\mathfrak{b}$	0·00009406	0·0000000661
Lateral, ...	$\mathfrak{h}$	0·00003405	0·0000000239

The following result is calculated from the experiments of M. Savart on the torsion of brass wire (*Ann. de Chim. et de Phys.*, August, 1829):—

		Kilog. on the Square Millim.	Lbs. on the Square Inch.
Coefficient of Rigidity, . .	C	3682	5237100
The difference being . . .		63·3	90000

Hence we see that the rigidity of wire-drawn brass, as determined directly by torsion, differs from that calculated from the longitudinal and cubic extensibilities by only *one-sixtieth part*, being a very small discrepancy in experiments of this kind.

The following table gives the means of the four sets of results, I., III., IV., V., for crystal:—

	Kilog. on the Square Millim.	Lbs. on the Square Inch.
C	1518	2159100
J	1438	2045300
A	5992	8522600
B	2956	4204400
$\frac{1}{\mathfrak{a}}$	4039	5746100
$\frac{1}{\mathfrak{d}}$	3968	5643800

	Per Kilog. on the Square Millim.	Per lb. on the Square Inch.
$\mathfrak{a}$	0·0002476	0·0000001740
$\mathfrak{d}$	0·0002520	0·0000001772
$\mathfrak{b}$	0·0000818	0·0000000575

It is obvious that the above mean values for crystal are not to be relied upon as equally accurate with those for brass; for the wide discrepancies between the results of the experiments on the five crystal tubes show that this substance, like every kind of glass, is subject to great variations in the physical properties of different specimens.

24. So far as I am aware, there is no substance whose elasticity varies in different directions, for which experimental data as yet exist, adequate to determine the three coefficients of rigidity, and the coefficient of fluid elasticity.

Supposing the three coefficients of rigidity of a substance of this kind to be known by experiments on torsion, the process of MM. Regnault and Wertheim would readily furnish data for calculating the fluid elasticity.

For example: let a tension R per unit of area be applied to the ends of a tube whose axis is one of the axes of elasticity, say that of x. Let $\frac{\Delta U}{U}$ be the fraction by which its volume is increased, as before. Then

$$\left.\begin{aligned}\frac{\Delta U}{R U} &= -\frac{N_1 + N_2 + N_3}{P_1} = \mathfrak{a}_1 - \mathfrak{b}_3 - \mathfrak{b}_2 \\ &= \frac{1}{K}\{8 C_1^2 - 2 C_1 (C_2 + C_3) - 6 (C_2 - C_3)^2\}\end{aligned}\right\} \quad (15.)$$

Let the above equation be abbreviated into

$$\frac{\Delta U}{R U} = \frac{\phi (C)}{K_0 + m J},$$

where $K = K_0 + m J$, as in equation (13). Then

$$J = \frac{1}{m} \left\{ \frac{R U}{\Delta U} \phi (C) - K_0 \right\} \qquad \text{(15A.)}$$

The formulæ corresponding to equation (15) for tubes whose axes are parallel to y and z, are easily found by permutations of the indices 1, 2, 3. The sum of the three values of $\frac{\Delta U}{R U}$ thus obtained is obviously $= \mathfrak{v}$.

[It may be remarked, with reference to Sect. 17 of the preceding paper, that the effect of alterations of direction in the lines joining pairs of particles is not taken into account in the investigation of the elastic forces arising from the states of strain which are there considered. It appears to me that this effect, except for particular laws of force, will be of the same order as that which depends on the alterations of the mutual distances between the particles; and that if it be taken into account, the demonstration of Theorem IV. fails.

This objection occurred to me after the whole of the paper was in type, and I immediately suggested it to the author; but, as he was not convinced of the correctness of my view, he desired that the paper should be published as it stands, reserving additional explanations or modifications, if necessary, for the next number of the *journal*.—W. T.]

Supplementary Paper to Section III., Article 17.

In the portion above referred to of my paper on the Elasticity of Solids, published in the *Cambridge and Dublin Mathematical Journal* for February, 1851, the theorem is laid down, that in a given plane in an elastic solid consisting entirely of atoms acting on each other by attractions and repulsions between their centres, the coefficients of rigidity and of lateral elasticity are equal.

The proof of this proposition depends on the principle that *the elastic force in such a solid, called into play by a strain, in which the relative displacements of the atoms are very small as compared with their distances apart, is sensibly the resultant of the variations of force due to the variations of distance only, the variations of relative direction producing no appreciable effect.* This principle being granted, it is easily shown that the portion of that resultant for each pair of atoms is the same for a given amount of strain in a given plane, whether lateral or transverse with respect to the plane on which elastic pressure is estimated.

In the paper referred to, I assumed this principle without demonstration. The editor of this journal, however, has since shown me, that my having done so may be considered as causing a defect in the chain of reasoning. I shall now, therefore, proceed to prove it.

Let it be possible for a solid to exist in an unstrained condition, consisting entirely of atomic centres of force acting on each other along the lines joining them, with forces which are functions of the lengths of these lines. Then must the pressure, estimated in any direction, on any portion of any plane in that solid be null. That pressure is the resultant, in the direction assumed, of the mutual actions of all the atoms whose lines of junction pass through the given portion of the given plane.

Let the given portion be indefinitely small, and let it be called ω, being situated in the arbitrarily-assumed plane $P \omega p$, which divides the solid into two portions, A and B. Let $- X \omega + X$ be an arbitrary axis along which pressure is to be estimated. The pressure exerted by the portion A upon the infinitesimal area ω of the portion B, is the resultant, reduced to the direction $- X \omega + X$, of all the forces exerted by the atoms in A on the atoms in B, in lines passing through ω; and the body being unstrained, this resultant must be null.

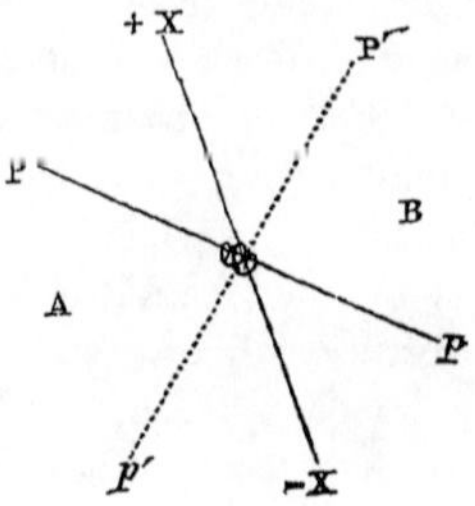

Assume a new position, $P' \omega p'$, for the plane of separation, making an equal angle $P' \omega + X = + X \omega P$ on the opposite side of the axis to the original position. The same letters applying to the two portions of the solid, the pressure of A on the area ω of B along $- X \omega X$ must still be null.

The two planes divide the solid into two pairs of opposite wedges. The action of A on B along X through ω in the original position of the plane, may be divided into two parts, viz.—

The resultant of the actions of the atoms in the wedge $P \omega p'$ on those in the opposite wedge $P' \omega p$;

The resultant of the actions of the atoms in the wedge $p \omega p'$ on those in the wedge $P \omega P'$.

In the new position of the plane, the pressure on ω is made up as follows:

The resultant of the actions of the atoms in the wedge $P \omega p'$ on those in the wedge $P' \omega p$, which is the same as in the original position of the plane;

The resultant of the actions of the atoms in the wedge $P \omega P'$ on those in the wedge $p' \omega p$, being identical in amount but *opposite in direction* to that of the atoms in $p' \omega p$ on those in $P \omega P'$, which formed part of the pressure in the original position of the plane.

Now the pressures in the two positions of the plane of separation cannot both be null, unless the resultant of the mutual actions of the atoms in each pair of opposite wedges is separately null; for we see that the action of a pair of wedges can be reversed in direction without affecting the nullity of the total resultant. The position of the pair of opposite wedges is arbitrary; so also is their angular magnitude, which may be indefinitely small.

Therefore, no mere change of angular position of a pair of opposite elementary wedges can produce a pressure.

Every strain in which the relative displacements of the particles are small as compared with their relative distances, may be reduced to angular displacements of pairs of opposite elementary wedges, and variations of the mutual distances of the particles contained in them. The angular displacements can produce no pressure of themselves; the variations of distance are therefore the sole cause of that portion of the pressure which is of the same order of small quantities with the strain: being the principle to be proved.

The combination of the angular displacements with the variations of distance will give rise to pressures of the second and higher orders of small quantities as compared with the strain; but for the small strains to which the present inquiry is limited, those are inappreciable, and may be neglected.

NOTE RESPECTING MR. CLERK MAXWELL'S PAPER "ON THE EQUILIBRIUM OF ELASTIC SOLIDS." (*Trans. Roy. Soc. Edin.*, VOL. XX., PART I.)

I HAVE already referred to the researches of Mr. Clerk Maxwell, of the general nature of which only I was aware at the time of the publication of my paper on this subject in the *Cambridge and Dublin Mathematical Journal* for February, 1851.

Since then I have had an opportunity of reading Mr. Maxwell's paper, so as to compare his notation with my own.

Mr. Maxwell's investigations relate to such solids only as are equally elastic in all directions. He expresses their elasticity by means of two coefficients, μ and m, having the following properties:

$$\mu = -\frac{1}{3} \cdot \frac{P_1 + P_2 + P_3}{\frac{d\xi}{dx} + \frac{d\eta}{dy} + \frac{d\zeta}{dz}},$$

$$m = -\frac{P_1 - P_2}{\frac{d\xi}{dx} - \frac{d\eta}{dy}} = -\frac{P_2 - P_3}{\frac{d\eta}{dy} - \frac{d\zeta}{dz}} = -\frac{P_3 - P_1}{\frac{d\zeta}{dz} - \frac{d\xi}{dx}}.$$

From which it is clear, that those coefficients have the following values in the notation of the paper which I have published.

$$\mu = \frac{1}{\mathfrak{d}} = \tfrac{5}{3}\,C + J = \text{reciprocal of the cubic compressibility,}$$

$$m = 2\,C = \text{twice the rigidity;}$$

consequently

$$C = \tfrac{1}{2}\,m, \quad J = \mu - \tfrac{5}{6}\,m.$$

The particular problems solved by Mr. Clerk Maxwell are of a very interesting character, especially those relative to the optical changes produced in transparent bodies by straining them.

ON THE LAWS OF ELASTICITY.*

SECTION VI.—ON THE APPLICATION OF THE METHOD OF VIRTUAL VELOCITIES TO THE THEORY OF ELASTICITY.

25. Lagrange's method of virtual velocities having been applied to the problems of the equilibrium and motion of elastic media by Mr. Green (*Camb. Trans.*, VII.) and by Mr. Haughton (*Trans. Royal Irish Acad.*, XXI., XXII.), it is my purpose in this and the following section to point out the mutual correspondence between the coefficients in the formulæ arrived at by these gentlemen, and the coefficients of elasticity which form the subject of the previous portion of this paper, and also to show how far the laws of relation between the nine coefficients of elasticity of a homogeneous body, which I originally proved by a method chiefly geometrical, are capable of being deduced symbolically from equations found according to Lagrange's method.

26. The principle of virtual velocities, as applied to molecular action, it

* Originally published in the *Cambridge and Dublin Mathematical Journal*, Nov., 1852.

as follows: Let X, Y, Z, denote the total accelerative forces applied to any particle, whose mass is m, of an elastic medium, through agencies distinct from molecular action (such as the attraction of gravitation); let u, v, w, be the components of the velocity of m; let δx, δy, δz, denote indefinitely small virtual variations of x, y, z; let S be the total accelerative molecular force applied to m, δs an indefinitely small virtual variation of the line along which it acts; then the following equation

$$\left.\begin{aligned}\Sigma\left[m\left\{\left(\mathrm{X}-\frac{du}{dt}\right)\delta x+\left(\mathrm{Y}-\frac{dv}{dt}\right)\delta y+\left(\mathrm{Z}-\frac{dw}{dt}\right)\delta z\right\}\right]\\ +\Sigma\,(m\,\mathrm{S}\,\delta s)=0\end{aligned}\right\}\quad(16.)$$

(the summation Σ being extended to all the particles of the medium), expresses at once all the conditions of equilibrium and motion of every particle of the medium.

27. In applying this principle to the theory of elastic media, both Mr. Green and Mr. Haughton assume the following postulates:

First. That in calculation we may treat each particle m as if it were a small rectangular space, $dx\ dy\ dz$, filled with matter of a certain density ρ: so that for the symbol Σm we may substitute that of a triple integration.

$$\int\!\!\int\!\!\int \rho\, dx\, dy\, dz.$$

Secondly. That the virtual moment $m\,\mathrm{S}\,\delta s$ of the total molecular force acting on any particle m, is capable of being expressed by the product of the small rectangular space $dx\ dy\ dz$ into the variation, $\delta\mathrm{V}$, of a certain function V of the relative position of m and the other particles of the body.*

Equation (16) is thus transformed into the following:—

$$\left.\begin{aligned}\int\!\!\int\!\!\int\rho\left\{\left(\mathrm{X}-\frac{du}{dt}\right)\delta x+\left(\mathrm{Y}-\frac{dv}{dt}\right)\delta y\right.\qquad\qquad\\ \left.+\left(\mathrm{Z}-\frac{dw}{dt}\right)\delta z\right\}dx\,dy\,dz+\int\!\!\int\!\!\int\delta\mathrm{V}\,dx\,dy\,dz=0\end{aligned}\right\}\quad(17.)$$

28. The term *elasticity* properly comprehends those molecular forces only whose variations are produced by, and tend to produce, variations in the volume and figure of bodies. There are, therefore, conceivable kinds of molecular force, which are not included in the term *elasticity*. For example, let us take the forces which Mr. MacCullagh ascribed to the particles of the medium which transmits light.

* This amounts, in fact, to the assumption that no part of the power developed by a variation of the relative positions of the particles is permanently converted into heat, or any other agency: in other words, that the body is perfectly elastic.

Let ξ, η, ζ, denote displacements of a point in the medium parallel respectively to x, y, z. Then Mr. MacCullagh supposes the molecular forces to be functions of

$$\frac{d\eta}{dz} - \frac{d\zeta}{dy}; \quad \frac{d\zeta}{dx} - \frac{d\xi}{dz}; \quad \frac{d\xi}{dy} - \frac{d\eta}{dx};$$

which are proportional to the rotations of an element $dx\,dy\,dz$ from its position of equilibrium about the three axes respectively. This amounts to ascribing to the particles of the medium a species of *polarity*, tending to place three orthogonal axes in each particle parallel respectively to the three corresponding axes in each of the other particles: the rotative force acting between the corresponding axes in each pair of particles being a function of the projection of the relative angular displacement of the axes on the plane passing through them, of the position of that plane, and of the distance between the particles.

A portion of a medium endowed with such molecular forces only would transmit oscillations; but it would not tend to preserve any definite bulk or figure, nor would it resist any change of bulk or figure. It would be a *medium* or *system*, but not a *body*. Molecular forces of this kind, therefore, are not comprehended under the term *elasticity;* and the limits of the present investigation exclude those forms of the function V which represent the laws of their action.

29. The inquiry being thus restricted to molecular forces dependent on the variations of the bulk and figure of bodies, there is to be introduced a

Third Postulate: That supposing the body to be divided mentally into small parts, which, in the undisturbed state of the body, are rectangular and of equal size, those parts, in the disturbed state, continue to be *sensibly* of equal bulk and similar figure, throughout a distance round each point at least equal to the greatest extent of appreciable molecular action.

This assumption has been made in all previous investigations, except those respecting the dispersion of light; and it seems, indeed, to be perfectly consistent with the real state of tangible bodies.

Its advantage in calculation is, that it enables us to treat the variations of the molecular forces acting on a given particle, as functions simply of the variations of bulk and figure of an originally rectangular element situated at that particle: seeing that the adjoining elements throughout the extent of appreciable molecular action continue always to undergo sensibly the same variations of bulk and figure as the element under consideration.

Let x_0, y_0, z_0 be the co-ordinates of any physical point in a homogeneous body in equilibrio, and whose particles are not operated upon by any extraneous forces, X, Y, Z. In this condition it is evident that

$$\delta V = 0$$

at every point, and that we may also make

$$V = 0.$$

In the disturbed condition, let ξ, η, ζ, be the displacements of the point whose undisturbed position is (x_0, y_0, z_0), so that

$$x = x_0 + \xi, \text{ \&c.}$$

Then all the variations of bulk and figure which can be undergone by an originally rectangular element, consistently with the third postulate, may be expressed by means of the following six quantities, which I have elsewhere called *strains:*

$$\frac{d\xi}{dx} = \alpha; \quad \frac{d\eta}{dy} = \beta; \quad \frac{d\zeta}{dz} = \gamma;$$

$$\frac{d\eta}{dz} + \frac{d\zeta}{dy} = \lambda; \quad \frac{d\zeta}{dx} + \frac{d\xi}{dz} = \mu; \quad \frac{d\xi}{dy} + \frac{d\eta}{dx} = \nu; *$$

of which α, β, γ, are longitudinal *extensions* if positive, *compressions* if negative, and λ, μ, ν, are *distortions* in the planes perpendicular to x, y, z, respectively.

Hence it appears that

$$V = \phi(\alpha, \beta, \gamma, \lambda, \mu, \nu). \qquad (18.)$$

30. The first assumption, that we may treat the body in calculation as composed of rectangular elements $\rho\, dx\, dy\, dz$, involves the consequence that we may express all the molecular forces which act on each such element by means of pressures, normal and tangential, exerted on its six faces. Taking yz, zx, xy, to denote the position of the faces of such an element, P to denote generally a normal pressure expressed in units of force per unit of area, and Q a tangential pressure similarly expressed, let the nine component pressures on unity of area of those faces be thus denoted:

* This notation is substituted for

$$N_1, N_2, N_3, 2T_1, 2T_2, 2T_3,$$

as being more convenient.

Position of Face.	Direction of Pressure.		
	x	y	z
yz	P_1	Q_3	Q_2'
zx	Q_3'	P_2	Q_1
xy	Q_2	Q_1'	P_3

By the definition of elasticity all pressures are excluded, except those whose variations produce and are produced by variations of volume and figure of the parts of the body. Hence the pressures

$$Q_1 - Q_1'; \; Q_2 - Q_2'; \; Q_3 - Q_3';$$

whose tendency is to make the element $dx\,dy\,dz$ rotate about its three axes respectively, without change of form, must be null; and therefore

$$Q_1 = Q_1'; \; Q_2 = Q_2'; \; Q_3 = Q_3'. \quad . \quad . \quad (19.)$$

Mr. Haughton correctly remarks that this often quoted theorem of Cauchy is not true for all conceivable media. It is not true, for instance, for a medium such as that which Mr. MacCullagh assumed to be the means of transmitting light. It is true, nevertheless, for all molecular pressures which properly fall under the definition of elasticity, if that term be confined to the forces which preserve the figure and volume of bodies.

Let us now express the sum of the virtual moments of the molecular forces acting on the element $dx\,dy\,dz$, in terms of the pressures P_1, &c.; to do which, we must multiply each pressure by the virtual variation of the effect which it tends to produce in its own direction. Thus we obtain the following result:—

$$\delta V = P_1\,\delta a + P_2\,\delta\beta + P_3\,\delta\gamma + Q_1\,\delta\lambda + Q_2\,\delta\mu + Q_3\,\delta\nu. \quad . \quad (20.)$$

Hence the function V bears the following relations to the normal and tangential pressures at the faces of a rectangular element:

$$\left.\begin{array}{l} P_1 = \dfrac{dV}{da}; \; P_2 = \dfrac{dV}{d\beta}; \; P_3 = \dfrac{dV}{d\gamma} \\[2ex] Q_1 = \dfrac{dV}{d\lambda}; \; Q_2 = \dfrac{dV}{d\mu}; \; Q_3 = \dfrac{dV}{d\nu} \end{array}\right\} \quad . \quad . \quad (21.)$$

31. A *Fourth Postulate*, generally assumed in investigations of this kind, is that the pressures are sensibly proportional simply to the strains with which they are connected. This assumption must be approximately true of any law of molecular action, when the pressures and strains are sufficiently small. It is known to be sensibly true for almost all bodies, so

long as the pressures and strains are not so great as to impair their power of recovering their original volume and figure.

According to this postulate, the pressures P_1, &c., are algebraic functions of the first order of the strains α, &c.; and consequently V is an algebraic function of the second order of those strains. The constant part of V, as we have already seen (Art. 29), is null.

Following the notation adopted by Mr. Haughton, let (α) denote the coefficient of α in V, (α^2) that of $-\frac{\alpha^2}{2}$, $(\beta\gamma)$ that of $-\beta\gamma$, &c. Then

$$\begin{aligned} V = {} & (\alpha)\,\alpha + (\beta)\,\beta + (\gamma)\,\gamma + (\lambda)\,\lambda + (\mu)\,\mu + (\nu)\,\nu \\ & - (\alpha^2)\frac{\alpha^2}{2} - (\beta^2)\frac{\beta^2}{2} - (\gamma^2)\frac{\gamma^2}{2} - (\lambda^2)\frac{\lambda^2}{2} - (\mu^2)\frac{\mu^2}{2} - (\nu^2)\frac{\nu^2}{2} \\ & - (\beta\gamma)\,\beta\gamma - (\gamma\alpha)\,\gamma\alpha - (\alpha\beta)\,\alpha\beta - (\mu\nu)\,\mu\nu - (\nu\lambda)\,\nu\lambda - (\lambda\mu)\,\lambda\mu \\ & - (\alpha\lambda)\,\alpha\lambda - (\alpha\mu)\,\alpha\mu - (\alpha\nu)\,\alpha\nu \\ & - (\beta\lambda)\,\beta\lambda - (\beta\mu)\,\beta\mu - (\beta\nu)\,\beta\nu \\ & - (\gamma\lambda)\,\gamma\lambda - (\gamma\mu)\,\gamma\mu - (\gamma\nu)\,\gamma\nu. \qquad (22.) \end{aligned}$$

The six coefficients of the terms of the first order in this equation obviously represent the pressures, uniform throughout the whole extent of the body, to which it is subjected when its particles are in those positions from which the displacements are reckoned: that is to say, when

$$\xi = 0; \quad \eta = 0; \quad \zeta = 0.$$

Let $P_{1,0}$ &c., denote those pressures. Then

$$\left.\begin{array}{l} (\alpha) = P_{1,0}; \quad (\beta) = P_{2,0}; \quad (\gamma) = P_{3,0} \\ (\lambda) = Q_{1,0}; \quad (\mu) = Q_{2,0}; \quad (\nu) = Q_{3,0} \end{array}\right\} \quad . \quad (23.)$$

The *twenty-one* coefficients of the terms of the second order are the *coefficients of elasticity* of the body, as referred to the three axes selected. The negative sign is prefixed to each, because it is essential to the stability of a body that molecular pressures should be opposite in direction to the strains producing them.

The transformation of the quantities in equation (22) for any set of rectangular axes, is effected by means of equation (2) of Sect. I., Art. 9, by making the following substitutions:—

for	P_1, P_2, P_3, $2\,Q_1$, $2\,Q_2$, $2\,Q_3$,
substitute	α, β, γ, λ, μ, ν,

and make similar substitutions for the accented symbols. By multiplying the six equations referred to together by pairs, twenty-one equations are obtained, serving to transform the squares and products of α, β, γ, λ, μ, ν. Formulæ similar to those which transform the strains α, &c., and their half-squares and products, serve also to transform the respective coefficients of those quantities in equation (22).*

It is shown by Mr. Haughton, that by properly selecting the axes of co-ordinates, the number of independent coefficients of elasticity may always be reduced to three less than when the axes are indefinite; and by Mr. Haughton and Mr. Green, that when the body has orthogonal axes of elasticity at each point, then, if those axes be taken as the axes of co-ordinates, the coefficients of elasticity are reduced to the first nine. The latter proposition is obvious, because if molecular action be symmetrical about three orthogonal planes, and those be taken for co-ordinate planes, then the value of that part of the function V which is of the second order cannot be altered by a change in the *sign* of either of the distortions λ, μ, ν; so that the coefficients of the last twelve terms of equation (22) must each be null.

The nine coefficients of elasticity of a body in those circumstances have the following values, in terms of the notation of the previous sections:

$$\left.\begin{array}{c}
\text{Coefficients of Longitudinal Elasticity.} \\
(\alpha^2) = A_1; \quad (\beta^2) = A_2; \quad (\gamma^2) = A_3. \\
\text{Coefficients of Lateral Elasticity.} \\
(\beta\gamma) = B_1; \quad (\gamma\alpha) = B_2; \quad (\alpha\beta) = B_3. \\
\text{Coefficients of Rigidity.} \\
(\lambda^2) = C_1; \quad (\mu^2) = C_2; \quad (\nu^2) = C_3.
\end{array}\right\} \quad . \quad (24.)$$

32. DEF. *Let the term* PERFECT FLUID *be used to denote the state of a body, which under a given uniform normal pressure, and at a given temperature, tends to preserve, and if disturbed to recover, a certain bulk; but offers no resistance to change of figure.*

In such a body, if the element whose original bulk was $dx\,dy\,dz$, becomes of the bulk $(1+\sigma)\ dx\,dy\,dz$ (σ being a small fraction), we shall have

$$\sigma = \alpha + \beta + \gamma,$$

and the function V must be of the form

$$V = P_0\,\sigma - (\sigma^2)\frac{\sigma^2}{2}, \quad . \quad . \quad . \quad (25.)$$

where P_0 is the uniform normal pressure when the particles are not displaced, and (σ^2) a coefficient of elasticity, whose value, in the notation of the previous sections, is:

* See the Note at the end of this paper.

Coefficient of Fluid Elasticity.

$$(\sigma^2) = (a^2) = (\beta^2) = (\gamma^2) = (\beta\gamma) = (\gamma a) = (a\beta) = \text{J}. \quad (26.)$$

The normal pressure in the disturbed state, which is the same in all directions, is obviously

$$\text{P}_1 = \text{P}_2 = \text{P}_3 = \text{P}_0 - \text{J}\,\sigma. \quad . \quad . \quad . \quad (27.)$$

The tangential pressures are each = 0.

SECTION VII.—ON THE PROOF OF THE LAWS OF ELASTICITY BY THE METHOD OF VIRTUAL VELOCITIES.

33. Having thus followed very nearly the steps of the researches of Mr. Haughton and Mr. Green, so as to compare their coefficients with those used in the previous part of this paper, I shall now investigate how far the method of Lagrange can be used to establish those relations between the coefficients of elasticity of different kinds in homogeneous solid bodies, which I have elsewhere deduced from geometrical and physical considerations.

The fluid elasticity considered in the last article cannot arise from the mutual actions of centres of force; for such actions would necessarily tend to preserve a certain arrangement amongst those centres, and would therefore resist change of figure. Fluid elasticity must arise either from the mutual actions of the parts of continuous matter, or from the centrifugal force of molecular motions, or from both those causes combined.

On the other hand, it is only by the mutual action of centres of force that resistance to change of figure and molecular arrangement can be explained, that property being inconceivable of a continuous body. The elasticity *peculiar* to solid bodies is, therefore, due to the mutual action on centres of force. Solid bodies may nevertheless possess, in addition, a portion of that species of elasticity which belongs to fluids.

The investigation is simplified by considering, in the first place, the elasticity of a solid body as arising from the mutual action of centres of force only, and afterwards adding the proper portion of fluid elasticity.

It is known that solid bodies are capable of preserving bulk and figure, although their surfaces are acted upon by no sensible pressure, normal or tangential. We may take the positions of the particles in this condition as points from which to measure their displacements. Thus we cause the coefficients of all the terms of the first order in equation (22) to vanish.

To investigate the properties of the coefficients of elasticity, the function δ V is to be expressed in a new form,—viz., as the sum of the virtual moments of the actions exerted upon each of the centres of force in the particle under consideration, by the centres of force in all the other particles. Mr. Haughton, in his first memoir, having performed this process, shows by means of its results, that in a body composed entirely of centres of force acting along the lines joining them, the number of independent coefficients of elasticity for any system of orthogonal axes is reduced to *fifteen*, which, by properly selecting those axes, may be reduced, for bodies in general, to *twelve*, and for those having axes of elasticity, to *six*.

I shall now endeavour to prove by the method of virtual velocities, what I have in the third section proved by other modes of reasoning, that in a homogeneous body constituted of centres of force only, the independent coefficients of elasticity are reducible to *three*, of which, and of the position of the axes, the twenty-one in equation (22) are functions.

A fourth independent coefficient is to be added in solids possessing a portion of fluid elasticity; that is to say, in all known solids.

34. It is known that a homogeneous solid can exist, with its particles in an unstrained condition, bounded by plane surfaces in any direction. In this condition, therefore, the total molecular action upon a particle situated at any bounding plane must be null. Conceive the bounding plane still to pass through the same particle, but to have its position shifted through any angle. The molecular action on the particle will still be null. Now the effect of the shifting of the bounding plane is to take away a wedge of matter from one side of the particle, and to substitute an equal and similarly constituted wedge, lying in a diametrically opposite direction. Hence, in the unstrained condition of a solid body, the action exercised upon any particle, by a wedge of matter bounded by any two planes passing through the particle, is null.

This shows that the action of a wedge of solid matter on a particle situated at its edge, is not altered by varying the angular position of the wedge; and consequently, that the molecular actions which produce elasticity are not *directly* functions of the relative angular positions of the centres of force which act on each other, but merely of their distances apart, so that if the actions of the several equal wedges into which a body may be conceived to be divided, round a given particle, are different, this does not arise directly from the angular positions of the wedges, but from the different distribution of their centres of force as to *distance* from those of the particle operated upon.

(I have proved this, in a manner slightly different in form, in a supplementary paper to Sect. III., Art. 17.) *See p.* 98.

This further shows, that the mutual action of two centres of force in a solid must be directed along the line joining them; for otherwise it would tend to bring that line into some definite angular position, and would be a function of the direction of the line.

It finally results, from what has been stated, that the action of an indefinitely slender pyramid of a solid body upon a particle at its apex must be a direct attraction or repulsion along the axis of the pyramid, which is a function of the several distances of the centres of force in the pyramid from those in the particle at the apex, and which, in the unstrained condition of the body, must be null.

The principles stated above have to a greater or less extent been taken for granted in previous investigations, but have not hitherto been demonstrated. They may all be regarded as the necessary consequences of the following:—

DEF. *Let the term* ELASTIC SOLID *be used to denote the condition of a body, which, when acted upon by any given system of pressures, or by none, and at a given temperature, tends to preserve, and if disturbed, to recover, a definite bulk and figure; and such that, if while in an unstrained condition it be cut into parts of any figure, those parts, when separate, will tend to preserve the same bulk and figure as they did when they formed one body.*

Experience informs us that bodies *sensibly* agreeing with this definition exist; its consequences are, therefore, applicable to them in practice.

35. Let r denote the distance apart of two centres of force in an unstrained solid, and let ϕr be proportional to their mutual action. Then

$$dx\,dy\,dz\,d^2\omega \,.\, \Sigma\,\phi r = 0$$

may be taken to represent the total action of an indefinitely slender pyramid which subtends the element of angular space $d^2\omega$ upon a particle at its apex $dx\,dy\,dz$.

In consequence of a strain, let each of the distances r become

$$(1 + \varepsilon)\,r,$$

ε being a very small fraction. Then the total action of the pyramid becomes

$$dx\,dy\,dz\,d^2\omega \,.\, \Sigma\,(\phi r + \varepsilon r \phi' r) = dx\,dy\,dz\,d^2\omega \,.\, \varepsilon\,\Sigma\, r\,\phi' r;$$

for by the third postulate, ε is uniform throughout the extent of appreciable molecular action.

The quantity which the force acting between two centres of force tends to vary, is their relative displacement along the line joining them, or εr. Hence the sum of the virtual moments of the actions of all the slender pyramids into which the solid is conceived to be divided, that is to say, the total virtual moment of its molecular action upon the particle $dx\,dy\,dz$, is

$$\delta V\, dx\,dy\,dz = dx\,dy\,dz \iint \varepsilon\,\delta\varepsilon \,.\, \Sigma\,(r^2\,\phi' r) \,.\, d^2\omega,$$

the double integration extending to all angular space. Consequently, we obtain as a new value of the function V,

$$\mathrm{V} = \iint \tfrac{1}{2}\,\varepsilon^2\, \Sigma\,(r^2\,\phi'\,r)\,.\,d^2\,\omega. \qquad (28.)$$

Let a, b, c be the direction-cosines of the axis of a given slender pyramid. Then it is easily seen that the strain ε along that axis has the following value, in terms of the six strains as referred to the axes of co-ordinates:

$$\varepsilon = \alpha\, a^2 + \beta\, b^2 + \gamma\, c^2 + \lambda\, b\, c + \mu\, c\, a + \nu\, a\, b,$$

and, consequently, that

$$\begin{aligned}\frac{\varepsilon^2}{2} = \frac{\alpha^2}{2}\, a^4 &+ \frac{\beta^2}{2}\, b^4 + \frac{\gamma^2}{2}\, c^4 + \frac{\lambda^2}{2}\, b^2\, c^2 + \frac{\mu^2}{2}\, c^2\, a^2 + \frac{\nu^2}{2}\, a^2\, b^2 \\ &+ \beta\gamma\, b^2\, c^2 + \gamma\alpha\, c^2\, a^2 + \alpha\beta\, a^2\, b^2 \\ &+ \mu\nu\, a^2\, b\, c + \nu\lambda\, a\, b^2\, c + \lambda\mu\, a\, b\, c^2 \\ &+ \alpha\lambda\, a^2\, b\, c + \alpha\mu\, a^3\, c + \alpha\nu\, a^3\, b \\ &+ \beta\lambda\, b^3\, c + \beta\mu\, a\, b^2\, c + \beta\nu\, a\, b^3 \\ &+ \gamma\lambda\, b\, c^3 + \gamma\mu\, a\, c^3 + \gamma\nu\, a\, b\, c^2.\end{aligned}$$

If this value of $\frac{1}{2}\,\varepsilon^2$ be substituted in equation (28), and the result compared with equation (22), it is at once obvious that the twenty-one coefficients of elasticity have the following values (putting $\Sigma\,(r^2\,\phi'\,r) = -\,\mathrm{R}$, which is negative that equilibrium may be stable):

$$\left.\begin{aligned}
(\alpha^2) &= \iint a^4\, \mathrm{R}\, d^2\,\omega \\
(\beta^2) &= \iint b^4\, \mathrm{R}\, d^2\,\omega \\
(\gamma^2) &= \iint c^4\, \mathrm{R}\, d^2\,\omega \\
(\lambda^2) = (\beta\gamma) &= \iint b^2\, c^2\, \mathrm{R}\, d^2\,\omega \\
(\mu^2) = (\gamma\,\alpha) &= \iint c^2\, a^2\, \mathrm{R}\, d^2\,\omega \\
(\nu^2) = (\alpha\beta) &= \iint a^2\, b^2\, \mathrm{R}\, d^2\,\omega \\
(\mu\nu) = (\alpha\,\lambda) &= \iint a^2\, b\, c\, \mathrm{R}\, d^2\,\omega \\
(\nu\lambda) = (\beta\mu) &= \iint a\, b^2\, c\, \mathrm{R}\, d^2\,\omega \\
(\lambda\,\mu) = (\gamma\,\nu) &= \iint a\, b\, c^2\, \mathrm{R}\, d^2\,\omega \\
(\beta\lambda) &= \iint b^3\, c\, \mathrm{R}\, d^2\,\omega \\
(\gamma\,\lambda) &= \iint b\, c^3\, \mathrm{R}\, d^2\,\omega \\
(\gamma\,\mu) &= \iint c^3\, a\, \mathrm{R}\, d^2\,\omega \\
(\alpha\,\mu) &= \iint c\, a^3\, \mathrm{R}\, d^2\,\omega \\
(\alpha\,\nu) &= \iint a^3\, b\, \mathrm{R}\, d^2\,\omega \\
(\beta\nu) &= \iint a\, b^3\, \mathrm{R}\, d^2\,\omega
\end{aligned}\right\} \quad . \quad (29.)$$

In the above equations, which agree with those given by Mr. Haughton, the number of independent coefficients is *fifteen.*

36. Their reduction to a smaller number arises from the nature of the function

$$R = - \Sigma (r^2 \phi' r).$$

This quantity is a function of the distances of the centres of force in a given indefinitely slender pyramid from those in a particle at its apex, and can vary with the direction-cosines a, b, c of the axis of the pyramid, solely because those distances vary with them. Now in a homogeneous solid, that is, one composed of a succession of similar and regularly placed groups of centres of force, those distances depend upon a quantity which may be called the *mean interval* between the centres of force in a given direction: a quantity of such a nature that the product of its three values for any three orthogonal directions is a constant quantity; being the space occupied by a centre of force, or by a definite group of such centres. To have this property, the mean interval must be a quantity of this form:

$$i = e^{f + g a^2 + h b^2 + k c^2 + l b c + m c a + n a b}; \quad . \quad . \quad (30.)$$

that is to say, its logarithm must be proportional to the reciprocal of the square of the radius of an ellipsoid, whose axes are those of molecular arrangement, and therefore of molecular action, and of elasticity.

Let the axes of this ellipsoid be taken as axes of co-ordinates. Then $l = 0$, $m = 0$, $n = 0$; and the above equation is reduced to

$$i = e^{f + g a^2 + h b^2 + k c^2}, \quad . \quad . \quad . \quad (30\text{A.})$$

and because the quantity

$$R = F(i) = \psi (f + g a^2 + h b^2 + k c^2), \quad . \quad (31.)$$

does not change its sign or value by any change of the signs of the cosines a, b, c, it follows that all the coefficients in (29) containing odd powers of those cosines, that is to say, all except the *first six,* disappear when the axes of molecular arrangement are taken for axes of co-ordinates.

These six, for all known homogeneous substances, are reducible to *three,* by the following reasoning:

Let us assume as a *Fifth Postulate,* what experience shows to be sensibly true of all known homogeneous substances—viz., that their elasticity varies very little in different directions. Those substances, such as timber, whose elasticity in different directions varies much, are not homogeneous, but composed of fibres, layers, and tubes of different substances.

If this be assumed, it follows that, in the expression (31), for the quantity R, the variable terms

$$g a^2 + h b^2 + k c^2,$$

are very small compared with the constant term f, and that R may be developed in the form

$$R = \psi(f) + \psi'(f) \cdot (g a^2 + h b^2 + k c^2) + \&c.$$

If this value of R be introduced into equation (29), and if small quantities of the second order be neglected, it is easily seen, on performing the integrations, that the following relations exist amongst the six coefficients already specified:

$$\left.\begin{array}{l}
(\beta^2) + (\gamma^2) = 6\,(\lambda^2) = 6\,(\beta\,\gamma) \\
(\gamma^2) + (a^2) = 6\,(\mu^2) = 6\,(\gamma\,a) \\
(a^2) + (\beta^2) = 6\,(\nu^2) = 6\,(a\,\beta) \\
\text{or, by transformation,} \\
(a^2) = 3\,\{(\mu^2) + (\nu^2) - (\lambda^2)\} \\
(\beta^2) = 3\,\{(\nu^2) + (\lambda^2) - (\mu^2)\} \\
(\gamma^2) = 3\,\{(\lambda^2) + (\mu^2) - (\nu^2)\}
\end{array}\right\} \quad . \quad (32.)$$

These equations reduce the number of independent coefficients of elasticity, arising from the actions of centres of force, to *three*. They are identical with the equations (7) and (8), embodied in the fifth theorem in Sect. III., although arrived at by a different process.

37. Let us suppose the solid under consideration to possess a portion of fluid elasticity, represented by the coefficient J. Then the coefficients of elasticity have evidently the following relations:

$$\left.\begin{array}{l}
(a^2) = 3\,\{(\mu^2) + (\nu^2) - (\lambda^2)\} + J \\
(\beta^2) = 3\,\{(\nu^2) + (\lambda^2) - (\mu^2)\} + J \\
(\gamma^2) = 3\,\{(\lambda^2) + (\mu^2) - (\nu^2)\} + J \\
(\beta\,\gamma) = (\lambda^2) + J \\
(\gamma\,a) = (\mu^2) + J \\
(a\,\beta) = (\nu^2) + J
\end{array}\right\} \quad . \quad . \quad (33.)$$

which are identical with the six equations (9) comprehended under the sixth theorem, in Sect. IV.

38. The laws of elasticity stated in this paper are the necessary consequences of the definitions of elasticity and of fluid and solid bodies, given in Arts. 28, 32, and 34, respectively, when taken in conjunction with five postulates or assumptions, which, however, may be summed up in two—viz.

First, That the variations of molecular force concerned in producing elasticity are sufficiently small to be represented by functions of the first order of the quantities on which they depend; and,

Secondly, That the integral calculus and the calculus of variations are applicable to the theory of molecular action. It is thus apparent that the science of elasticity is, to a great extent, one of deduction *à priori*.

The functions of perceptive experience in connection with it are two-fold: first, by observation, to inform us of the existence of substances, agreeing to a greater or less degree of approximation with the definitions and postulates; and, secondly, by experiment, to ascertain the numerical value of the coefficients of elasticity of each substance.

Note to Sections VI. and VII. of Preceding Paper.

On the Transformation of Coefficients of Elasticity, by the aid of a Surface of the Fourth Order.

(The following note contains no original principle, and is designed merely to put on record, for the sake of convenient reference, a series of equations which will be found useful in future investigations.)

It has been pointed out by Mr. Haughton, in his first paper, that if we take into consideration that part only of the elasticity of a solid which arises from the mutual actions of centres of force, so that the function V shall contain at most but fifteen unequal coefficients (viz., those whose values are given in equation 29), and if, with those fifteen coefficients, we construct a surface of the fourth order, whose equation is the following,

$$\begin{aligned} U = & (a^2)\, x^4 + (\beta^2)\, y^4 + (\gamma^2)\, z^4 \\ & + 6\,(\lambda^2)\, y^2 z^2 + 6\,(\mu^2)\, z^2 x^2 + 6\,(\nu^2)\, x^2 y^2 \\ & + 12\,(a\lambda)\, x^2 y z + 12\,(\beta\mu)\, x y^2 z + 12\,(\gamma\nu)\, x y z^2 \\ & + 4\,(a\iota)\, x^3 y + 4\,(a\mu)\, x^3 z \\ & + 4\,(\beta\lambda)\, y^3 z + 4\,(\beta\nu)\, y^3 x \\ & + 4\,(\gamma\mu)\, z^3 x + 4\,(\gamma\lambda)\, z^3 y \\ = & 1 \; . \quad . \quad . \quad . \quad . \quad . \quad . \quad . \quad . \quad . \quad . \qquad \text{(A.)} \end{aligned}$$

then will U be the same function of the six quantities

$$x^2, \quad y^2, \quad z^2, \quad 2yz, \quad 2zx, \quad 2xy,$$

that $-2V$ is of the six strains

$$\alpha, \quad \beta, \quad \gamma, \quad \lambda, \quad \mu, \quad \nu,$$

which are known to be transformed by the same equations with the above functions of the second order of x, y, z; and consequently, the same equations which serve to transform the coefficients of the surface $U = 1$, into those suitable for a new set of rectangular co-ordinates, will also serve to transform the coefficients of elasticity in the function V.

Now, it is obvious that if the equation

$$\phi(x, y, z) = \psi(x', y', z')$$

be true for two sets of rectangular co-ordinates having the same origin, then must the equation

$$\phi\left(\frac{d}{dx}, \frac{d}{dy}, \frac{d}{dz}\right) = \psi\left(\frac{d}{dx'}, \frac{d}{dy'}, \frac{d}{dz'}\right) \quad . \quad \text{(B.)}$$

be true also.

It follows that the fifteen coefficients of elasticity (a^2), &c., which are proportional to the differential coefficients of U of the fourth order with respect to x, y, z, are transformable by means of the same equations which serve to transform the fifteen algebraical functions of the fourth order of x, y, z, by which they are respectively multiplied in the value of U.

The following is the investigation of those fifteen equations of the fourth order, as well as of the six equations of the second order, from which they are formed by multiplication.

Let the relative direction-cosines of the two sets of rectangular axes be expressed as follows:

Original Axes.	*New Axes.* x'	y'	z'	
x	a_1	b_1	c_1	cosines.
y	a_2	b_2	c_2	
z	a_3	b_3	c_3	

Let the following notation be used for functions of those cosines. (It is the same which is employed by Mr. Haughton.)

$$p_1 = b_1 c_1; \quad p_2 = b_2 c_2; \quad p_3 = b_3 c_3;$$
$$q_1 = c_1 a_1; \quad q_2 = c_2 a_2; \quad q_3 = c_3 a_3;$$
$$r_1 = a_1 b_1; \quad r_2 = a_2 b_2; \quad r_3 = a_3 b_3;$$

$$l_1 = b_2 c_3 + b_3 c_2; \quad l_2 = b_3 c_1 + b_1 c_3; \quad l_3 = b_1 c_2 + b_2 c_1;$$
$$m_1 = c_2 a_3 + c_3 a_2; \quad m_2 = c_3 a_1 + c_1 a_3; \quad m_3 = c_1 a_2 + c_2 a_1;$$
$$n_1 = a_2 b_3 + a_3 b_2; \quad n_2 = a_3 b_1 + a_1 b_3; \quad n_3 = a_1 b_2 + a_2 b_1;$$

then the following are the six equations of transformation of the second order for the surface U = 1,

$$x'^2 = x^2 a_1^2 + y^2 a_2^2 + z^2 a_3^2 + 2\,y\,z\,a_2 a_3 + 2\,z\,x\,a_3 a_1 + 2\,x\,y\,a_1 a_2,$$

(for y'^2, z'^2, similar equations in b, c, respectively).

$$y' z' = x^2 p_1 + y^2 p_2 + z^2 p_3 + y\,z\,l_1 + z\,x\,l_2 + x\,y\,l_3,$$

(for $z' x'$, a similar equation in q and m),

(for $x' y'$, a similar equation in r and n), . . . (C.)

Those equations are made applicable to the transformation of strains by the following substitutions:

for x^2, y^2, z^2, $2\,y\,z$, $2\,z\,x$, $2\,x\,y$,

substitute α, β, γ, λ, μ, ν;

and to that of pressures, by the following:

for x^2, y^2, z^2, $y\,z$, $z\,x$, $x\,y$,

substitute P_1, P_2, P_3, Q_1, Q_2, Q_3,

and similar substitutions for the accented symbols.

The following are the fifteen equations of transformation of the fourth order:

$$x'^4 = x^4 a_1^4 + y^4 a_2^4 + z^4 a_3^4 + 6\,y^2 z^2 a_2^2 a_3^2 + 6 z^2 x^2 a_3^2 a_1^2 + 6\,x^2 y^2 a_1^2 a_2^2$$
$$+ 12\,x^2 y\,z\,a_1^2 a_2 a_3 + 12\,x\,y^2 z\,a_1 a_2^2 a_3 + 12\,x\,y\,z^2 a_1 a_2 a_3^2$$
$$+ 4\,x^3 y\,a_1^3 a_2 + 4\,x^3 z\,a_1^3 a_3 + 4\,y^3 z\,a_2^3 a_3 + 4\,y^3 x\,a_2^3 a_1 + 4\,z^3 x\,a_3^3 a_1$$
$$+ 4\,z^3 y\,a_3^3 a_2;$$

(for y'^4, z'^4, similar equations in b, c, respectively).

$$y'^2 z'^2 = x^4 p_1^2 + y^4 p_2^2 + z^4 p_3^2 + y^2 z^2 (l_1^2 + 2\,p_2 p_3) + z^2 x^2 (l_2^2 + 2\,p_3 p_1)$$
$$+ x^2 y^2 (l_3^2 + 2\,p_1 p_2) + 2\,x^2 y\,z\,(p_1 l_1 + l_2 l_3) + 2\,x\,y^2 z\,(p_2 l_2 + l_3 l_1)$$
$$+ 2\,x\,y\,z^2 (p_3 l_3 + l_1 l_2) + 2\,x^3 y\,p_1 l_3 + 2\,x^3 z\,p_1 l_2$$
$$+ 2\,y^3 z\,p_2 l_1 + 2\,y^3 x\,p_2 l_3 + 2\,z^3 x\,p_3 l_2 + 2\,z^3 y\,p_3 l_1;$$

(for $z'^2 x'^2$, a similar equation in q and m);

(for $x'^2 y'^2$, a similar equation in r and n);

$$
\begin{aligned}
x'^2 y' z' = {} & x^4 a_1^2 p_1 + y^4 a_2^2 p_2 + z^4 a_3^2 p_3 \\
& + y^2 z^2 (a_2^2 p_3 + a_3^2 p_2 + 2 a_2 a_3 l_1) \\
& + z^2 x^2 (a_3^2 p_1 + a_1^2 p_3 + 2 a_3 a_1 l_2) \\
& + x^2 y^2 (a_1^2 p_2 + a_2^2 p_1 + 2 a_1 a_2 l_3) \\
& + x^2 y z (a_1^2 l_1 + 2 a_2 a_3 p_1 + 2 a_3 a_1 l_3 + 2 a_1 a_2 l_2) \\
& + x y^2 z (a_2^2 l_2 + 2 a_3 a_1 p_2 + 2 a_1 a_2 l_1 + 2 a_2 a_3 l_3) \\
& + x y z^2 (a_3^2 l_3 + 2 a_1 a_2 p_3 + 2 a_2 a_3 l_2 + 2 a_3 a_1 l_1) \\
& + x^3 y (a_1^2 l_3 + 2 a_1 a_2 p_1) + x^3 z (a_1^2 l_2 + 2 a_1 a_3 p_1) \\
& + y^3 z (a_2^2 l_1 + 2 a_2 a_3 p_2) + y^3 x (a_2^2 l_3 + 2 a_2 a_1 a_2) \\
& + z^3 x (a_3^2 l_2 + 2 a_3 a_1 p_3) + z^3 y (a_3^2 l_1 + 2 a_3 a_2 p_3);
\end{aligned}
$$

(for $x' y'^2 z'$, a similar equation in b, q, m);

(for $x' y' z'^2$, a similar equation in c, r, n);

$$
\begin{aligned}
x'^3 y' = {} & x^4 a_1^2 r_1 + y^4 a_2^2 r_2 + z^4 a_3^2 r_3 \\
& + 3 y^2 z^2 a_2 a_3 n_1 \\
& + 3 z^2 x^2 a_3 a_1 n_2 \\
& + 3 x^2 y^2 a_1 a_2 n_3 \\
& + 3 x^2 y z (a_1^2 n_1 + 2 a_2 a_3 r_1) \\
& + 3 x y^2 z (a_2^2 n_2 + 2 a_3 a_1 r_2) \\
& + 3 x y z^2 (a_3^2 n_3 + 2 a_1 a_2 r_3) \\
& + x^3 y (a_1^2 n_3 + 2 a_1 a_2 r_1) + x^3 z (a_1^2 n_2 + 2 a_3 a_1 r_1) \\
& + y^3 z (a_2^2 n_1 + 2 a_2 a_3 r_2) + y^3 x (a_2^2 n_3 + 2 a_1 a_2 r_2) \\
& + z^3 x (a_3^2 n_2 + 2 a_3 a_1 r_3) + z^3 y (a_3^2 n_1 + 2 a_2 a_3 r_3);
\end{aligned}
$$

(for $x'^3 z'$, a similar equation in a, q, and m);

(for $y'^3 z'$, a similar equation in b, p, and l);

(for $y'^3 x'$, a similar equation in b, r, and n);

(for $z'^3 x'$, a similar equation in c, q, and m);

(for $z'^3 y'$, a similar equation in c, p, and l); (D.)

The above equations are made applicable to the transformation of the coefficients of elasticity arising from the mutual actions of centres of force only, by the following substitutions:

for x^4, y^4, z^4, $y^2 z^2$, $z^2 x^2$, $x^2 y^2$,

substitute (α^2), (β^2), (γ^2), $(\beta \gamma) = (\lambda^2)$, $(\gamma \alpha) = (\mu^2)$, $(\alpha \beta) = (\nu^2)$;

for $x^2 y z$, $x y^2 z$, $x y z^2$,

substitute $(\alpha \lambda) = (\mu \nu)$, $(\beta \mu) = (\nu \lambda)$, $(\gamma \nu) = (\lambda \mu)$;

$$\text{for } x^3 y, \quad x^3 z, \quad y^3 z, \quad y^3 x, \quad z^3 x, \quad z^3 y,$$
$$\text{substitute } (\alpha \nu), \ (\alpha \mu), \ (\beta \lambda), \ (\beta \nu), \ (\gamma \mu), \ (\gamma \lambda);$$

and similar substitutions for the accented symbols.

Should the substance under consideration be endowed with a portion of fluid elasticity in addition to that which arises from the mutual action of centres of force, the coefficient of that fluid elasticity J must be subtracted from the coefficients into which it enters, viz.—

$$(\alpha^2), \ (\beta^2), \ (\gamma^2), \ (\beta \gamma) = (\lambda^2) + \mathrm{J}, \ (\gamma \alpha) = (\mu^2) + \mathrm{J}, \ (\alpha \beta) = (\nu^2) + \mathrm{J},$$

before effecting the transformation.

The results of the transformation for those six coefficients, being increased by the same quantity J which was previously subtracted, will give their entire values for the new axes.

If the original axes of co-ordinates are those of elasticity, each of the fifteen equations of transformation is reduced to its first six terms, in which the following substitutions are to be made for the unaccented symbols:

$$\text{for} \quad x^4, \quad y^4, \quad z^4, \quad y^2 z^2, \quad z^2 x^2, \quad x^2 y^2,$$
$$\text{substitute } \mathrm{A}_1 - \mathrm{J}, \ \mathrm{A}_2 - \mathrm{J}, \ \mathrm{A}_3 - \mathrm{J}, \ \mathrm{B}_1 - \mathrm{J} = \mathrm{C}_1, \ \mathrm{B}_2 - \mathrm{J} = \mathrm{C}_2, \ \mathrm{B}_3 - \mathrm{J} = \mathrm{C}_3.$$

VI.—ON AXES OF ELASTICITY AND CRYSTALLINE FORMS.*

SECTION 1.—GENERAL DEFINITION OF AXES OF ELASTICITY.

As originally understood, the term "axes of elasticity" was applied to the intersections of three orthogonal planes at a given point of an elastic medium, with respect to each of which planes the molecular actions causing elasticity were conceived to be symmetrical.

If the elasticity of solids arose either wholly from the mutual attractions and repulsions of centres of force, such attractions and repulsions being functions of the mutual distances of those centres, or partly from such mutual actions and partly from an elasticity like that of a fluid, resisting change of volume only, it is easy to prove that there would be three such orthogonal planes of symmetry of molecular action in every homogeneous solid.

But there is now no doubt that the elastic forces in solid bodies are not such as can be analysed into fluid elasticity and mutual attractions between centres simply; and though there are, as will presently be shown, orthogonal planes of symmetry for certain kinds of elastic forces, those planes are not necessarily the same for all kinds of elastic forces in a given solid.

The term "*axes of elasticity*," therefore, may now be taken in a more extended sense, to signify *all directions with respect to which certain kinds of elastic forces are symmetrical;* or speaking algebraically, *directions for which certain functions of the coefficients of elasticity are null or infinite.*

The theory of axes and coëfficients of elasticity is specially connected with that branch of the calculus of forms which relates to linear transformations, and which has recently been so greatly advanced by the researches of Mr. Sylvester, Mr. Cayley, and Mr. Boole. In such applications of that calculus as occur in this paper, the nomenclature of Mr. Sylvester is followed; † and by the adoption of the "*Umbral*

* Read before the Royal Society of London, on June 21, 1855.

† See *Cambridge and Dublin Mathematical Journal*, Vol. VII.; and *Philosophical Transactions*, 1853.

Notation" of that author immense advantages are gained in conciseness and simplicity.*

SECTION 2.—STRAINS, STRESSES, POTENTIAL ENERGY, AND COEFFICIENTS OF ELASTICITY.

In this paper, the word "*Strain*" will be used to denote the change of volume and figure constituting the deviation of a molecule of a solid from that condition which it preserves when free from the action of external forces; and the word "*Stress*" will be used to denote the force, or combination of forces, which such a molecule exerts in tending to recover its free condition, and which, for a state of equilibrium, is equal and opposite to the combination of external forces applied to it.

In framing a nomenclature for quantities connected with the theory of elasticity, θλίψις is adopted to denote *strain*, and τάσις to denote *stress*.

It is well known that the condition of *strain* at any given point in the interior of a molecule may be completely expressed by means of the following six *elementary strains*, in which ξ, η, ζ are the components of the molecular displacement parallel to three rectangular axes, x, y, z.

Elongations, . $\frac{d\xi}{dx} = \alpha; \; \frac{d\eta}{dy} = \beta; \; \frac{d\zeta}{dz} = \gamma;$

Distortions, . $\frac{d\zeta}{dy} + \frac{d\eta}{dz} = \lambda; \; \frac{d\xi}{dz} + \frac{d\zeta}{dx} = \mu; \; \frac{d\eta}{dx} + \frac{d\xi}{dy} = \nu.$

It is also well known that the condition of *stress* at a given point may be completely expressed, relatively to the three rectangular co-ordinate planes, by means of six *elementary stresses*, viz.—

Normal Pressures, . . . P_1, P_2, P_3,

Tangential Pressures, . . Q_1, Q_2, Q_3;

these quantities being estimated in units of force per unit of surface.

Let each elementary stress be integrated with respect to the elementary strain which it tends directly to diminish, *from* the actual amount of that strain, *to* the condition of freedom; the sum of the integrals is the *potential* energy of elasticity of the molecule $dx\,dy\,dz$, expressed in units of work per unit of volume; viz.—

$$\begin{aligned} U = \int_\alpha^0 P_1\, d\alpha + \int_\beta^0 P_2\, d\beta + \int_\gamma^0 P_3\, d\gamma \\ + \int_\lambda^0 Q_1\, d\lambda + \int_\mu^0 Q_2\, d\mu + \int_\nu^0 Q_3\, d\nu. \quad . \quad . \quad (1.) \end{aligned}$$

* See the Note at the end of the paper.

The condition that the function U shall have the same value, in what order soever the variations of the different elementary strains take place, amounts to supposing that no transformation of energy of the kind well distinguished by Professor Thomson as *frictional* or *irreversible*, takes place during such variations; in other words, that the substance is *perfectly elastic*.

Each of the elementary stresses being sensibly a linear function of the six elementary strains, the potential energy of elasticity is, as Mr. Green first showed, a function of those strains of the second degree, having twenty-one constant coefficients, which are the coefficients of elasticity of the body, and will in this paper be called the *Tasinomic Coefficients;* that is to say, adopting Mr. Green's notation for such coefficients,—

$$\begin{aligned} \mathrm{U} = (\alpha^2)\frac{\alpha^2}{2} + (\beta^2)\frac{\beta^2}{2} + (\gamma^2)\frac{\gamma^2}{2} + (\lambda^2)\frac{\lambda^2}{2} + (\mu^2)\frac{\mu^2}{2} + (\nu^2)\frac{\nu^2}{2} \\ + (\beta\gamma)\,\beta\gamma + (\gamma\alpha)\,\gamma\alpha + (\alpha\beta)\,\alpha\beta \\ + (\mu\nu)\,\mu\nu + (\nu\lambda)\,\nu\lambda + (\lambda\mu)\,\lambda\mu \\ + (\alpha\lambda)\,\alpha\lambda + (\beta\mu)\,\beta\mu + (\gamma\nu)\,\gamma\nu \\ + (\beta\lambda)\,\beta\lambda + (\gamma\mu)\,\gamma\mu + (\alpha\nu)\,\alpha\nu \\ + (\gamma\lambda)\,\gamma\lambda + (\alpha\mu)\,\alpha\mu + (\beta\nu)\,\beta\nu \end{aligned} \qquad (2.)$$

From a theorem of Mr. Sylvester it follows, that every such function as U is reducible by linear transformations to the sum of six positive squares, each multiplied by a coefficient. The nature and meaning of this reduction have been discussed by Professor William Thomson.

The following classification of the tasinomic coefficients will be used in the sequel:—

Designation of Coefficients.		Elasticities.	Symbols.		
Orthotatic	Euthytatic,	Direct or Longitudinal,	(α^2)	(β^2)	(γ^2)
	Platytatic,	Lateral, . . .	$(\beta\gamma)$	$(\gamma\alpha)$	$(\alpha\beta)$
	Goniotatic,	Rigidities, . . .	(λ^2)	(μ^2)	(ν^2)
Plagiotatic, . . .		Unsymmetrical, . .	$(\mu\nu)$, &c., &c.		

The twenty-one equations of transformation by which the values of these coefficients, being known for any one set of orthogonal axes, are found for any other, are founded on the following principles.

It is well known, that for rectangular transformations, the operations

$$\frac{d}{dx},\ \frac{d}{dy},\ \frac{d}{dz}$$

are respectively covariant with

$$x, \ y, \ z,$$

from which it is easily deduced, that because the displacements

$$\xi, \ \eta, \ \zeta$$

are respectively covariant with

$$x, \ y, \ z,$$

therefore the elementary strains,

$$\alpha, \ \beta, \ \gamma, \ \lambda, \ \mu, \ \nu$$

the operations,

$$\frac{d}{d\alpha}, \ \frac{d}{d\beta}, \ \frac{d}{d\gamma}, \ 2\frac{d}{d\lambda}, \ 2\frac{d}{d\mu}, \ 2\frac{d}{d\nu}$$

and the strains

$$P_1, \ P_2, \ P_3, \ 2Q_1, \ 2Q_2, \ 2Q_3,$$

must be respectively covariant with the squares and products,

$$x^2, \ y^2, \ z^2, \ 2yz, \ 2zx, \ 2xy.$$

Section 3.—Thlipsimetric and Tasimetric Surfaces and Invariants.

Isotropic functions of the elementary strains and stresses, which may be called respectively *Thlipsimetric* and *Tasimetric Invariants,* are easily deduced from the principle, that the strains may be represented by the coefficients of the following *Thlipsimetric Surface,*

$$\alpha x^2 + \beta y^2 + \gamma z^2 + \lambda yz + \mu zx + \nu xy = 1, \qquad (3.)$$

and the stresses by the coefficients of the *Tasimetric Surface,*

$$P_1 x^2 + P_2 y^2 + P_3 z^2 + 2Q_1 yz + 2Q_2 zx + 2Q_3 xy = 1. \qquad (4.)$$

These surfaces, and others deduced from them, have been fully discussed by M. Cauchy and M. Lamé.

The invariants in question may all be deduced from the following pair of contragredient matrices :—

$$
\text{(5.)}\left\{\begin{array}{ccc}
\multicolumn{3}{c}{\text{For Strains.}}\\
\alpha & \frac{\nu}{2} & \frac{\mu}{2} \\
\frac{\nu}{2} & \beta & \frac{\lambda}{2} \\
\frac{\mu}{2} & \frac{\lambda}{2} & \gamma
\end{array}\right.
\qquad\qquad
\left.\begin{array}{ccc}
\multicolumn{3}{c}{\text{For Stresses.}}\\
P_1 & Q_3 & Q_2 \\
Q_3 & P_2 & Q_1 \\
Q_2 & Q_1 & P_3
\end{array}\right\}\text{(5 A.)}
$$

The following are the primitive thlipsimetric invariants, from which an indefinite number of others may be deduced by involution, multiplication, addition, and subtraction:—

$$
\left.\begin{array}{l}
\alpha + \beta + \gamma = \theta_1 \text{ (the cubic dilatation);} \\
\beta\gamma + \gamma\alpha + \alpha\beta - \frac{1}{4}(\lambda^2 + \mu^2 + \nu^2) = \theta_2; \\
\alpha\beta\gamma + \frac{1}{4}\lambda\mu\nu - \frac{1}{4}(\alpha\lambda^2 + \beta\mu^2 + \gamma\nu^2 = \theta_3.
\end{array}\right\} \quad . \quad (6.)
$$

The potential energy U is what Mr. Sylvester calls a "Universal Mixed Concomitant," its value being

$$
U = -\tfrac{1}{2}(P_1\alpha + P_2\beta + P_3\gamma + Q_1\lambda + Q_2\mu + Q_3\nu). \qquad (7.)
$$

SECTION 4.—TASINOMIC FUNCTIONS, SURFACES, AND UMBRÆ.

If, in any isotropic function of the co-ordinates and the elementary strains, there be substituted for each square or product of elementary strains that tasinomic coefficient which is covariant with it, the result will be an isotropic function of the co-ordinates and tasinomic coefficients, called a *Tasinomic Function.*

The following Table of Covariants is readily deduced from the principles stated at the end of Sect. 2:—

$$
\left.\begin{array}{l}
\text{Co-variant}\left\{\begin{array}{ll}
\text{Squares of Strains,} & \alpha^2,\ \ \beta^2,\ \ \gamma^2,\ \ \lambda^2,\ \ \mu^2,\ \ \nu^2, \\
\text{Tasinomic Coefficients} & (\alpha^2),\ (\beta^2),\ (\gamma^2),\ 4(\lambda^2),\ 4(\mu^2),\ 4(\nu^2);
\end{array}\right. \\[2ex]
\text{Co-variant}\left\{\begin{array}{ll}
\text{Products of Strains,} & \beta\gamma,\ \ \gamma\alpha,\ \ \alpha\beta,\ \ \mu\nu,\ \ \nu\lambda,\ \ \lambda\mu, \\
\text{Tasinomic Coefficients} & (\beta\gamma),\ (\gamma\alpha),\ (\alpha\beta),\ 4(\mu\nu),\ 4(\nu\lambda),\ 4(\lambda\mu),
\end{array}\right. \\[2ex]
\alpha\lambda,\ \ \alpha\mu,\ \ \alpha\nu,\ \ \beta\lambda,\ \ \beta\mu,\ \ \beta\nu,\ \ \gamma\lambda,\ \ \gamma\mu,\ \ \gamma\nu, \\
2(\alpha\lambda), 2(\alpha\mu), 2(\alpha\nu), 2(\beta\lambda), 2(\beta\mu), 2(\beta\nu), 2(\gamma\lambda), 2(\gamma\mu), 2(\gamma\nu).
\end{array}\right\}(8.)
$$

Each tasinomic function being equated to a constant, forms the equation of a *Tasinomic Surface;* and on the geometrical properties of such surfaces depend many of the laws of coefficients and axes of elasticity.

A convenient and expeditious mode of forming tasinomic functions is obtained by the aid of an *Umbral Notation* analogous to that introduced by Mr. Sylvester in the calculus of forms.

Let each tasinomic coefficient be regarded as compounded of two *Tasinomic Umbræ*, those umbræ being expressed by the following notation:

$$(\alpha),\quad (\beta),\quad (\gamma),\quad (\lambda),\quad (\mu),\quad (\nu);$$

then the following equation, deduced from that of the thlipsimetric surface (3), by substituting umbræ for elementary strains according to the following Table of Covariance,

Strains, . . .	α,	β,	γ,	λ,	μ,	ν,
Umbrae, . . .	(α),	(β),	(γ),	$2(\lambda)$,	$2(\mu)$,	$2(\nu)$,

is the equation of the *Tasinomic Umbral Ellipsoid*, from which, by elimination, multiplication, involution, addition, subtraction, and differentiation, various tasinomic functions may be deduced,

$$(\alpha)x^2 + (\beta)y^2 + (\gamma)z^2 + 2(\lambda)yz + 2(\mu)zx + 2(\nu)xy = (\phi) = 1. \quad (8\text{A.})$$

Section 5.—Tasinomic Invariants and Spheres.

Tasinomic invariants are constant isotropic functions of the tasinomic coefficients, which are deduced, either by substitution from thlipsimetric invariants, or directly from the *Umbral Matrix*,

$$\left.\begin{matrix} (\alpha) & (\nu) & (\mu) \\ (\nu) & (\beta) & (\lambda) \\ (\mu) & (\lambda) & (\gamma) \end{matrix}\right\} \quad . \quad . \quad . \quad . \quad (9.)$$

The following invariant is umbral of the first order:—

$$\left(\frac{d^2}{dx^2} + \frac{d^2}{dy^2} + \frac{d^2}{dz^2}\right) \cdot (\phi) = (\alpha) + (\beta) + (\gamma) = (\theta_1). \quad (9a.)$$

Invariants of the second order in umbræ are real quantities of the first order, viz.—

$$(\alpha^2) + (\beta^2) + (\gamma^2) + 2(\beta\gamma) + 2(\gamma\alpha) + 2(\alpha\beta) = (\theta_1)^2 \text{ (the \textit{cubic elasticity})}$$

$$(\beta\gamma) + (\gamma\alpha) + (\alpha\beta) - (\lambda^2) - (\mu^2) - (\nu^2) = (\theta_2)$$

$$(\alpha^2) + (\beta^2) + (\gamma^2) + 2(\lambda^2) + 2(\mu^2) + 2(\nu^2) = (\theta_1)^2 - 2(\theta_2). \quad . \quad (10).$$

The equation of a *Tasinomic Sphere* is formed by multiplying a tasinomic invariant by

$$x^2 + y^2 + z^2,$$

or any power of that quantity, and equating the result to a constant.

SECTION 6.—OF TWO TASINOMIC ELLIPSOIDS, AND THEIR AXES, ORTHOTATIC AND HETEROTATIC.

The equations of two ellipsoids with tasinomic coefficients are derived from that of the umbral ellipsoid (8A.), in one case by multiplying each term by the umbral invariant (θ), and in the other by substituting for each umbra in the function (ϕ_1), the contravariant component of the *Inverse* to the umbral matrix (9). The results are as follows:—

ORTHOTATIC ELLIPSOID.

$$\begin{aligned}(\theta_1)\times(\phi)=\{(a^2)+(a\beta)+(\gamma a)\}x^2+\{(\beta^2)+(\beta\gamma)+(\gamma a)\}\,y^2\\ +\{(\gamma^2)+(\gamma a)+(\beta\gamma)\}\,z^2\\ +2\{(a\lambda)+(\beta\lambda)+(\gamma\lambda)\}\,yz+2\{(a\mu)+(\beta\mu)+(\gamma\mu)\}\,zx\\ +2\{(a\nu)+(\beta\nu)+(\gamma\nu)\}\,xy=1. \quad . \quad . \quad (11).\end{aligned}$$

HETEROTATIC ELLIPSOID.

$$\begin{aligned}\{(\beta\gamma)-(\lambda^2)\}x^2+\{(\gamma a)-(\mu^2)\}y^2+\{(a\beta)-(\nu^2)\}z^2\\ +2\{(\mu\nu)-(a\lambda)\}yz+2\{(\nu\lambda)-(\beta\mu)\}zx+2\{(\lambda\mu)-(\gamma\nu)\}xy=1. \quad (12.)\end{aligned}$$

The three *Orthotatic Axes* are three rectangular directions for which the following sums of plagiotatic coefficients are null:—

$$\left.\begin{aligned}(a\,\lambda)+(\beta\,\lambda)+(\gamma\,\lambda)=0;\quad (a\,\mu)+(\beta\,\mu)+(\gamma\,\mu)=0;\\ (a\,\nu)+(\beta\,\nu)+(\gamma\,\nu)=0.\end{aligned}\right\} \quad (13.)$$

It was proved by Mr. Haughton, in a paper published in the *Transactions of the Royal Irish Academy*, Vol. III., Part 2, that there are three rectangular directions having this property in a solid whose elasticity arises solely from the mutual actions of physical points, and which has but fifteen independent coefficients of elasticity. The present investigation shows that there are three such axes at each point of every solid, independently of all hypothesis. The physical meaning of this result is expressed by the following

Theorem as to Orthotatic Axes.

At each point of an elastic solid there is one position in which a cubical molecule may be cut out, such that a uniform dilatation or condensation of that molecule by equal elongations or equal compressions of its three dimensions, shall produce no tangential stress on the faces of the molecule.

The properties of the *Heterotatic Axes* are expressed by the following equations :—

$$(\mu\,\nu) - (\alpha\,\lambda) = 0\,;\ (\nu\,\lambda) - (\beta\,\mu) = 0\,;\ (\lambda\,\mu) - (\gamma\,\nu) = 0\,; \qquad (14.)$$

or by the following

Theorem as to Heterotatic Axes.

At each point of an elastic solid there is one position in which a cubical molecule may be cut out, such that if there be a distortion of that molecule round x (x being any one of its three axes), *and an equal distortion round* y (y being either of its other two axes), *the normal stress on the faces normal to* x *arising from the distortion round* x *shall be equal to the tangential stress round* z *arising from the distortion round* y.

The six coefficients of the heterotatic ellipsoid may be called the *Heterotatic Differences.* For a solid whose elasticity is wholly due to the mutual attractions and repulsions of physical points, each of those differences is necessarily null; therefore, they represent a part of the elasticity which is necessarily irreducible to such attractions and repulsions. There is reason to believe that part at least of the elasticity of every substance is of this kind.

If this part of the elasticity of a solid be, as suggested in a series of papers in the *Cambridge and Dublin Mathematical Journal,* for 1851–52, a species of *fluid elasticity,* resisting change of volume only, the solid may be said to be *heterotatically isotropic.* The equations (14) will be fulfilled for all directions of axes, and also the following equations :—

$$(\beta\,\gamma) - (\lambda^2) = (\gamma\,\alpha) - (\mu^2) = (\alpha\,\beta) - (\nu^2)\,; \quad . \quad . \quad (15.)$$

that is to say, the excess of the platytatic above the goniotatic coefficient will be the same in every plane.

In a substance *orthotatically isotropic,* the equations (13) are fulfilled for all directions, and also the following :—

$$(\alpha^2)+(\alpha\beta)+(\gamma\alpha)=(\beta^2)+(\beta\gamma)+(\alpha\beta)=(\gamma^2)+(\gamma\alpha)+(\beta\gamma), \quad (16.)$$

that is to say, a uniform compression in all directions produces a uniform normal stress in all directions, and no tangential stress.

The equations (16) may be reduced to the following form:—

$$(\alpha^2) - (\beta\,\gamma) = (\beta^2) - (\gamma\,\alpha) = (\gamma^2) - (\alpha\,\beta). \quad . \quad (17.)$$

In a substance which is at once *orthotatically* and *heterotatically isotropic*, there may still be eleven independent quantities amongst the tasinomic coefficients, viz.—

$$\left.\begin{array}{ll} \text{Three Euthytatic Coefficients,} & (\alpha^2), (\beta^2), (\gamma^2), \\ \text{The Isotropic excess, .} \quad . & (\alpha^2) - (\beta\,\gamma), \\ \text{The Isotropic excess, .} \quad . & (\beta\,\gamma) - (\lambda^2), \\ \text{Six Plagiotatic Coefficients,} & (\beta\lambda), (\gamma\lambda), (\gamma\mu), (\alpha\mu), (\alpha\nu), (\beta\nu), \end{array}\right\} (18.)$$

Such a substance may, therefore, be far from being completely isotropic with respect to elasticity.

SECTION 7.—BIQUADRATIC TASINOMIC SURFACE; HOMOTATIC COEFFICIENTS; EUTHYTATIC AXES DEFINED.

If the equation (8A.) of the umbral ellipsoid be squared, there is obtained the following equation of a *Biquadratic Tasinomic Surface.*

$$\begin{aligned} (\phi)^2 &= (\alpha^2)\,x^4 + (\beta^2)\,y^4 + (\gamma^2)\,z^4 \\ &+ 2\{(\beta\gamma) + 2(\lambda^2)\}y^2z^2 + 2\{(\gamma\alpha) + 2(\mu^2)\}z^2x^2 + 2\{(\alpha\beta) + 2(\nu^2)\}x^2y^2 \\ &+ 4\{2(\mu\nu) + (\alpha\lambda)\}x^2yz + 4\{2(\nu\lambda) + (\beta\mu)\}xy^2z + 4\{2(\lambda\mu) + (\gamma\nu)\}xyz^2 \\ &+ 4(\beta\lambda)y^3z + 4(\gamma\lambda)yz^3 + 4(\gamma\mu)z^3x + 4(\alpha\mu)zx^3 + 4(\alpha\nu)x^3y + 4(\beta\nu)xy^3 = 1 \quad (19) \end{aligned}$$

The fifteen coefficients of this surface (which will be called the *Homotatic Coefficients*) are covariant respectively with the fifteen biquadratic powers and products of the co-ordinates, with proper numerical factors.

It is obvious, that when the fifteen homotatic coefficients, and the six heterotatic differences, are known for any set of orthogonal axes, the twenty-one tasinomic coefficients are completely determined.

Mr. Haughton, in the paper previously referred to, discovered the biquadratic surface for a solid constituted of centres of force. It is here shown to exist for all solids, independently of hypotheses.

Those diameters of the biquadratic surface which are normal to that surface, are *axes of maximum and minimum direct elasticity*, and have also this property, that a direct elongation along one of them produces, on a plane perpendicular to it, a normal stress, and no tangential stress; so that they may be called *Euthytatic Axes.* Though such axes sometimes form orthogonal systems, their complete investigation requires the use of oblique co-ordinates, and is therefore deferred till after the eighteenth section of this paper, which relates to such co-ordinates.

SECTION 8.—ORTHOGONAL AXES OF THE BIQUADRATIC SURFACE. METATATIC AXES, ORTHOGONAL AND DIAGONAL.

By rectangular linear transformations, it is always possible to make three of the terms with odd exponents, or three functions of such terms, vanish from the equation of the biquadratic surface. Thus are ascertained sets of orthogonal axes having special properties.

To exemplify this, let the rectangular transformation be such as to make the following functions vanish :—

$$\{(\beta\lambda)-(\gamma\lambda)\}(y^2-z^2)yz;\ \{(\gamma\mu)-(a\mu)\}(z^2-x^2)zx;\ \{(a\nu)-(\beta\nu)\}(x^2-y^2)xy.$$

A cubical molecule having its faces normal to the axes fulfilling this condition has the following property :—*if there be a linear elongation along* y, *and an equal linear compression along* z (or *vice versâ*), *no tangential stress will result round* x *on planes normal to* y *and* z; and similarly of other pairs of axes.

This set of axes may be called the *Orthogonal* or *Principal Metatatic Axes*, and their planes, *Metatatic Planes.*

Let the suffix 1 designate co-ordinates and coefficients referred to these axes. Let Oy, Oz be any new pair of orthogonal axes in the plane $y_1 z_1$. Then since $(\beta\lambda)-(\gamma\lambda)$ is covariant with $(y^2-z^2)yz$, it follows that

$$(\beta\lambda)-(\gamma\lambda)=\{2(\beta\gamma)_1+4(\lambda^2)_1-(\beta^2)_1-(\gamma^2)_1\}\cdot\frac{\sin 4\omega}{4} \quad (20.)$$

$$(\text{where } \omega = \angle y_1 O y),$$

a quantity which is $=0$ for all values of ω which are multiples of 45°. There are of course similar equations for the other metatatic planes. Hence it appears that *in each of the three Metatatic Planes there is a pair of Diagonal Metatatic Axes, bisecting the right angles formed by the Principal Metatatic Axes.*

Each pair of diagonal axes is metatatic for that plane only in which it is situated.

Thus there are in all *nine* metatatic axes, three orthogonal axes, and three pairs of diagonal axes. The diagonal axes are normal to the faces of a regular rhombic dodecahedron.

Let Oy, Oz be a pair of rectangular axes in *any plane whatsoever;* Oy', Oz' any other pair of rectangular axes in the same plane; and let

$$\angle y O y' = \omega';$$

then

$$(\beta\lambda)' - (\gamma\lambda)' = \{2(\beta\gamma) + 4(\lambda^2) - (\beta^2) - (\gamma^2)\}\frac{\sin 4\omega'}{4} + \{(\beta\lambda) - (\gamma\lambda)\}\cos 4\omega', \quad . \quad . \quad (21.)$$

a quantity which is null for eight values of ω', differing from each other by multiples of 45°. Hence, *in each plane in an elastic solid, there is a system of two pairs of axes metatatic for that plane, and forming with each other eight equal angles of* 45°.

In equation (21), make

$$\omega' = -\omega$$

$$(\beta\lambda)' - (\gamma\lambda)' = (\beta\lambda)_1 - (\gamma\lambda)_1 = 0;$$

then from equations (20) and (21), it is easily seen that

$$2(\beta\gamma) + 4(\lambda^2) - (\beta^2) - (\gamma^2) = \{2(\beta\gamma)_1 + 4(\lambda^2)_1 - (\beta^2)_1 - (\gamma^2)_1\} . \cos 4\omega. \quad . \quad (22.)$$

The trigonometrical factor $\cos 4\omega$ is $+1$ for all values of ω which are even multiples of 45°, -1 for all odd multiples of 45°, and $= 0$ for all odd multiples of $22\frac{1}{2}$°. Hence, in every plane in an elastic solid, the quantity (22), which may be called the *Metatatic Difference*, is a maximum for one of the two pairs of metatatic axes, a minimum of equal amount and negative sign for the other, and null for the eight intermediate directions.

SECTION 9.—OF METATATIC ISOTROPY.

A solid is *Metatatically Isotropic* when, if a cubical molecule, cut out in any position whatsoever, undergoes simultaneously an elongation along one axis, and an equal and opposite linear compression along another axis, no tangential stress will result on the faces of that molecule.

For such a substance, the metatatic differences must be null for all sets of axes, viz. :—

$$\left.\begin{aligned} 2(\beta\gamma) + 4(\lambda^2) - (\beta^2) - (\gamma^2) = 0; \\ 2(\gamma\alpha) + 4(\mu^2) - (\gamma^2) - (\alpha^2) = 0; \\ 2(\alpha\beta) + 4(\nu^2) - (\alpha^2) - (\beta^2) = 0. \end{aligned}\right\} . \quad . \quad (23.)$$

In a paper in the *Cambridge and Dublin Mathematical Journal*, Vol. VI., this theorem was alleged of all homogeneous solids, it having been, in fact, tacitly taken for granted, that homogeneity involves metatatic isotropy, as above defined.

SECTION 10.—OF ORTHOTATIC SYMMETRY.

If it be taken for granted that symmetrical action with respect to a certain set of axes, between the parts of a body under one kind of strain, involves symmetrical action with respect to the same axes under all kinds of strains, then one and the same set of orthogonal axes will be at once orthotatic, heterotatic, metatatic, and euthytatic, and for them the whole twelve plagiotatic coefficients will vanish at once, and the independent tasinomic coefficients be reduced to the nine orthotatic coefficients enumerated in Sect. 2. As long as the rigidity of solid bodies was ascribed wholly to mutual attractions and repulsions between centres of force, it is difficult to see how, with respect to homogeneous substances, the above assumption could be avoided. It is probable that there exist substances for which it is true. Such substances may be said to be *Orthotatically Symmetrical.*

Orthotatic symmetry requires that the equation (19) of the biquadratic surface should be reducible by rectangular transformations to its first six terms, and that the axes so found should also be those of the heterotatic ellipsoid. The conditions which must be fulfilled in order that a biquadratic function of three variables may be reducible by rectangular transformations to its first six terms, have been investigated by Mr. Boole.*

SECTION 11.—OF CYBOTATIC SYMMETRY.

Let a substance be conceived which is not only orthotatically symmetrical, but for which the three kinds of orthotatic coefficients are equal for the three orthotatic axes, viz.—

$$(\alpha^2) = (\beta^2) = (\gamma^2);\ (\beta\,\gamma) = (\gamma\,\alpha) = (\alpha\,\beta);\ (\lambda^2) = (\mu^2) = (\nu^2). \quad (24.)$$

Then, for such a substance the metatatic difference may be expressed by

$$2\,(\beta\,\gamma) + 4\,(\lambda^2) - 2\,(\alpha^2); \quad . \quad . \quad . \quad (25.)$$

and if the body be not metatatically isotropic, this difference will have equal maxima or minima for the three orthogonal axes, normal to the faces of a cube, and, conversely, equal minima or maxima for the six diagonal axes, normal to the faces of a regular rhombic dodecahedron.

Symmetry of this kind may be called *Cybotatic,* from its analogy to that of crystals of the tessular system.

* *Cambridge and Dublin Mathematical Journal,* Vol. VI.

SECTION 12.—OF PANTATIC ISOTROPY.

When a body fulfils the conditions of cybotatic symmetry, and at the same time those of metatatic isotropy, it is completely isotropic with respect to elasticity, or pantatically isotropic. It has but three tasinomic coefficients—viz., the euthytatic, platytatic, and goniotatic coefficients, which are equal for all sets of axes, and are connected by the following equation, expressing the condition of metatatic isotropy:

$$(a^2) = (\beta\,\gamma) + 2\,(\lambda^2). \qquad (26.)$$

The properties of such bodies have been fully investigated by various authors.

SECTION 13.—OF THLIPSINOMIC COEFFICIENTS.

If the six elementary strains, a, &c., at a given point in an elastic solid, be expressed as linear functions of the six elementary stresses, P_1, &c., these expressions will contain twenty-one coefficients of compressibility, extensibility, and pliability, which are the second differential coefficients of the potential energy of elasticity with respect to the six elementary stresses; that energy being represented as follows:—

$$U = (a^2)\frac{P_1^2}{2} + (b^2)\frac{P_2^2}{2} + (c^2)\cdot\frac{P_3^2}{2} + (l^2)\cdot\frac{Q_1^2}{2} + (m^2)\frac{Q_2^2}{2} + (n^2)\frac{Q_3^2}{2}$$

$$+(bc)P_2P_3+(ca)P_3P_1+(ab)P_1P_2+(mn)Q_2Q_3+(nl)Q_3Q_1+(lm)Q_1Q_2$$

$$+\{(al)\,P_1+(bl)\,P_2+(cl)\,P_3\}\,Q_1$$

$$+\{(am)P_1+(bm)P_2+(cm)\,P_3\}\,Q_2$$

$$+\{(an)\,P_1+(bn)P_2+(cn)\,P_3\}\,Q_3. \qquad (27.)$$

The twenty-one coefficients in the above equation may be comprehended under the general term *Thlipsinomic*, and classified as follows:—

Designations of Coefficients.		Properties expressed by them.	Symbols.
Orthothliptic	Euthythliptic,	Longitudinal Extensibilities,	(a^2), (b^2), (c^2)
	Platythliptic,	Lateral Extensibilities, . .	(bc), (ca), (ab),
	Goniothliptic,	Pliabilities, ...	(l^2), (m^2), (n^2),
Plagiothliptic, ...		Unsymmetrical Pliabilities,	(mn), &c., &c.

SECTION 14.—OF THLIPSINOMIC TRANSFORMATIONS, UMBRÆ, SURFACES, AND INVARIANTS.

The equations of transformation of the thlipsinomic coefficients are easily deduced from the principle, that the operations

$$\frac{d}{d\,P_1}, \quad \frac{d}{d\,P_2}, \quad \frac{d}{d\,P_3}, \quad \frac{d}{d\,Q_1}, \quad \frac{d}{d\,Q_2}, \quad \frac{d}{d\,Q_3}$$

are respectively covariant with

$$P_1, \quad P_2, \quad P_3, \quad 2\,Q_1, \quad 2\,Q_2, \quad 2\,Q_3,$$

and these with

$$x^2, \quad y^2, \quad z^2, \quad 2\,y\,z, \quad 2\,z\,x, \quad 2\,x\,y.$$

We may regard the thlipsinomic coefficients, like the tasinomic coefficients, as binary compounds of the following six *Umbræ*,

$$(a), \quad (b), \quad (c), \quad (l), \quad (m), \quad (n),$$

which being respectively substituted for

$$P_1, \quad P_2, \quad P_3, \quad 2\,Q_1, \quad 2\,Q_2, \quad 2\,Q_3,$$

in the equation of the tasimetric surface (4), produce the following equation of the *Umbral Thlipsinomic Ellipsoid*,

$$(a)\,x^2 + (b)\,y^2 + (c)\,z^2 + (l)\,yz + (m)\,zx + (n)\,x\,y = 1, \quad (28.)$$

from which, by involution, multiplication, and other operations exactly analogous to those performed on the umbral tasinomic ellipsoid, there may be deduced the equations of *Thlipsinomic Surfaces* exactly corresponding to the tasinomic surfaces already described; while, from the umbral matrix,

$$\left.\begin{matrix} (a) & \frac{1}{2}(n) & \frac{1}{2}(m) \\ \frac{1}{2}(n) & (b) & \frac{1}{2}(l) \\ \frac{1}{2}(m) & \frac{1}{2}(l) & (c) \end{matrix}\right\} \quad . \quad . \quad . \quad (29.)$$

may be formed *Thlipsinomic Invariants* corresponding to the tasinomic invariants.

Hence it appears, that every function of the tasinomic coefficients is converted into a function of the thlipsinomic coefficients with analogous properties, by the substitution of thlipsinomic for tasinomic umbræ according to the following table:—

Tasinomic Umbræ, . .	(a), (β), (γ), (λ), (μ), (ν),
Thlipsinomic Umbræ, . .	(a), (b), (c), $\frac{1}{2}(l)$, $\frac{1}{2}(m)$, $\frac{1}{2}(n)$.

Amongst the thlipsinomic invariants may be distinguished the *Cubic Compressibility*, which is formed by squaring the umbral invariant $(a) + (b) + (c)$, and has the following value:

$$(a^2) + (b^2) + (c^2) + 2\,(b\,c) + 2\,(c\,a) + 2\,(a\,b).$$

SECTION 15.—THLIPSINOMIC AND TASINOMIC CONTRAGREDIENT SYSTEMS.

Let the following square matrices be formed with the tasinomic and thlipsinomic coefficients respectively:—

$$\left.\begin{array}{cccccc}
(a^2) & (a\beta) & (\gamma a) & (a\lambda) & (a\mu) & (a\nu) \\
(a\beta) & (\beta^2) & (\beta\gamma) & (\beta\lambda) & (\beta\mu) & (\beta\nu) \\
(\gamma a) & (\beta\gamma) & (\gamma^2) & (\gamma\lambda) & (\gamma\mu) & (\gamma\nu) \\
(a\lambda) & (\beta\lambda) & (\gamma\lambda) & (\lambda^2) & (\lambda\mu) & (\nu\lambda) \\
(a\mu) & (\beta\mu) & (\gamma\mu) & (\lambda\mu) & (\mu^2) & (\mu\nu) \\
(a\nu) & (\beta\nu) & (\gamma\nu) & (\nu\lambda) & (\mu\nu) & (\nu^2)
\end{array}\right\} \quad . \quad (30.)$$

$$\left.\begin{array}{cccccc}
(a^2) & (a\,b) & (c\,a) & (a\,l) & (a\,m) & (a\,n) \\
(a\,b) & (b^2) & (b\,c) & (b\,l) & (b\,m) & (b\,n) \\
(c\,a) & (b\,c) & (c^2) & (c\,l) & (c\,m) & (c\,n) \\
(a\,l) & (b\,l) & (c\,l) & (l^2) & (l\,m) & (n\,l) \\
(a\,m) & (b\,m) & (c\,m) & (l\,m) & (m^2) & (m\,n) \\
(a\,n) & (b\,n) & (c\,n) & (n\,l) & (m\,n) & (n^2)
\end{array}\right\} \quad . \quad (31.)$$

Then will these matrices be mutually *inverse*, the two systems of coefficients arrayed in them, with their respective systems of functions mutually *contragredient*, and each coefficient or function belonging to one system *contravariant* to the corresponding coefficient or function belonging to the other system.

The values of the coefficients in either of those matrices are expressed in terms of those in the other matrix, in Mr. Sylvester's umbral notation, by twenty-one equations, of which the following are examples :—

$$\left.\begin{aligned}(a^2) &= \begin{vmatrix}(\beta), & (\gamma), & (\lambda), & (\mu), & (\nu)\\ (\beta), & (\gamma), & (\lambda), & (\mu), & (\nu)\end{vmatrix} \div \begin{vmatrix}(a), & (\beta), & (\gamma), & (\lambda), & (\mu), & (\nu)\\ (a), & (\beta), & (\gamma), & (\lambda), & (\mu), & (\nu)\end{vmatrix};\\ (ab) &= \begin{vmatrix}(\beta), & (\gamma), & (\lambda), & (\mu), & (\nu)\\ (a), & (\gamma), & (\lambda), & (\mu), & (\nu)\end{vmatrix} \div \begin{vmatrix}(a), & (\beta), & (\gamma), & (\lambda), & (\mu), & (\nu)\\ (a), & (\beta), & (\gamma), & (\lambda), & (\mu), & (\nu)\end{vmatrix};\end{aligned}\right\} \quad (32.)$$

SECTION 16.—OF THLIPSINOMIC AXES.

If, under given conditions, any symmetrical system or function of the constituents of one of the above matrices be null, then under the same conditions will the contravariant system or function of the constituents of the inverse matrix be null or infinite. Therefore, *Systems of Thlipsinomic Axes coincide with the corresponding systems of Tasinomic Axes.*

SECTION 17.—PLATYTHLIPTIC COEFFICIENTS ARE NEGATIVE.

It may be observed, as a matter of fact, that in consequence of the largeness of the euthytatic coefficients (a^2), (β^2), (γ^2), as compared with the other tasinomic coefficients, the platythliptic coefficients $(b\,c)$, $(c\,a)$, $(a\,b)$, are generally, if not always, negative.

To illustrate this, the case of pantatic isotropy may be taken, for which the two matrices have the following forms :—

$$\left.\begin{matrix}(a^2) & (\beta\gamma) & (\beta\gamma) & 0 & 0 & 0\\ (\beta\gamma) & (a^2) & (\beta\gamma) & 0 & 0 & 0\\ (\beta\gamma) & (\beta\gamma) & (a^2) & 0 & 0 & 0\\ 0 & 0 & 0 & (\lambda^2) & 0 & 0\\ 0 & 0 & 0 & 0 & (\lambda^2) & 0\\ 0 & 0 & 0 & 0 & 0 & (\lambda^2)\end{matrix} \;\middle\|\; \begin{matrix}(a^2) & (b\,c) & (b\,c) & 0 & 0 & 0\\ (b\,c) & (a^2) & (b\,c) & 0 & 0 & 0\\ (b\,c) & (b\,c) & (a^2) & 0 & 0 & 0\\ 0 & 0 & 0 & (l^2) & 0 & 0\\ 0 & 0 & 0 & 0 & (l^2) & 0\\ 0 & 0 & 0 & 0 & 0 & (l^2)\end{matrix}\right\} \quad (33.)$$

from which it is easily seen that the sole platythliptic coefficient has the following value :

$$(b\,c) = \frac{-(\beta\gamma)}{(a^2)^2 + (a^2)(\beta\gamma) - 2(\beta\gamma)^2}. \quad (33\text{A.})$$

The denominator of this fraction is always positive so long as (a^2)

exceeds $(\beta\gamma)$; a condition invariably fulfilled by solid bodies, and, in fact, necessary to their existence.

SECTION 18.—OF OBLIQUE CO-ORDINATES AND CONTRA-ORDINATES.

As there are, in the relations between two systems of oblique co-ordinates, or between a system of oblique co-ordinates and a system of rectangular co-ordinates, six independent constants of transformation, it is possible, by referring the equation of the biquadratic surface (19) to oblique co-ordinates, to make the six terms vanish which contain the cubes of the co-ordinates.

The conception of the physical meaning of such a transformation is much facilitated by the employment of a system of three auxiliary variables, which will be designated as *Contraordinates.*

The relations between co-ordinates and contraordinates are as follows:—

Through an origin O let any three axes pass, right or oblique. Let R be any point, and let

$$\overline{\mathrm{O\,R}} = r.$$

Through R draw three planes, parallel respectively to the three co-ordinate planes, and intersecting the axes respectively in the points X, Y, Z. Also, on O R, as a diameter, describe a sphere, intersecting the axes respectively in U, V, W. Then will

$$\mathrm{O\,X} = x, \ \mathrm{O\,Y} = y, \ \mathrm{O\,Z} = z,$$

be the *co-ordinates* of R, as usual, and

$$\mathrm{O\,U} = u, \ \mathrm{O\,V} = v, \ \mathrm{O\,W} = w,$$

its *contra-ordinates,* being, in fact, the projections of O R on the three axes.

For rectangular axes, co-ordinates and contra-ordinates are identical.

Co-ordinates and contra-ordinates are connected by the following equation:—

$$r^2 = u\,x + v\,y + w\,z. \qquad . \quad . \quad . \quad (34.)$$

In the language of Mr. Sylvester, a system of co-ordinates, and the concomitant system of contra-ordinates, are mutually *Contragredient;* and the square of the radius vector is their *universal mixed concomitant.*

Let the cosines of the angles made by the axes with each other be denoted as follows:

$$\cos y\,\mathrm{O}\,z = c_1; \ \cos z\,\mathrm{O}\,x = c_2; \ \cos x\,\mathrm{O}\,y = c_3;$$

then the contra-ordinates of a given point are the following functions of the co-ordinates:

$$\left.\begin{aligned} u &= x + c_3 y + c_2 z \\ v &= c_3 x + y + c_1 z \\ w &= c_2 x + c_1 y + z \end{aligned}\right\} \quad . \quad . \quad . \quad (35.)$$

Also let

$$\left.\begin{aligned} &\begin{vmatrix} 1, & c_3, & c_2 \\ c_3, & 1, & c_1 \\ c_2, & c_1, & 1 \end{vmatrix} = 1 - c_1^2 - c_2^2 - c_3^2 + 2\, c_1\, c_2\, c_3 = \mathrm{C}; \\ &\frac{1 - c_1^2}{\mathrm{C}} = h_1; \quad \frac{1 - c_2^2}{\mathrm{C}} = h_2; \quad \frac{1 - c_3^2}{\mathrm{C}} = h_3; \\ &\frac{c_1 - c_2\, c_3}{\mathrm{C}} = k_1; \; \frac{c_2 - c_3\, c_1}{\mathrm{C}} = k_2; \; \frac{c_3 - c_1\, c_2}{\mathrm{C}} = k_3; \end{aligned}\right\}$$

then the co-ordinates are the following linear functions of the contra-ordinates :—

$$\left.\begin{aligned} x &= h_1 u - k_3 v - k_2 w; \\ y &= - k_3 u + h_2 v - k_1 w; \\ z &= - k_2 u - k_1 v + h_3 w; \end{aligned}\right\} \quad . \quad . \quad (36.)$$

Also,

$$r^2 = x^2 + y^2 + z^2 + 2\, c_1\, y\, z + 2\, c_2\, z\, x + 2\, c_3\, x\, y \quad . \quad . \quad (37.)$$

$$= h_1 u^2 + h_2 v^2 + h_3 w^2 - 2\, k_1\, v\, w - 2\, k_2\, w\, u - 2\, k_3\, u\, v. \quad (37\text{A}.)$$

Differentiations with respect to the contra-ordinates are obviously covariant with the co-ordinates, and *vice versa;* that is to say,

$$\left.\begin{array}{lcccccc} \text{the operations} \;\; . & \dfrac{d}{dx}, & \dfrac{d}{dy}, & \dfrac{d}{dz}, & \dfrac{d}{du}, & \dfrac{d}{dv}, & \dfrac{d}{dw} \\ \text{are respectively covariant with} & u, & v, & w, & x, & y, & z. \end{array}\right\} \quad . \quad . \quad (38.)$$

By making substitutions according to the above law of covariance in the equations (34), (37), (37A), three equivalent symbols of operation are obtained, which, being applied to isotropic functions of the second degree, produce invariants of the first degree.

SECTION 19.—OF MOLECULAR DISPLACEMENTS AND STRAINS AS REFERRED TO OBLIQUE AXES.

If the displacement of a particle from its free position be resolved into three components ξ, η, ζ, parallel respectively to three oblique axes, O x, O y, O z, those components are evidently covariant respectively with the co-ordinates x, y, z.

It is now necessary to find a method of expressing the strain at any particle in an elastic solid by a system of six elementary strains, which shall be covariant respectively with the squares and doubled-products of these oblique co-ordinates. This condition is fulfilled by considering the elementary strains as being constituted by the variations of the components of the molecular displacement with respect to the distances of the strained particle from three planes passing through the origin, and normal respectively to the three axes; that is to say, with respect to the *contra-ordinates* of the particle, as expressed in the following equations:—

$$\left.\begin{array}{ll} \text{Elongations,} \quad . & \alpha=\frac{d\xi}{du};\ \beta=\frac{d\eta}{dv};\ \gamma=\frac{d\zeta}{dw}; \\ \text{Quasi-Distortions,} & \lambda=\frac{d\zeta}{dv}+\frac{d\eta}{dw};\ \mu=\frac{d\xi}{dw}+\frac{d\zeta}{du};\ \nu=\frac{d\eta}{du}+\frac{d\xi}{dv} \end{array}\right\} \quad (39.)$$

The six elementary strains, as above defined, are obviously covariant with the squares and doubled-products of the co-ordinates, according to the following table:

$$\left.\begin{array}{cccccc} \alpha, & \beta, & \gamma, & \lambda, & \mu, & \nu, \\ x^2, & y^2, & z^2, & 2yz, & 2zx, & 2xy. \end{array}\right\} \quad . \quad . \quad (40.)$$

SECTION 20.—OF STRESSES, AS REFERRED TO OBLIQUE AXES.

It is next required to express the stress at any particle of an elastic solid by means of a system of six elementary stresses, which shall be contragredient to the system of six elementary strains defined in the preceding section. This is accomplished in the following manner.

It is known that the total stress at any point may be resolved into three normal stresses on the three principal planes of the tasimetric surface. Let the direction and sign of any one of those three *principal* stresses be represented by those of a line O R, and its magnitude, as reduced to unity of area of the plane normal to that direction, by the square of that line.

$$\overline{\mathrm{OR}}^2 = r^2.$$

Let u, v, w be the contraordinates of R, as referred to the oblique axes O X, O Y, O Z. Then will the stresses on unity of area of planes normal to those axes, in the direction O R, be represented respectively by

$$u r, \quad v r, \quad w r.$$

Let the *Elementary Stresses* be defined to be, *the projections on the three axes of co-ordinates of the total stresses on unity of area of the three pairs of faces of a parallelopiped, normal to the three axes respectively:* then, if we take S to denote the summation of three terms arising from the three principal stresses, the elementary stresses will be expressed as follows:—

$$\left.\begin{array}{l}
\text{Normal stresses on the faces normal to} \\
\qquad\qquad x, \qquad\qquad y, \qquad\qquad z, \\
\qquad P_1 = S\,.\,u^2; \quad P_2 = S\,.\,v^2; \quad P_3 = S\,.\,w^2; \\
\text{Oblique stresses on the faces normal to} \\
\text{In the directions} \quad \underbrace{\begin{matrix} y & z \\ z & y \end{matrix}} \quad \underbrace{\begin{matrix} z & x \\ x & z \end{matrix}} \quad \underbrace{\begin{matrix} x & y \\ y & x \end{matrix}} \\
\qquad Q_1 = S.vw; \quad Q_2 = S.wu; \quad Q_3 = S\,.\,uv.
\end{array}\right\} \quad . \quad (41.)$$

These expressions fulfil the condition of making the elementary stresses

$$P_1, \quad P_2, \quad P_3, \quad Q_1, \quad Q_2, \quad Q_3$$

contravariant respectively to the elementary strains

$$\alpha, \quad \beta, \quad \gamma, \quad \lambda, \quad \mu, \quad \nu,$$

so that for oblique axes, as for rectangular axes, the potential energy of elasticity is represented by

$$U = -\tfrac{1}{2}\,(P_1\,\alpha + P_2\,\beta + P_3\,\gamma + Q_1\,\lambda + Q_2\,\mu + Q_3\,\nu),$$

the universal concomitant; and may be expressed either by a homogeneous quadratic function of the six elementary strains (as in equation 2) with twenty-one tasinomic coefficients, or by a homogeneous quadratic function of the six elementary stresses, as in equation (27), with twenty-one thlipsinomic coefficients, forming a system contragredient to that of the tasinomic coefficients.

Section 21.—Of Tasinomic and Thlipsinomic Umbræ for Oblique Axes.

The *tasinomic* coefficients for oblique axes may be regarded as compounded of umbræ

$$(\alpha), \quad (\beta), \quad (\gamma), \quad (\lambda), \quad (\mu), \quad (\nu),$$

contravariant respectively to the elementary strains

$$\alpha, \quad \beta, \quad \gamma, \quad \tfrac{1}{2}\lambda, \quad \tfrac{1}{2}\mu, \quad \tfrac{1}{2}\nu,$$

and consequently *covariant* with the squares and products of the *contra-ordinates*

$$u^2, \quad v^2, \quad w^2, \quad v\,w, \quad w\,u, \quad u\,v;$$

and the *thlipsinomic* coefficients for oblique axes may be regarded as compounded of umbræ

$$(a), \quad (b), \quad (c), \quad (l), \quad (m), \quad (n),$$

contravariant respectively to the stresses

$$P_1, \quad P_2, \quad P_3, \quad 2\,Q_1, \quad 2\,Q_2, \quad 2\,Q_3,$$

and consequently *covariant* with the squares and products of the *co-ordinates*

$$x^2, \quad y^2, \quad z^2, \quad 2\,y\,z, \quad 2\,z\,x, \quad 2\,x\,y.$$

Section 22.—Of the Biquadratic Surface, and of Principal Euthytatic Axes.

For oblique as well as for rectangular axes of co-ordinates, the characteristic function of the biquadratic tasinomic surface is represented by equation (19); and the fifteen homotatic coefficients are covariant respectively with suitable multiples of the fifteen biquadratic powers and products of the contraordinates.

If by linear transformations a system of three axes, oblique or rectangular, be found which reduces the characteristic function of the biquadratic surface to the canonical form, consisting of not more than nine terms, viz.—

$$\begin{aligned}(\phi)^2 &= (\alpha^2)x^4+(\beta^2)y^4+(\gamma^2)z^4\\ &+2\{(\beta\gamma)+2(\lambda^2)\}y^2z^2+2\{(\gamma\alpha)+2(\mu^2)\}z^2x^2+2\{(\alpha\beta)+2(\nu^2)\}x^2y^2\\ &+4\{2(\mu\,\nu)+(\alpha\,\lambda)\}x^2y\,z+4\{2(\nu\,\lambda)+(\beta\,\mu)\}x\,y^2z+4\{2(\lambda\,\mu)\\ &\qquad+(\gamma\nu)\}xyz^2=1\,; \quad . \quad . \quad . \quad . \quad . \quad (42)\end{aligned}$$

then, for that system of axes, the following six plagiotatic coefficients are null,

$$(\beta\lambda)=0;\ (\gamma\lambda)=0;\ (\gamma\mu)=0;\ (\alpha\mu)=0;\ (\alpha\nu)=0;\ (\beta\nu)=0;\ \quad (43.)$$

and each of those axes is EUTHYTATIC, according to the definition in Sect. 7, that is to say, is a direction of maximum or minimum direct elasticity (absolute or relative), and also a direction in which a direct elongation or compression produces a simply normal stress.

There are necessarily *three* euthytatic axes at least in every solid—viz., the three *Principal Euthytatic Axes*, as above described, which are normal to the faces of a hexahedron, right or oblique, as the case may be; but in special cases of symmetry there are *additional* or *secondary* euthytatic axes, of which examples will now be given.

SECTION 23.—OF RHOMBIC AND HEXAGONAL SYMMETRY.

When a solid has three oblique principal euthytatic axes making equal angles with each other round an axis of symmetry, and having equal systems of homotatic coefficients corresponding to them, viz.—

$$\left.\begin{aligned}(\alpha^2)=(\beta^2)=(\gamma^2);\ (\beta\gamma)+2(\lambda^2)=(\gamma\alpha)+2(\mu^2)=(\alpha\beta)+2(\nu^2)\\ 2(\mu\nu)+(\alpha\lambda)=2(\nu\lambda)+(\beta\mu)=2(\lambda\mu)+(\gamma\nu)\end{aligned}\right\}\ (43\text{A.})$$

it may be said to possess rhombic symmetry, because the three oblique axes are normal to the faces of one rhombohedron, and to the edges of another belonging to the same series, crystallographically speaking. It is evident in this case, that the axis of symmetry must be a *fourth Euthytatic Axis.*

In the limiting case, when the three oblique axes make with each other equal angles of 120°, they lie in the same plane, normal to the axis of symmetry, and are normal to the faces of one hexagonal prism and the edges of another.

Let Oy_1 denote the longitudinal axis of symmetry of the prism; Oz_1 any one of the three transverse axes perpendicular to Oy_1. The equation of a section of the biquadratic surface by the *Plane of Hexagonal Symmetry* y_1z_1, is as follows:—

$$(\beta^2)_1 y_1^4+(\gamma^2)_1 z_1^4+2\{(\beta\gamma)_1+2(\lambda^2)_1\} y_1^2 z_1^2=1. \quad . \quad (44.)$$

The equation of the same section, referred to any other pair of orthogonal axes Oy, Oz, in the plane of y_1z_1, is as follows:—

$$(\beta^2)\cdot y^4+(\gamma^2)\cdot z^4+2\{(\beta\gamma)+2(\lambda^2)\}y^2z^2+4\{(\beta\lambda)y^2+(\gamma\lambda)z^2\}yz=1.\ (44\text{A.})$$

From considerations of symmetry, it is evident that the coefficient $(\beta\nu)$ must be null for every direction of the axis $O\,y$, in the plane of $y_1 z_1$; consequently, every direction $O\,y$ in that plane, for which $(\beta\lambda)=0$, is an euthytatic axis.

To ascertain whether, and under what conditions, there are other euthytatic axes in the planes of hexagonal symmetry besides the longitudinal and transverse axes, it is to be considered, that for rectangular co-ordinates $(\beta\lambda)$ is covariant with $y^3 z$; hence, let

$$\angle y_1 O\, y = \omega,$$

then

$$(\beta\lambda)=\frac{\sin 2\omega}{4}\cdot\Big[\{2\,(\beta\,\gamma)_1 + 4\,(\lambda^2)_1 - (\beta^2)_1 - (\gamma^2)_1\}\cos 2\,\omega - (\beta^2)_1 + (\gamma^2)_1\Big] \quad . \quad . \quad . \quad (45.)$$

The first factor of the above expression is null for the longitudinal and transverse axes only. The condition of there being additional euthytatic axes in the plane $y_1 z_1$ is, that the second factor shall vanish; that is to say, that

$$\cos 2\,\omega = \frac{(\beta^2)_1 - (\gamma^2)_1}{2\,(\beta\,\gamma)_1 + 4\,(\lambda^2)_1 - (\beta^2)_1 - (\gamma^2)_1} \quad . \quad (46.)$$

and that the value of ω which makes it vanish shall neither be 0° nor 90°; that is to say, that the second member of the above equation (46) shall lie between $+1$ and -1; in which case the equation is satisfied by equal values of ω with opposite signs. Hence are deduced the following theorems, which are stated in such a form as to be applicable to planes of symmetry, whether hexagonal or otherwise:

If, in any plane of tasinomic symmetry containing a pair of orthogonal euthytatic axes, the difference of the euthytatic coefficients for these axes be equal to or greater than the metatatic difference, there are no additional euthytatic axes in that plane.

If, on the other hand, the difference of such euthytatic coefficients be less than the metatatic difference, there are, in such plane of symmetry, a pair of additional euthytatic axes making with each other a pair of angles bisected by the orthogonal euthytatic axes.

$2\,\omega$ is the angle bisected by the axis $O\,y_1$.

In the case of hexagonal symmetry, the additional axes thus found are normal to the faces of one pyramidal dodecahedron and the edges of another.

Section 24.—Of Orthorhombic Symmetry.

Let a solid have one of the three principal euthytatic axes, $O z_1$, normal to the other two, $O x_1$, $O y_1$; let the last two be oblique to each other, and have equal sets of homotatic coefficients, viz.—

$$(a^2)_1 = (\beta^2)_1; \quad (\beta\gamma)_1 + 2(\lambda^2)_1 = (\gamma a)_1 + 2(\mu^2)_1;$$
$$2(\mu\nu) + (a\lambda) = 2(\nu\lambda) + (\beta\mu), \qquad (47.)$$

then, that solid may be said to have *Orthorhombic Symmetry*, its principal euthytatic axes being normal to the faces of a right rhombic prism.

The existence or non-existence, and the position, of a pair of additional euthytatic axes in the longitudinal planes of $y_1 z_1$, $z_1 x_1$, is to be determined as in the preceding section. When such axes exist, they are normal to the faces of an *octahedron with a rhombic base.*

Section 25.—Of Orthogonal Symmetry.

If the three principal euthytatic axes be orthogonal, they are normal to the faces of a *right rectangular or square prism,* and to the edges of a *right rhombic or square prism.* The existence or non-existence, and position, of a pair of additional euthytatic axes in each of the principal planes of such a solid, are determined as in Sect. 23.

If there be a pair of such additional axes in each of the three principal planes, they are normal to the faces of an *irregular rhombic dodecahedron,* and to the edges of a *rhombic octahedron.*

If there be a pair of such additional axes in two of the three principal planes, those axes are normal to the faces of an *octahedron with a rectangular or square base,* and to the edges of an *octahedron with a rhombic or square base.*

If there be a pair of such additional axes in one of the planes of orthotatic symmetry only, those axes are normal to the lateral faces of a *right rhombic prism.*

Section 26.—Of Cyboïd Symmetry.

The case of *Cyboïd Symmetry* is that in which the homotatic coefficients are equal for three orthogonal axes, viz.—

$$(a^2) = (\beta^2) = (\gamma^2); \quad (\beta\gamma) + 2(\lambda^2) = (\gamma a) + 2(\mu^2) = (a\beta) + 2(\nu^2);$$
$$2(\mu\nu) + (a\lambda) = 2(\nu\lambda) + (\beta\mu) = 2(\lambda\mu) + (\gamma\nu) = 0. \qquad (48.)$$

In this case, the principal metatatic axes coincide with the principal euthytatic axes, which are normal to the faces of a cube; the diagonal metatatic axes normal to the faces of a regular rhombic dodecahedron, are euthytatic also; and there are, besides, four additional euthytatic axes symmetrically situated between the first nine, and normal to the faces of a regular octahedron, making in all *thirteen euthytatic axes.*

SECTION 27.—OF MONAXAL ISOTROPY.

Monaxal Isotropy denotes the case in which the homotatic coefficients are completely isotropic round one axis only. In this case, the principal euthytatic axes are, the axis of isotropy, and every direction perpendicular to it; and when there are additional axes, determined as in the preceding sections, they are normal to the surface of a cone.

SECTION 28.—OF COMPLETE ISOTROPY.

In the case of complete isotropy of the homotatic coefficients, every direction is a euthytatic axis.

SECTION 29.—PROBABLE RELATIONS BETWEEN EUTHYTATIC AXES AND CRYSTALLINE FORMS.

In the preceding sections it has been shown what must be the nature of the relations between the fifteen homotatic coefficients, for various solids, having systems of euthytatic axes normal to the faces and edges of the several *Primitive Forms* known in crystallography.

It is probable that the normals to *Planes of Cleavage* are euthytatic axes of minimum elasticity.

It may also be considered probable, that in some cases, especially in the tessular system, which corresponds to cyboïd symmetry, and in the case of the pyramidal summits of crystals of the rhombohedral system, euthytatic axes correspond to symmetrical summits of crystalline forms. In the icositetrahedral crystals of leucite and analcime, and the tetracontaoctahedral crystals of diamond, there are twenty-six symmetrical summits, one pair corresponding to each of the thirteen axes of cyboïd symmetry.

The following is a synoptical table of the various possible systems of euthytatic axes, arranged according to their degrees and kinds of symmetry, and of the crystalline forms to the faces and edges of which such systems of axes are respectively normal.

Systems of Euthytatic Axes.	Crystalline Forms. Faces.	Edges.
I. Asymmetry.	Tetarto-prismatic System.	
1. Three unequal oblique axes, .	Oblique hexahedron.	
II. Symmetry about One Plane.	Hemiprismatic System.	
2. Two unequal oblique axes, and one rectangular axis, . .	Right rhomboidal prism,	Oblique rhombic prism.
3. Two equal and one unequal oblique axis, . . .	Oblique rhombic prism,	Right rhomboidal prism.
III. Rhombic and Hexagonal Symmetry.	Rhombohedral System.	
4. Three equi-oblique principal axes round one axis of symmetry, ...	Rhombohedron, . .	Rhombohedron.
5. Three equi-oblique principal axes in one plane, normal to axis of symmetry, . .	Hexagonal prism, .	Hexagonal prism.
6. Three pairs of secondary axes in planes of symmetry, .	Pyramidal dodecahedron, . . .	Pyramidal dodecahedron.
IV. Orthorhombic Symmetry.	Prismatic and Pyramidal Systems.	
7. Two equal oblique transverse axes normal to one longitudinal axis, . . .	Right rhombic prism, .	Rectangular prism.
8. Two pairs of secondary axes in longitudinal planes, . .	Octahedron with rhombic base, . . .	Octahedron with rectangular base.
V. Orthogonal Symmetry.		
9. Three orthogonal axes, not all equal, ...	Rectangular and square prisms, . . .	Right rhombic and square prisms.
10. Three pairs of secondary axes in principal planes, . .	Irregular rhombic dodecahedron, . . .	Octahedron with rhombic base and rectangular prism.
11. Two pairs of secondary axes, .	Octahedron with square or rectangular base, .	Octahedron with square or rhombic base.
Same with 7. One pair of secondary axes, ...	Right rhombic prism, .	Rectangular prism.
VI. Cyboïd Symmetry.	Tessular System.	
12. Three equal orthogonal axes, .	Cube.	
13. Six diagonal axes, . . .	Regular rhombic dodecahedron, . . .	Cube and regular octahedron.
14. Four symmetrical intermediate axes, ...	Regular octahedron, .	Rhombic dodecahedron.

VII. MONAXAL ISOTROPY.

15. One axis of isotropy, . .	Isotropic laminæ.
16. Innumerable transverse axes, .	Istropic fibres.
17. Innumerable equi-oblique axes,	Conical cleavage.

VIII. COMPLETE ISOTROPY.

18. Innumerable axes of isotropy, .	Amorphism.

SECTION 30.—MUTUAL INDEPENDENCE OF THE EUTHYTATIC AND HETEROTATIC AXES, AND OF THE HOMOTATIC AND HETEROTATIC COEFFICIENTS.

The fifteen homotatic coefficients of the biquadratic surface, on which the euthytatic axes depend, and the six heterotatic differences, coefficients of the heterotatic ellipsoid, constitute twenty-one independent quantities; so that the euthytatic axes may possess any kind or degree of symmetry or asymmetry, and the heterotatic axes any other kind or degree, in the same solid.

Hence, if it be true that crystalline form depends on the arrangement of euthytatic axes, it follows that two substances may be exactly alike in crystalline form, and yet differ materially in the laws of their elasticity, owing to differences in their respective heterotatic coefficients.

It may be observed, however, that this complete independence of those two systems of axes and coefficients is *mathematical* only, and that their physical dependence or independence is a question for experiment.

SECTION 31.—ON REAL AND ALLEGED DIFFERENCES BETWEEN THE LAWS OF THE ELASTICITY OF SOLIDS, AND THOSE OF THE LUMINIFEROUS FORCE.

For every conceivable system of tasinomic coefficients in a solid, the *plane of polarisation* of a wave of distortion is that which includes the direction of the molecular vibration and the direction of its propagation, being, in fact, the plane of distortion.

On the other hand, it appears to be impossible to avoid concluding, from the laws of the diffraction of polarised light, as discovered by Professor Stokes, and from those of the more minute phenomena of the reflexion of light, as investigated theoretically by M. Cauchy and experimentally by M. Jamin, that in plane-polarised light the plane of polarisation is perpendicular to the direction of vibration, or rather (to avoid hypothetical language) to the direction of some physical phenomenon whose laws of communication are to a certain extent analogous to those of a vibratory movement.

This constitutes an essential difference between the laws of the elastic forces in a solid, and those of the luminiferous force.

In order to frame, in connection with the wave theory of light, a mechanical hypothesis which should take that difference into account, it has been proposed to consider the *elasticity* of the luminiferous medium to be the same in all substances, and for all directions, or *Pantatically Isotropic;* and to ascribe the various retardations of light to variations in the *inertia* of the mass moved in luminiferous waves, in different substances, and for different directions of motion.*

Another essential difference between the laws of solid elasticity and those of the luminiferous force is, that under no conceivable system of tasinomic coefficients in a homogeneous solid, would the plane of distortion in a wave be rotated continuously round the direction of propagation.

Much has been written, both recently and in former times, concerning an alleged difficulty in the theories of waves, both of sound and of light, arising from the physical impossibility of the actual divergence of waves from, or their convergence to, a mathematical point. This impossibility must be admitted; but the supposed difficulty to which it gives rise in the theories of waves is completely overcome in Mr. Stokes' paper "On the Dynamical Theory of Diffraction,"† in which that author proves, that waves spreading from a focal space, or origin of disturbance, of finite magnitude, and of any figure, sensibly agree in all respects with waves spreading from an imaginary focal point, so soon as they have attained a distance from the focal space which is large as compared with the dimensions of that space; so that the equations of the propagation of waves spreading from imaginary focal points may be applied, without sensible error, to all those cases of actual waves to which it is usual to apply them.

The physical impossibility of focal points applies to light independently of all hypotheses; for at such points the intensity would be infinite. It appears to be worthy of consideration, whether this impossibility may not be connected with the appearance of spurious disks of fixed stars in the foci of telescopes.

SECTION 32.—ON THE ACTION OF CRYSTALS ON LIGHT.

If we set aside those actions on light to which there is nothing analogous in the phenomena of the elasticity of homogeneous solids, the laws of

* *Philosophical Magazine,* June, 1851, December, 1853.

† *Cambridge Transactions,* Vol. IX., Part 1.

the refractive action of a crystal on light are in general of a more symmetrical kind, or depend on fewer quantities than those of its elasticity.

Thus, the elasticity of a homogeneous solid depends on twenty-one quantities; its crystalline form, on fifteen (the homotatic coefficients), while its refractive action on homogeneous light in most cases is expressible by means of the magnitudes and directions of the orthogonal axes of Fresnel's wave-surface, making in all six quantities. Crystals which possess only rhombic or hexagonal symmetry in their euthytatic axes, are usually monaxally isotropic in their action on light; while crystals which possess only cyboïd symmetry in their euthytatic axes, are completely isotropic in their action on light.

From these remarks, however, there are exceptions, as in the case of the extraordinary optical properties discovered by Sir David Brewster in analcime, which, in its refraction as well as in its form, is cyboïdally symmetrical without being isotropic.

NOTE REFERRED TO AT PAGE 120, ON SYLVESTRIAN UMBRÆ.

Without attempting to enter into the abstract theory of the umbral method, it may here be useful to explain the particular case of its application which is employed in this paper.

Let U be a quantity having an absolute value, constant or variable, (such, for example, as any physical magnitude), and $u, v, \ldots$ &c. a set of quantities, m in number, such that U is of them a homogeneous rational function of the nth degree. There are an indefinite number of possible sets of m quantities satisfying this condition; and the quantities of each set are related to those of each other set by m equations of the first degree, called equations of *linear transformation.* Let

$$u_1, v_1, \ldots$$
$$u_2, v_2, \ldots$$

be two such sets.

Let $C_{a, b, \ldots}$ denote the coefficient of $u^a v^b \ldots$ in the development of

$$(u + v + \ldots)^n,$$

and let

$$U = \Sigma\{C_{a, b, \ldots} A_{1, a, b, \ldots} u_1^a v_1^b \ldots\}$$
$$= \Sigma\{C_{a, b, \ldots} A_{2, a, b, \ldots} u_2^a v_2^b \ldots\}.$$

The two sets of coefficients A_1, A_2, are connected by linear equations of transformation, the investigation of which is much facilitated by the following process.

Let two sets, each of m symbols, a_1, β_1, &c. . . . a_2, β_2, . . . &c., be assumed such that

$$a_1 u_1 + \beta_1 v_1 + \quad . \ . \ . \quad = a_2 u_2 + \beta_2 v_2 + \quad . \ . \ .$$

and that, consequently,

$$(a_1 u_1 + \beta_1 v_1 + \ . \ . \ .)^n = \Sigma\{C_{a,b}, \ . \ . \ . \ a_1{}^a \beta_1{}^b \ . \ . \ . \ u_1{}^a v_1{}^b \ . \ . \ .\}$$
$$= (a_2 u_2 + \beta_2 v_2 + \ . \ . \ .)^n = \Sigma\{C_{a,b}, \ . \ . \ . \ a_2{}^a \beta_2{}^b \ . \ . \ . \ u_2{}^a v_2{}^b \ . \ . \ .\}.$$

Then, if the m equations of transformation between the two sets of symbols a_1, β_1 . . . and a_2, β_2 . . . be formed, and if from them be deduced the equations between the two sets of products $a_1{}^a \beta_1{}^b$ . . . , and $a_2{}^a \beta_2{}^b$ . . . . , &c., and if in the latter system of equations, there be substituted for each product $a^a \beta^b$ . . . the corresponding coefficient $A_{a,b}$, ... , the result will be the system of equations sought. Also, if any function of the products $a^a \beta^b$ . . . be *invariant* (*i. e.*, a function whose value, like that of the original function U, is not altered by the transformation), the corresponding function of the coefficients A will be invariant.

The symbols a, β, &c., with reference to their relation to the coefficients A, are called *umbræ;* that is, *factors of symbols, whose equations of transformation are similar to those of the coefficients* A. In the *umbral notation,* umbræ are usually distinguished from symbols denoting actual quantities by being enclosed in brackets thus:

$$(a), \ (\beta), \ \&c. \ . \ . \ .$$

and each coefficient A is represented by enclosing in brackets that product of umbræ with which it is *covariant;* thus:

$$A_{a,b}, \ . \ . \ . \ = (a^a \beta^b \ . \ . \ .).$$

The umbral notation is applied to abbreviate the expression of determinants in a manner of which the following are examples:—

$$\left| \begin{matrix} a, & \beta, & \gamma, & \&c. \\ a, & \beta, & \gamma, & \&c. \end{matrix} \right| \text{ denotes } \left| \begin{matrix} (a^2) & (a\beta) & (a\gamma) & \&c. \\ (a\beta) & (\beta^2) & (\beta\gamma) & \&c. \\ (a\gamma) & (\beta\gamma) & (\gamma^2) & \&c. \\ \&c. & \&c. & \&c. & \&c. \end{matrix} \right|$$

$$\begin{vmatrix} \alpha, & \gamma, & \delta, & \&c. \\ \beta, & \gamma, & \delta, & \&c. \end{vmatrix} \text{ denotes } \boxed{\begin{matrix} (\alpha\beta) & (\beta\gamma) & (\beta\delta) & \&c. \\ (\alpha\gamma) & (\gamma^2) & (\gamma\delta) & \&c. \\ (\alpha\delta) & (\gamma\delta) & (\delta^2) & \&c. \\ \&c. & \&c. & \&c. & \&c. \end{matrix}}$$

VII.—ON THE VIBRATIONS OF PLANE-POLARISED LIGHT.*

1. THE important question, whether the direction of vibration in plane-polarised light is normal or parallel to the plane of polarisation, is equivalent to this:—whether the velocity of propagation of a rectilinear transverse vibratory movement in the medium which transmits light in crystallised bodies is a function of the *direction of vibration* only, or of the *position of the plane, which includes the direction of vibration and the direction of transmission.*

The former of these views was adopted by Fresnel, as being necessary to explain the phenomena of polarisation by reflexion; and in Mr. Green's investigation of the laws of these phenomena, in which the conclusions of Fresnel are shown to agree either exactly or approximately with the consequences of strictly mechanical principles, the same supposition is adopted.

But if we follow the generally received theory, that the different velocities of differently polarised rays in crystalline bodies are due solely to the different degrees of elasticity possessed by the vibrating medium in different directions, Fresnel's supposition must be abandoned, and the opposite one adopted. For if there is any proposition more certain than others respecting the laws of elasticity, it is this:—that the transverse elasticity of a medium, or the elasticity which resists *distortion* of the particles, depends upon the position of the *plane of distortion,* being the same for all directions of distortion in a given plane. This law is implicitly involved in the researches of Poisson, of M. Cauchy, of Mr. Green, and others on elasticity; and in a memoir read to the British Association at Edinburgh, in 1850, and published in the *Cambridge and Dublin Mathematical Journal* for February, 1851, I have shown that it is true independently of all hypotheses respecting the constitution of matter, being a necessary consequence of the conception of an elastic medium. Now a wave of plane-polarised light is a wave of distortion: the plane of distortion is the plane which includes the direction of transmission and the direction of vibration: the elasticity called into play depends on the position of this plane; therefore, if the velocity of propagation depends upon elasticity alone, the plane of distortion must be the plane of polarisation; and if a normal be drawn to that plane, the velocity of propagation

* Read before the Royal Society of Edinburgh, on December 2, 1850, and published in the *Philosophical Magazine,* June, 1851.

will be a function of the position of that normal, and not, as supposed by Fresnel, of the direction of vibration itself.

2. Up to a very recent period, no experimental data existed adequate to determine which of these suppositions is supported by facts; for the phenomena of double refraction are consistent with either; and the theory of polarisation by reflexion is not regarded as sufficiently certain to afford the means of deciding this question. At length, however, the *experimentum crucis* has been made by Professor Stokes, and the result is conclusive in favour of the supposition of Fresnel.

In his paper on Diffraction (*Cambridge Transactions*, Vol. IX., Part 1), Professor Stokes has shown, that on any conceivable theory of the propagation of undulations of light, vibrations *normal to the plane of diffraction* must be transmitted round the edge of an opaque body with less diminution of intensity than vibrations *in* that plane. Therefore, when light, of which the vibrations are oblique to the plane of diffraction, is so transmitted, the plane of vibration will be *more nearly perpendicular* to the plane of diffraction in the diffracted ray than in the incident ray. He has found by experiment, that when light of which the plane of polarisation is oblique to the plane of diffraction is transmitted round the edge of an opaque body, the plane of polarisation is *more nearly parallel to* the plane of diffraction in the diffracted than in the incident ray. The necessary conclusion is, that the direction of vibration in plane-polarised light is normal to the plane of polarisation; in other words, that the *velocity of light in crystallised media depends on the direction of vibration*, as conjectured by Fresnel.

This result of experiment is at variance with the necessary consequences of the supposition, that the velocity of light depends on elasticity alone; therefore, that supposition is inadequate to explain the phenomena of polarised light.

3. Having considered what modifications must be introduced into our hypothetical conceptions of the nature of the medium which transmits light, to make them adequate to explain the facts which have thus been established, I now offer the following suggestions.

In a paper read to the Royal Society of Edinburgh, and published in their *Transactions*, Vol. XX., Part 1, I proposed, as a foundation for the theory of heat, and of the elasticity of gases and vapours, an hypothesis called that of molecular vortices; and in a subsequent paper, already referred to, I deduced from the same hypothesis some principles relative to the elasticity of solids. I shall now show that Fresnel's conjecture as to the direction of vibration in plane-polarised light is a natural consequence of that hypothesis.

The fundamental suppositions of the hypothesis of molecular vortices are the following:—

First. That each atom of matter consists of a nucleus or central physical point enveloped by an elastic atmosphere, which is retained round it by attraction; so that the elasticity of bodies is made up of two parts—one arising from the diffused portion of the atmospheres, and resisting change of volume only; the other arising from the mutual actions of the nuclei, and of the portions of atmosphere condensed round them, and resisting not only change of volume, but also change of figure.

Secondly. That the changes of elasticity due to heat arise from the centrifugal force of revolutions or oscillations among the particles of the atomic atmospheres, diffusing them to a greater distance from their nuclei, and thus increasing the elasticity which resists change of volume only, at the expense of that which resists change of figure also.

Thirdly. That the medium which transmits light and radiant heat consists of the *nuclei* of the atoms vibrating independently, or almost independently, of their atmospheres; *absorption* being the transference of motion from the nuclei to the atmospheres, and *emission* its transference from the atmospheres to the nuclei.

This last supposition is peculiar to my own researches, the first two having more or less resemblance to ideas previously entertained by others.

If an indefinitely extended vibrating medium, equally elastic in all directions, consists of a system of atomic nuclei, tending to preserve a certain configuration in consequence of their mutual attractions and repulsions, it is well known that such a medium is capable of transmitting two sorts of vibrations only, longitudinal and transverse, the latter alone being supposed to be concerned in the phenomena of light. It is also well known, that the square of the velocity of propagation of transverse vibrations is directly proportional to a quantity called the transverse elasticity of the medium, arising from the mutual actions of the nuclei, and inversely proportional to its density—that is, to the sum of the masses of all the nuclei contained in unity of volume. To account for the immense velocity of light, the masses of the atomic nuclei must be supposed to be very small as compared with the mutual forces exerted by them.

In stating the third supposition of the hypothesis, the nuclei are said to vibrate independently, *or almost independently*, of their atmospheres; for the absolute independence of their vibrations is probably an ideal case, not realised in nature, though approached very nearly in the celestial space, where the atomic atmospheres must be inconceivably rarefied.

As a pendulum is known to be accompanied in its oscillations by a portion of the air in which it swings, so the nuclei probably in all cases carry along with them in their vibrations a small portion of their atmospheres, which acts as a *load*, increasing the vibrating mass without increasing in the same proportion the elasticity, and consequently retarding the velocity of transmission. The amount of this load must depend on the density of the atomic atmosphere; and, accordingly, we find that, generally speaking, the most dense substances are those in which the velocity of light is least.

Now, if we assume, what is extremely probable, that in crystallised media the atomic atmospheres are not similarly diffused in all directions round their nuclei, but are more dense in certain directions than in others, we must at once conclude that in such media the velocity of propagation of vibratory movement depends *on the direction of vibration;* for upon that direction depends the load of atmosphere which each nucleus carries along with it.

4. Having thus shown that the conjecture of Fresnel, which has been confirmed by the experiments of Professor Stokes, is a natural consequence of the hypothesis of molecular vortices, I shall now prove that that hypothesis leads to those mathematical laws of the transmission of light in crystalline media which Fresnel discovered.

Considering it desirable in this paper to avoid lengthened algebraical analysis, I shall with that view state, in the first place, certain known geometrical properties of the ellipsoid, to which it will be necessary for me to refer.

I. If a curved surface be described about a centre, such that the sum of the reciprocals of the squares of any three orthogonal diameters is a constant quantity, that surface, if no diameter is infinite, is an ellipsoid.

II. Every function of direction round a centre, whose variation from a given amount varies as the reciprocal of the square of the diameter of an ellipsoid described about that centre, is itself proportional to the reciprocal of the square of the diameter of another ellipsoid described about the same centre with the first, and having the directions of its axes the same.

III. It follows from the last proposition, that if there be a function of direction round a centre which is proportional to the reciprocal of the square of the diameter of an ellipsoid of small excentricities, so that the range of variation of the function is small as compared with its amount, then any function of that function, whose range of variation is small also, may be represented *approximately* by the reciprocal of the square of the diameter of another ellipsoid, having its centre and the directions of its axes the same with those of the first.

The square of the velocity of propagation of transverse vibrations is proportional to the transverse elasticity of the medium divided by the mean density; that is, by the sum of all the vibrating masses in unity of volume. That sum is the sum of the masses of the nuclei, added to the masses of atmosphere with which they are loaded. The atmospheric load of each nucleus depends on, or is a function of, the density of the atmosphere adjoining the nucleus along the line in which the latter vibrates. The mode of distribution of the atmospheres depends on the attraction of the nuclei upon them, and therefore on the mode of arrangement of the nuclei. The mode of arrangement of the nuclei, when it is symmetrical and uniform, may be expressed by means of their *mean intervals.*

The mean interval of the nuclei is a function of direction, of such a nature that its three values for any three orthogonal directions being multiplied together give a constant result—viz., the space occupied (not *filled*) by one nucleus, or the quotient of a given space by the number of nuclei contained in it. Hence the sum of the values of the logarithm of the mean interval for any three orthogonal directions is a constant quantity; and that logarithm, therefore, is proportional to the reciprocal of the square of the diameter of an ellipsoid, whose three axes may be called the *axes of atomic distribution.* Therefore, the mean interval, the atmospheric load of the nuclei, and the square of the velocity of propagation, for a given direction of transverse vibration, are all functions of the reciprocal of the square of the diameter of an ellipsoid, and have maxima and minima corresponding to its three axes, which are those of atomic distribution.

Now, in all known crystalline media, the range of variation of these quantities for different directions is very small compared with their amount. Therefore, each of them may be *approximately* represented by the reciprocal of the square of the diameter of an ellipsoid whose axes are parallel to those of atomic distribution.

If, then, the directions of vibration in a given crystal which correspond to the greatest and least velocities of transmission are known, let these directions (which are at right angles to each other), and a third direction at right angles to them both, be taken for the axes of an ellipsoid, the lengths of those axes being inversely proportional to the corresponding velocities of transmission. Then will the velocity of transmission of any transverse vibratory movement be sensibly proportional to the reciprocal of a diameter of that ellipsoid, drawn parallel to the direction of vibration. And if a plane be drawn through the centre of the ellipsoid parallel to a series of plane-waves, the two axes of the elliptic section so made will represent, in magnitude, the reciprocals of the greatest and least normal velocities of transmission of waves parallel to that plane, and, in direction, the corresponding directions of vibration.

This agrees exactly with the construction given by Fresnel, on which his entire theory of double refraction is founded.

The degree of symmetry and uniformity of arrangement of the atoms which is necessary in order that the mean interval may have a definite value, and that three axes of distribution may exist, is the same which is necessary to the existence of rectangular axes of elasticity in a solid body. It must extend round each point, throughout a space which is large as compared with the sphere of appreciable molecular action.

The experiments of Sir David Brewster and of Fresnel on the action of compressed glass on polarised light, show that rays polarised in a plane normal to the direction of compression—that is to say, vibrations parallel to that direction—are accelerated. This indicates that the atmospheric load on each vibrating nucleus in that direction is diminished, probably by the displacement of a portion of the atmosphere out of that line.

5. Though I have assumed, in the course of this investigation, that the luminiferous medium is equally elastic in all directions, I by no means intend to assert that it is necessarily so in all substances; but merely that, in most known crystalline media, an atmospheric load on the vibrating nuclei is the *predominant* cause of variation in the velocity of transmission with the direction of transverse vibration.

It is remarkable that Fresnel, in his theory of the intensity of reflected and refracted light, speaks of the particles of the luminiferous medium as being more or less *loaded* in substances of greater or less refractive power. He did not, however, apply this idea to double refraction, although he adopted a theory, which, as we have seen, results from it.

The principles laid down in this paper are not compatible with the prevalent idea of a luminiferous ether enveloping ponderable particles. The fundamental idea from which they spring is the converse: that the luminiferous medium is a system of atomic nuclei or centres of force, whose office it is to give *form* to matter; while the atmospheres by which they are surrounded give, of themselves, merely *extension*.

VIII.—GENERAL VIEW OF AN OSCILLATORY THEORY OF LIGHT.*

SECTION I.—DIFFICULTIES OF THE PRESENT HYPOTHESIS.

NOTWITHSTANDING the perfection to which the geometrical part of the undulatory theory of light has been brought, it is admitted that great difficulty exists in framing, to serve as a basis for the theory, a physical hypothesis which shall at once be consistent with itself and with the known properties of matter.

The present paper is a summary of the results of an attempt to diminish that difficulty. All the conclusions stated have been deduced by means of strict mathematical analysis; and although it is impossible to read the investigations before the British Association in detail, their results can easily be verified by every mathematician who is familiar with the undulatory theory in its present form.

It may be considered as established, that if we assume the supposition that plane-polarised light (out of the varieties of which all other light can be compounded) consists in the wave-like transmission of a state of motion, the nature and magnitude of which are functions of the direction and length of a line transverse to the direction of propagation, we can deduce from this supposition, with the aid of experimental data, and of certain auxiliary hypotheses, the laws of the phenomena of the interference of light, of its propagation in crystalline and uncrystalline substances, of diffraction, of single and double refraction, of dispersion by refraction, and of partial and total reflexion.

It has hitherto been always assumed, that the kind of motion which constitutes light is a vibration from side to side, transmitted from particle to particle of the luminiferous medium, by means of forces acting between the particles. In order to account for the transmission of such transverse vibrations, the luminiferous medium has been supposed to possess a kind of elasticity which resists distortion of its parts, like that of an elastic solid; and in order to account for the non-appearance in ordinary cases of effects which can be ascribed to longitudinal vibrations, it has been found

* Read before the British Association at Hull, September 10, 1853, and published in the *Philosophical Magazine*, December, 1853.

necessary to suppose further, that this medium resists compression with an elasticity immensely greater than that with which it resists distortion; the latter species of elasticity being, nevertheless, sufficiently great to transmit one of the most powerful kinds of physical energy through interstellar space, with a speed in comparison with which that of the swiftest planets of our system in their orbits is appreciable, but no more.

It seems impossible to reconcile these suppositions with the fact, that the luminiferous medium in interstellar space offers no sensible resistance to the motions of the heavenly bodies.

A step towards the solution of this difficulty was made by Mr. MacCullagh. The equations which he used to express the laws of the propagation of light, when interpreted physically, denote the condition of a medium whose molecules tend to range themselves in straight lines, and when disturbed, to return to those lines with a force depending on the curvature of the lines into which they have been moved. But even this hypothesis requires the assumption that the elasticity of the luminiferous medium to resist compression is immensely greater than the elasticity which transmits transverse vibrations.

The difficulty just referred to arises from a comparison of the hypothesis of transverse vibrations with the observed phenomena of the world.

Another difficulty arises within the hypothesis itself. Fresnel originally assumed, that in crystalline media, where the velocity of light varies with the position of the plane of polarisation, the direction of vibration is perpendicular to that plane. This is equivalent to the supposition, that the velocity with which a state of rectilinear transverse vibration is transmitted through such a medium, is a function simply of the direction in which the particles vibrate. From this hypothesis he deduced the form of that wave-surface which expresses completely the law of the propagation of plane-polarised light through crystalline media, and he obtained also a near approximation to the laws of the intensity of plane-polarised light reflected from singly refracting substances.

But it was afterwards demonstrated, that the elastic forces which propagate a transverse movement in any medium must necessarily be functions, not merely of the direction of the movement, but jointly of this direction and the direction of propagation; that is to say, of the position of the plane containing these two directions. Consequently, if the various velocities of variously polarised light in a doubly refracting medium, arise from variations of elasticity in different directions, the direction of vibration is in the plane of polarisation, contrary to the hypothesis of Fresnel. Fresnel's wave-surface, and his approximate formulæ for the intensity of reflected light, are deducible from this supposition as readily as from his original hypothesis; and Mr. MacCullagh obtained from it formulæ for the intensity of light reflected from doubly refracting substances, agreeing closely with the experiments of Sir David Brewster.

On the other hand, the formulæ of M. Cauchy, and those of Mr. Green, as modified by Mr. Haughton,* expressing the effects of reflexion on the intensity and phase of polarised light, all of which are founded on the supposition that the direction of vibration is perpendicular to the plane of polarisation, have been shown to be capable, by the introduction of proper constants, of giving results agreeing closely with those of the important experiments of M. Jamin (*Annales de Chimie et de Physique*, 3rd Series, Vol. XXIX., 1850); and it is difficult, if not impossible, to see how such formulæ could have been deduced from the opposite supposition.

But the true crucial experiment on this subject has been furnished by the researches of Professor Stokes on the Diffraction of Polarised Light (*Camb. Trans.*, Vol. IX.). Whatsoever may be the nature of the motion that constitutes light, if it can be expressed by a function of the direction and length of a line perpendicular to the direction of propagation (which may be called a *transversal*), it is certain that this motion will be more abundantly communicated round the edge of an obstacle, when its transversal is parallel than when it is perpendicular to that edge; so that the effect of diffraction is, to bring every oblique transversal into a position more nearly *parallel* to the diffracting edge. But it has been shown by the experiments of Professor Stokes, that the effect of diffraction upon every ray of light polarised in a plane oblique to the diffracting edge, is to bring the plane of polarisation into a position more nearly *perpendicular* to the diffracting edge. Therefore, the transversal of a ray of plane-polarised light (which, if light consists in linear vibratory movement, is the direction of vibration) is perpendicular to the plane of polarisation.

Hence it follows, that, in a crystalline medium, the velocity of light depends simply on the direction of the transversal characteristic of the movement propagated, and not on the direction of propagation.

This conclusion is opposed to the laws of the propagation of transverse vibrations through a crystalline elastic solid, or through any medium in which the velocity of propagation depends on elasticity varying in different directions. Therefore, the velocity of light depends on something not analogous to the variations of elasticity in such a medium.

To solve this difficulty, the author of this paper some time since suggested the hypothesis, that the luminiferous medium consists of particles forming the nuclei of atmospheres of ordinary matter; that it transmits transverse vibrations by means of an elasticity which is the same in all substances and in all directions; and that the variations in the velocity of the transmission of vibrations arise from variations in the atmospheric load, which the luminiferous particles carry along with them in their vibrations,

**Philosophical Magazine*, August, 1853.

and which is a function of the nature of the substance, and, in a crystalline body, of the direction of vibration.

But although this hypothesis removes the inconsistency just pointed out as existing within the theory itself, it leaves undiminished the difficulty of conceiving a medium pervading all space, and possessed of an elasticity of figure, at once so strong as to transmit the powerful energy of light with its enormous velocity, and so feeble as to exercise no direct appreciable effect on the motions of visible bodies.

Section II.—Statement of the Proposed Hypothesis of Oscillations.

The hypothesis now to be proposed as a groundwork for the undulatory theory of light, consists mainly in conceiving that the luminiferous medium is constituted of detached atoms or nuclei distributed throughout all space, and endowed with a peculiar species of polarity, in virtue of which three orthogonal axes in each atom tend to place themselves parallel respectively to the corresponding axes in every other atom; and that plane-polarised light consists in a small oscillatory movement of each atom round an axis transverse to the direction of propagation.

Such a movement would be transmitted through such a medium with a velocity proportional, directly, to the square root of the total rotative force exercised by the luminiferous atoms in a given small space, upon those in a given adjacent small space lying in the direction of propagation, in consequence of a given amount of relative angular displacement round the axis of oscillation; and inversely, to the square root of the sum of the moments of inertia round the axes of oscillation of the atoms contained in a given space, loaded with such portions of molecular atmospheres surrounding them as they may carry along with them in their oscillations.

Then, denoting by

h, the velocity, in a given direction of plane-waves, of oscillation round transverse axes parallel to a given line;

C, a coefficient of polarity or rotative force for the given directions of propagation and of axes;

M, a coefficient of moment of inertia for the given direction of axes;
the above principle may be represented by this equation,

$$h^2 = \frac{C}{M}.$$

The coefficient of polarity in question is proper only to an axis of oscillation transverse to the direction of propagation. To account for the stability of direction of the axes of the atoms, and also for the non-appearance, in ordinary cases, of phenomena capable of being ascribed to oscillations round axes parallel to the direction of propagation, it is necessary to suppose the corresponding coefficient for the latter species of oscillations to be much greater than the coefficient for transverse axes of oscillation.

It is evident, that how powerful soever the polarity may be which is here ascribed to the atoms of the luminiferous medium, it is a kind of force which must be absolutely destitute of direct influence on resistance to change of volume or change of figure in the parts of that medium, or of any body of which that medium may form part; and that, consequently, the difficulty, which in the hypothesis of vibrations arises from the necessity of ascribing to the luminiferous medium properties like those of an elastic solid, has no existence in the hypothesis of oscillations now proposed.

The luminiferous atoms may now be supposed to be diffused throughout all space, and, as molecular nuclei, throughout all bodies; the distribution and motion of their centres being regulated by forces wholly independent of that species of polarity which is the means of transmitting a state of oscillation round those centres.

Section III.—Of the Diffraction of Plane-polarised Light, and the Relation of Axes of Oscillation to Planes of Polarisation.

In the diffraction of an oscillatory movement round transverse axes past the edge of an obstacle, a law holds good exactly analogous to that demonstrated by Professor Stokes for a transverse vibratory movement, substituting only the axis of oscillation for the direction of vibration—that is to say:

The direction of the axes of oscillation in the diffracted wave is the projection of that of the axes of oscillation in the incident wave on a plane tangent to the front of the diffracted wave.

Consequently, oscillations in the incident wave, round axes oblique to the diffracting edge, give rise to oscillations in the diffracted wave round axes *more nearly parallel* to the diffracting edge.

But the experiments of Professor Stokes have proved, that light polarised in a plane oblique to the diffracting edge, becomes, after diffraction, polarised in a plane *more nearly perpendicular* to the diffracting edge.

Therefore, the axes of oscillation in plane-polarised light are perpendicular to the plane of polarisation.

Therefore, the velocity of transmission of oscillations round transverse axes, through the luminiferous medium in a crystalline body, is a function simply of the direction of the axes of oscillation.

Now, if the variations of the velocity of transmission arose from variations of the coefficient of transverse polarity (denoted by C), they would depend on the direction of propagation as well as upon that of the axes of oscillation, so that the plane of polarisation would be that which contains these two directions. Since the velocity of transmission depends on the direction of the axes of oscillation only, it follows that its variations in a given crystalline medium arise wholly from variations of the moment of inertia of the luminiferous atoms, together with their loads of extraneous matter.

Consequently, the coefficient of polarity, C, for transverse axes of oscillation is the same for all directions in a given substance.

To account for the known laws of the intensity and phase of reflected and refracted light consistently with the hypothesis of oscillations, it is necessary to suppose, also, that this coefficient is the same for all substances; so that the variations of the velocities of light, and indices of refraction for different media, depend solely on those of the moments of inertia of the loaded luminiferous atoms.

There is reason to anticipate, that, upon further investigation, it will appear that this condition is necessary to the stability of the luminiferous atoms.

SECTION IV.—OF THE WAVE-SURFACE IN CRYSTALLINE BODIES.

Let the axes of co-ordinates be those of molecular symmetry in a crystalline medium.

Let M_1, M_2, M_3, be coefficients proportional to the moments of inertia of the luminiferous atoms with their loads of extraneous matter, round axes parallel to x, y, z, respectively.

Let r be a radius vector of the diverging wave-surface in the direction (α, β, γ).

Then the equation of that surface for polar co-ordinates is,

$$\frac{1}{r^4} - \frac{1}{r^2} \cdot \frac{1}{C} \{(M_2 + M_3) \cos^2 \alpha + (M_3 + M_1) \cos^2 \beta + (M_1 + M_2) \cos^2 \gamma\}$$

$$+ \frac{1}{C^2} \{M_2 M_3 \cos^2 \alpha + M_3 M_1 \cos^2 \beta + M_1 M_2 \cos^2 \gamma\} = 0;$$

and for rectangular co-ordinates,

$$\frac{1}{C^2}(x^2+y^2+z^2).(M_2 M_3 x^2+M_3 M_1 y^2+M_1 M_2 z^2)$$

$$-\frac{1}{C}\{(M_2+M_3)x^2+(M_3+M_1)y^2+(M_1+M_2)x^2\}=1.$$

The above equations are exactly those of Fresnel's wave-surface, with the following semi-axes:—

Directions.	Semi-axes.	
x,	$\sqrt{\frac{C}{M_2}}$,	$\sqrt{\frac{C}{M_3}}$;
y,	$\sqrt{\frac{C}{M_3}}$,	$\sqrt{\frac{C}{M_1}}$;
z,	$\sqrt{\frac{C}{M_1}}$,	$\sqrt{\frac{C}{M_2}}$;

the squares of the semi-axes of the wave-surface along each axis of co-ordinates being inversely proportional to the moments of inertia of the loaded luminiferous atoms in a given space round the other two axes of co-ordinates.

The plane of polarisation at each point of the wave-surface is perpendicular to the direction of greatest declivity.

The equation of the index-surface, whose radius in any direction is inversely proportional to the normal velocity of the wave, is formed from that of the wave-surface by substituting respectively,

$$C,\ \frac{1}{M_1},\ \frac{1}{M_2},\ \frac{1}{M_3},$$

for

$$\frac{1}{C},\ M_1,\ M_2,\ M_3,$$

These equations are obtained on the supposition that the coefficient of polarity for axes of oscillation parallel to the direction of propagation, (which we may call A), is either very large, or very small, compared with that for transverse axes. By treating the ratio of these quantities as finite, there is obtained an equation of the sixth order, representing a wave-surface of three sheets, differing somewhat from that of the propagation of vibrations in an elastic crystalline solid; inasmuch as the former has always three circular sections, while the latter has none, unless it is symmetrical all round one axis at least. By increasing the ratio $\frac{A}{C}$

without limit, this equation is made to approximate indefinitely to the product of the equation of Fresnel's wave-surface by the following—

$$\frac{M_1}{A}x^2 + \frac{M_2}{A}y^2 + \frac{M_3}{A}z^2 = 1;$$

which represents a very large ellipsoidal wave of oscillations round axes parallel to the direction of propagation.

SECTION V.—OF REFLEXION AND REFRACTION.

According to the proposed hypothesis of oscillations, the laws of the phase and intensity of light reflected and refracted at the bounding surface of two transparent substances, are to be determined by conditions analogous to those employed in the hypothesis of vibrations by M. Cauchy and Mr. Green. They are the consequences of the principle, that if we have two sets of formulæ expressing the nature and magnitude of the oscillations in the two substances respectively, then either of those formulæ, being applied to a particle at the bounding surface, ought to give the same results.

According to this principle, the following six quantities for a particle at the bounding surface must be the same at every instant, when computed by either of the two sets of formulæ :—

The three angular displacements round the three axes of co-ordinates,

The three rotative forces round the same three axes.

There is, generally speaking, a change of phase when light undergoes refraction or reflexion. It is known that we may express this change of phase by subdividing each reflected or refracted disturbance into two, of suitable intensities and signs; one synchronous in phase with the corresponding incident disturbance, and the other retarded by a quarter of an undulation. There are thus twelve quantities to be found—viz., the amplitudes of the six components of the reflected disturbance, and those of the six components of the refracted disturbance. To determine these quantities there are twelve conditions—viz., the equality at every instant, according to the formulæ for either medium, of the total angular displacements, and of the total rotative forces, round each of the three axes of co-ordinates, for the set of waves composed of the incident wave and those synchronous with it, and for the set of waves retarded by one quarter of an undulation.

The results of these conditions have been investigated in detail for singly refracting substances.

The indices of refraction of such substances are proportional to the square roots of the moments of inertia of the loaded luminiferous atoms in a given space. Thus, if the coefficients M′, M″ are proportional to these moments in two given substances respectively, then the index of refraction of the second substance relatively to the first is

$$\mu = \sqrt{\frac{M''}{M'}}.$$

In the case of light incident on a plane surface between two such media, the axes of co-ordinates may be assumed respectively perpendicular to the reflecting surface, perpendicular to the plane of reflexion, and along the intersection of those two planes; and oscillations round axes normal and parallel to the plane of reflexion may be considered separately.

When the axes of oscillation are normal to the plane of reflexion—that is to say, when the light is polarised in that plane—the formulæ for the intensities of the reflected and refracted light agree exactly with those of Fresnel. When the reflexion takes place in the rarer medium, the reflected light is retarded by half an undulation; when in the denser, there is no change of phase, unless the reflexion is total, when there is a certain acceleration of phase depending on the angle of incidence. In the last case, the disturbance in the second medium is an *evanescent wave*, analogous to those introduced into the vibratory theory by M. Cauchy and Mr. Green—that is to say, a wave in which the amplitude of oscillation diminishes in proportion to an exponential function of the distance from the bounding surface (called by M. Cauchy the *modulus*), and which travels along that surface with a velocity less than the velocity of an ordinary wave; the square of the negative exponent of the modulus being proportional to the difference of the squares of those velocities, divided by the square of the velocity of an ordinary wave.

This is an evanescent wave of oscillation round transverse axes.

How large soever the coefficient of polarity for oscillations round longitudinal axes may be, an evanescent wave of such oscillations may travel along the bounding surface of a medium with any velocity, however slow, provided the negative exponent of the modulus is made large enough. Consequently, in framing the formulæ to represent oscillations round axes parallel to the plane of incidence, we must introduce in each medium two such evanescent waves of suitable exponents and indeterminate amplitudes; one travelling along the surface with the incident wave, and the other a quarter of an undulation behind it. The maximum amplitudes of oscillation in these evanescent waves constitute four unknown quantities; the amplitudes in the two ordinary reflected waves, and the two ordinary refracted waves, differing by one quarter of an undulation, constitute four more unknown quantities, making eight in all: four conditions having been fulfilled by the waves polarised in the plane of incidence, there remain to be fulfilled eight conditions—viz., the identity, as calculated by the formulæ for the first and second substance respectively, of the following eight functions at the bounding surface; the angular displacement, and the rotative forces, round each of the two axes in the plane of incidence, for the incident wave, and the set of waves synchronous with it, and for the set of waves retarded by one quarter of an undulation. These conditions are sufficient to determine the unknown quantities, and to complete the solution of the problem.

The following is a general statement of the results of the solution when the second medium is the denser. They agree with the results of the experiments of M. Jamin, and are, in every respect, analogous to those deduced from the hypothesis of vibrations by M. Cauchy, Mr. Green, and Mr. Haughton.

Light polarised in a plane perpendicular to the plane of incidence, suffers by reflexion at a perpendicular incidence no alteration of phase.

At a grazing incidence (or when the angle of incidence differs insensibly from 90°), the phase, like that of light polarised in the plane of incidence, is retarded by half an undulation.

The variation of phase with the angle of incidence is, in fact, continuous; but it is, generally speaking, not appreciable by observation, except in the immediate neighbourhood of an angle, called by M. Jamin the *principal incidence*, where the retardation of phase is a quarter of an undulation.

This angle differs by a very small amount, appreciable only in certain substances, from the *polarising angle*, at which the intensity of light polarised in a plane at right angles to the plane of incidence is a minimum.

The "law of Brewster," that *the tangent of the polarising angle is equal to the index of refraction*, is, theoretically, only approximately true; but the error is quite inappreciable.

When the second medium is the less dense, the phase of the reflected light is half an undulation in advance of its value when the second medium is the denser.

In either case, light polarised in planes perpendicular to the plane of incidence is less retarded—that is to say, is accelerated in phase—as compared with light polarised in that plane, according to the following table:—

	Angle of Incidence.	Relative Acceleration.
	°	
Perpendicular incidence, . .	0	$\frac{1}{2}$ undulation.
Principal incidence, . . .	...	$\frac{1}{4}$ undulation.
Grazing incidence, . . .	90	0·

In the case of total reflexion, light polarised in planes perpendicular to the plane of incidence, has its phase more accelerated than light polarised in that plane, by an amount to which the formulæ of Fresnel give a close approximation.

The proposed hypothesis has not yet been applied to reflexion from doubly refracting crystals; but there can be little doubt that it will be found to represent the phenomena correctly.

Section VI.—Of Circular and Elliptic Polarisation.

Light polarised in a plane oblique to the angle of incidence is, generally speaking, elliptically polarised after reflexion, the plane-polarised components of the disturbance being in different phases.

According to the hypothesis of oscillations, circularly and elliptically polarised light, being compounded of oscillations in different phases round two transverse axes, consist in a sort of *nutation* of the longitudinal axis of each luminiferous atom. The direction of this nutation, and the form of the circle or ellipse described by the ends of the longitudinal axes, serve to define the character of the light. The ellipse of nutation has its axes in the same proportion with, but perpendicular in position to, those of the elliptic orbit supposed to be described by each atom according to the hypothesis of vibrations.

The molecular mechanism by which certain media transmit right and left-handed circularly or elliptically polarised light with different velocities, is still problematical, according to either hypothesis. The laws of the phenomena, however, may be represented by means of the assumption, that in the substances in question the extraneous load on the luminiferous atoms is a function of the direction of nutation.

Section VII.—Of Dispersion.

If we assume the extent of sensible direct action of the polarity of the luminiferous atoms to be appreciable, as compared with the length of a wave, the velocity of propagation (precisely as with the vibratory hypothesis) is found to consist of a constant quantity, diminished by the sum of a series in terms of the reciprocal of the square of the length of a wave.

It may be doubted, however, whether this supposition is of itself adequate to explain the phenomena of dispersion, and whether it may not be necessary to assume, also, that the load upon the luminiferous atoms is a function of the time of oscillation, as well as of the nature of the substance and the position of the axes of oscillation.

In conclusion, it may be affirmed, that, as a mathematical system, the proposed theory of oscillations round axes represents the laws of all the phenomena which have hitherto been reduced to theoretical principles, as well, at least, as the existing theory of vibrations; while, as a physical hypothesis, it is free from the principal objections to which the hypothesis of vibrations is liable.

IX.—ON THE VELOCITY OF SOUND IN LIQUID AND SOLID BODIES OF LIMITED DIMENSIONS, ESPECIALLY ALONG PRISMATIC MASSES OF LIQUID.*

Introductory Remarks.

1. The velocity of sound in elastic substances of different kinds, solid, liquid, and gaseous, has been made the subject of numerous and careful experiments, most of which are well known. The object of this investigation is to determine to what extent our present knowledge of the condition and properties of elastic bodies, and of the laws of elasticity, enables us to use those experiments as data for calculating the elasticity of the materials; and, also, to point out circumstances which, so far as I am aware, have been insufficiently attended to, if not altogether overlooked, in previous theoretical researches, and which must limit our power of drawing definite conclusions from those experiments, until our knowledge of molecular forces shall be in a more advanced state.

2. If it were possible for us to ascertain by experiment the velocities of transmission of vibratory movements along the axes of elasticity of an indefinitely extended mass of any substance, we could at once calculate the coefficients of elasticity of that material; for in such a mass we can assign the direction of vibratory movement corresponding to each given direction of transmission, and consequently the nature of the molecular forces which are called into play, and whose intensity is indicated by the velocity of transmission. In an uncrystallised medium, for instance, the direction of vibration must either be exactly longitudinal or exactly transverse with respect to the direction of transmission, so that we can calculate from the velocity of transmission the longitudinal or the transverse elasticity, as the case may be. In a crystalline medium having rectangular axes of elasticity, the directions of vibration, though not always exactly longitudinal or transverse, unless the direction of transmission coincides with an axis, have still certain definite positions.

3. It is only in air and water, however, that such experiments are possible. For other substances, the best experiments which it is practicable for us to make, are those upon the transmission of nearly longitudinal vibrations along prismatic or cylindrical bodies. Were we able

* Read before the British Association at Ipswich, July 3, 1851, and published in the *Cambridge and Dublin Mathematical Journal*, Nov., 1851.

to ensure that the vibrations of those prisms and cylinders should be exactly longitudinal, we might compute from their velocity of transmission, as from that of such vibrations in an unlimited mass, the true longitudinal elasticity. This we can do for gaseous substances, as M. Wertheim has proved (*Ann. de Chim. et de Phys.*, Sér. III., tom. xxiii.), by making the organ-pipes in which they vibrate of proper construction.

In liquid and solid columns, on the other hand, it is impossible to prevent a certain amount of lateral vibration of the particles, the effect of which is to diminish the velocity of transmission in a ratio depending on circumstances in the molecular condition of the superficial particles, which are yet almost entirely unknown.

4. It has, indeed, been sometimes supposed, that the coefficient of elasticity, as calculated from the vibrations of a solid rod, is that called the *weight of the modulus of elasticity*—that is to say, the reciprocal of the fraction by which the length of a rod is increased by a tension applied to its ends of unity of weight upon unity of area; that coefficient being less than the true coefficient of longitudinal elasticity, because the lateral collapsing of the particles enables them to yield more in a longitudinal direction to a given force, than if their displacements were wholly longitudinal.

This conjecture, however, is inconsistent with the mechanics of vibratory movement; and, accordingly, experiment has shown that the elasticity corresponding to the velocity of sound in a rod agrees neither with the modulus of elasticity, nor with the true longitudinal elasticity; although it is, in some cases, nearly equal to the former of those quantities, and in others to the latter.

5. In liquids, it has been shown by the experiments of M. Wertheim (*Ann. de Chim. et de Phys.*, Sér. III., tom. xxiii.), that the velocity of sound in a mass contained in a trough, and set in motion through an organ-pipe, bears to that in an unlimited mass the ratio of $\sqrt{2}$ to $\sqrt{3}$. This has led him to form the conjecture, that liquids possess a momentary rigidity for very small molecular displacements as great in comparison with their other elastic forces as that of solids. This conjecture, paradoxical as it may seem, would indeed be necessary to account for the facts, if the supposition I have already mentioned were true, that the velocity of sound in a rod depends upon the modulus of elasticity. I shall show, however, in the sequel, that if we suppose that at the free surface of every mass of liquid, an atmosphere of its own vapour is retained by molecular attraction under certain conditions of equilibrium, the ratio $\sqrt{2} : \sqrt{3}$ between the velocities of sound in a prism and an unlimited mass, is a consequence of the equations of motion in all cases in which the liquid has any rigidity whatsoever, even although so small as to be insensible by any means of observation; so that the supposition of a

rigidity for small displacements equal to that of solids becomes unnecessary.

6. With respect to solids, all that theory is yet adequate to show us is, that the velocity of sound along a rod must be less than in an unlimited mass, a conclusion in accordance with experiment. The precise ratio depends on properties of the superficial particles yet unknown.

General Equations of Vibratory Movement in Homogeneous Bodies.

7. Having now stated generally the objects of this paper, I shall proceed, in the first place, to the mathematical investigation of the integrals of the general differential equations of vibratory movement in homogeneous bodies; because, although those equations have already been integrated by many mathematicians, it will be necessary in this paper to introduce functions into the integrals which have hitherto been almost totally neglected in such researches; having been applied only to the theory of waves rolling by the influence of gravity, to that of total reflexion, by Mr. Green (*Camb. Trans.*, Vol. VI.), and by Professor Stokes to represent the gradual extinction of sound by its conversion into heat.

8. Let g represent the accelerating force of gravity:

D the weight of unity of volume of a homogeneous substance, having orthogonal axes of elasticity whose directions are the same throughout its extent;

A_1, A_2, A_3, the coefficients of longitudinal elasticity for the axes of x, y, z, respectively:

B_1, B_2, B_3, the coefficients of lateral elasticity; and

C_1, C_2, C_3, those of rigidity for the planes of yz, zx, xy, respectively;

ξ, η, ζ, the displacements of a particle parallel to x, y, z, respectively.

Then, it is well known that the differential equations of small vibratory movements are the following, when small quantities of the second order are neglected:

$$\left.\begin{aligned} 0 &= \left(-\frac{D}{g}\cdot\frac{d^2}{dt^2} + A_1\frac{d^2}{dx^2} + C_3\frac{d^2}{dy^2} + C_2\frac{d^2}{dz^2}\right)\xi \\ &\quad + (B_3 + C_3)\frac{d^2\eta}{dx\,dy} + (B_2 + C_2)\frac{d^2\zeta}{dz\,dx} \\ 0 &= \left(-\frac{D}{g}\cdot\frac{d^2}{dt^2} + C_3\frac{d^2}{dx^2} + A_2\frac{d^2}{dy^2} + C_1\frac{d^2}{dz^2}\right)\eta \end{aligned}\right\} \quad (1.)$$

$$
\left.\begin{aligned}
&+ (B_1 + C_1)\frac{d^2 \zeta}{dy\,dz} + (B_3 + C_3)\frac{d^2 \xi}{dx\,dy} \\
0 = &\left(-\frac{D}{g}\cdot\frac{d^2}{dt^2} + C_2\frac{d^2}{dx^2} + C_1\frac{d^2}{dy^2} + A_3\frac{d^2}{dz^2}\right)\zeta \\
&+ (B_2 + C_2)\frac{d^2 \xi}{dz\,dx} + (B_1 + C_1)\frac{d^2 \eta}{dy\,dz},
\end{aligned}\right\} \quad . \quad (1.)
$$

of which the integrals are

$$
\left.\begin{aligned}
\xi &= \Sigma\{L_1 \phi(\sqrt{\varepsilon}.t + \alpha x + \beta y + \gamma z + \kappa)\} \\
\eta &= \Sigma\{L_2 \phi(\sqrt{\varepsilon}.t + \alpha x + \beta y + \gamma z + \kappa)\} \\
\zeta &= \Sigma\{L_3 \phi(\sqrt{\varepsilon}.t + \alpha x + \beta y + \gamma z + \kappa)\}.
\end{aligned}\right\} \quad . \quad (2.)
$$

The form of the function ϕ being arbitrary, subject to a restriction to be afterwards referred to, and Σ extending to any number of terms, the coefficients of which fulfil the following conditions. Let

$$
\left.\begin{aligned}
\omega_1 &= A_1\alpha^2 + C_3\beta^2 + C_2\gamma^2 \\
\omega_2 &= C_3\alpha^2 + A_2\beta^2 + C_1\gamma^2 \\
\omega_3 &= C_2\alpha^2 + C_1\beta^2 + A_3\gamma^2 \\
\rho_1 &= (B_1 + C_1)\beta\gamma \\
\rho_2 &= (B_2 + C_2)\gamma\alpha \\
\rho_3 &= (B_3 + C_3)\alpha\beta \\
\frac{\varepsilon D}{g} &= E.
\end{aligned}\right\} \quad . \quad . \quad (a.)
$$

Then the following equations must be satisfied by the coefficients of each set of terms in equation (2):

$$
\left.\begin{aligned}
0 &= L_1(\omega_1 - E) + L_2\rho_3 + L_3\rho_2 \\
0 &= L_1\rho_3 + L_2(\omega_2 - E) + L_3\rho_1 \\
0 &= L_1\rho_2 + L_2\rho_1 + L_3(\omega_3 - E).
\end{aligned}\right\} \quad . \quad . \quad (3.)
$$

By elimination we transform those equations as follows: let

$$
\left.\begin{aligned}
G &= \omega_1 + \omega_2 + \omega_3, \\
H &= \omega_2\omega_3 + \omega_3\omega_1 + \omega_1\omega_2 - \rho_1^2 - \rho_2^2 - \rho_3^2, \\
K &= \omega_1\omega_2\omega_3 + 2\rho_1\rho_2\rho_3 - \omega_1\rho_1^2 - \omega_2\rho_2^2 - \omega_3\rho_3^2.
\end{aligned}\right\} \quad (b.)
$$

Then for each set of values of α, β, γ, E has three values which are the roots of the cubic equation,

$$0 = \mathrm{E}^3 - \mathrm{G}\,\mathrm{E}^2 + \mathrm{H}\,\mathrm{E} - \mathrm{K}\,; \quad . \quad . \quad . \quad (4.)$$

so that $\sqrt{\varepsilon}$ has six values, three positive and three negative, of equal arithmetical amount.

The absolute values of L_1, L_2, L_3, are arbitrary, but their mutual ratios are fixed by the following equations:

$$\left.\begin{aligned} \mathrm{L}_1\{(\omega_1-\mathrm{E})\rho_1-\rho_2\rho_3\} &= \mathrm{L}_2\{(\omega_2-\mathrm{E})\rho_2-\rho_3\rho_1\} \\ &= \mathrm{L}_3\{(\omega_3-\mathrm{E})\rho_3-\rho_1\rho_2\}\,; \end{aligned}\right\} \quad (5.)$$

consequently, they have in general three sets of ratios for each set of values of α, β, γ, corresponding to the three values of E.

9. The condition that the motions of the particles of the body must be *small oscillations* restricts the variations of the displacements ξ, η, ζ, within certain limits. Now, as the time t increases *ad infinitum*, this can be fulfilled only when each of those quantities is either a periodical circular function of t, or a function developable into a sum or definite integral of such functions. We may, therefore, make each of the functions ϕ a trigonometrical function of t. This being the case, those functions must be either trigonometrical or exponential with respect to x, y, and z, or compounded of both, being trigonometrical so far as α, β, γ, are real, and exponential so far as they are imaginary.

If we suppose each of these coefficients to consist of a real and an imaginary part, then each of their functions which enters into the equations of condition, will also consist of a real and an imaginary part. Each of the equations of condition thus becomes divided into two, which must be separately satisfied.

Thus we arrive at the following results:

For the symbol $\phi\ \{\ \}$, put $e^{2\pi\sqrt{-1}\{\ \}}$; so as to make ξ, &c., trigonometrical with respect to t. Let λ be a line of such a length that

$$a^2 + b^2 + c^2 = 1,$$

and let

$$\alpha = \frac{1}{\lambda}(\mp a - a'\sqrt{-1}),$$

$$\beta = \frac{1}{\lambda}(\mp b - b'\sqrt{-1}),$$

$$\gamma = \frac{1}{\lambda}(\mp c - c'\sqrt{-1}):$$

also, let

$$L_1 = l \mp l'\sqrt{-1},\ L_2 = m \mp m'\sqrt{-1},\ L_3 = n \mp n'\sqrt{-1};$$

so that the displacements become

$$\left.\begin{aligned} \xi &= \Sigma\{(l \mp l'\sqrt{-1})\,e^{\frac{2\pi}{\lambda}\{a'x+b'y+c'z\pm\sqrt{-1}(\sqrt{\epsilon}\,.\,t-ax-by-cz)\}} \\ \eta &= \Sigma\{\text{terms in } m,\, m'\},\quad \zeta = \Sigma\{\text{terms in } n,\, n'\}. \end{aligned}\right\} \quad (7.)$$

Let the quantities in the equations of condition be thus represented:

$$\omega_1 = p_1 \pm p'_1\sqrt{-1},\ \&c.;\quad \rho_1 = q_1 \pm q'_1\sqrt{-1},\ \&c.;$$

$$G = \mathfrak{g} \pm \mathfrak{g}'\sqrt{-1};\ H = \mathfrak{h} \pm \mathfrak{h}'\sqrt{-1};\ K = \mathfrak{k} \pm \mathfrak{k}'\sqrt{-1}.$$

The equations of notation now become

$$\left.\begin{aligned}
p_1 &= A_1(a^2-a'^2) + C_3(b^2-b'^2) + C_2(c^2-c'^2) \\
p_2 &= C_3(a^2-a'^2) + A_2(b^2-b'^2) + C_1(c^2-c'^2) \\
p_3 &= C_2(a^2-a'^2) + C_1(b^2-b'^2) + A_3(c^2-c'^2) \\
q_1 &= (B_1+C_1)(bc-b'c') \\
q_2 &= (B_2+C_2)(ca-c'a') \\
q_3 &= (B_3+C_3)(ab-a'b') \\
p_1' &= 2(A_1aa' + C_3bb' + C_2cc') \\
p_2' &= 2(C_3aa' + A_2bb' + C_1cc') \\
p_3' &= 2(C_2aa' + C_1bb' + A_3cc') \\
q_1' &= (B_1+C_1)(bc'+b'c) \\
q_2' &= (B_2+C_2)(ca'+c'a) \\
q_3' &= (B_3+C_3)(ab'+a'b) \\
\frac{\epsilon D}{g} &= E \text{ as before, or } \epsilon = \frac{Eg}{D} \\
\mathfrak{g} &= p_1+p_2+p_3,\quad \mathfrak{g}' = p_1'+p_2'+p_3' \\
\mathfrak{h} &= p_2p_3+p_3p_1+p_1p_2-q_1^2-q_2^2-q_3^2 \\
&\quad -p_2'p_3'-p_3'p_1'-p_1'p_2'+q_1'^2+q_2'^2+q_3'^2 \\
\mathfrak{h}' &= p_2p_3'+p_2'p_3+p_3p_1'+p_3'p_1+p_1p_2'+p_1'p_2 \\
&\quad -2q_1q_1'-2q_2q_2'-2q_3q_3' \\
\mathfrak{k} &= p_1p_2p_3+2q_1q_2q_3-p_1q_1^2-p_2q_2^2-p_3q_3^2
\end{aligned}\right\} \quad (c.)$$

$$
\left.\begin{aligned}
&-p_1 p_2' p_3' - p_1' p_2 p_3' - p_1' p_2' p_3 \\
&-2\,(q_1 q_2' q_3' + q_1' q_2 q_3' + q_1' q_2' q_3) \\
&+p_1 q_1'^2 + p_2 q_2'^2 + p_3 q_3'^2 + 2\,(p_1' q_1 q_1' + p_2' q_2 q_2' + p_3' q_3 q_3') \\
\mathfrak{k}' = {} &p_1' p_2' p_3' + 2\, q_1' q_2' q_3' - p_1' q_1'^2 - p_2' q_2'^2 - p_1' q_3'^2 \\
&-p_1' p_2 p_3 - p_1 p_2' p_3 - p_1 p_2 p_3' \\
&-2\,(q_1' q_2 q_3 + q_1 q_2' q_3 + q_1 q_2 q_3') \\
&+p_1' q_1^2 + p_2' q_2^2 + p_3' q_3^2 + 2\,(p_1 q_1 q_1' + p_2 q_2 q_2' + p_3 q_3 q_3'). \\
&\text{Also, let} \\
\mathfrak{x}_1 &= (p_1 - \mathrm{E})\, q_1 - p_1' q_1' - q_2 q_3 + q_2' q_3' \\
\mathfrak{x}_2 &= (p_2 - \mathrm{E})\, q_2 - p_2' q_2' - q_3 q_1 + q_3' q_1' \\
\mathfrak{x}_3 &= (p_3 - \mathrm{E})\, q_3 - p_3' q_3' - q_1 q_2 + q_1' q_2' \\
\mathfrak{x}_1' &= (p_1 - \mathrm{E})\, q_1' + p_1' q_1 - q_2 q_3' - q_2' q_3 \\
\mathfrak{x}_2' &= (p_2 - \mathrm{E})\, q_2' + p_2' q_2 - q_3 q_1' - q_3' q_1 \\
\mathfrak{x}_3' &= (p_3 - \mathrm{E})\, q_3' + p_3' q_3 - q_1 q_2' - q_1' q_2
\end{aligned}\right\} \quad \text{(c.)}
$$

Then the equations of condition relative to the coefficients become the following :—

$$0 = \mathrm{E}^3 - \mathfrak{g}\,\mathrm{E}^2 + \mathfrak{h}\,\mathrm{E} - \mathfrak{k}, \quad . \quad . \quad . \quad (8.)$$

$$0 = \mathfrak{g}'\,\mathrm{E}^2 - \mathfrak{h}'\,\mathrm{E} + \mathfrak{k}', \quad . \quad . \quad . \quad . \quad (9.)$$

$$
\left.\begin{aligned}
l\,\mathfrak{x}_1 + l'\,\mathfrak{x}_1' &= m\,\mathfrak{x}_2 + m'\,\mathfrak{x}_2' = n\,\mathfrak{x}_3 + n'\,\mathfrak{x}_3' \\
l\,\mathfrak{x}_1' - l'\,\mathfrak{x}_1 &= m\,\mathfrak{x}_2' - m'\,\mathfrak{x}_2 = n\,\mathfrak{x}_3' - n'\,\mathfrak{x}_3.
\end{aligned}\right\} \quad . \quad . \quad (10.)
$$

The three original equations of condition are transformed into the following six, to which (8), (9), (10) are equivalent :—

$$
\left.\begin{aligned}
0 &= l\,(p_1 - \mathrm{E}) + l' p_1' + m\, q_3 + m' q_3' + n\, q_2 + n' q_2' \\
0 &= l'(p_1 - \mathrm{E}) - l\, p_1' + m' q_3 - m\, q_3' + n' q_2 - n\, q_2' \\
0 &= l\, q_3 + l' q_3' + m\,(p_2 - \mathrm{E}) + m' p_2' + n\, q_1 + n' q_1' \\
0 &= l' q_3 - l\, q_3' + m'(p_2 - \mathrm{E}) - m\, p_2' + n' q_1 - n\, q_1' \\
0 &= l\, q_2 + l' q_2' + m\, q_1 + m' q_1' + n\,(p_3 - \mathrm{E}) + n' p_3' \\
0 &= l' q_2 - l\, q_2' + m' q_1 - m\, q_1' + n'(p_3 - \mathrm{E}) - n\, p_3'
\end{aligned}\right\} \quad (10\,\mathrm{A.})
$$

To give an intelligible result, the terms of the series in equation (7) must be taken in pairs, with the imaginary exponents in each pair of equal arithmetical value and opposite signs.

Hence equations (7) are equivalent to the following:—

$$\left.\begin{array}{l}\xi=\Sigma\left[e^{\frac{2\pi}{\lambda}(a'x+b'y+c'z)}\left\{l.\cos\frac{2\pi}{\lambda}(\sqrt{\varepsilon}.t-ax-by-cz)\right.\right.\\ \qquad\qquad\left.\left.+l'.\sin\frac{2\pi}{\lambda}(\sqrt{\varepsilon}.t-ax-by-cz)\right\}\right]\\ \eta=\Sigma\{\text{terms in } m, m'\},\quad \zeta=\Sigma\{\text{terms in } n, n'\}\end{array}\right\}\quad(11.)$$

The above equations (11), together with the equations of condition (8), (9), (10), or their equivalent (10A), and the equations of notation (c), contain the complete representation of the laws of small molecular oscillations in a homogeneous body of any dimensions and figure; it being understood that in the symbol of summation Σ are included as many definite integrations as the problem may require with respect to independent variables of which the coefficients λ, $\sqrt{\varepsilon}, a, b, c, a', b', c', l, m, n, l', m', n'$, are functions.

As there are fourteen coefficients, connected by seven equations—viz., $a^2+b^2+c^2=1$, and the six equations of condition, the greatest number of independent variables is limited to seven; therefore, in the most general case, the symbol $\Sigma\ \{\ldots\}$ in equations (11) may be replaced by

$$\Sigma\int\!\!\int\!\!\int\!\!\int\!\!\int\!\!\int\!\!\int \mathrm{F}(\theta_1,\theta_2,\theta_3,\theta_4,\theta_5,\theta_6,\theta_7)\{\ldots\}d\theta_1, d\theta_2, d\theta_3, d\theta_4, d\theta_5, d\theta_6, d\theta_7, \quad (12.)$$

θ_1, &c., being variables of which the coefficients are functions, and F an arbitrary function.*

10. Let us consider the physical meaning of a single set of terms of the sums in equations (11), containing but one set of values of the coefficients. It represents a system of plane waves, the wave surfaces, or planes of equal phase, in which are normal to the line whose direction-cosines are a, b, c. λ is the *length of a wave* measured along that line.

$$\left.\begin{array}{l}\frac{2\pi}{\lambda}(\sqrt{\varepsilon}.t-ax-by-cz)+\tan^{-1}\frac{l}{l'}\\ \frac{2\pi}{\lambda}(\sqrt{\varepsilon}.t-ax-by-cz)+\tan^{-1}\frac{m}{m'}\\ \frac{2\pi}{\lambda}(\sqrt{\varepsilon}.t-ax-by-cz)+\tan^{-1}\frac{n}{n'}\end{array}\right\}\text{are the \textit{phases} of vibration for}\left\{\begin{array}{l}\xi,\\ \eta,\\ \zeta.\end{array}\right.$$

* To make the functions in equations (11) satisfy the conditions of equilibrium, instead of those of oscillation, it is only necessary to make $\varepsilon=0$, and to substitute $k=0, k'=0$, for equations (8) and (9). Some additional functions, however, are necessary in order to complete the values of ξ, η, ζ.

$\sqrt{\varepsilon} = \sqrt{\left(\frac{\mathrm{E}g}{\mathrm{D}}\right)}$ is the normal velocity of propagation, $\frac{\lambda}{\sqrt{\varepsilon}}$ is the periodic time of an oscillation of a particle

$$\sqrt{(l^2 + l'^2)} \cdot e^{\frac{2\pi}{\lambda}(a'x + b'y + c'z)},$$

and the corresponding expressions in m and n are the semi-amplitudes of vibration parallel to x, y, z, respectively; a', b', c', are proportional to the direction-cosines of a normal to a series of planes of equal amplitude of vibration.

The trajectory of each particle affected by a single series of plane waves is in general an ellipse, the position and magnitude of which are found as follows. Let ϕ_0 denote the value of $\frac{2\pi}{\lambda}(\sqrt{\varepsilon} \cdot t - a x - b y - c z)$, which makes the total displacement $\sqrt{(\xi^2 + \eta^2 + \zeta^2)}$ a maximum or minimum. It is easily seen that

$$\tan \phi = -\frac{s}{2} \pm \sqrt{\left(1 + \frac{s^2}{4}\right)}, \quad . \quad . \quad . \quad (13.)$$

where

$$s = \frac{l^2 + m^2 + n^2 - l'^2 - m'^2 - n'^2}{l\,l' + m\,m' + n\,n'}.$$

The values of ξ, η, ζ, calculated from ϕ_0 by equations (11), are the co-ordinates of the extremities of the axes of the elliptic trajectory, referred to the natural position of the particle as origin.

The processes of summation and definite integration denote the representation of an arbitrary manner of oscillation by the combination of a definite or indefinite number of systems of plane waves.

Case of an Indefinitely Extended Medium.

11. Let the medium, in the first place, be supposed to be indefinitely extended in all directions. This case having been thoroughly investigated by MM. Poisson, Cauchy, Green, MacCullagh, Haughton, Stokes, and others, I shall give merely an outline of the general results. The condition that the motion shall consist of small oscillations, here makes it necessary that the exponential factor in the displacements should in all cases be equal to unity, and therefore that

$$a' = 0; \quad b' = 0; \quad c' = 0;$$

and, consequently, each of the accented symbols in equations $(c) = 0$. Equation (9) vanishes, and the normal velocity of propagation for each set of direction-cosines a, b, c, has, generally speaking, three values, corresponding to the three values of E, roots of equation (8). Equations (10) become

$$l : m : n :: l' : m' : n' :: \frac{1}{\mathfrak{r}_1} : \frac{1}{\mathfrak{r}_2} : \frac{1}{\mathfrak{r}_3}, \quad . \quad . \quad (14.)$$

consequently, the phases of ξ, η, ζ, are simultaneous; so that $\sqrt{(l^2+l'^2)}$, $\sqrt{(m^2+m'^2)}$, $\sqrt{(n^2+n'^2)}$, are proportional to the direction-cosines of a rectilinear vibratory movement of the semi-amplitude $\sqrt{(l^2+l'^2+m^2+m'^2+n^2+n'^2)}$, which cosines have in general three sets of values corresponding to the three values of E. It is easily shown that those three directions are at right angles to each other. The number of coefficients being in this case reduced to eleven, connected by six equations—viz., $a^2+b^2+c^2=1$, equation (8), and the proportional equation (14), which is equivalent to four, the greatest number of definite integrations in the operation (12) is restricted to five.

Thus it appears that the velocity of transmission of vibratory movement through an indefinitely extended mass, has a set of definite values, not exceeding three, for each position of plane waves. When the direction of propagation coincides with an axis of elasticity, we find those values to be:

For vibrations parallel to	Velocity of propagation along		
	x	y	z
x . .	$\sqrt{\left(\frac{A_1 g}{D}\right)}$	$\sqrt{\left(\frac{C_3 g}{D}\right)}$	$\sqrt{\left(\frac{C_2 g}{D}\right)}$
y . .	$\sqrt{\left(\frac{C_3 g}{D}\right)}$	$\sqrt{\left(\frac{A_2 g}{D}\right)}$	$\sqrt{\left(\frac{C_1 g}{D}\right)}$
z . .	$\sqrt{\left(\frac{C_2 g}{D}\right)}$	$\sqrt{\left(\frac{C_1 g}{D}\right)}$	$\sqrt{\left(\frac{A_3 g}{D}\right)}$

(15.)

When the substance is equally elastic in all directions, we have simply, Velocities of propagation in any direction for longitudinal vibrations

$$\sqrt{\left(\frac{A g}{D}\right)}.$$

For transverse vibrations, in any direction perpendicular to that of propagation,

$$\sqrt{\left(\frac{\mathrm{C}\,g}{\mathrm{D}}\right)}.$$

Hence, experiments on the velocity of sound in an indefinitely extended mass, or one so large as to be practically such, afford the means of directly calculating the coefficients of elasticity.

General Case of a Body of Limited Dimensions.

12. It is not so, however, in a body of limited dimensions; for the coefficients a', b', c', in the exponents of the exponential factors, are no longer necessarily null, but have values which must depend on the molecular condition of the external surface of the body, and on the forces applied to it. The velocity of propagation is no longer a function of the direction-cosines a, b, c, alone, but also of the coefficients a', b', c'. It has in general but one value, corresponding to the common root of the equations (8) and (9). By substituting successively the two roots of equation (9)—viz.,

$$\frac{\mathfrak{h}'}{2\,\mathfrak{g}'} \pm \sqrt{\left(\frac{\mathfrak{h}'^2}{4\,\mathfrak{g}'^2} - \frac{\mathfrak{k}'}{\mathfrak{g}'}\right)},$$

for E in equation (8), the latter is converted into two *alternative* equations between the six quantities a, b, c, a', b', c', showing the relations they must have in order that equations (8) and (9) may have a common root. In the only particular problems, however, of which I shall here give the solutions, those relations are obvious without going through that process, for they belong to a class of cases in which the three quantities $\mathfrak{g}'$, $\mathfrak{h}'$, $\mathfrak{k}'$, have a common factor; which being made $=0$, the necessary conditions are fulfilled.

It is obvious that in all cases the effect of the coefficients a', b', c', is to diminish the velocity of propagation.

13. The following are the values of the three components of the velocity of a particle:

$$\left.\begin{aligned} \frac{d\xi}{dt} = \Sigma\Big[\frac{2\pi}{\lambda}\sqrt{\varepsilon}\,.\,e^{\frac{2\pi}{\lambda}(a'x+b'y+c'z)}\Big\{ & -l\sin\frac{2\pi}{\lambda}(\sqrt{\varepsilon}\,.\,t-ax-by-cz) \\ & + l'\cos\frac{2\pi}{\lambda}(\sqrt{\varepsilon}\,.\,t-ax-by-cz)\Big\}\Big] \\ \frac{d\eta}{dt} = \Sigma\,\{\text{terms in } m, m'\}\,; \quad \frac{d\zeta}{dt} = \Sigma\,\{\text{terms in } n, n'\}. & \end{aligned}\right\} \quad (16.)$$

The *strains*, or coefficients of relative molecular displacement, are as follows:

Longitudinal Strains.

Along

$$
\left.\begin{array}{l|l}
x & \dfrac{d\xi}{dx} = \Sigma\Big[\dfrac{2\pi}{\lambda} e^{\frac{2\pi}{\lambda}(a'x+b'y+c'z)}\Big\{(la'-l'a)\cos\dfrac{2\pi}{\lambda}(\sqrt{\varepsilon}.t-ax-by-cz) \\
 & \qquad\qquad + (la+l'a')\sin\dfrac{2\pi}{\lambda}(\sqrt{\varepsilon}.t-ax-by-cz)\Big\}\Big] \\
y & \dfrac{d\eta}{dy} = \Sigma\{\text{terms in } (mb'-m'b),\ (mb+m'b')\} \\
z & \dfrac{d\zeta}{dz} = \Sigma\{\text{terms in } (nc'-n'c),\ (nc+n'c')\}
\end{array}\right\} \quad (17.)
$$

Distortions.

In the plane

$$
\left.\begin{array}{l|l}
yz & \dfrac{d\eta}{dz}+\dfrac{d\zeta}{dy} = \Sigma\Big[\dfrac{2\pi}{\lambda} e^{\frac{2\pi}{\lambda}(a'x+b'y+c'z)} \\
 & \Big\{(mc'-m'c+nb'-n'b)\cos\dfrac{2\pi}{\lambda}(\sqrt{\varepsilon}.t-ax-by-cz) \\
 & +(mc+m'c'+nb+n'b')\sin\dfrac{2\pi}{\lambda}(\sqrt{\varepsilon}.t-ax-by-cz)\Big\}\Big] \\
zx & \dfrac{d\zeta}{dx}+\dfrac{d\xi}{dz} = \Sigma\{\text{terms in } (na'-n'a+lc'-l'c), \\
 & \qquad (na+n'a'+lc+l'c')\} \\
xy & \dfrac{d\xi}{dy}+\dfrac{d\eta}{dx} = \Sigma\{\text{terms in } (lb'-l'b+ma'-m'a), \\
 & \qquad (lb+l'b'+ma+m'a')\}
\end{array}\right\} \quad (17A.)
$$

The pressures on the co-ordinate planes, arising from those strains (using the notation of my paper on the Laws of Elasticity, *Cambridge and Dublin Mathematical Journal*, February, 1851), are the following:—

Normal.

On

$$yz \quad . \quad . \quad . \quad P_1 = -A_1\frac{d\xi}{dx} - B_3\frac{d\eta}{dy} - B_2\frac{d\zeta}{dz}$$

$$= -\Sigma\left[\frac{2\pi}{\lambda}e^{\frac{2\pi}{\lambda}(a'x+b'y+c'z)}\left\{\Big(A_1(la'-l'a)\right.\right.$$

$$+ B_3(mb'-m'b) + B_2(nc'-n'c)\Big) \cdot \cos\frac{2\pi}{\lambda}(\sqrt{\varepsilon}\,.\,t - ax - by - cz)$$

$$+\Big(A_1(la+l'a') + B_3(mb+m'b') + B_2(nc+n'c')\Big)$$

$$\left.\left.\sin\frac{2\pi}{\lambda}(\sqrt{\varepsilon}\,.\,t - ax - by - cz)\right\}\right] \quad . \quad . \quad (18.)$$

The other normal pressures are found by substituting symbols according to the following table :—

Plane.	Pressure.	Coefficients.
yz . . .	P_1 . . .	A_1, B_3, B_2,
zx . . .	P_2 . . .	B_3, A_2, B_1,
xy . . .	P_3 . . .	B_2, B_1, A_3.

Tangential Pressures.

Plane of Distortion	Along	On the plane	
yz	y, z	xy, zx	$Q_1 = -C_1\left(\frac{d\eta}{dz}+\frac{d\zeta}{dy}\right)$
zx	z, x	yz, xy	$Q_2 = -C_2\left(\frac{d\zeta}{dx}+\frac{d\xi}{dz}\right)$
xy	x, y	zx, yz	$Q_3 = -C_3\left(\frac{d\xi}{dy}+\frac{d\eta}{dx}\right)$

(18A.)

CASE OF AN UNCRYSTALLISED MEDIUM.

14. I shall now take the particular case of an uncrystallised medium, in which the coefficients of elasticity are the same for all axes, and may be represented thus :—

rigidity $= C$; fluid elasticity $= J$;

longitudinal elasticity $A = 3C + J$,

lateral elasticity $B = C + J = A - 2C$.

The position of the axes being in this case arbitrary, I shall take the direction of propagation as the axis of x, so as to make

$$a = 1, \quad b = 0, \quad c = 0.$$

To fulfil the condition that equations (8) and (9) shall have common roots, we must make

$$a' = 0,$$

being in this case a common factor of $\mathfrak{g}'$, $\mathfrak{h}'$, $\mathfrak{k}'$.

The equations of notation (c) now become

$$\left.\begin{aligned}
p_1 &= \mathrm{A} - \mathrm{C}(b'^2 + c'^2)\\
p_2 &= -\mathrm{A}\,b'^2 + \mathrm{C}(1 - c'^2)\\
p_3 &= -\mathrm{A}\,c'^2 + \mathrm{C}(1 - b'^2)\\
q_1 &= -(\mathrm{A} - \mathrm{C})b'c'; \; q_2 = 0; \; q_3 = 0\\
p_1' &= 0; \; p_2' = 0, \; p_3' = 0\\
q_1' &= 0; \; q_2' = (\mathrm{A} - \mathrm{C})c'; \; q_3' = (\mathrm{A} - \mathrm{C})b';\\
\mathfrak{g} &= (\mathrm{A} + 2\,\mathrm{C})(1 - b'^2 - c'^2)\\
\mathfrak{h} &= (2\,\mathrm{A}\,\mathrm{C} + \mathrm{C}^2)(1 - b'^2 - c'^2)^2\\
\mathfrak{k} &= \mathrm{A}\,\mathrm{C}^2(1 - b'^2 - c'^2)^3\\
\mathfrak{g}' &= 0; \; \mathfrak{h}' = 0; \; \mathfrak{k}' = 0\\
\mathfrak{r}_1 &= (p_1 - \mathrm{E})q_1 + q_2'q_3'; \; \mathfrak{r}_2 = 0; \; \mathfrak{r}_3 = 0;\\
\mathfrak{r}_1' &= 0; \; \mathfrak{r}_2' = (p_2 - \mathrm{E})q_2' - q_3'q_1; \; \mathfrak{r}_3' = (p_3 - \mathrm{E})q_3' - q_1 q_2'.
\end{aligned}\right\} \text{(d.)}$$

Hence it appears, that for an uncrystallised medium, equation (8) has three roots, viz.,—

$$\left.\begin{aligned}
&\text{one root} \quad . \quad . \quad \mathrm{E} = \mathrm{A}(1 - b'^2 - c'^2),\\
&\text{two equal roots, each } \mathrm{E} = \mathrm{C}(1 - b'^2 - c'^2).
\end{aligned}\right\} \quad . \quad (19.)$$

So that the velocity of propagation is less than that in an unlimited mass, in the ratio $\sqrt{(1 - b'^2 - c'^2)} : 1$. Equation (9) disappears.

Equations (10) become

$$\left.\begin{aligned}
l\,\mathfrak{r}_1 &= m'\mathfrak{r}_2' = n'\mathfrak{r}_3'\\
-l'\mathfrak{r}_1 &= m\,\mathfrak{r}_2' = n\,\mathfrak{r}_3'\\
&\text{or}\\
l : m' : n' :: -l' &: m : n :: \frac{1}{\mathfrak{r}_1} : \frac{1}{\mathfrak{r}'_2} : \frac{1}{\mathfrak{r}'_3}
\end{aligned}\right\} \quad . \quad (20.)$$

Equations (10A) become

$$\left.\begin{aligned} 0 &= l\,(p_1 - \mathrm{E}) + m' q_3' + n' q_2' \\ 0 &= l'(p_1 - \mathrm{E}) - m\, q_3' - n\, q_2' \\ 0 &= l' q_3' + m\,(p_2 - \mathrm{E}) + n\, q_1 \\ 0 &= -\,l\, q_3' + m'\,(p_2 - \mathrm{E}) + n' q_1 \\ 0 &= l' q_2' + m\, q_1 + n\,(p_3 - \mathrm{E}) \\ 0 &= -\,l\, q_2' + m' q_1 + n'\,(p_3 - \mathrm{E}) \end{aligned}\right\} \qquad (20\mathrm{A}.)$$

15. It may be shown that the vibrations corresponding to the roots C $(1-b'^2-c'^2)$ cannot take place in a body of which the surface is free, unless $b'=0$, $c'=0$, in which case they are reduced to ordinary transverse vibrations. (*See Appendix, No. II.*)

Nearly-longitudinal Vibrations in an Uncrystallised Medium.

16. For the present, therefore, I shall confine the investigation to the root

$$\mathrm{E} = \mathrm{A}\,(1 - b'^2 - c'^2),$$

corresponding to the velocity of propagation

$$\sqrt{\varepsilon} = \sqrt{\left\{\frac{\mathrm{A}g}{\mathrm{D}}(1 - b'^2 - c'^2)\right\}}. \quad . \quad . \quad (21.)$$

The vibrations to which this root is applicable may be called *nearly-longitudinal;* because in them the longitudinal component predominates, and their velocity of transmission is a function of the longitudinal elasticity A.

This value being substituted for E in the expressions for $\mathfrak{r}_1$, &c., gives

$$\left.\begin{aligned} m' &= -\,b' l; \quad n' = -\,c' l \\ m &= b' l'; \quad n' = c' l'. \end{aligned}\right\} \quad . \quad . \quad (22.)$$

Which values being substituted in equations (11), (16), (17), (18), give the following results:—

For brevity's sake, let

$$\frac{2\,\pi}{\lambda}(b' y + c' z) = \psi; \quad \frac{2\,\pi}{\lambda}(\sqrt{\varepsilon}\,.\,t - x) = \phi;$$

also, let

$$\Phi = \Sigma\left\{\frac{\lambda}{2\,\pi}e^{\psi}(l' \cos\phi - l \sin\phi)\right\}.$$

Then the displacements are:

$$\left.\begin{aligned}\xi &= \Sigma\,\{e^{\psi}\,(l\cos\phi + l'\sin\phi)\} &&= \frac{d\,\Phi}{d\,x}\\ \eta &= \Sigma\,\{b'\,e^{\psi}\,(l'\cos\phi - l\sin\phi)\} &&= \frac{d\,\Phi}{d\,y}\\ \zeta &= \Sigma\,\{c'\,e^{\psi}\,(l'\cos\phi - l\sin\phi)\} &&= \frac{d\,\Phi}{d\,z}\end{aligned}\right\} \quad (23.)$$

The velocities of the particles are:

$$\left.\begin{aligned}\frac{d\,\xi}{d\,t} &= \Sigma\left\{\frac{2\,\pi}{\lambda}\sqrt{\varepsilon}\,.\,e^{\psi}\,(l'\cos\phi - l\sin\phi)\right\}\\ \frac{d\,\eta}{d\,t} &= -\,\Sigma\left\{\frac{2\,\pi}{\lambda}\sqrt{\varepsilon}\,.\,b'\,e^{\psi}\,(l\cos\phi + l'\sin\phi)\right\}\\ \frac{d\,\zeta}{d\,t} &= -\,\Sigma\left\{\frac{2\,\pi}{\lambda}\sqrt{\varepsilon}\,.\,c'\,e^{\psi}\,(l\cos\phi + l'\sin\phi)\right\}\end{aligned}\right\} \quad (24.)$$

The longitudinal strains:

$$\frac{d\,\xi}{d\,x} = -\,\Sigma\left\{\frac{2\,\pi}{\lambda}\,e^{\psi}\,(l'\cos\phi - l\sin\phi)\right\}$$

$$\frac{d\,\eta}{d\,y} = \Sigma\left\{\frac{2\,\pi}{\lambda}\,b'^2\,e^{\psi}\,(l'\cos\phi - l\sin\phi)\right\}$$

$$\frac{d\,\zeta}{d\,z} = \Sigma\left\{\frac{2\,\pi}{\lambda}\,c'^2\,e^{\psi}\,(l'\cos\phi - l\sin\phi)\right\} \quad (25.)$$

The total change of volume:

$$\frac{d\,\xi}{d\,x} + \frac{d\,\eta}{d\,y} + \frac{d\,\zeta}{d\,z} = -\,\Sigma\left\{\frac{2\,\pi}{\lambda}(1 - b'^2 - c'^2)\,e^{\psi}\,(l'\cos\phi - l\sin\phi)\right\}$$

The distortions:

$$\left.\begin{aligned}\frac{d\,\eta}{d\,z} + \frac{d\,\zeta}{d\,y} &= 2\,\Sigma\left\{\frac{2\,\pi}{\lambda}\,b'\,c'\,e^{\psi}\,(l'\cos\phi - l\sin\phi)\right\}\\ \frac{d\,\zeta}{d\,x} + \frac{d\,\xi}{d\,z} &= 2\,\Sigma\left\{\frac{2\,\pi}{\lambda}\,c'\,e^{\psi}\,(l\cos\phi + l'\sin\phi)\right\}\\ \frac{d\,\xi}{d\,y} + \frac{d\,\eta}{d\,x} &= 2\,\Sigma\left\{\frac{2\,\pi}{\lambda}\,b'\,e^{\psi}\,(l\cos\phi + l'\sin\phi)\right\}\end{aligned}\right\} \quad (25.)$$

The pressures due to the displacements are as follows:

Normal Pressures.

$$\left.\begin{aligned} P_1 &= \Sigma\left[\frac{2\pi}{\lambda} e^{\psi}\{A(1-b'^2-c'^2)+2\,C(b'^2+c'^2)\}(l'\cos\phi-l\sin\phi)\right] \\ P_2 &= \Sigma\left[\frac{2\pi}{\lambda} e^{\psi}\{A(1-b'^2-c'^2)-2\,C(1-c'^2)\}(l'\cos\phi-l\sin\phi)\right] \\ P_3 &= \Sigma\left[\frac{2\pi}{\lambda} e^{\psi}\{A(1-b'^2-c'^2)-2\,C(1-b'^2)\}(l'\cos\phi-l\sin\phi)\right] \end{aligned}\right\} \quad (26.)$$

The tangential pressures Q_1, Q_2, Q_3, are found by multiplying the distortions by $-$ C.

Let R_1, R_2, R_3, be the three components of the pressure exerted by the particles of the body, in consequence of the molecular displacements, at any part of its external surface, the normal to which makes with the axes the angles α, β, γ. Then

$$\left.\begin{aligned} R_1 &= P_1\cos\alpha + Q_3\cos\beta + Q_2\cos\gamma \\ R_2 &= Q_3\cos\alpha + P_2\cos\beta + Q_1\cos\gamma \\ R_3 &= Q_2\cos\alpha + Q_1\cos\beta + P_3\cos\gamma \end{aligned}\right\} \quad . \quad (27.)$$

Should there be any surface along which the particles are constrained to slide, it is obvious that at that surface the following condition must be fulfilled:

$$0 = \xi\cos\alpha + \eta\cos\beta + \zeta\cos\gamma;$$

or if $z_1 = f(x, y)$ be the equation of the surface,

$$0 = \xi\frac{dz_1}{dx} + \eta\frac{dz_1}{dy} - \zeta. \qquad (28.)$$

Were we acquainted with the laws which determine the superficial pressures in vibrating bodies, equations (27) would enable us to determine the values which b' and c' must have, in virtue of those laws, during the transmission of sound in a limited mass of an uncrystallised material, and thence the ratio $\sqrt{(1-b'^2-c'^2)}$: 1, in which the velocity of sound in such a body is less than in an unlimited mass of the same material. Those laws, however, are as yet a matter of conjecture only.

Transmission of a Definite Musical Tone.

17. When the body transmits one or more definite musical tones (which is the case in all experiments capable of yielding useful results), the

velocity of propagation must be the same for all the elementary vibrations into which the motion may be resolved: that is to say, $1-b'^2-c'^2$ must have the same value in all the terms of the sums Σ. This affords the means of simplifying the equations. Let

$$b'^2 + c'^2 = h^2; \quad b' = h \cos\theta; \quad c' = h \sin\theta;$$

h being the same for all the terms in the sums Σ. Then the velocity of propagation is

$$\sqrt{\varepsilon} = \sqrt{\left\{\frac{\mathrm{A}\,g}{\mathrm{D}}(1-h^2)\right\}} \quad . \quad . \quad . \quad (29.)$$

and this factor may be removed outside the sign of summation.

When but one musical tone is transmitted, the factor $\frac{2\pi}{\lambda}$ also may be removed outside that sign, and for $\Sigma\ \{\ \}$ may be substituted a definite integration,

$$\left.\begin{array}{l} \Sigma \int \mathrm{F}\,\theta\,\{\ .\ .\ .\ \}\,d\theta, \\ \text{F } \theta \text{ being arbitrary.} \\ \text{We have also} \\ e^{\psi} = e^{\frac{2\pi}{\lambda} h (y \cos\theta + z \sin\theta)} \end{array}\right\} \quad . \quad . \quad (30.)$$

in which $\frac{2\pi}{\lambda}$ h, y, and z, are independent of θ, and may be treated as constants in the definite integration.

Introducing these modifications into equations (23), &c., we find

$$\Phi = \frac{\lambda}{2\pi}(l' \cos\phi - l \sin\phi)\,\Sigma \int e^{\psi}\,\mathrm{F}\,\theta\,d\theta$$

Displacements.

$$\xi = (l \cos\phi + l' \sin\phi)\,\Sigma \int e^{\psi}\,\mathrm{F}\,\theta\,d\theta$$

$$\eta = (l' \cos\phi - l \sin\phi)\,h.\,\Sigma \int \cos\theta\, e^{\psi}\,\mathrm{F}\,\theta\,d\theta$$

$$\zeta = (l' \cos\phi - l \sin\phi)\,h.\,\Sigma \int \sin\theta\, e^{\psi}\,\mathrm{F}\,\theta\,d\theta \quad (31.)$$

Velocities of the Particles.

$$\frac{d\xi}{dt} = \frac{2\pi}{\lambda}\sqrt{\varepsilon}\,.\,(l' \cos\phi - l \sin\phi)\,\Sigma \int e^{\psi}\,\mathrm{F}\,\theta\,d\theta$$

$$\frac{d\eta}{dt} = -\frac{2\pi}{\lambda}\sqrt{\varepsilon}\,.\,(l \cos\phi + l' \sin\phi)\,h.\,\Sigma \int \cos\theta\, e^{\psi}\,\mathrm{F}\,\theta\,d\theta$$

$$\frac{d\zeta}{dt} = -\frac{2\pi}{\lambda}\sqrt{\varepsilon}.(l \cos\phi + l' \sin\phi)\, h.\Sigma\int \sin\theta\, e^{\psi}\, F\theta\, d\theta$$

Longitudinal Strains.

$$\frac{d\xi}{dx} = -\frac{2\pi}{\lambda}(l' \cos\phi - l \sin\phi)\,\Sigma\int e^{\psi}\, F\theta\, d\theta$$

$$\frac{d\eta}{dy} = \frac{2\pi}{\lambda}(l' \cos\phi - l \sin\phi)\, h^2.\Sigma\int \cos^2\theta\, e^{\psi}\, F\theta\, d\theta$$

$$\frac{d\zeta}{dz} = \frac{2\pi}{\lambda}(l' \cos\phi - l \sin\phi)\, h^2.\Sigma\int \sin^2\theta\, e^{\psi}\, F\theta\, d\theta.$$

Cubic Dilatation.

$$\frac{d\xi}{dx} + \frac{d\eta}{dy} + \frac{d\zeta}{dz} = -\frac{2\pi}{\lambda}(l' \cos\phi - l \sin\phi)\,(1 - h^2)\,\Sigma\int e^{\psi}\, F\theta\, d\theta.$$

Distortions.

$$\frac{d\eta}{dz} + \frac{d\zeta}{dy} = \frac{4\pi}{\lambda}(l' \cos\phi - l \sin\phi)\, h^2\,\Sigma\int \cos\theta \sin\theta\, e^{\psi}\, F\theta\, d\theta$$

$$\frac{d\zeta}{dx} + \frac{d\xi}{dz} = \frac{4\pi}{\lambda}(l \cos\phi + l' \sin\phi)\, h\,\Sigma\int \sin\theta\, e^{\psi}\, F\theta\, d\theta \qquad (31.)$$

$$\frac{d\xi}{dy} + \frac{d\eta}{dx} = \frac{4\pi}{\lambda}(l \cos\phi + l' \sin\phi)\, h\,\Sigma\int \cos\theta\, e^{\psi}\, F\theta\, d\theta.$$

Which, being multiplied by $-$ C, give the tangential pressures Q_1, Q_2, Q_3, on the co-ordinate planes.

Normal Pressures on the Co-ordinate Planes, due to the Displacements.

$$P_1 = \frac{2\pi}{\lambda}(l' \cos\phi - l \sin\phi)\,\{A(1 - h^2) + 2Ch^2\}\,\Sigma\int e^{\psi}\, F\theta\, d\theta$$

$$P_2 = \frac{2\pi}{\lambda}(l' \cos\phi - l \sin\phi)\,[\{A(1 - h^2) - 2C\}\,\Sigma\int e^{\psi}\, F\theta\, d\theta + 2Ch^2\,\Sigma\int \sin^2\theta\, e^{\psi}\, F\theta\, d\theta]$$

$$P_3 = \frac{2\pi}{\lambda}(l' \cos\phi - l \sin\phi)\,[\{A(1 - h^2) - 2C\}\,\Sigma\int e^{\psi}\, F\theta\, d\theta + 2Ch^2\,\Sigma\int \cos^2\theta\, e^{\psi}\, F\theta\, d\theta].$$

Let R_1, R_2, R_3, be the components of the pressure exerted by the body, in consequence of the molecular displacements, at a point of its surface normal to the direction (α, β, γ). Also, let

$$\cos\beta = \sin\alpha\cos\chi,$$
$$\cos\gamma = \sin\alpha\sin\chi,$$

so as to make x the axis of polar co-ordinates, and xy the plane from which longitudes χ are measured. Then,

$$\left.\begin{aligned}
R_1 &= \frac{2\pi}{\lambda}[\cos\alpha(l'\cos\phi - l\sin\phi)\{A(1-h^2)+2Ch^2\}\,\Sigma\int e^{\psi}F\theta\,d\theta \\
&\qquad - 2\sin\alpha\,(l\cos\phi + l'\sin\phi)\,C\,h\,\Sigma\int\cos(\theta-\chi)\,e^{\psi}F\theta\,d\theta] \\
R_2 &= \frac{2\pi}{\lambda}[-2\cos\alpha\,(l\cos\phi + l'\sin\phi)\,C\,h\,\Sigma\int\cos\theta\,e^{\psi}F\theta\,d\theta \\
&+\sin\alpha\,(l'\cos\phi - l\sin\phi)\,\{\cos\chi\left(A(1-h^2) - 2\cdot C\right)\Sigma\int e^{\psi}F\theta\,d\theta \\
&\qquad + 2\,C\,h^2\,\Sigma\int\sin\theta\sin(\theta-\chi)\,e^{\psi}F\theta\,d\theta\}] \\
R_3 &= \frac{2\pi}{\lambda}[-2\cos\alpha\,(l\cos\phi + l'\sin\phi)\,C\,h\,\Sigma\int\sin\theta\,e^{\psi}F\theta\,d\theta \\
&+\sin\alpha\,(l'\cos\phi - l\sin\phi)\,\{\sin\chi\left(A(1-h^2) - 2\,C\right)\Sigma\int e^{\psi}F\theta\,d\theta \\
&\qquad - 2\,C\,h^2\,\Sigma\int\cos\theta\sin(\theta-\chi)\,e^{\psi}F\theta\,d\theta\}]
\end{aligned}\right\} \quad (32.)$$

Let P′ represent the normal pressure at the given point of the surface due to molecular displacements: then

$$\left.\begin{aligned}
P' &= R_1\cos\alpha + \sin\alpha\,(R_2\cos\chi + R_3\sin\chi) \\
&= P_1\cos^2\alpha + P_2\sin^2\alpha\cos^2\chi + P_3\sin^2\alpha\sin^2\chi \\
&+ 2\,Q_1\sin^2\alpha\cos\chi\sin\chi + 2\,Q_2\cos\alpha\sin\alpha\sin\chi \\
&\qquad + 2\,Q_3\cos\alpha\sin\alpha\cos\chi \\
&= \frac{2\pi}{\lambda}[(l'\cos\phi - l\sin\phi)\,\{\Big(A(1-h^2) \\
&\qquad + 2\,C\,(h^2\cos^2\alpha - \sin^2\alpha)\Big)\Sigma\int e^{\psi}F\theta\,d\theta \\
&\qquad + 2\,C\,h^2\sin^2\alpha\,\Sigma\int\sin^2(\theta-\chi)\,e^{\psi}F\theta\,d\theta\} \\
&- 4\,(l\cos\phi + l'\sin\phi)\,C\,h\cos\alpha\sin\alpha\,\Sigma\int\cos(\theta-\chi)\,e^{\psi}F\theta\,d\theta
\end{aligned}\right\} \quad (32A)$$

Propagation of Sound by Nearly-longitudinal Vibrations along a Horizontal Prism of Liquid contained in a Rectangular Trough, investigated according to a Peculiar Hypothesis.

18. I shall now suppose the vibrating body to be a rectangular horizontal prism of liquid contained in a trough of some substance so dense, hard, and smooth, that the particles at the sides and bottom of the trough are constrained to slide along those surfaces, and that the vertical ends of the trough are capable of perfectly reflecting a wave of sound travelling horizontally; so that the propagation of that wave may take place as if in a trough of indefinite length; and I shall investigate the velocity of such a wave according to a peculiar hypothetical view of the molecular condition of the upper surface of the liquid.

The axis of x being the horizontal axis of the trough, and parallel to the direction of propagation, let that of y be transverse, and that of z vertical. Let the middle of the bottom of the trough be the origin of co-ordinates, $2y_1$ being its breadth, and z_1 the depth of liquid in it.

The conditions to be fulfilled at the bottom are, when

$$z = 0, \quad \alpha = \tfrac{1}{2}\pi, \text{ and } \chi = -\tfrac{1}{2}\pi.$$

Let

$$\Sigma \int \sin \theta \, e^{\psi} \mathrm{F}\,\theta \, d\theta = \Sigma \int \sin \theta \, e^{\frac{2\pi}{\lambda} h y \cos\theta} \mathrm{F}\,\theta\, d\theta = 0$$

at the sides, when

$$y = \pm\, y_1, \quad \alpha = \tfrac{1}{2}\pi, \text{ and } \chi = 0 \text{ or } \pi.$$

Let

$$\Sigma \int \cos \theta \, e^{\psi} \mathrm{F}\,\theta\, d\theta = \Sigma \int \cos \theta \, e^{\frac{2\pi}{\lambda} h(\pm y_1 \cos\theta + z \sin\theta)} \mathrm{F}\,\theta\, d\theta = 0;$$

which conditions are fulfilled by making

$$\cos \theta = 0, \quad \sin \theta = \pm\, 1,$$

and putting for $\Sigma \int \mathrm{F}\,\theta\, d\theta$ a summation of two terms in which the signs of the exponent are respectively positive and negative.

Thus we obtain

$$\left.\begin{aligned} \xi &= (l \cos \phi + l' \sin \phi)\left(e^{\frac{2\pi}{\lambda}hz} + e^{-\frac{2\pi}{\lambda}hz}\right) \\ \eta &= 0 \\ \zeta &= (l' \cos \phi - l \sin \phi) h \left(e^{\frac{2\pi}{\lambda}hz} - e^{-\frac{2\pi}{\lambda}hz}\right) \end{aligned}\right\} \quad (33.)$$

The trajectory of each particle is an ellipse in a vertical longitudinal plane; the motion being *direct* in the upper part of the ellipse, because the sign of $\frac{d\xi}{dt}$ is the same with that of ζ. The axes are vertical and horizontal respectively, and have the following values :—

$$\text{Horizontal axis} = 2\sqrt{(l^2 + l'^2)} \, . \left(e^{\frac{2\pi}{\lambda}hz} + e^{-\frac{2\pi}{\lambda}hz}\right),$$

$$\text{Vertical axis} \quad = 2\sqrt{(l^2 + l'^2)} \, . \, h \, . \left(e^{\frac{2\pi}{\lambda}hz} - e^{-\frac{2\pi}{\lambda}hz}\right);$$

so that the motion is analogous to that of waves propagated by gravitation, being entirely horizontal at the bottom of the trough, and elliptical elsewhere, the ellipse being larger and less eccentric as the height above the bottom increases. The ratio of the axes, however, instead of approaching equality as the depth of the trough increases (which is the case with waves of gravitation), approaches $1 : h$.

19. To determine this ratio, upon which the velocity of sound along such a mass of liquid must depend, I shall assume the following hypothetical principles respecting the state of the particles at the upper surface:

First, That (as laid down in a previous paper, *Cambridge and Dublin Mathematical Journal,* February, 1851) the elasticity of bodies is due partly to the mutual actions of atomic centres producing elasticity both of volume and figure, and partly to a mere fluid elasticity resisting change of volume only, and exerted by atmospheres surrounding those centres; and that the effect of the mutual actions of the atomic centres in producing pressure is very small in liquids, and absolutely inappreciable in gases and vapours.

Secondly, That every liquid maintains at its surface, by molecular attraction, an atmosphere of its own vapour, under these conditions— that the total pressures of the liquid and vapour, and also their fluid pressures, shall be equal at the bounding surface. (From this hypothesis I have already deduced the form of an approximate equation between the pressure and temperature of vapour at saturation.) The total pressure of the vapour on the liquid is sensibly equal to its fluid pressure; the total pressure of the liquid on the vapour consists of its fluid pressure, and a pressure due to atomic centres; the latter quantity must, therefore, be null.

Thirdly, That the pressure of the vapour follows that of the liquid throughout its variations during the propagation of sound; so that the portion of the pressure of the liquid on the vapour, due to atomic centres, must continue null throughout these variations.

Let ω be the mutual pressure of the liquid and its vapour in a state of rest, then $\omega + P'$ is their momentary mutual pressure during the passage of a wave of sound horizontally along the trough. The portion of P′ depending on the coefficient of rigidity C being made $= 0$, we shall obtain an equation from which the value of h may be deduced.

Making the proper substitutions in equation (32A), viz.—

$$\cos a = 0,\ \sin a = 1,\ \cos \chi = 0,\ \sin \chi = 1,\ \psi = \pm \frac{2\pi}{\lambda} h z,$$

$$\cos \theta = 0,\ \sin \theta = \pm 1,\ F\theta = 1,\ z = z_1,\ \&c.,$$

we find

$$\omega + P' = \omega + \frac{2\pi}{\lambda}(l' \cos \phi - l \sin \phi)\{A(1 - h^2) - 2C\}\left(e^{\frac{2\pi}{\lambda}hz_1} + e^{-\frac{2\pi}{\lambda}hz_1}\right).$$

The part of this depending on mere fluid elasticity, in which the liquid is followed by the vapour, is

$$\omega - J\left(\frac{d\xi}{dx} + \frac{d\zeta}{dz}\right)$$

$$= \omega + \frac{2\pi}{\lambda}(l' \cos \phi - l \sin \phi) . J(1 - h^2)\left(e^{\frac{2\pi}{\lambda}hz_1} + e^{-\frac{2\pi}{\lambda}hz_1}\right),$$

which, being subtracted, there remains for the part depending on atomic centres,

$$0 = \frac{2\pi}{\lambda}(l' \cos \phi - l \sin \phi)\, C(1 - 3h^2)\left(e^{\frac{2\pi}{\lambda}hz_1} + e^{-\frac{2\pi}{\lambda}hz_1}\right).$$

Consequently,

$$1 - 3h^2 = 0, \text{ or } h = \sqrt{\tfrac{1}{3}} \quad . \quad . \quad . \quad (33.)$$

is the equation of condition sought, arising from the state of the free surface; and this equation is independent of the amount of rigidity of the liquid, requiring only that it shall be *something*, however small, while that of the vapour is null.

It follows from this equation, that the velocity of propagation of sound along a trough of liquid of the density D, and longitudinal elasticity A, is

$$\sqrt{\varepsilon} = \sqrt{\left\{\frac{Ag}{D}(1 - h^2)\right\}} = \sqrt{\left(\frac{2}{3} \cdot \frac{Ag}{D}\right)}, \quad . \quad (34.)$$

or less than the velocity in an unlimited mass in the ratio of $\sqrt{2}$ to $\sqrt{3}$.

20. This is precisely the result arrived at by M. Wertheim from a comparison of his numerous experiments on the propagation of sound in water at various temperatures, from 15° to 60° Centigrade, in solutions of various salts, in alcohol, turpentine, and ether (*Ann. de Chim.*, Sér. III.,

tom. xxiii.), with those of M. Grassi on the compressibility of the same substances (*Comptes Rendus*, XIX., p. 153), and with the experiments of MM. Colladon and Sturm on the velocity of sound in an expanse of water.

M. Wertheim having given this comparison in detail, I shall quote one example only.

The velocity of sound in an unlimited mass of water, at the temperature of 16° Centigrade, as ascertained by MM. Colladon and Sturm, was 1435 mètres per second.

That of sound in water contained in a trough, the vibrations of which were regulated by an organ-pipe, was found by M. Wertheim, at 15° Centigrade, to be 1173·4 mètres per second.

The ratio of the squares of those quantities is 0·6686 : 1, differing from $\frac{2}{3}$ by 0·0009 only.

Remarks on the Propagation of Sound along Solid Rods.

21. I refrain from giving, in the body of this paper, detailed investigations of particular problems respecting the propagation of sound along a solid prism or cylinder; for, in the present state of our knowledge of the condition of the superficial particles of such bodies, the conclusions would be almost entirely speculative and conjectural.

I may mention briefly, however, the following general results. If we adopt for solids the same hypothesis as for liquids, then the ratio of the velocity of sound in a rod of an uncrystallised material to that in an unlimited mass has the following values:

For a rectangular prismatic rod, the lateral vibrations of the particles of which are confined to planes parallel to one pair of faces of the prism, but are perfectly free in other respects, the ratio is $\sqrt{2} : \sqrt{3}$, being the same as for a liquid.

For a cylindrical rod, the surface being perfectly free, the ratio has various values, from $\sqrt{\frac{1}{2}} : \sqrt{\frac{2}{3}}$, approaching the less value as the diameter of the rod diminishes, and the greater as it increases *ad infinitum*. (*See Appendix, No. I.*)

22. These conditions, however, cannot be realised in practice; and the lateral vibrations being more or less confined by the means used in fixing the rods, we find that the ratio generally exceeds $\sqrt{2} : \sqrt{3}$, and sometimes approaches equality.

The following table illustrates this fact. I have selected, in the first place, the experiments of M. Wertheim on tubes of crystal (*Ann. de Chim.*, Sér. III., tom. xxiii.), because in them the coefficients of elasticity and the velocities of sound were ascertained by experiments on the same pieces of the material. To these I have added a calculation, founded on a comparison of the experiments of M. Wertheim on the elasticity of brass, with those of M. Savart on the velocity of sound in it, as being the only other data of the kind now existing from which a satisfactory conclusion can be drawn.

The coefficients of longitudinal elasticity, calculated by myself from M. Wertheim's experiments, are extracted from my paper on elasticity in the *Cambridge and Dublin Mathematical Journal* for February, 1851. The quantities $\frac{\varepsilon D}{g}$ for crystal are given as calculated by M. Wertheim. For brass I have used the following data:

$\sqrt{\varepsilon}$ = velocity of sound in brass rods; mean of many experiments by M. Savart = 3560 mètres per second.

D = density, in kilogrammes per cubic mètre, 8395.

TABLE.

Crystal.	Longitudinal Elasticity A Kilogrammes per square millimètre.	$\frac{\varepsilon D}{g}$ Kilogrammes per square millimètre.	Ratio $1 - h^2$
Tube No. I., . .	5514·2	5354·0	0·970,
„ „ III., . .	5611·0	5476·7	0·976,
„ „ IV., . .	6183·1	5597·3	0·905,
„ „ V., . .	6659·9	5489·8	0·824,
Brass, . .	15625	10847	0·694.

Concluding Remarks.

23. The chief positive results arrived at in this paper may be summed up as follows:—

(I.) In liquid and solid bodies of limited dimensions, the freedom of lateral motion possessed by the particles causes vibrations to be propagated less rapidly than in an unlimited mass.

(II.) The symbolical expressions for vibrations in limited bodies are distinguished by containing exponential functions of the co-ordinates as factors; and the retardation referred to depends on the coefficients of the co-ordinates in the exponents of those functions, which coefficients depend on the molecular condition of the body's surface—a condition yet imperfectly understood.

(III.) If we adopt the hypothetical principle, *that at the free surface of a vibrating mass of liquid the normal pressure depending on the actions of atomic centres is always null*, then we deduce from theory that the ratio of the velocity of sound along a mass of liquid contained in a rectangular trough to that in an unlimited mass is $\sqrt{2} : \sqrt{3}$, that ratio being independent of the specific rigidity of the liquid—a conclusion agreeing with our present experimental knowledge.

24. I do not put forward the hypothetical part of these researches as more than a probable conjecture; nor should I be justified in so doing in the present state of our knowledge of molecular forces. I think, however, that these investigations are sufficient to prove that we are not warranted in concluding from M. Wertheim's experiments (as he is disposed to do) that liquids possess a momentary rigidity as great as that of solids, seeing that any amount of rigidity, howsoever small, will account for the phenomena if we adopt certain suppositions as to molecular forces; and to show that our knowledge of those forces is not yet sufficiently advanced to enable us to use experiments on sound as a means of determining the coefficients of elasticity of solids.

Appendix.—No. I.

Propagation of Sound by Nearly-Longitudinal Vibrations along a Cylindrical Uncrystallised Rod.

Let the vibrating body be cylindrical round the axis of x, and let the vibrations of all particles in a given circle round that axis be assumed to be equal and simultaneous. Let r represent the distance of any particle from the axis of x, and χ, the angle $\widehat{y\,r}$.

Then

$$e^{\psi} = e^{\frac{2\pi}{\lambda} r h \cos(\theta - \chi)} \quad . \quad . \quad . \quad (35.)$$

To make the results of the definite integrations $\Sigma \int \mathrm{F}\,\theta\, d\,\theta$ independent of the angle χ, we must have $\mathrm{F}\,\theta =$ constant, and the limits of integration 0 and 2π.

The following are the definite integrals which enter into the solution of this problem.

Let

$$\frac{2\pi}{\lambda} h r = k,$$

$$\left.\begin{aligned}
\Theta &= \int_0^{2\pi} e^{k\cos\theta}\, d\theta = 4\int_0^{\frac{1}{2}\pi} \frac{e^{k\cos\theta} + e^{-k\cos\theta}}{2}\, d\theta \\
&= 4\int_0^{\frac{1}{2}\pi} d\theta \left\{ 1 + \Sigma\left(\frac{k^{2n}\cos^{2n}\theta}{\Gamma(2n+1)}\right) \right\} \\
&= 2\pi\left[1 + \Sigma\left\{ \frac{k^{2n}}{2^{2n}(\Gamma(n+1))^2} \right\} \right] \\
\Theta' = \frac{d\Theta}{dk} &= \int_0^{2\pi} \cos\theta\, e^{k\cos\theta}\, d\theta \\
&= 2\pi \,.\, \Sigma\left\{ \frac{k^{2n-1}}{2^{2n-1}\,.\,\Gamma(n)\,.\,\Gamma(n+1)} \right\} \\
\Theta'' = \frac{d^2\Theta}{dk^2} &= \int_0^{2\pi} \cos^2\theta\, e^{k\cos\theta}\, d\theta \\
&= 2\pi\left[\tfrac{1}{2} + \Sigma\left\{ \frac{(2n+1)k^{2n}}{2^{2n+1}\Gamma(n+1)\Gamma(n+2)} \right\} \right];
\end{aligned}\right\} \qquad (36.)$$

the values of n comprehending all integers from 1 inclusive.

Those series have the following properties:

(I.) The term (n) of Θ = term $(n-1) \times \frac{k^2}{4n^2}$; therefore, this series always becomes convergent at the term for which $n > \frac{1}{2}k$.

(II.) Term (n) of Θ' = term $(n-1) \times \frac{k^2}{4(n-1)n}$; therefore, this series becomes convergent when $n^2 - n > \frac{1}{4}k^2$.

(III.) Term (n) of Θ'' = term $(n-1) \times \frac{(2n+1)k^2}{4(2n-1).n.(n+1)}$; therefore, it begins to converge when $n^2 - \frac{n}{2n-1} > \frac{k^2}{4}$.

(IV.) $\Theta' = k(\Theta - \Theta'')$.

(V.) Term (n) of Θ'' = term (n) of $\Theta \times \frac{2n+1}{2n+2}$; a ratio which is $\frac{1}{2}$ for the first term $(n = 0)$, and approaches equality as n increases; therefore, when $\frac{1}{4}k^2$ is an inappreciably small fraction, $\frac{\Theta''}{\Theta} = \frac{1}{2}$ sensibly.

And the larger k is, the more nearly is $\frac{\Theta''}{\Theta} = 1$.

The following table of a few numerical results illustrates this :—

$\frac{k^2}{4}$	$\frac{\Theta}{2\pi}$	$\frac{\Theta''}{2\pi}$	$\frac{\Theta''}{\Theta}$
0	1·0000	0·5000	0·5000
$\frac{1}{4}$	1·2661	0·7010	0·5537
$\frac{1}{3}$	1·3622	0·7741	0·5683
$\frac{1}{2}$	1·5661	0·9302	0·5490
1	2·2796	1·4843	0·6511
2	4·2523	3·0550	0·7160
3	7·1590	5·4238	0·7576
4	11·3019	9·3620	0·8284

The displacements in this case are as follows :—

$$\left.\begin{aligned} \xi &= (l \cos\phi + l' \sin\phi)\,\Theta \\ \eta &= (l' \cos\phi - l \sin\phi)h\,\Theta' \cos\chi \\ \zeta &= (l' \cos\phi - l \sin\phi)h\,\Theta' \sin\chi; \end{aligned}\right\} \quad . \quad . \quad (37.)$$

whence it appears that the two transverse displacements η and ζ compose a radial displacement,

$$\rho = (l' \cos\phi - l \sin\phi)h\,\Theta'. \quad . \quad . \quad (37\,\text{A.})$$

Therefore, the trajectory of each particle is an ellipse, in a plane passing through the axis of the cylinder; and the axes of the ellipse are longitudinal and radial, and have the following values:

$$\left.\begin{aligned} &\text{Longitudinal axis,} = 2\sqrt{(l^2 + l'^2)}\;.\;\Theta \\ &\text{Radial axis,} \quad . \quad = 2\sqrt{(l^2 + l'^2)}\;.\;h\,\Theta' \end{aligned}\right\} \quad . \quad (38.)$$

If we now adopt the same hypotheses with respect to the outer surface of the cylinder that have been used in the problem respecting liquids, we shall have for the mutual pressure of the solid and its atmosphere of vapour

$$\omega + \mathrm{P}' = \omega + \frac{2\pi}{\lambda}(l' \cos\phi - l \sin\phi)\{(\mathrm{A} - 2\,\mathrm{C})(1 - h^2)\Theta_1 - 2\,\mathrm{C}\,h^2\,\Theta_1''\},$$

Θ_1, Θ_1'' being the values of those integrals corresponding to the radius of the cylinder.

The portion of this pressure depending on mere fluid elasticity is

$$\omega - J\left(\frac{d\,\xi}{d\,x} + \frac{d\,\eta}{d\,y} + \frac{d\,\zeta}{d\,z}\right) = \omega + \frac{2\,\pi}{\lambda}(l' \cos\phi - l \sin\phi)\, J\,(1 - h^2)\,.\,\Theta_1,$$

which, being subtracted, leaves

$$0 = \frac{2\,\pi}{\lambda}(l' \cos\phi - l \sin\phi)\, C\,\{\Theta_1 - h^2(\Theta_1 + 2\,\Theta_1'')\};$$

therefore, according to the hypothesis adopted,

$$h^2 = \frac{1}{1 + 2\dfrac{\Theta_1''}{\Theta_1}}: \quad . \quad . \quad . \quad . \quad (39.)$$

and the velocity of sound along the cylinder is

$$\sqrt{\varepsilon} = \sqrt{\left\{\frac{A\,g}{D}(1 - h^2)\right\}} = \sqrt{\left(\frac{A\,g}{D}\cdot\frac{2}{2 + \dfrac{\Theta_1}{\Theta_1''}}\right)} \quad . \quad (40.)$$

Now, the limits of the ratios in the above formulæ are the following:

Limits of	$\frac{2\,\pi\,h\,r_1}{\lambda} = k_1$	0	∞,
„	$\frac{\Theta_1}{\Theta_1''}$	2	1,
„	h	$\sqrt{\frac{1}{2}}$	$\sqrt{\frac{1}{3}}$,
„	$\sqrt{(1 - h^2)}$	$\sqrt{\frac{1}{2}}$	$\sqrt{\frac{2}{3}}$.

That is to say, if the hypothesis already explained with reference to liquids is applicable to a solid cylinder of an uncrystallised material, the velocity of sound along such a cylinder, when its surface is perfectly free, will be less than that in an unlimited mass in some ratio between $\sqrt{\frac{1}{2}}$ and $\sqrt{\frac{2}{3}}$.

APPENDIX.—No. II.

General Equations of Nearly-Transverse Vibrations.

The two equal roots of equation (8) in uncrystallised bodies, viz.

$$E = C(1 - b'^2 - c'^2),$$

correspond to what may be called *nearly-transverse* vibrations, propagated with the velocity

$$\sqrt{\varepsilon} = \sqrt{\left\{\frac{Cg}{D}(1 - b'^2 - c'^2)\right\}}. \quad . \quad . \quad (41.)$$

Equations (20) in this case give no result; but equations (20A) are reducible to the following two:

$$\left.\begin{array}{l} l = -m'b' - n'c' \quad . \quad . \quad . \quad . \\ l' = mb' + nc' \quad . \quad . \quad . \quad . \quad . \end{array}\right\} \quad . \quad . \quad (42);$$

the ratios $m:n$ and $m':n'$ are arbitrary.

Equations (11), (16), (17), (18), become the following:

Displacements.

$$\xi = \Sigma\left[e^{\psi}\{-(m'b' + n'c')\cos\phi + (mb' + nc')\sin\phi\}\right]$$

$$\eta = \Sigma\{e^{\psi}(m\cos\phi + m'\sin\phi)\}$$

$$\zeta = \Sigma\{e^{\psi}(n\cos\phi + n'\sin\phi)\}$$

Velocities of the Particles.

$$\frac{d\xi}{dt} = \Sigma\left[\frac{2\pi}{\lambda}\sqrt{\varepsilon}\,.\,e^{\psi}\{(m'b' + n'c')\sin\phi + (mb' + nc')\cos\phi\}\right]$$

$$\frac{d\eta}{dt} = \Sigma\left\{\frac{2\pi}{\lambda}\sqrt{\varepsilon}\,.\,e^{\psi}(-m\sin\phi + m'\cos\phi)\right\}$$

$$\frac{d\zeta}{dt} = \Sigma\left\{\frac{2\pi}{\lambda}\sqrt{\varepsilon}\,.\,e^{\psi}(-n\sin\phi + n'\cos\phi)\right\}$$

Longitudinal Strains.

$$\frac{d\xi}{dx} = -\Sigma\left[\frac{2\pi}{\lambda}e^{\psi}\{(mb' + nc')\cos\phi + (m'b' + n'c')\sin\phi\}\right]$$

(43.)

$$\frac{d\eta}{dy} = \Sigma\left\{\frac{2\pi}{\lambda} e^{\psi} (m b' \cos\phi + m' b' \sin\phi)\right\}$$

$$\frac{d\zeta}{dz} = \Sigma\left\{\frac{2\pi}{\lambda} e^{\psi} (n c' \cos\phi + n' c' \sin\phi)\right\}$$

Cubic Dilatation.

$$\frac{d\xi}{dx} + \frac{d\eta}{dy} + \frac{d\zeta}{dz} = 0.$$

Distortions.

$$\frac{d\eta}{dz} + \frac{d\zeta}{dy} = \Sigma\left[\frac{2\pi}{\lambda} e^{\psi} \{(m c' + n b') \cos\phi + (m' c' + n' b') \sin\phi\}\right]$$

$$\frac{d\zeta}{dx} + \frac{d\xi}{dz} = \Sigma\left[\frac{2\pi}{\lambda} e^{\psi} \{-\left(n' (1 + c')^2 + m' b' c'\right) \cos\phi + \left(n (1 + c')^2 + m b' c'\right) \sin\phi\}\right] \qquad (43.)$$

$$\frac{d\xi}{dy} + \frac{d\eta}{dx} = \Sigma\left[\frac{2\pi}{\lambda} e^{\psi} \{-\left(m' (1 + b'^2) + n' b' c'\right) \cos\phi + \left(m (1 + b'^2) + n b' c'\right) \sin\phi\}\right]$$

Which, being multiplied by $-$ C, give the tangential pressures Q_1, Q_2, Q_3.

Normal Pressures on the Co-ordinate Planes due to the Displacements.

$$P_1 = 2\,C.\,\Sigma\left[\frac{2\pi}{\lambda} e^{\psi} \{(m b' + n c') \cos\phi + (m' b' + n' c') \sin\phi\}\right]$$

$$P_2 = -\,2\,C.\,\Sigma\left[\frac{2\pi}{\lambda} e^{\psi} \{m b' \cos\phi + m' b' \sin\phi\}\right]$$

$$P_3 = -\,2\,C.\,\Sigma\left[\frac{2\pi}{\lambda} e^{\psi} \{n c' \cos\phi + n' c' \sin\phi\}\right]$$

$$P_1 + P_2 + P_3 = 0.$$

The normal pressure due to the displacements at any point of the surface of a prism or cylinder described round x is

$$P' = -2\,C\,.\,\Sigma\left[\frac{2\pi}{\lambda}e\psi\left\{\left(m\,b'\cos^2\chi + n\,c'\sin^2\chi + (m\,c' + n\,b')\cos\chi\sin\chi\right)\cos\phi + \left(m'\,b'\cos^2\chi + n'\,c'\sin^2\chi + (m'\,c' + n'\,b')\cos\chi\sin\chi\right)\sin\phi\right\}\right] \quad (44.)$$

If this pressure is to be null at all points of the surface, we must have $b'=0$, $c'=0$, and, consequently, $l=0$, $l'=0$; and the motion is restricted to common exactly-transverse vibrations, for which

$$E = C \text{ and } \sqrt{\varepsilon} = \sqrt{\left(\frac{C\,g}{D}\right)}.$$

Nearly-transverse vibrations, therefore, cannot be transmitted along a cylindrical or prismatic uncrystallised body whose surface is absolutely free.

PART II.

PAPERS RELATING TO ENERGY AND ITS TRANSFORMATIONS, THERMODYNAMICS, MECHANICAL ACTION OF HEAT IN THE STEAM ENGINE, &c.

PART II.

PAPERS RELATING TO ENERGY AND ITS TRANSFORMATIONS, THERMODYNAMICS, MECHANICAL ACTION OF HEAT IN THE STEAM ENGINE, &c.

X.—ON THE RECONCENTRATION OF THE MECHANICAL ENERGY OF THE UNIVERSE.*

THE following remarks have been suggested by a paper by Professor William Thomson † of Glasgow, on the tendency which exists in nature to the dissipation or indefinite diffusion of mechanical energy originally collected in stores of power.

The experimental evidence is every day accumulating, of a law which has long been conjectured to exist,—that all the different kinds of physical energy in the universe are mutually convertible; that the total amount of physical energy, whether in the form of visible motion and mechanical power, or of heat, light, magnetism, electricity, or chemical agency, or in other forms not yet understood, is unchangeable; the transformations of its different portions from one of those forms of power into another, and their transference from one portion of matter to another, constituting the phenomena which are the objects of experimental physics.

Professor William Thomson has pointed out the fact, that there exists (at least in the present state of the known world), a predominating tendency to the conversion of all the other forms of physical energy into heat, and to the uniform diffusion of all heat throughout all matter. The form in which we generally find energy originally collected, is that of a store of chemical power, consisting of uncombined elements. The combination of these elements produces energy in the form known by the name of electric currents, part only of which can be employed in analysing

* Read before the British Association at Belfast, on September 2, 1852, and published in the *Philosophical Magazine* for November, 1852.

† Now Sir William Thomson.

compounds, and thus reconverted into a store of chemical power; the remainder is necessarily converted into heat: a part only of this heat can be employed in analysing compounds, or in reproducing electric currents. If the remainder of the heat be employed in expanding an elastic substance, it may be entirely converted into visible motion, or into a store of visible mechanical power (by raising weights, for example), provided the elastic substance is enabled to expand until its temperature falls to the point which corresponds to absolute privation of heat; but unless this condition be fulfilled, a certain proportion only of the heat, depending upon the range of temperature through which the elastic body works, can be converted, the rest remaining in the state of heat. On the other hand, all visible motion is of necessity ultimately converted entirely into heat by the agency of friction. There is thus, in the present state of the known world, a tendency towards the conversion of all physical energy into the sole form of heat.

Heat, moreover, tends to diffuse itself uniformly by conduction and radiation, until all matter shall have acquired the same temperature.

There is, consequently, Professor Thomson concludes, so far as we understand the present condition of the universe, a tendency towards a state in which all physical energy will be in the state of heat, and that heat so diffused that all matter will be at the same temperature; so that there will be an end of all physical phenomena.

Vast as this speculation may seem, it appears to be soundly based on experimental data, and to represent truly the present condition of the universe, so far as we know it.

My object now is to point out how it is conceivable that, at some indefinitely distant period, an opposite condition of the world may take place, in which the energy which is now being diffused may be reconcentrated into foci, and stores of chemical power again produced from the inert compounds which are now being continually formed.

There must exist between the atmospheres of the heavenly bodies a material medium capable of transmitting light and heat; and it may be regarded as almost certain that this interstellar medium is perfectly transparent and diathermanous; that is to say, that it is incapable of converting heat, or light (which is a species of heat), from the radiant into the fixed or conductible form.

If this be the case, the interstellar medium must be incapable of acquiring any temperature whatsoever; and all heat which arrives in the conductible form at the limits of the atmosphere of a star or planet, will there be totally converted, partly into ordinary motion, by the expansion of the atmosphere, and partly into the radiant form. The ordinary motion will again be converted into heat, so that *radiant heat* is the ultimate form to which all physical energy tends; and in this form it is, in the present condition of the world, diffusing itself from the heavenly bodies through the interstellar medium.

Let it now be supposed, that, in all directions round the visible world, the interstellar medium has bounds beyond which there is empty space.

If this conjecture be true, then on reaching those bounds the radiant heat of the world will be totally reflected, and will ultimately be reconcentrated into foci. At each of these foci the intensity of heat may be expected to be such, that should a star (being at that period an extinct mass of inert compounds) in the course of its motions arrive at that part of space, it will be vaporised and resolved into its elements; a store of chemical power being thus reproduced at the expense of a corresponding amount of radiant heat.

Thus it appears, that although, from what we can see of the known world, its condition seems to tend continually towards the equable diffusion, in the form of radiant heat, of all physical energy, the extinction of the stars, and the cessation of all phenomena; yet the world, as now created, may possibly be provided within itself with the means of reconcentrating its physical energies, and renewing its activity and life.

For aught we know, these opposite processes may go on together; and some of the luminous objects which we see in distant regions of space may be, not stars, but foci in the interstellar ether.

XI.—ON THE GENERAL LAW OF THE TRANSFORMATION OF ENERGY.*

ACTUAL, or SENSIBLE ENERGY, is a measurable, transmissible, and transformable condition, whose presence causes a substance to tend to change its state in one or more respects. By the occurrence of such changes, actual energy disappears, and is replaced by

POTENTIAL OR LATENT ENERGY; which is measured by the product of a change of state into the resistance against which that change is made.

(The *vis viva* of matter in motion, thermometric heat, radiant heat, light, chemical action, and electric currents, are forms of actual energy; amongst those of potential energy are the mechanical powers of gravitation, elasticity, chemical affinity, statical electricity, and magnetism).

The law of the *Conservation of Energy* is already known—viz., that the sum of all the energies of the universe, actual and potential, is unchangeable.

The object of the present paper is to investigate the law according to which all *transformations of energy*, between the actual and potential forms, take place.

Let V be the magnitude of a measurable state of a substance;

U, the species of potential energy which is developed when the state V increases;

P, the common magnitude of the tendency of the state V to increase, and of the equal and opposite resistance against which it increases; so that—

$$d\,U = P\,d\,V;\ \text{and}\ P = \frac{d\,U}{d\,V}. \qquad \text{(A.)}$$

Let Q be the quantity which the substance possesses, of a species of actual energy whose presence produces a tendency of the state V to increase.

It is required to find how much energy is transformed from the actual form Q to the potential form U, during the increment $d\,V$; that is to say, the magnitude of the portion of $d\,U$, the potential energy developed, which is due to the disappearance of an equivalent portion of actual energy of the species Q.

The development of this portion of potential energy is the immediate

* Read before the Philosophical Society of Glasgow, on January 5, 1853, and published in the *Proceedings* of that Society, Vol. III., No. V.

effect of the presence in the substance of the total quantity Q of actual energy.

Let this quantity be conceived to be divided into indefinitely small equal parts dQ. As those parts are not only equal, but altogether alike in nature and similarly circumstanced, their effects must be equal; therefore, the effect of the total energy Q must be equal simply to the effect of one of its small parts dQ, multiplied by the ratio $\frac{Q}{dQ}$.

But the effect of the indefinitely small part dQ in causing development of potential energy of the species U, during the increment of state dV, is represented by—

$$\frac{dP}{dQ} dQ\, dV;$$

whence it follows, that the effect of the presence of the total actual energy Q, in causing transformation of energy from the actual form Q to the potential form U, is expressed by the following formula:—

$$Q \cdot \frac{dP}{dQ} \cdot dV, \quad . \quad . \quad . \quad . \quad (1.)$$

which is the solution required, and is the symbolical expression of the GENERAL LAW OF THE TRANSFORMATION OF ENERGY:—

The effect of the whole actual energy present in a substance, in causing transformation of energy, is the sum of the effects of all its parts.

The difference between this quantity and the potential energy developed, viz:—

$$\left(P - Q \cdot \frac{dP}{dQ}\right) dV,$$

represents a portion of potential energy, due to causes different from the actual energy Q. This difference is null, when the resistance $\left(P = \frac{dU}{dV}\right)$ against which the state V increases, is simply proportional to the total actual energy Q.

It is next proposed to find the quantity of actual energy of the form Q, which must be transmitted to the substance from without, in order that its total actual energy may receive the increment dQ, and its state V at the same time the increment dV.

This quantity is composed of three parts—viz., actual energy, which preserves its form, dQ; actual energy which transforms itself to some unknown form, in consequence of the resistance which is offered to the increase of the total actual energy, $L\,dQ$; actual energy, already determined, which transforms itself into potential energy of the form U, $Q \cdot \frac{P}{dQ} \cdot dV$; the sum of these parts being—

$$d.Q = (1 + L)\,dQ + Q \cdot \frac{dP}{dQ} \cdot dV, \quad . \quad . \quad (2.)$$

in which nothing remains to be determined except the function L.

If we subtract from the above formula the total potential energy developed during the increment dV, viz:—

$$P.\,dV,$$

we obtain the algebraical sum of the energies, actual and potential, received and developed by the substance during the changes dQ, dV; which is thus expressed:—

$$d\Psi = d.Q - d.U = (1 + L)\,dQ + \left(Q\frac{d}{dQ} - 1\right)P.\,dV. \quad (B.)$$

This quantity must be the exact differential of a function of Q and V; for otherwise it would be possible, by varying the order of the increments dQ, dV, to change the sum of the energies of the universe.

It follows that—

$$\frac{dL}{dV} = \frac{d}{dQ}\left(Q\frac{d}{dQ} - \right)P = Q \cdot \frac{d^2}{dQ^2}P;$$

and, consequently, that

$$L = f'(Q) + Q \cdot \frac{d^2}{dQ^2}\int . P\,dV,$$

where $f'(Q)$ is a function of Q and constants, the first derivative of $f'(Q)$.

We find at length the following equation—

$$d\Psi = d.Q - d.U = \left(1 + f'(Q) + Q \cdot \frac{d^2}{dQ^2}\int P\,dV\right)dQ + \left(Q\frac{d}{dQ} - 1\right)$$

$$= d.\left\{Q + f(Q) + \left(Q\frac{d}{dQ} - 1\right)\int P\,dV\right\} \qquad (3.)$$

which represents the algebraical sum of the energy, actual and potential, received and developed by a substance, when the total actual energy of the species Q, and the state V, receive respectively the increments dQ, dV.

It is to be observed, that in the last equation, the symbol $\int P.\,dV$ denotes a *partial integral*, taken in treating the particular value of Q, to which it corresponds as a constant quantity; while $d.U$ represents the real magnitude of the potential energy developed.

The application of the general law of the transformation of energy may be extended to any number of kinds of energy, actual and potential, by means of the following equation:

$$d.\Psi = \Sigma d.Q - \Sigma d.U.$$

$$= \Sigma \left\{ \left(1 + f'(Q) + Q.\Sigma \frac{d^2}{dQ^2} \int P\,dV \right) dQ \right\} + \Sigma \left\{ \left(\Sigma\, Q \frac{d}{dQ} - 1 \right) P\,dV \right\}$$

$$= d \left\{ \Sigma Q + \Sigma f(Q) + \Sigma (\Sigma.Q \frac{d}{dQ} - 1) \int P\,dV \right\} \quad . \quad (4.)$$

This equation is the complete expression of the general law of the transformation of energy of all possible kinds, known and unknown. It affords the means, so soon as the necessary experimental data have been obtained, of analysing every development of potential energy, and referring its several portions to the species of actual energy from which they have been produced.

Amongst the consequences of this law, the author deduces that which may be called the general principle of the maximum effect of engines.

An engine consists essentially in a substance whose changes of state, and of actual energy, between given limits, are so regulated as to produce a permanent transformation of energy.

Let Q_1 be the given superior limit of actual energy; Q_2, the inferior limit.

To produce the maximum permanent transformation of energy from the actual to the potential form, the substance must undergo a cycle of four operations, viz:—

First Operation.

The substance, preserving the constant quantity Q_1 of actual energy, passes from the state V_A to the state V_B, receiving from without the following quantity of actual energy, which is converted into potential energy:—

$$H_1 = Q_1 . \frac{d}{dQ} \int_{V_A}^{V_B} P.dV.$$

Second Operation.

The substance passes from the superior limit of actual energy Q_1, to the inferior limit Q_2. Let V_C be the value of the state V at the end of this operation.

Third Operation.

The substance, preserving the constant quantity Q_2 of actual energy, passes from the state V_C to the state V_D, transmitting to external substances the following quantity of actual energy, produced by the disappearance of potential energy:—

$$H_2 = Q_2 \frac{d}{dQ} \int_{V_D}^{V_C} P \, dV.$$

Fourth Operation.

The substance is brought back to its original actual energy Q_1, and state V_A, thus completing the cycle of operations.

In order that the second and fourth operations may be performed without expenditure of energy, the following condition must be fulfilled.—

$$\frac{d}{dQ} \int_{V_D}^{V_C} P \, dV \text{ (for } Q = Q_2) = \frac{d}{dQ} \int_{V_A}^{V_B} P \, dV \text{ (for } Q = Q_1).$$

This being the case, the total expenditure of energy during a cycle of operations will be H_1, being the quantity converted from the actual to the potential form during the first operation; the energy lost will be H_2, the quantity reconverted to the actual form, and transmitted to external substances, during the third operation; and the quantity of energy permanently transformed from the actual to the potential form, that is to say, the work done by the engine, will be—

$$H_1 - H_2 = (Q_1 - Q_2) \frac{d}{dQ} \int_{V_A}^{V_B} P \, dV \text{ (for } Q = Q_1) \quad . \quad (6.)$$

The ratio of this work to the total expenditure of energy is

$$\frac{H_1 - H_2}{H_1} = \frac{Q_2 - Q_2}{Q_1} \quad . \quad . \quad . \quad (7.)$$

This principle is applicable to all possible engines, known and unknown.

In the sequel of the paper, the author gives some examples of the application of the general principles of the transformation of energy to the theory of heat, and to that of electro-magnetism; and deduces from them, as particular cases, several laws already known through specific researches.

The details of the application of these principles to the theory of heat are contained in the sixth section of a memoir read before the Royal Society of Edinburgh, "On the Mechanical Action of Heat."

The actual energy produced by an electric pile in unity of time is expressed by—

$$Q = M u,$$

where M is the electro-motive force, and u, the strength of the current.

The actual energy of an electric circuit is expressed by—

$$R u^2,$$

where R is the resistance of the circuit. This energy is immediately and totally transformed into sensible heat.

The proportion of the actual energy produced in the pile, which is transformed into mechanical work by an electro-dynamic machine, is represented by—

$$\frac{Q_1 - Q_2}{Q_2} = \frac{M - Ru}{M}$$

The strength of the current is known to be found by means of the equation—

$$u = \frac{M - N}{R}$$

where N is the negative or inverse electro-motive force of the apparatus by means of which electricity is transformed into mechanical work. Hence,

$$\frac{Q_1 - Q_2}{Q_1} = \frac{N}{M}.$$

The above particular forms of the general equation agree with formulæ already deduced from special researches by Mr. Joule and Professor William Thomson.

XII.—OUTLINES OF THE SCIENCE OF ENERGETICS.*

SECTION I.—WHAT CONSTITUTES A PHYSICAL THEORY.

AN essential distinction exists between two stages in the process of advancing our knowledge of the laws of physical phenomena; the first stage consists in observing the relations of phenomena, whether of such as occur in the ordinary course of nature, or of such as are artificially produced in experimental investigations, and in expressing the relations so observed by propositions called formal laws. The second stage consists in reducing the formal laws of an entire class of phenomena to the form of a science; that is to say, in discovering the most simple system of principles, from which all the formal laws of the class of phenomena can be deduced as consequences.

Such a system of principles, with its consequences methodically deduced, constitutes the PHYSICAL THEORY of a class of phenomena.

A physical theory, like an abstract science, consists of definitions and axioms as first principles, and of propositions, their consequences; but with these differences:—First, That in an abstract science, a definition assigns a name to a class of notions derived originally from observation, but not necessarily corresponding to any existing objects of real phenomena; and an axiom states a mutual relation amongst such notions, or the names denoting them: while in a physical science, a definition states properties common to a class of existing objects, or real phenomena; and a physical axiom states a general law as to the relations of phenomena. And, secondly, That in an abstract science, the propositions first discovered are the most simple; whilst in a physical theory, the propositions first discovered are in general numerous and complex, being formal laws, the immediate results of observation and experiment, from which the definitions and axioms are subsequently arrived at by a process of reasoning differing from that whereby one proposition is deduced from another in an abstract science, partly in being more complex and difficult, and partly in being, to a certain extent, *tentative*—that is to say, involving the trial of conjectural principles, and their acceptance or rejection, according as their consequences are found to agree or disagree with the formal laws deduced immediately from observation and experiment.

* Read before the Philosophical Society of Glasgow on May 2, 1855, and published in the *Proceedings* of that Society, Vol. III., No. VI.

SECTION II.—THE ABSTRACTIVE METHOD OF FORMING A PHYSICAL THEORY DISTINGUISHED FROM THE HYPOTHETICAL METHOD.

Two methods of framing a physical theory may be distinguished, characterised chiefly by the manner in which classes of phenomena are defined. They may be termed, respectively, the ABSTRACTIVE and the HYPOTHETICAL methods.

According to the ABSTRACTIVE method, a class of objects or phenomena is defined by describing, or otherwise making to be understood, and assigning a name or symbol to, that assemblage of properties which is common to all the objects or phenomena composing the class, as perceived by the senses, without introducing anything hypothetical.

According to the HYPOTHETICAL method, a class of objects or phenomena is defined, according to a conjectural conception of their nature, as being constituted, in a manner not apparent to the senses, by a modification of some other class of objects or phenomena whose laws are already known. Should the consequences of such a hypothetical definition be found to be in accordance with the results of observation and experiment, it serves as the means of deducing the laws of one class of objects or phenomena from those of another.

The conjectural conceptions involved in the hypothetical method may be distinguished into two classes, according as they are adopted as a probable representation of a state of things which may really exist, though imperceptible to the senses, or merely as a convenient means of expressing the laws of phenomena; two kinds of hypotheses, of which the former may be called *objective*, and the latter *subjective*. As examples of objective hypotheses may be taken, that of vibrations or oscillations in the theory of light, and that of atoms in chemistry; as an example of a subjective hypothesis, that of magnetic fluids.

SECTION III.—THE SCIENCE OF MECHANICS CONSIDERED AS AN ILLUSTRATION OF THE ABSTRACTIVE METHOD.

The principles of the science of mechanics, the only example yet existing of a complete physical theory, are altogether formed from the data of experience by the abstractive method. The class of *objects* to which the science of mechanics relates—viz., material bodies—are defined by means of those sensible properties which they all possess—viz., the property of occupying space, and that of resisting change of motion. The two classes of *phenomena* to which the science of mechanics relates are distinguished by two words, *motion* and *force*—*motion* being a word denoting that which is common to the fall of heavy bodies, the flow of streams, the tides, the winds, the vibrations of sonorous bodies, the revolutions of the stars, and, generally, to all phenomena involving change of the portions of space occupied by bodies; and *force*, a word denoting that which is common to the mutual attractions and repulsions of bodies, distant or near, and of the parts of bodies, the mutual pressure or stress of bodies in contact, and of the parts of bodies, the muscular exertions of animals, and, generally, to all phenomena tending to produce or to prevent motion.

The laws of the composition and resolution of motions, and of the composition and resolution of forces, are expressed by propositions which are the consequences of the definitions of motion and force respectively. The laws of the relations between motion and force are the consequences of certain axioms, being the most simple and general expressions for all that has been ascertained by experience respecting those relations.

SECTION IV.—MECHANICAL HYPOTHESES IN VARIOUS BRANCHES OF PHYSICS.

The fact that the theory of motions and motive forces is the only complete physical theory, has naturally led to the adoption of *mechanical hypotheses* in the theories of other branches of physics; that is to say, hypothetical definitions, in which classes of phenomena are defined conjecturally as being constituted by some kind of motion or motive force not obvious to the senses (called *molecular* motion or force), as when light and radiant heat are defined as consisting in molecular vibrations, thermometric heat in molecular vortices, and the rigidity of solids in molecular attractions and repulsions.

The hypothetical motions and forces are sometimes ascribed to *hypothetical bodies*, such as the luminiferous ether; sometimes to *hypothetical parts*, whereof tangible bodies are conjecturally defined to consist, such as atoms, atomic nuclei with elastic atmospheres, and the like.

A mechanical hypothesis is held to have fulfilled its object, when, by applying the known axioms of mechanics to the hypothetical motions and forces, results are obtained agreeing with the observed laws of the classes of phenomena under consideration; and when, by the aid of such a hypothesis, phenomena previously unobserved are predicted, and laws anticipated, it attains a high degree of probability.

A mechanical hypothesis is the better the more extensive the range of phenomena whose laws it serves to deduce from the axioms of mechanics; and the perfection of such a hypothesis would be, if it could, by means of one connected system of suppositions, be made to form a basis for all branches of molecular physics.

SECTION V.—ADVANTAGES AND DISADVANTAGES OF HYPOTHETICAL THEORIES.

It is well known that certain hypothetical theories, such as the wave theory of light, have proved extremely useful, by reducing the laws of a various and complicated class of phenomena to a few simple principles, and by anticipating laws afterwards verified by observation.

Such are the results to be expected from well-framed hypotheses in every branch of physics, when used with judgment, and especially with that caution which arises from the consideration, that even those hypotheses whose consequences are most fully confirmed by experiment never can, by any amount of evidence, attain that degree of certainty which belongs to observed facts.

Of mechanical hypotheses in particular, it is to be observed, that their tendency is to combine all branches of physics into one system, by making the axioms of mechanics the first principles of the laws of all phenomena—an object for the attainment of which an earnest wish was expressed by Newton.*

In the mechanical theories of elasticity, light, heat, and electricity, considerable progress has been made towards that end.

The neglect of the caution already referred to, however, has caused some hypotheses to assume, in the minds of the public generally, as well as in those of many scientific men, that authority which belongs to facts alone; and a tendency has, consequently, often evinced itself to explain away, or set aside, facts inconsistent with these hypotheses, which facts, rightly appreciated, would have formed the basis of true theories. Thus, the fact of the production of heat by friction, the basis of the true theory of heat, was long neglected, because inconsistent with the hypothesis of caloric; and the fact of the production of cold by electric currents, at certain metallic junctions, the key (as Professor William Thomson recently showed) to the true theory of the phenomena of thermo-electricity, was, from inconsistency with prevalent assumptions respecting the so-called "electric fluid," by some regarded as a thing to be explained away, and by others as a delusion.

Such are the evils which arise from the misuse of hypotheses.

* Utinam cætera naturæ phænomena ex principiis mechanicis eodem argumentandi genere derivare liceret.—(*Phil. Nat. Prin. Math.; Præf.*)

SECTION VI.—ADVANTAGES OF AN EXTENSION OF THE ABSTRACTIVE METHOD OF FRAMING THEORIES.

Besides the perfecting of mechanical hypotheses, another and an entirely distinct method presents itself for combining the physical sciences into one system; and that is, by an *extension of the* ABSTRACTIVE PROCESS in framing theories.

The abstractive method has already been partially applied, and with success, to special branches of molecular physics, such as heat, electricity, and magnetism. We are now to consider in what manner it is to be applied to physics generally, considered as one science.

Instead of supposing the various classes of physical phenomena to be constituted, in an occult way, of modifications of motion and force, let us distinguish the properties which those classes possess in common with each other, and so define more extensive classes denoted by suitable terms. For axioms, to express the laws of those more extensive classes of phenomena, let us frame propositions comprehending as particular cases the laws of the particular classes of phenomena comprehended under the more extensive classes. So shall we arrive at a body of principles, applicable to physical phenomena in general, and which, being framed by induction from facts alone, will be free from the uncertainty which must always attach, even to those mechanical hypotheses whose consequences are most fully confirmed by experiment.

This extension of the abstractive process is not proposed in order to supersede the hypothetical method of theorising; for in almost every branch of molecular physics it may be held, that a hypothetical theory is necessary, as a preliminary step, to reduce the expression of the phenomena to simplicity and order, before it is possible to make any progress in framing an abstractive theory.

SECTION VII.—NATURE OF THE SCIENCE OF ENERGETICS.

Energy, or the capacity to effect changes, is the common characteristic of the various states of matter to which the several branches of physics relate; if, then, there be general laws respecting energy, such laws must be applicable, *mutatis mutandis,* to every branch of physics, and must express a body of principles as to physical phenomena in general.

In a paper read before the Philosophical Society of Glasgow, on the 5th of January, 1853 (*see p. 203*), a first attempt was made to investigate such principles by defining *actual energy* and *potential energy*, and by demonstrating a general law of the mutual transformations of those kinds of energy, of which one particular case is a previously known law of the mechanical action of heat in elastic bodies, and another, a subsequently demonstrated law which forms the basis of Professor William Thomson's theory of thermo-electricity.

The object of the present paper is to present, in a more systematic form, both these and some other principles, forming part of a science whose subjects are, material bodies and physical phenomena in general, and which it is proposed to call the SCIENCE OF ENERGETICS.

SECTION VIII.—DEFINITIONS OF CERTAIN TERMS.

The peculiar terms which will be used in treating of the Science of Energetics are purely abstract; that is to say, they are not the names of any particular object, nor of any particular phenomena, nor of any particular notions of the mind, but are names of very comprehensive *classes* of objects and phenomena. About such classes it is impossible to think or to reason, except by the aid of examples or of symbols. General terms are symbols employed for this purpose.

Substance.

The term "*substance*" will be applied to all bodies, parts of bodies, and systems of bodies. The parts of a substance may be spoken of as distinct substances, and a system of substances related to each other may be spoken of as one complex substance. Strictly speaking, the term should be "*material substance;*" but it is easily borne in mind, that in this essay none but material substances are referred to.

Property.

The term "*property*" will be restricted to *invariable* properties; whether such as always belong to all material substances, or such as constitute the invariable distinctions between one kind of substance and another.

Mass.

Mass means "*quantity of substance.*" Masses of one kind of substance may be compared together by ascertaining the numbers of equal parts which they contain; masses of substances of different kinds are compared by means to be afterwards referred to.

Accident.

The term "*accident*" will be applied to every variable state of substances, whether consisting in a condition of each part of a substance, how small soever, (which may be called an *absolute accident*), or in a physical relation between parts of substances, (which may be called a *relative accident*). Accidents to be the subject of scientific inquiry, must be capable of being measured and expressed by means of quantities. The quantity, even of an absolute accident, can only be expressed by means of a mentally-conceived relation.

The whole condition or state of a substance, so far as it is variable, is a *complex accident;* the independent quantities which are at once *necessary* and *sufficient* to express completely this complex accident, are *independent accidents.* To express the same complex accident, different systems of independent accidents may be employed; but the number of independent accidents in each system will be the same.

Examples.—The variable thermic condition of an elastic fluid is a *complex accident*, capable of being completely expressed by *two independent accidents*, which may be any two out of these three quantities—the *temperature*, the *density*, the *pressure*—or any two independent functions of these quantities.

The condition of strain at a point in an elastic solid, is a *complex accident*, capable of being completely expressed by *six independent accidents*, which may be the three elongations of the dimensions and the three distortions of the faces of a molecule originally cubical, or the lengths and directions of the axes of the ellipsoidal figure assumed by a molecule originally spherical; or any six independent functions of either of those systems of quantities.

The distinction of accidents into absolute and relative is, to a certain extent, arbitrary; thus, the figure and dimensions of a molecule may be regarded as absolute accidents when it is considered as a whole, or as relative accidents when it is considered as made up of parts. Most kinds of accidents are necessarily relative; but some kinds can only be considered as relative accidents when some hypothesis is adopted as to the occult condition of the substances which they affect, as when heat is ascribed hypothetically to molecular motions; and such suppositions are excluded from the present inquiry.

Accidents may be said to be *homogeneous* when the quantities expressing them are capable of being put together, so that the result of the combination of the different accidents shall be expressed by one quantity. The number of heterogeneous kinds of accidents is evidently indefinite.

Effort, or Active Accident.

The term "*effort*" will be applied to every cause which varies, or tends to vary, an accident. This term, therefore, comprehends not merely *forces* or *pressures*, to which it is usually applied, but all causes of variation in the condition of substances.

Efforts may be homogeneous or heterogeneous.

Homogeneous efforts are compared by balancing them against each other.

An effort being a condition of the parts of a substance, or a relation between substances, is itself an accident, and may be distinguished as an "*active accident.*"

With reference to a given limited substance, *internal efforts* are those which consist in actions amongst its parts; *external efforts* those which consist in actions between the given substance and other substances.

Passive Accident.

The condition which an effort tends to vary may be called a "*passive accident*," and when the word "accident" is not otherwise qualified, "passive accident" may be understood.

Radical Accident.

If there be a quantity such that it expresses at once the magnitude of the passive accident caused by a given effort, and the magnitude of the active accident or effort itself, let the condition denoted by that quantity be called a "*radical accident.*"

[The velocity of a given mass is an example of a radical accident, for it is itself a passive accident, and also the measure of the kind of effort called accelerative force, which, acting for unity of time, is capable of producing that passive accident.]

[The strength of an electric current is also a radical accident.]

Effort as a Measure of Mass.

Masses, whether homogeneous or heterogeneous, may be compared by means of the efforts required to produce in them variations of some particular accident. The accident conventionally employed for this purpose is *velocity.*

Work.

"*Work*" is the variation of an accident by an effort, and is a term comprehending all phenomena in which physical change takes place. *Quantity of work* is measured by the product of the variation of the passive accident by the magnitude of the effort, when this is constant; or by the integral of the effort, with respect to the passive accident, when the effort is variable.

Let x denote a passive accident;

X an effort tending to vary it;

W the work performed in increasing x from x_0 to x_1: then,

$$\left.\begin{aligned} W &= \int_{x_0}^{x_1} X\,dx, \text{ and} \\ W &= X\,(x_1 - x_0), \text{ if X is constant.} \end{aligned}\right\} \quad . \quad . \quad (1.)$$

Work is represented geometrically by the area of a curve, whereof the abscissa represents the passive accident, and the ordinate, the effort.

Energy, Actual and Potential.

The term *"energy"* comprehends every state of a substance which constitutes a capacity for performing work. *Quantities of energy* are measured by the quantities of work which they constitute the means of performing.

"Actual energy" comprehends those kinds of capacity for performing work which consist in particular states of each part of a substance, how small soever; that is, in an *absolute accident,* such as heat, light, electric current, *vis viva.* Actual energy is essentially positive.

"Potential energy" comprehends those kinds of capacity for performing work which consist in relations between substances, or parts of substances; that is, in *relative accidents.* To constitute potential energy there must be a *passive accident* capable of variation, and an *effort* tending to produce such variation; the integral of this effort, with respect to the *possible variation* of the passive accident, is *potential energy,* which differs in work from this—that in work the change *has been effected,* which, in potential energy, is *capable of being effected.*

Let x denote an accident; x_1, its actual value; X, an effort tending to vary it; x_0, the value to which the effort tends to bring the accident; then

$$\int_{x_1}^{x_0} X\,dx = U, \text{ denotes potential energy.}$$

Examples of potential energy are, the chemical affinity of uncombined elements; the energy of gravitation, of magnetism, of electrical attraction and repulsion, of electro-motive force, of that part of elasticity which arises from actions between the parts of a body, and, generally, of all mutual actions of bodies, and parts of bodies.

Potential energy may be passive or negative, according as the effort in question is of the same sign with the variation of the passive accident, or of the opposite sign; that is, according as X is of the same sign with dx, or of the opposite sign.

It is to be observed, that the states of substances comprehended under the term *actual energy*, may possess the characteristics of potential energy also; that is to say, may be accompanied by a tendency or effort to vary relative accidents; as heat, in an elastic fluid, is accompanied by a tendency to expand; that is, an effort to increase the volume of the receptacle containing the fluid.

The states to which the term *potential energy* is especially applied, are those which are solely due to mutual actions.

To put a substance into a state of energy, or to increase its energy, is obviously a *kind of work*.

SECTION IX.—FIRST AXIOM.

All kinds of Work and Energy are Homogeneous.

This axiom means, that *any kind of energy may be made the means of performing any kind of work.* It is a fact arrived at by induction from experiment and observation, and its establishment is more especially due to the experiments of Mr. Joule.

This axiom leads, in many respects, to the same consequences with the hypothesis that all those kinds of energy which are not sensibly the results of motion and motive force are the results of occult modifications of motion and motive force.

But the axiom differs from the hypothesis in this, that the axiom is simply the generalised allegation of the facts proved by experience, while the hypothesis involves conjectures as to objects and phenomena which never can be subjected to observation.

It is the truth of this axiom which renders a science of energetics possible.

The efforts and passive accidents to which the branches of physics relate are varied and heterogeneous; but they are all connected with *energy*, a uniform species of quantity which pervades every branch of physics.

This axiom is also equivalent to saying, that *energy is transformable and transferable* (an allegation which, in the previous paper referred to, was included in the definition of energy); for, to *transform energy*, means to employ energy depending on accidents of one kind in putting a substance into a state of energy depending on accidents of another kind; and to *transfer energy*, means to employ the energy of one substance in putting another substance into a state of energy, both of which are kinds of work, and may, according to the axiom, be performed by means of any kind of energy.

SECTION X.—SECOND AXIOM.

The Total Energy of a Substance cannot be altered by the Mutual Actions of its Parts.

Of the truth of this axiom there can be no doubt; but some difference of opinion may exist as to the evidence on which it rests. There is ample experimental evidence from which it might be proved; but independently of such evidence, there is the argument, that the law expressed by this axiom is essential to the stability of the universe, such as it exists.

The special application of this law to mechanics is expressed in two ways, which are virtually equivalent to each other, the principle of *vis viva*, and that of the equality of action and reaction. The latter principle is demonstrated by Newton, from considerations connected with the stability of the universe (*Principia*, Scholium to the Laws of Motion); for he shows, that but for the equality of action and reaction, the earth, with a continually accelerated velocity, would fly away through infinite space.

It follows, from the second axiom, that *all work consists in the transfer and transformation of energy alone;* for otherwise the total amount of energy would be altered. Also, that the energy of a substance can be varied by *external efforts alone.*

SECTION XI.—EXTERNAL POTENTIAL EQUILIBRIUM.

The entire condition of a substance, so far as it is variable, as explained in Sect. VIII., under the head of *accident*, is a complex accident, which may be expressed in various ways by means of different systems of quantities denoting independent accidents; but the number of independent accidents in each system must be the same.

The quantity of work required to produce any change in the condition of the substance, that is to say, the potential energy received by it from without during that change, may in like manner be expressed in different ways by the sums of different systems of integrals of external efforts, each integrated with respect to the independent accident which it tends to augment; but the number of integrals in each system, and the number of efforts, like the number of independent accidents, must be the same; and so also must the sums of the integrals, each sum representing the same quantity of work in a different way.

The different systems of efforts which correspond to different systems of independent accidents, each expressing the same complex accident, may be called *equivalent systems of efforts;* and the finding of a system of efforts equivalent to another may be called conversion of efforts.*

When the law of variation of potential energy, by a change of condition of a substance, is known, the system of external efforts corresponding to any system of independent accidents is found by means of this principle:

Each effort is equal to the rate of variation of the potential energy with respect to the independent accident which that effort tends to vary; or, symbolically,

$$X = \frac{dU}{dx}. \qquad \qquad (2.)$$

EXTERNAL POTENTIAL EQUILIBRIUM *of a substance takes place, when the external effort to vary each of the independent accidents is null;* that is to say, *when the rate of variation of the potential energy of the substance with the variation of each independent accident is null.*

For a given substance there are as many conditions of equilibrium, of the form

$$\frac{dU}{dx} = 0, \qquad \qquad (3.)$$

as there are independent accidents in the expression of its condition.

The special application of this law to motion and motive force constitutes the *principle* of *virtual velocities,* from which the whole science of statics is deducible.

SECTION XII.—INTERNAL POTENTIAL EQUILIBRIUM.

The internal potential equilibrium of a substance consists in the equilibrium of each of its parts, considered separately; that is to say, in the nullity of the rate of variation of the potential energy of each part with respect to each of the independent accidents on which the condition of such part depends.

Examples of particular cases of this principle are, the laws of the equilibrium of elastic solids, and of the distribution of statical electricity.

SECTION XIII.—THIRD AXIOM.

The Effort to Perform Work of a Given Kind, caused by a Given Quantity of Actual Energy, is the Sum of the Efforts caused by the Parts of that Quantity.

A law equivalent to this axiom, under the name of the "GENERAL

* The conversion of efforts in physics is connected with the theory of lineal transformations in algebra.

LAW OF THE TRANSFORMATION OF ENERGY," formed the principal subject of the previous paper already referred to. (*See p. 203.*)

This axiom appears to be a consequence of the definition of actual energy, as a capacity for performing work possessed by each part of a substance independently of its relations to other parts, rather than an independent proposition.

Its applicability to natural phenomena arises from the fact, that there are states of substances corresponding to the definition of actual energy.

The mode of applying this third axiom is as follows:—

Let a homogeneous substance possess a quantity Q, of a particular kind of actual energy, uniformly distributed, and let it be required to determine the amount of the effort arising from the actual energy, which tends to perform a particular kind of work W, by the variation of a particular passive accident x.

The total effort to perform this kind of work is represented by the rate of its increase relatively to the passive accident, viz.,—

$$X = \frac{dW}{dx}.$$

Divide the quantity of actual energy Q into an indefinite number of indefinitely small parts δQ; the portion of the effort X due to each of those parts will be

$$\delta Q \frac{dX}{dQ},$$

and adding these partial efforts together, the effort caused by the whole quantity of actual energy will be

$$Q \frac{dX}{dQ} = Q \frac{d^2 W}{dQ\, dx}. \qquad (4.)$$

If this be equal to the *effective effort* X, then that effort is simply proportional to, and wholly caused by, the actual energy Q. This is the case of the pressure of a perfect gas, and the centrifugal force of a moving body.

If the effort caused by the actual energy differs from the effective effort, their difference represents, when the former is the less, an additional effort,

$$\left(1 - Q \frac{d}{dQ}\right) X,$$

and when the former is the greater, a counter effort (5.)

$$\left(Q \frac{d}{dQ} - 1\right) X,$$

due to some other cause or causes.

SECTION XIV.—RATE OF TRANSFORMATION; METAMORPHIC FUNCTION.

The effort to augment a given accident x, caused by actual energy of a given kind Q, may also be called the "*rate of transformation*" of the given kind of actual energy, with increase of the given accident; for the limit of the amount of actual energy which disappears in performing work by an indefinitely small augmentation dx, of the accident, is

$$d\mathrm{H} = \mathrm{Q}\frac{d\mathrm{X}}{a\mathrm{Q}}dx \quad . \quad . \quad . \quad . \quad (6.)$$

$$= \mathrm{Q}\frac{d^2\mathrm{W}}{d\mathrm{Q}\,dx}\,dx = \mathrm{Q}\,d\frac{d\mathrm{W}}{d\mathrm{Q}}$$

The *last* form of the above expression is obviously applicable when the work W is the result of the variation of any number of independent accidents, each by the corresponding effort. For example, let x, y, z, &c., be any number of independent accidents, and X, Y, Z, &c., the efforts to augment them; so that

$$d\mathrm{W} = \mathrm{X}\,dx + \mathrm{Y}\,dy + \mathrm{Z}\,dz + \&c.$$

Then,

$$d\mathrm{H} = \mathrm{Q}\left\{\frac{d\mathrm{X}}{d\mathrm{Q}}dx + \frac{d\mathrm{Y}}{d\mathrm{Q}}dy + \frac{d\mathrm{Z}}{d\mathrm{Q}}dz + \&c.\right\} \quad . \quad (7.)$$

$$= \mathrm{Q}\,d\frac{d\mathrm{W}}{d\mathrm{Q}}, \text{ as before.}$$

The function of actual energy, efforts, and passive accidents, denoted by

$$\frac{d\mathrm{W}}{d\mathrm{Q}} = \int\frac{d\mathrm{H}}{\mathrm{Q}} = \mathrm{F}, \quad . \quad . \quad . \quad . \quad (8.)$$

whose variation, multiplied by the actual energy, gives the amount of actual energy transformed in performing the work $d\mathrm{W}$, may be called the "METAMORPHIC FUNCTION" of the kind of actual energy Q, relatively to the kind of work W.

When this metamorphic function is known for a given homogeneous substance, the quantity H of actual energy of the kind Q transformed to the kind of work W, during a given operation, is found by taking the integral

$$\mathrm{H} = \int \mathrm{Q}\,d\mathrm{F}. \quad . \quad . \quad . \quad . \quad (9.)$$

The transformation of actual energy into work by the variation of passive accidents is a *reversible operation;* that is to say, if the passive accidents be made to vary to an equal extent in an opposite direction, potential energy will be exerted upon the substance, and transformed into actual energy: a case represented by the expression (9) becoming negative.

The metamorphic function of heat relatively to expansive power, was first employed in a paper on the Economy of Heat in Expansive Machines, read before the Royal Society of Edinburgh in April, 1851. (*Trans. Roy. Soc. Edin.*, Vol. XXI.)

The metamorphic function of heat relatively to electricity was employed by Professor William Thomson, in a paper on Thermo-Electricity, read before the Royal Society of Edinburgh in May, 1854 (*Trans. Roy. Soc. Edin.*, Vol. XXI.), and was the means of anticipating some most remarkable laws, afterwards confirmed by experiment.

Section XV.—Equilibrium of Actual Energy; Metabatic Function.

It is known by experiment, that a state of actual energy is directly transferable; that is to say, the actual energy of a particular kind (such as heat), in one substance, may be diminished, the sole work performed being an equal augmentation of the same kind of actual energy in another substance.

Equilibrium of actual energy of a particular kind Q between substances A and B, takes place when the tendency of B to transfer this kind of energy to B is equal to the tendency of B to transfer the same kind of energy to A.

Laws respecting the equilibrium of particular kinds of actual energy have been ascertained by experiment, and in some cases anticipated by means of mechanical hypotheses, according to which all actual energy consists in the *vis viva* of motion.

The following law will now be proved, respecting the equilibrium of actual energy of all possible kinds :—

Theorem.—If equilibrium of actual energy of a given kind take place between a given pair of substances, possessing respectively quantities of actual energy of that kind in a given ratio, then that equilibrium will subsist for every pair of quantities of actual energy bearing to each other the same ratio.

Demonstration.—The tendency of one substance to transfer actual energy of the kind Q to another, must depend on some sort of effort, whose nature and laws may be known or unknown. Let Y_A be this effort for the substance A, Y_B the corresponding effort for the substance B. Then a condition of equilibrium of actual energy is

$$Y_A = Y_B. \quad . \quad . \quad . \quad . \quad (10.)$$

The effort Y may or may not be proportionate to the actual energy Q multiplied by a quantity independent of Q.

Case first.—If it is so proportional, let

$$Y = \frac{1}{K} Q,$$

K being independent of Q; then the condition of equilibrium becomes

$$\frac{1}{K_A} Q_A = \frac{1}{K_B} Q_B,$$

or

$$\frac{Q_B}{Q_A} = \frac{K_B}{K_A},$$

a ratio independent of the absolute amounts of actual energy.

Case second.—If the effort Y is not simply proportional to the actual energy Q, the portion of it caused by that actual energy, according to the principle of Sect. XIII., deduced from the third axiom, is, for each substance,

$$Q \frac{dY}{dQ},$$

and a second condition of equilibrium of actual energy is furnished by the equation

$$Q_A \frac{dY}{dQ_A} = Q_B \frac{dY_B}{dQ_B}. \quad . \quad . \quad . \quad . \quad (11.)$$

In order that this condition may be fulfilled simultaneously with the condition (10), it is necessary that

$$\frac{dQ_A}{Q_A} = \frac{dQ_B}{Q_B},$$

that is to say, that the ratio of the quantities of actual energy in the two substances should be independent of those quantities themselves; a condition expressed, as before, by

$$\frac{Q_B}{Q_A} = \frac{K_B}{K_A}. \quad . \quad . \quad . \quad . \quad . \quad (11.)$$

Q.E.D.

This ratio is a quantity to be ascertained by experiment, and may be called the ratio of the SPECIFIC ACTUAL ENERGIES of the substances A and B, for the kind of energy under consideration.

The function

$$\frac{Q_A}{K_A} = \frac{Q_B}{K_B} = \theta, \quad . \quad . \quad . \quad . \quad (12.)$$

whose identity for the two substances expresses the condition of equilibrium of the actual energy Q between them, may be called the "METABATIC FUNCTION" for that kind of energy.

In the science of thermo-dynamics, the metabatic function is *absolute temperature;* and the factor K is *real specific heat.* The theorem stated above, when applied to heat, amounts to this: *that the real specific heat of a substance is independent of its temperature.*

SECTION XVI.—USE OF THE METABATIC FUNCTION; TRANSFORMATION OF ENERGY IN AN AGGREGATE.

From the mutual proportionality of the actual energy Q, and the metabatic function θ, it follows that the operations

$$Q \frac{d}{dQ}, \quad \theta \frac{d}{d\theta}$$

are equivalent; and that the latter may be substituted for the former in all the equations expressing the laws of the transformation of energy. We have therefore

$$Q \frac{dX}{dQ} = \theta \frac{dX}{d\theta} = \theta \frac{d^2 W}{d\theta \, dx}, \quad . \quad . \quad . \quad (13.)$$

for the effort to transform actual energy of the kind Q into work of the kind W, when expressed in terms of the metabatic function; and

$$dH = \theta \, d \frac{dW}{d\theta}, \quad . \quad . \quad . \quad . \quad (14.)$$

for the limit of the indefinitely small transformation produced by an indefinitely small variation of the accidents on which the kind of work W depends.

There is also a form of *metamorphic function,*

$$\phi = \frac{dW}{d\theta} = \int \frac{dH}{\theta} = K F, \quad . \quad . \quad . \quad (15.)$$

suited for employment along with the metabatic function, in order to find, by the integration

$$H = \int \theta \, d \phi, \quad . \quad . \quad . \quad . \quad (16.)$$

the quantity of actual energy of a given kind Q transformed to the kind of work W during any finite variation of accidents.

The advantage of the above expressions is, that they are applicable not merely to a homogeneous substance, but to any *heterogeneous substance or aggregate*, which is internally in a state of equilibrium of actual and potential energy; for throughout all the parts of an aggregate in that condition, the metabatic function θ is the same, and each of the efforts X, &c., is the same, and consequently the metamorphic function ϕ is the same.

"*Carnôt's function*" in thermo-dynamics is proportional to the reciprocal of the metabatic function of heat.

Section XVII.—Efficiency of Engines.

An engine is, a contrivance for transforming energy, by means of the periodical repetition of a cycle of variations of the accidents of a substance.

The *efficiency* of an engine is the proportion which the energy permanently transformed to a useful form by it, bears to the whole energy communicated to the working substance.

In a *perfect engine* the cycle of variations is thus:—

I. The metabatic function is increased, say from θ_0 to θ_1.

II. The metamorphic function is increased by the amount $\Delta \phi$.

III. The metabatic function is diminished from θ_1 back to θ_0.

IV. The metamorphic function is diminished by the amount $\Delta \phi$.

During the second operation, the energy received by the working substance, and transformed from the actual to the potential form is $\theta_1 \Delta \phi$. During the fourth operation energy is transformed back, to the amount $\theta_0 \Delta \phi$. So that the energy permanently transformed during each cycle is $(\theta_1 - \theta_0) \Delta \phi$; and the efficiency of the engine $\frac{\theta_1 - \theta_0}{\theta_1}$.

Section XVIII.—Diffusion of Actual Energy; Irreversible or Frictional Operations.

There is a tendency in every substance, or system of substances, to the *equable diffusion* of actual energy; that is to say, to its transfer between the parts of the substance or system, until the value of the *metabatic function* becomes uniform.

This process is *not directly reversible;* that is to say, there is no such operation as a direct concentration of actual energy through a tendency of the metabatic function to become unequal in different parts of a substance or system.

Hence arises the impossibility of using the energy reconverted to the actual form at the lower limit of the metabatic function in an engine.

There is an analogy in respect of this property of *irreversibility,* between the diffusion of one kind of actual energy and certain irreversible transformations of one kind of actual energy to another, called by Professor William Thomson, "Frictional Phenomena"—viz., the production of heat by rubbing, and agitation, and by electric currents in a homogeneous substance at a uniform temperature.

In fact, a conjecture may be hazarded, that immediate diffusion of the actual energy produced in frictional phenomena, is the circumstance which renders them irreversible; for, suppose a small part of a substance to have its actual energy increased by the exertion of some kind of work upon it, then, if the increase of actual energy so produced be immediately diffused amongst other parts, so as to restore the uniformity of the metabatic function, the whole process will be irreversible. This speculation, however, is, for the present, partly hypothetical; and, therefore, does not, strictly speaking, form part of the science of energetics.

Section XIX.—Measurement of Time.

The general relations between energy and time must form an important branch of the science of energetics; but for the present, all that I am prepared to state on this subject is the following DEFINITION OF EQUAL TIMES:—

Equal times are the times in which equal quantities of the same kind of work are performed by equal and similar substances, under wholly similar circumstances.

Section XX.—Concluding Remarks.

It is to be observed, that the preceding articles are not the results of a new and hitherto untried speculation, but are the generalised expression of a method of reasoning which has already been applied with success to special branches of physics.

In this brief essay, it has not been attempted to do more than to give an outline of some of the more obvious principles of the science of energetics, or the abstract theory of physical phenomena in general; a science to which the maxim, true of all science, is specially applicable—that its subjects are boundless, and that they never can, by human labours, be exhausted, nor the science brought to perfection.

XIII.—ON THE PHRASE "POTENTIAL ENERGY," AND ON THE DEFINITIONS OF PHYSICAL QUANTITIES.*

1. In the course of an essay by Sir John Herschel "On the Origin of Force," which appeared some time ago in the *Fortnightly Review*, and has lately been republished in a volume, entitled *Familiar Lectures on Scientific Subjects*, the opinion is expressed that the phrase "Potential Energy" is "unfortunate, inasmuch as it goes to substitute a truism for the announcement of a great dynamical fact" (*Familiar Lectures*, page 469).

2. There is here no question as to the reality of the class of relations amongst bodies to which that phrase is applied, nor as to any matter of fact concerning those relations, but as to the convenient and appropriate use of language. This is a sort of question in the discussion of which authority has much weight; and when an objection to the appropriateness of a term is made by an author who is not less eminent as a philosopher than as a man of science, and whose skill in the art of expressing scientific truth in clear language is almost unparalleled, it becomes the duty of those who use that term to examine carefully their grounds for doing so.

3. As the phrase "Potential Energy," now so generally used by writers on physical subjects, was first proposed by myself in a paper "On the General Law of the Transformation of Energy,"† read before the Philosophical Society of Glasgow, on the 5th of January, 1853 (*see p. 203*), I feel that the remark of Sir John Herschel makes it incumbent upon me to explain the reasons which led me, after much consideration, to adopt that phrase for the purpose of denoting all those relations amongst bodies, or the parts of bodies, which consist in a power of doing work dependent on mutual configurations.

4. The kind of quantity now in question forms part of the subject of the thirty-ninth proposition of Newton's *Principia;* but it is there represented by the area of a figure, or by symbols only, and not designated by a name; and such is also the case in many subsequent mathematical writings.

5. The application of the word "force" to that kind of quantity is open

* Read before the Philosophical Society of Glasgow on Jan. 23, 1867, and published in the *Proceedings* of that Society, Vol. VI., No. III.

† Viz.,—that the effect of the presence of a quantity of actual energy, in causing transformation of energy between the actual and the potential forms, is the sum of the effects of all the parts of that quantity.

to the objection, that when "force" is taken in the sense in which Newton defines "vis motrix," the power of performing work is not simply force, but force multiplied by space. To make such an application of the word "force," therefore, would have been to designate a product by the name properly belonging to one of its factors, and would have added to the confusion which has already arisen from the ambiguous employment of that word.

6. The word "power," though at first sight it might seem very appropriate, was already used in mechanics in at least three different senses:—viz., *first*, the power of an engine, meaning the rate at which it performs work, and being the product of force and space divided by time; *secondly*, the power, in the sense of effort or pressure, which drives a machine; and *thirdly*, "mechanical powers," meaning certain elementary machines. Thus "power" was open to the same sort of objection with "force."

7. About the beginning of the present century, the word "energy" had been substituted by Dr. Thomas Young for "vis viva," to denote the capacity for performing work due to velocity; and the application of the same word had at a more recent time been extended by Sir William Thomson to capacity of any sort for performing work. There can be no doubt that the word "energy" is specially suited for that purpose; for not only does the meaning to be expressed harmonise perfectly with the etymology of *ενέργεια*, but the word "energy" has never been used in precise scientific writings in a different sense; and thus the risk of ambiguity is avoided.

8. It appeared to me, therefore, that what remained to be done, was to qualify the noun "energy" by appropriate adjectives, so as to distinguish between energy of activity and energy of configuration. The well-known pair of antithetical adjectives, "actual" and "potential," seemed exactly suited for that purpose; and I accordingly proposed the phrases "actual energy" and "potential energy," in the paper to which I have referred.

9. I was encouraged to persevere in the use of those phrases, by the fact of their being immediately approved of and adopted by Sir William Thomson; a fact to which I am disposed to ascribe, in a great measure, the rapid extension of their use in the course of a period so short in the history of science as fourteen years.* I had also the satisfaction of receiving a very strong expression of approval from the late Professor Baden Powell.

10. Until some years afterwards I was not aware of the fact, that the idea of a phrase equivalent to "potential energy," in its purely mechanical sense, had been anticipated by Carnot, who, in an essay on machines in general, employed the term "force vive virtuelle," of which "potential

* Sir William Thomson and Professor Tait have lately substituted the word "kinetic" for "actual."

energy" might be supposed to be almost a literal translation. That coincidence shows how naturally the phrase "potential energy," or something equivalent, occurs to one in search of words appropriate to denote that power of performing work which is due to configuration, and not to activity.

11. Having explained the reasons which led me to propose the use of the phrase "potential energy," I have next to make some observations on the objection made by Sir John Herschel to that phrase, that "it goes to substitute a truism for a great dynamical fact."

12. It must be admitted that the use of the term "potential energy" tends to make the statement of the law of the conservation of energy wear, to a certain extent, the appearance of a truism. It seems to me, however, that such must always be the effect of denoting physical relations by words that are specially adapted to express the properties of those relations; or, what amounts virtually to the same thing, of drawing up precise and complete definitions of physical terms. Let A and B denote certain conceivable relations, and let them be precisely and completely defined; then, from the definitions follows the proposition, that A and B are related to each other in a certain way; and that proposition wears the appearance of a truism, and is virtually comprehended in the definitions. But it is not a bare truism; for when with the definitions are conjoined the two facts, ascertained by experiment and observation, that there are relations amongst real bodies corresponding to the definition of A, and that there are also relations amongst real bodies corresponding to the definition of B, the proposition as to relation between A and B becomes not a bare truism, but a physical fact. In the present case, for example, "actual energy" and "potential energy" are defined in such a way as to make the proposition: That what a body or a system of bodies gains in one form of energy through mutual actions, it loses in the other form—in other words, that the sum of actual and potential energies is "conserved"—follow from the definitions, so as to sound like a truism; but when it is proved by experiment and observation that there are relations amongst real bodies agreeing with the definitions of "actual energy" and "potential energy," that which otherwise would be a truism becomes a fact.

13. A definition cannot be true or false; for it makes no assertion, but says, "let such a word or phrase be used in such a sense;" but it may be *real* or *fantastic*, according as the description contained in it corresponds, or not, to real objects and phenomena; and when, by the aid of experiment and observation, a set of definitions have been framed which possess reality, precision, and completeness, the investing of a physical fact with the appearance of a truism is often an unavoidable consequence of the use of the term so defined.

14. In the case of physical *quantities* in particular, the definition involves a rule for measuring the quantity; and the proof of the reality of the

definition is the fact, that the application of the rule to the same quantity under different circumstances gives consistent results, which it would not do if the definition were fantastic; and hence the definitions of a set of physical quantities necessarily involve mathematical relations amongst those quantities, which, when expressed as propositions and compared with the definitions, wear the appearance of truisms, and are at the same time statements of fact.

15. In illustration of the foregoing principles, it may be pointed out that there is a certain set of definitions of the measurement of time, force, and mass, which reduce the laws of motion to the form of truisms, thus—

I. Let "*equal times*" mean the times in which a moving body, under the influence of no force, describes equal spaces. This definition is proved to be real by the fact, that times which are equal when compared by means of the free motion of one body, are equal when compared by means of the free motion of any other body. If the definition were fantastic, times might be equal as measured by the free motion of one body, and unequal as measured by that of another.

II. Let "*force*" mean a relation between a pair of bodies such that their relative velocity changes, or tends to change, in magnitude or direction, or both; and let "*equal forces*" mean those which act when equal changes of the relative velocity of a given pair of bodies occur in equal times. This definition is proved to be real by the fact, that the comparative measurements of forces made in different intervals of time are consistent with each other, which would not be the case if the definition were fantastic.

III. Let the "*mass*" of a body mean a quantity inversely proportional to the change of velocity impressed on that body in a given time by a given force. This definition is proved to be real by the fact, that the ratio of the masses of two given bodies is found experimentally to be always the same, when those masses are compared by means of the velocities impressed on them by different forces, and in different times; and is also the same, whether each of the masses is measured as a whole or as the sum of a set of parts.

Assuming those definitions as merely verbal, without reference to their reality, the laws of motion take the form of verbal truisms; but when experiment and observation inform us that permanent relations exist amongst real bodies and real events corresponding to the definitions, those apparent truisms become statements of fact.

16. One of the chief objects of mathematical physics is to ascertain, by the help of experiment and observation, what physical quantities or functions are "conserved." Such quantities or functions are, for example—

I. The *mass* of every particle of matter, conserved at all times and under all circumstances.

II. The *resultant momentum* of a body, or system of bodies, conserved so long as internal forces act alone.

III. The *resultant angular momentum* of a body or system of bodies, conserved so long as internal forces act alone.

IV. The *total energy* of a body, or system of bodies, conserved so long as internal forces act alone.

V. The *thermo-dynamic function*, conserved in a body while it neither receives nor gives out heat.

In defining such physical quantities as those, it is almost, if not quite, impossible to avoid making the definition imply the property of conservation; so that when the fact of conservation is stated, it has the form of a truism.

17. In conclusion, it appears to me that the making of a physical law wear the appearance of a truism, so far from being a ground of objection to the definition of a physical term, is rather a proof that such definition has been framed in strict accordance with reality.

XIV.—ON THE MECHANICAL ACTION OF HEAT, ESPECIALLY IN GASES AND VAPOURS.*

INTRODUCTION—SUMMARY OF THE PRINCIPLES OF THE HYPOTHESIS OF MOLECULAR VORTICES, AND ITS APPLICATION TO THE THEORY OF TEMPERATURE, ELASTICITY, AND REAL SPECIFIC HEAT.

THE ensuing paper forms part of a series of researches respecting the consequences of an hypothesis called that of Molecular Vortices, the object of which is, to deduce the laws of elasticity, and of heat as connected with elasticity, by means of the principles of mechanics, from a physical supposition consistent and connected with the theory which deduces the laws of radiant light and heat from the hypothesis of undulations. Those researches were commenced in 1842, and after having been laid aside for nearly seven years, from the want of experimental data, were resumed in consequence of the appearance of the experiments of M. Regnault on gases and vapours.

The investigation which I have now to describe, relates to the mutual conversion of heat and mechanical power by means of the expansion and contraction of gases and vapours.

In the introduction, which I here prefix to it, I purpose to give such a summary of the principles of the hypothesis as is necessary to render the subsequent investigation intelligible.

The fundamental suppositions are the following:—

First, That each atom of matter consists of a nucleus, or central physical point, enveloped by an elastic atmosphere, which is retained in its position by forces attractive towards the nucleus or centre.

Suppositions similar to this have been brought forward by Franklin, Æpinus, Mossotti, and others. They have in general, however, conceived the atmosphere of each nucleus to be of variable mass. I have treated it, on the contrary, as an essential part of the atom. I have left the question indeterminate, whether the nucleus is a small body of a character distinct from that of the atmosphere, or merely a portion of the atmosphere in a highly condensed state, owing to the mutual attraction of its parts.

According to this first supposition, the boundary between two con-

* Read before the Royal Society of Edinburgh on Feb. 4, 1850, and published in the *Transactions* of that Society, Vol. XX., Part I. (*See also p. 16.*)

tiguous atoms of a body is an imaginary surface at which the attractions of all the atomic centres of the body balance each other; and the elasticity of the body is made up of two parts: *First*, the elasticity of the atomic atmospheres at the imaginary boundaries of the atoms, which I shall call the superficial-atomic elasticity; and, *secondly*, the force resulting from the mutual actions of distinct atoms. If the atmospheres are so much condensed round their nuclei or centres, that the superficial-atomic elasticity is insensible, and that the resultants of the mutual actions of all parts of the distinct atoms are forces acting along the lines joining the nuclei or centres, then the body is a perfect solid, having a tendency to preserve not only a certain bulk, but a certain figure; and the elasticity of figure, or *rigidity*, bears certain definite relations to the elasticity of volume.

If the atmospheres are less condensed about their centres, so that the mutual actions of distinct atoms are not reducible to a system of forces acting along the lines joining the atomic centres, but produce merely a cohesive force sufficient to balance the superficial-atomic elasticity, then the condition is that of a *perfect liquid;* and the intermediate conditions between this and perfect solidity constitute the gelatinous, plastic, and viscous states.

When the mutual actions of distinct atoms are very small as compared with the superficial-atomic elasticity, the condition is that of *gas* or *vapour;* and when the substance is so far rarefied that the influence of the atomic nuclei or centres in modifying the superficial elasticity of their atmospheres is insensible, it is then in the state of *perfect gas.*

So far as our experimental knowledge goes, the elasticity of a perfect gas, at a given temperature, varies simply in proportion to its density. I have therefore assumed this to be the law of the elasticity of the atomic atmospheres, ascribing a specific coefficient of elasticity to each substance.

The *second supposition*, being that from which the hypothesis of molecular vortices derives its name, is the following:—*That the elasticity due to heat arises from the centrifugal force of revolutions or oscillations among the particles of the atomic atmospheres; so that quantity of heat is the* vis viva *of those revolutions or oscillations.*

This supposition appears to have been first definitely stated by Sir Humphry Davy. It has since been supported by Mr. Joule, whose valuable experiments to establish the convertibility of heat and mechanical power are well known. So far as I am aware, however, its consequences have not hitherto been mathematically developed.

To connect this hypothesis with the undulatory theory of radiation, I have introduced a *third supposition: That the medium which transmits light and radiant heat consists of the* nuclei *of the atoms, vibrating independently, or almost independently, of their atmospheres;* so that the absorption of light and of radiant heat, is the transference of motion from the nuclei to their atmospheres; and the emission of light and of radiant heat, the transference of motion from the atmospheres to their nuclei.

Although in all undulations of sensible length and amplitude, such as those of sound, the nuclei must carry their atmospheres along with them, and vibrating thus loaded, produce a comparatively slow velocity of propagation; yet, in all probability, the minute vibrations of light and radiant heat may be performed by the atomic nuclei in transparent and diathermanous bodies, without moving the atmospheres more than by that amount which constitutes absorption; and those vibrations will therefore be transmitted according to the laws of the elasticity of *perfect solids*, and with a rapidity corresponding to the extreme smallness of the masses set in motion, as compared with the mutual forces exerted by them.

This supposition is peculiar to my own view of the hypothesis, and is, in fact, the converse of the idea hitherto adopted, of an ether surrounding ponderable particles.

The second and third suppositions involve the assumption, that motion can be communicated between the nuclei and their atmospheres, and between the different parts of the atmospheres; so that there is a tendency to produce some permanent condition of motion, which constitutes equilibrium of heat. It is now to be considered what kind of motion is capable of producing increase of elasticity, and what are the conditions of permanency of that motion.

It is obvious, that the parts of the atomic atmospheres may have motions of alternate expansion and contraction, or of rectilinear oscillation about a position of equilibrium, without affecting the superficial atomic elasticity, except by small periodical changes. Should they have motions, however, of *revolution* about centres, so as to form a group of *vortices*, the centrifugal force will have the effect of increasing the density of the atmosphere at what I have called the bounding surfaces of the atoms, and thus of augmenting the elasticity of the body.

In this summary, I shall not enter into the details of mathematical analysis, but shall state results only. The following, then, are the conditions which must be fulfilled, in order that a group of vortices, of small size as compared with the bulk of an atom, and of various diameters, may permanently co-exist, whether side by side, or end to end, in the atomic atmospheres of one substance, or of various substances mixed.

First, The mean elasticity must vary continuously, which involves the condition, that at the surface of contact of two vortices of different substances, side by side, or end to end, the respective densities at each point of contact must be inversely proportional to the coefficients of elasticity. Hence, *the specific gravities of the* atmospheric parts of *all substances, under precisely similar circumstances as to heat and molecular forces* (a condition realised in perfect gases at the same pressure and temperature), *are inversely proportional to the coefficients of atmospheric elasticity.* Therefore, let μ represent the mass of the atmosphere of one atom of any substance, b its coefficient of elasticity, and n the number of atoms which, in the state of perfect gas, occupy unity of volume under unity of pressure at the temperature of melting ice;—then

$$n \mu b \quad . \quad . \quad . \quad . \quad . \quad \text{(I.)}$$

is a constant quantity for all substances.

Secondly, The superficial elasticity of a vortex must *not* be a function of its diameter: to fulfil which condition, the linear velocity of revolution must be equal throughout all parts of each individual vortex.

Thirdly, In all contiguous vortices of the same substance, the velocities of revolution must be equal; and in contiguous vortices of different substances, the squares of the velocities must be proportional to the coefficients of elasticity of the molecular atmospheres.

The second and third conditions are those of equilibrium of heat, and are equivalent to this law:

TEMPERATURE *is a function of the square of the velocity of revolution in the molecular vortices, divided by the coefficient of elasticity of the atomic atmospheres;* or

$$\text{Temperature} = \phi\left(\frac{w^2}{b}\right), \quad . \quad . \quad . \quad \text{(II.)}$$

where w represents that velocity.

The *mean* elasticity which a vortex exerts endways is not affected by its motion, being equal to

$$b \rho, \quad . \quad . \quad . \quad . \quad . \quad . \quad \text{(III.)}$$

where ρ is its mean density. The superficial elasticity at its lateral surfaces, however, is expressed by

$$\frac{w^2 \rho}{2 g} + b \rho. \quad . \quad . \quad . \quad . \quad . \quad \text{(IV.)}$$

The additional elasticity $\frac{w^2 \rho}{2 g}$, being that which is due to the motion, is independent of the diameter. The divisor g (the force of gravity) is introduced, on the supposition of the density ρ being measured by weight.

Supposing the atmosphere of an atom to be divided into concentric spherical layers, it may be shown that the effect of the co-existence of a great number of small vortices in one of those layers whose radius is r, and mean density ρ, is to give it a centrifugal force, expressed by

$$\frac{w^2}{g r}, \quad . \quad . \quad . \quad . \quad . \quad . \quad \text{(V.)}$$

which tends to increase the density and elasticity of the atmosphere at the surface, which I have called the boundary of the atom. The layer is also acted upon by the difference between the mean elasticities at its two surfaces, and by the attraction towards the atomic centre; and these three forces must balance each other.

I have integrated the differential equation which results from this condition, for substances in the gaseous state, in which the forces that interfere with the centrifugal force and atmospheric elasticity are comparatively small; and the result is

$$P = b \frac{\mu}{M} D \left(\frac{w^2}{3 g b} + 1 \right) (1 - F) + f(D). \qquad \text{(VI.)}$$

P is the entire elasticity of the gas, and D its mean density. M represents the total mass of an atom, measured by weight, and μ that of its atmospheric part; so that $\frac{\mu}{M} D$ is the mean density of the atomic atmospheres.

$f(D)$ denotes the effect of the mutual actions of separate atoms.

The first term represents the superficial-atomic elasticity. F denotes the effect of the attraction of the nucleus in modifying that elasticity, and can be represented approximately by a converging series, in terms of the negative powers of $\frac{w^2}{3 g b} + 1$, commencing with the inverse square, the coefficients being functions of the density D.

By using the first term of such a series, and determining its coefficient and the quantity $f(D)$ empirically, I have obtained formulæ agreeing closely with the results of M. Regnault's experiments on the expansion of atmospheric air, carbonic acid, and hydrogen.

In a perfect gas, the above expression is reduced to

$$P = b \frac{\mu}{M} D \left(\frac{w^2}{3 g b} + 1 \right). \qquad \text{(VII.)}$$

Let n, as before, denote the number of atoms of a substance which, in the state of perfect gas, occupy unity of volume under unity of pressure, at the temperature of melting ice, so that $n M$ is its specific gravity in that state: then,

$$P = \frac{D}{n M} n \mu b \left(\frac{w^2}{3 g b} + 1 \right). \qquad \text{(VIII.)}$$

The factor by which $\frac{D}{n M}$ is here multiplied, fulfils the condition of being a function of $\frac{w^2}{b}$, and of constants which are the same for all substances, and is, therefore, fitted for a measure of temperature. It obviously varies proportionally to the pressure of a perfect gas of a given density, or its volume under a given pressure.

Let τ, therefore, denote temperature, as measured from an *imaginary zero*, C degrees of the scale adopted, below the temperature of melting ice, at which

$$\frac{w^2}{3\,g\,b} + 1 = 0;$$

Then, for all substances

$$\tau = C\,n\,\mu\,b\left(\frac{w^2}{3\,g\,b} + 1\right),$$

and in perfect gases

$$P = \frac{\tau}{C}\,\frac{D}{n\,M}. \quad \ldots \quad \text{(IX.)}$$

τ may be termed *absolute temperature*, and the point from which it is measured, the *absolute zero of temperature*. This, as I have observed, is an imaginary point, being lower than the absolute zero of heat by the quantity $C\,n\,\mu\,b$, which is the same for all substances.

The value of C, or the absolute temperature of melting ice, as determined from M. Regnault's experiments, is

$$274^\circ{\cdot}6 \text{ Centigrade},$$

being the reciprocal of

$$0{\cdot}00364166 \text{ per Centigrade degree},$$

the value to which the coefficients of dilatation of gases at the temperature of melting ice approximate as they are rarefied.

For Fahrenheit's scale $C = 494^\circ{\cdot}28$.

In the sequel I shall represent temperatures measured from that of melting ice by

$$T = \tau - C.$$

We have now to consider the absolute quantity of heat, or of molecular *vis viva* which corresponds to a given temperature in a given substance. It is obvious that

$$\frac{\mu\,w^2}{2\,g}$$

represents, in terms of gravity, the portion of *vis viva*, in one atom, due to the molecular vortices; but besides the vortical motion, there may be oscillations of expansion and contraction, or of rectilinear vibration about a position of equilibrium. The velocity with which these additional motions are performed will be in a permanent condition, when the mean value of its square, independent of small periodic changes, is equal throughout the atomic atmosphere. We may therefore represent by

$$\frac{\mu v^2}{2g} = k \frac{\mu w^2}{2g}, \quad . \quad . \quad . \quad . \quad (X.)$$

the total *vis viva* of the atomic atmosphere. To this we have to add that of the nucleus, raising the quantity of heat in one atom to

$$\frac{M v^2}{2g} = q,$$

while the quantity of heat in unity of weight is . . . (XI.)

$$\frac{v^2}{2g} = Q.$$

The coefficient k (which enters into the value of specific heat) being the ratio of the *vis viva* of the entire motion impressed on the atomic atmospheres by the action of their nuclei, to the *vis viva* of a peculiar kind of motion, may be conjectured to have a specific value for each substance depending, in a manner as yet unknown, on some circumstance in the constitution of its atoms. Although it varies in some cases for the same substance in the solid, liquid, and gaseous states, there is no experimental evidence that it varies for the same substance in the same condition. In the investigation which follows, therefore, I have treated it as sensibly constant.

The following, then, are the expressions for quantity of heat in terms of temperature: in one atom,—

$$q = \frac{v^2}{2g} M = \frac{3kM}{2Cn\mu}(\tau - Cn\mu b).$$

In unity of weight,— (XII.)

$$Q = \frac{v^2}{2g} = \frac{3k}{2Cn\mu}(\tau - Cn\mu b).$$

Real specific heat is, consequently, expressed by the following equations:—

For one atom,—

$$\frac{dq}{d\tau} = \frac{3\,k\,\mathrm{M}}{2\,\mathrm{C}\,n\,\mu}.$$

For unity of weight,—

$$\frac{d\mathrm{Q}}{d\tau} = \frac{3\,k}{2\,\mathrm{C}\,n\,\mu}. \qquad \text{(XIII.)}$$

For so much of a perfect gas as occupies unity of volume under unity of pressure at the temperature of melting ice,—

$$n\frac{dq}{d\tau} = \frac{3\,k\,\mathrm{M}}{2\,\mathrm{C}\,\mu}.$$

The laws established experimentally by Dulong, that the specific heats of simple atoms, and of certain groups of compound atoms, bear to each other simple ratios, generally that of equality, and that the specific heats of equal volumes of all simple gases are equal, show that the specific factor $\frac{k\,\mathrm{M}}{\mu}$ depends on the chemical constitution of the atom, and thus confirm the conjecture I have stated respecting the coefficient k.

As I shall have occasion, in the investigation which follows, to refer to and to use the equation for the elasticity of vapours in contact with their liquids, which I published in the *Edinburgh New Philosophical Journal* for July, 1849, I shall here state generally the nature of the reasoning from which it was deduced.

The equilibrium of a vapour in contact with its liquid depends on three conditions:

First, The total elasticity of the substance in the two states must be the same.

Secondly, The superficial atomic elasticity must vary continuously; so that if at the surface which reflects light there is an abrupt change of density (which seems almost certain), there must there be two densities corresponding to the same superficial-atomic elasticity.

Thirdly, The two forces which act on each stratum of vapour parallel to the surface of the liquid—namely, the preponderance of molecular attraction towards the liquid, and the difference of the superficial-atomic elasticities at the two sides of the stratum—must be in equilibrio.

Close to the surface of the liquid, therefore, the vapour is highly condensed. The density diminishes rapidly as the distance from the liquid increases, and at all appreciable distances has a sensibly uniform value, which is a function of the temperature and of certain unknown molecular forces.

The integration of a differential equation representing the third condition of equilibrium, indicates the *form* of the approximate equation,

$$\text{Log P} = a - \frac{\beta}{\tau} - \frac{\gamma}{\tau^2}, \quad . \quad . \quad . \quad \text{(XIV.)}$$

the coefficients of which have been determined empirically by three experimental data for each fluid. For proofs of the extreme closeness with which the formulæ thus obtained agree with experiment, I refer to the Journal in which they first appeared.

I annex a table of the coefficients for water, alcohol, ether, turpentine, petroleum, and mercury, in the direct equation, and also in the inverse formula,

$$\frac{1}{\tau} = \sqrt{\frac{a - \log \text{P}}{\gamma} + \frac{\beta^2}{4\gamma^2}} - \frac{\beta}{2\gamma}, \quad . \quad . \quad \text{(XV.)}$$

by which the temperature of vapour at saturation may be calculated from the pressure. In the ninth and tenth columns are stated the limits between which the formulæ have been compared with experiment.

For turpentine, petroleum, and mercury, the formula consists of two terms only.

$$\text{Log P} = a - \frac{\beta}{\tau}, \quad . \quad . \quad . \quad \text{(XVI.)}$$

the small range of the experiments rendering the determination of γ impossible.

The following are some additional values of the constant a for steam, corresponding to various units of pressure used in practice.

Units of Pressure.	Values of a.
ATMOSPHERES of 760 millimètres of mercury,	
=29·922 inches of mercury,	
=14·7 lbs. on the square inch,	
=1·0333 kilogrammes on the square centimètre,	4·950433
ATMOSPHERES of 30 inches of mercury,	
=761·99 millimètres,	
=14·74 lbs. on the square inch,	
=1·036 kilogrammes on the square centimètre,	4·949300
Kilogrammes on the square centimètre, ...	4·964658
Kilogrammes on the circular centimètre, ...	4·859748
Pounds avoirdupois on the square inch, ...	6·117662
Pounds avoirdupois on the circular inch, ...	6·012752
Pounds avoirdupois on the square foot, ...	8·276025

All the numerical values of the constants are for common logarithms.

TABLE OF THE CONSTANTS IN THE FORMULÆ FOR THE ELASTICITIES OF VAPOURS IN CONTACT WITH THEIR LIQUIDS.

(1.) Names of the Fluids.	(2.) Scale of Pressures.	(3.) Scale of Temperatures.	(4.) a.	(5.) Log. β.	(6.) Log. γ.	(7.) $\frac{\beta}{2\gamma}$	(8.) $\frac{\beta^2}{4\gamma^2}$	(9.) Range of Temperatures.	(10.) Range of Pressures.
Water, ...	Millims. of mercury.	Centigrade.	7·831247	3·1851091	5·0827176	·0063294	·00004006	Centigrade. −30° to + 230°	Millimètres. 0·35 to 20945
Water, ...	Inches of mercury.	Fahrenheit.	6·426421	3·4403816	5·5932626	·0035163	·000012364	Fahrenheit. −22° to +446°	Inches. 0·014 to 824·63
Alcohol, spec. gr. 0·813.	Inches of mercury.	Fahrenheit.	6·16620	3·3165220	5·7602709	·0017998	·000003239	+32° to 264°	0·41 to 165·58
Ether, boiling at 105° F.	Inches of mercury.	Fahrenheit.	5·33590	3·2084573	5·5119893	·0024856	·000006178	105° to 210°	30·00 to 163·27
Ether, boiling at 104° F.	Inches of mercury.	Fahrenheit.	5·44580	3·2571312	5·3962460	·0036296	·000013174	34° to 104°	6·20 to 30·00
Turpentine, . .	Inches of mercury.	Fahrenheit.	5·98187	3·5380701	...	...	...	304° to 362°	30·00 to 62·24
Petroleum, . . .	Inches of mercury.	Fahrenheit.	6·19451	3·5648490	...	...	...	316° to 375°	30·00 to 64·50
Mercury, . . .	Millims. of mercury.	Centigrade.	7·5305	3·4685511	...	...	...	Centigrade. 72°·74 to 358°	Millimètres. 0·115 to 760
Mercury, . . .	Inches of mercury.	Fahrenheit.	6·1259	3·7238236	...	...	...	Fahrenheit 162°·93 to 676°·4	Inches. ·0046 to 29·92

SECTION I.—OF THE MUTUAL CONVERSION OF HEAT AND EXPANSIVE POWER.

1. The quantity of heat in a given mass of matter, according to the hypothesis of molecular vortices, as well as every other hypothesis which ascribes the phenomena of heat to motion, is measured by the mechanical power to which that motion is equivalent, that being a quantity the total amount of which, in a given system of bodies, cannot be altered by their mutual actions, although its distribution and form may be altered. This is expressed in equation XII. of the introduction, where the quantity of heat in unity of weight, Q, is represented by the height $\frac{v^2}{2g}$, from which a body must fall in order to acquire the velocity of the molecular oscillations. This height, being multiplied by the weight of a body, gives the mechanical power to which the oscillations constituting its heat are equivalent. The real specific heat of unity of weight, as given in equation (XIII.) of the introduction,

$$\frac{dQ}{d\tau} = \frac{3k}{2Cn\mu},$$

represents the *depth of fall*, which is equivalent to *one degree of rise of temperature* in any given weight of the substance under consideration.

We know, to a greater or less degree of precision, the ratios of the specific heats of many substances to each other, and they are commonly expressed by taking that of water at the temperature of melting ice as unity; but their actual mechanical values have as yet been very imperfectly ascertained, and, in fact, the data necessary for their determination are incomplete.

2. Mr. Joule, indeed, has made several very interesting series of experiments, in order to ascertain the quantity of heat developed in various substances by mechanical power employed in different ways—viz., by electric currents excited by the rotation of a magnet, by the forcing of water through narrow tubes, by the agitation of water and oil with a paddle, by the compression of air, and by the friction of air rushing through a narrow orifice. The value of the depth of fall equivalent to a rise of one degree of Fahrenheit's scale in the temperature of a mass of water, as determined by that gentleman, varies, in the different series of experiments, between the limits of 760 feet and 890 feet, the value in which Mr. Joule appears to place the greatest confidence being about 780 feet.

Although the smallness of the differences of temperature measured in those experiments renders the numerical results somewhat uncertain, it appears to me that, as evidence of the convertibility of heat and mechanical power, they are unexceptionable. Nevertheless, there is reason to believe that the true mechanical equivalent of heat is considerably less than any of the values deduced from Mr. Joule's experiments; for in all of them there are causes of loss of power the effect of which it is impossible to calculate. In all machinery, a portion of the power which disappears is carried off by waves of condensation and expansion, along the supports of the machine, and through the surrounding air: this portion cannot be estimated, and is, of course, not operative in producing heat within the machine. It is also impossible to calculate, where friction is employed to produce heat, what amount of it has been lost in the production of electricity, a power which is, no doubt, convertible into heat, but which, in such experiments, probably escapes without undergoing that conversion. To make the determination of the mechanical equivalent of heat by electro-magnetic experiments correct, it is necessary that the whole of the mechanical power should be converted into magnetic power, the whole of the magnetic power into what are called electric currents, and the whole of the power of the electric currents into heat, not one of which conditions is likely to be exactly fulfilled. Even in producing heat by the compression of air, it must not be assumed that the whole of the mechanical power is expended in raising the temperature.

3. The best means of determining the mechanical equivalent of heat are furnished by those experiments in which no machinery is employed. Of this kind are experiments on the velocity of sound in air and other gases, which, according to the received and well-known theory of Laplace, is accelerated by the heat developed by the compression of the medium.

The accuracy of this theory has lately been called in question. There can be no doubt that it deviates from absolute exactness, in so far that the magnitude of the displacements of the particles of air is neglected in comparison with the length of a wave. It appears to me, however, that the Astronomer-Royal, in his remarks on the subject in the *London and Edinburgh Philosophical Magazine* for July, 1849, has shown, in a satisfactory manner, that although the effect of the appreciable magnitude of those displacements, as compared with the length of a wave of sound, is to alter slowly the form of the function representing the wave, still that effect is not sufficiently great to make Laplace's theory practically erroneous. I have, therefore, in the sequel, adhered to the experiments of Dulong, and to those quoted by Poisson, on the velocity of sound, as the best data for determining the mechanical equivalent of heat.

4. The expression already given for the real specific heat of unity of weight of a given substance may be resolved into two factors, thus :—

$$\frac{d\,Q}{d\,\tau} = \frac{1}{C\,n\,M} \times \frac{3\,k\,M}{2\,\mu}. \quad . \quad . \quad . \quad (1.)$$

The first factor, $\frac{1}{C\,n\,M}$, may be considered in general as a known quantity; for C represents, as already stated, 274·6 Centigrade degrees, the absolute temperature of melting ice, and n M the theoretical weight, in the perfectly gaseous state, of unity of volume of the substance, under unity of pressure, at that temperature; or what is the same thing, $\frac{1}{n\,M}$ is the height of an imaginary column of the substance, of uniform density, and at the temperature of melting ice, whose pressure by weight upon a given area of base is equal to its pressure by elasticity, supposing it to be perfectly gaseous. The determination of the ratio $\frac{3\,k\,M}{2\,\mu}$ is necessary to complete the solution of the problem.

5. The relation now to be investigated between heat and mechanical power, is that which exists between the power expended in compressing a body into a smaller volume, and the increase of heat in consequence of such a compression; and, conversely, between the heat which disappears, or, as it is said, becomes *latent* during the expansion of a body to a greater volume, and the mechanical power gained or developed by that expansion. Those phenomena, according to the hypothesis now under consideration, as well as every hypothesis which ascribes heat to motion, are simply the transformation of mechanical power from one shape into another.

It is obvious, in the first place, without the aid of algebraical symbols, that the general effect of the compression of an oscillating atomic atmosphere, or molecular vortex, must be to accelerate its motion, and of its dilatation, to retard its motion; for every portion of such an atmosphere is urged toward the nucleus or atomic centre by a centripetal force equal to the centrifugal force arising from the oscillation; so that when, by compression, each portion of the atmosphere is made to *approach* the centre by a given distance, the *vis viva* of its motion will be *increased* by the amount corresponding to the centripetal force acting through that distance; and, conversely, when by expansion each portion of the atmosphere is made to *retreat* from the centre, the *vis viva* of its motion will be *diminished* by a similar amount.

It is not, however, to be taken for granted, that *all* the power expended in compressing a body appears in the form of heat. More or less power may be consumed or developed by changes of molecular arrangement, or of the internal distribution of the density of the atomic atmospheres; and changes of molecular arrangement or distribution may develop or consume heat, independently of changes of volume.

6. We shall now investigate, according to the hypothesis of mole-

cular vortices, the amount of heat produced by an indefinitely small compression of one atom of a body in that state of perfect fluidity which admits of the bounding surface of the atom being treated as if it were spherical: its radius being denoted by R, and the radius of any internal spherical layer of the atmosphere by multiplying R by a fraction u.

I shall denote by the ordinary symbol of differentiation d, such changes as depend on the various positions of portions of the atomic atmosphere relatively to each other, when changes of volume and temperature are not taken into consideration; while by the symbol δ of the calculus of variations, I shall represent such changes as arise from the variations of volume and temperature.

Let us consider the case of an indefinitely thin spherical layer of the atomic atmosphere, whose distance from the nucleus is Ru, its thickness $\mathrm{R}\,du$, its area $4\,\pi\,\mathrm{R}^2 u^2$, and its density $\frac{\mu}{\mathrm{M}}\mathrm{D}\,\psi\,(u, \mathrm{D}, \tau)$.

The weight, then, of this layer is

$$4\,\pi\,\mathrm{R}^3\,\frac{\mu}{\mathrm{M}}\mathrm{D}\,u^2\,\psi\,(u, \mathrm{D}, \tau)\,d\,u.$$

Its velocity of oscillation is v, and having, in virtue of that velocity, a mean centrifugal force, as explained in the introduction (Equation V.), equal to

$$\text{its weight} \times \left(\frac{v^2}{g\,k\,\mathrm{R}\,u} = \frac{2\,\mathrm{Q}}{k\,\mathrm{R}\,u}\right)$$

it is kept in equilibrio by an equal and opposite centripetal force, arising from attraction and elastic pressure, which is consequently represented by

$$4\,\pi\,\mathrm{R}^2\,\frac{\mu}{\mathrm{M}}\,\frac{v^2}{g\,k}\mathrm{D}\,u\,\psi\,(u, \mathrm{D}, \tau)\,d\,u$$

$$= 8\,\pi\,\mathrm{R}^2\,\frac{\mu}{k\,\mathrm{M}}\,\mathrm{Q}\,\mathrm{D}\,u\,\psi\,(u, \mathrm{D}, \tau)\,d\,u.$$

Let the mean density of the atom now be increased by the indefinitely small quantity $\delta\,\mathrm{D}$. Then the layer will approach the nucleus through the distance $-\delta\,(\mathrm{R}u) = -u\,\delta\,\mathrm{R} - \mathrm{R}\,\delta\,u$, and being acted upon through that distance by the centripetal force already stated, the *vis viva* of oscillation will be increased by a quantity corresponding to the mechanical power (that is to say, the *heat*), represented by the product of that distance by that force, or by

$$-\,8\,\pi\,\mathrm{R}^2\,\frac{\mu}{k\,\mathrm{M}}\,\mathrm{Q}\,\mathrm{D}\,u\,\psi\,(u, \mathrm{D}, \tau)\,d\,u \times \delta\,(\mathrm{R}\,u)$$

$$= -8\pi R^3 \frac{\mu}{k M} Q D \psi(u, D, \tau) u^2 \left(\frac{\delta R}{R} + \frac{\delta u}{u}\right) du$$

which, because

$$\frac{\delta R}{R} = -\frac{1}{3} \cdot \frac{\delta D}{D}, \text{ and } \frac{4\pi R^3 D}{3} = M, \text{ is equal to}$$

$$+ Q M . \frac{2\mu}{k M} \psi(u, D, \tau) u^2 \left(\frac{\delta D}{D} - 3\frac{\delta u}{u}\right) du.$$

We must suppose that the velocity of oscillation is equalised throughout the atomic atmosphere, by a propagation of motion so rapid as to be practically instantaneous.

Then, if the above expression be integrated with respect to du, from $u=0$ to $u=1$, the result will give the whole increase of heat in the atom arising from the condensation δD; and dividing that integral by the atomic weight M, we shall obtain the corresponding development of heat in unity of weight. This is expressed by the following equation:—

$$\delta Q' = 2 Q \frac{\mu}{k M} \left\{ \frac{\delta D}{D} \int_0^1 du . u^2 \psi(u, D, \tau) - 3 \int_0^1 du . u \,\delta u \,\psi(u, D, \tau) \right\} \quad . \quad (2.)$$

The letter Q′ is here introduced to denote, when negative, that heat which is consumed in producing changes of volume and of molecular arrangement; and when positive, as in the above equation, the heat which is produced by such changes.

The following substitutions have to be made in equation (1) of this Section :

For Q is to be substituted its value, according to equation XII. of the introduction; or abbreviating $C n \mu b$ into κ,—

$$Q = \frac{3k}{2 C n \mu}(\tau - \kappa). \quad . \quad . \quad . \quad (3.)$$

The value of the first integral in equation (2) of this Section is

$$\int_0^1 du . u^2 \psi(u, D, \tau) = \frac{1}{3}.$$

The value of the second integral,

$$-3 \int_0^1 du . u \,\delta u \,\psi(u, D, \tau)$$

remains to be investigated. The first step in this inquiry is given by the condition, that whatsoever changes of magnitude a given spherical layer undegoes, the portion of atmosphere between it and the nucleus is invariable. This condition is expressed by the equation

$$0 = \left(\delta u \frac{d}{du} + \delta \tau \frac{d}{d\tau} + \delta \mathrm{D} \frac{d}{d\mathrm{D}}\right) \int_0^u du \, . \, u^2 \psi (u, \mathrm{D}, \tau), \qquad (4.)$$

from which it follows that

$$\delta u = - \frac{1}{u^2 \psi (u, \mathrm{D}, \tau)} \left(\delta \tau \frac{d}{d\tau} + \delta \mathrm{D} \frac{d}{d\mathrm{D}}\right) \int_0^u du \, . \, u^2 \psi (u, \mathrm{D}, \tau),$$

and, consequently, that

$$- 3 \int_0^1 du \, . \, u \, \delta u \, \psi (u, \mathrm{D}, \tau) =$$

$$+ \left(\delta \tau \, . \, \frac{d}{d\tau} + \delta \mathrm{D} \frac{d}{d\mathrm{D}}\right) 3 \int_0^1 \frac{du}{u} \int_0^u du \, . \, u^2 \psi (u, \mathrm{D}, \tau).$$

Hence, making

$$9 \int_0^1 \frac{du}{u} \int_0^u du \, . \, u^2 \psi (u, \mathrm{D}, \tau) = \mathrm{U}. \quad . \quad . \quad (5.)$$

The second integral in equation (2) is transformed into

$$+ \frac{1}{3} \left(\delta \tau \frac{d}{d\tau} + \delta \mathrm{D} \frac{d}{d\mathrm{D}}\right) \mathrm{U}.$$

By means of those substitutions we obtain for the mechanical value of the heat developed in unity of weight of a fluid by any indefinitely small change of volume or of molecular distribution—

$$\delta \mathrm{Q}' = \frac{\tau - \kappa}{\mathrm{C} n \mathrm{M}} \left(\delta \mathrm{D} \left(\frac{1}{\mathrm{D}} + \frac{d \mathrm{U}}{d \mathrm{D}}\right) + \delta \tau \frac{d \mathrm{U}}{d\tau}\right)$$

or taking $\mathrm{V} = \frac{1}{\mathrm{D}}$ to denote the volume of unity of weight of the substance, . (6.)

$$\delta \mathrm{Q}' = - \frac{\tau - \kappa}{\mathrm{C} n \mathrm{M}} \left(\delta \mathrm{V} \left(\frac{1}{\mathrm{V}} - \frac{d \mathrm{U}}{d \mathrm{V}}\right) - \delta \tau \frac{d \mathrm{U}}{d\tau}\right)$$

Of this expression, the portion $\frac{\tau - \kappa}{\mathrm{C} n \mathrm{M}} \cdot \frac{\delta \mathrm{D}}{\mathrm{D}} = - \frac{\tau - \kappa}{\mathrm{C} n \mathrm{M}} \cdot \frac{\delta \mathrm{V}}{\mathrm{V}}$ represents the variation of heat arising from mere change of volume.

$\frac{\tau - \kappa}{\mathrm{C} n \mathrm{M}} \delta \mathrm{V} \frac{d \mathrm{U}}{d \mathrm{V}} = \frac{\tau - \kappa}{\mathrm{C} n \mathrm{M}} \delta \mathrm{D} \frac{d \mathrm{U}}{d \mathrm{D}}$, denotes the variation of heat produced by change of molecular distribution dependent on change of volume.

$\frac{\tau - \kappa}{C n M} \delta \tau \frac{d U}{d \tau}$ expresses the variation of heat due to change of molecular distribution dependent on change of temperature.

7. The function U is one depending on molecular forces, the nature of which is as yet unknown. The only case in which it can be calculated directly is that of a perfect gas. Without giving the details of the integration, it may be sufficient to state, that in this case

$$U = \frac{\kappa}{\tau},$$

and, therefore, that

$$\left.\begin{aligned} U &= \frac{\kappa}{\tau}, \\ \frac{d U}{d \tau} &= -\frac{\kappa}{\tau^2}; \quad \frac{d U}{d V} = 0. \end{aligned}\right\} \quad . \quad . \quad . \quad (7.)$$

In all other cases, however, the value of this function can be determined indirectly, by introducing into the investigation the principle of the *conservation of vis viva.*

Suppose a portion of any substance, of the weight *unity,* to pass through a variety of changes of temperature and volume, and at length to be brought back to its primitive volume and temperature. Then the absolute quantity of heat in the substance, and the molecular arrangement and distribution, being the same as at first, the effect of their changes is eliminated; *and the algebraical sum of the vis viva expanded and produced, whether in the shape of expansion and compression, or in that of heat, must be equal to zero:* that is to say, if, on the whole, any mechanical power has appeared, and been given out from the body, in the form of expansion, an equal amount must have been communicated to the body, and must have disappeared in the form of heat; and if any mechanical power has appeared and been given out from the body in the form of heat, an equal amount must have been communicated to the body, and must have disappeared in the form of compression. This principle expressed symbolically is

$$\Delta \Pi + \Delta Q' = 0, \quad . \quad . \quad . \quad . \quad (8.)$$

where Π, when positive, represents expansive power given out, when negative, compressive power absorbed; and Q' represents, when positive, heat given out, when negative, heat absorbed.

To take the simplest case possible, let the changes of temperature and of volume be supposed to be indefinitely small, and to occur during distinct intervals of time, so that τ and V are independent variables; let the initial absolute temperature be τ, the initial volume V, and the initial total elasticity P; and let the substance go through the following four changes:

First, Let its temperature be raised from τ to $\tau + \delta\tau$, the volume remaining unchanged. Then the quantity of heat absorbed is

$$-\delta\tau\left(\frac{dQ}{d\tau} - \frac{\tau - \kappa}{C n M}\frac{dU}{d\tau}\right),$$

and there is no expansion nor compression.

Secondly, Let the body expand, without change of temperature, from the volume V to the volume $V + \delta V$. Then the quantity of heat absorbed is

$$-\delta V \,.\, \frac{\tau + \delta\tau - \kappa}{C n M}\left(\frac{1}{V} - \frac{d}{dV}\,.\,(U + \frac{dU}{d\tau}\delta\tau)\right),$$

while the power given out by expansion is

$$\delta V\,(P + \frac{dP}{d\tau}\delta\tau).$$

Thirdly, Let the temperature fall from $\tau + \delta\tau$, to its original value τ, the volume $V + \delta V$ continuing unchanged; then the heat given out is

$$+\delta\tau\left(\frac{dQ}{d\tau} - \frac{\tau - \kappa}{C n M}\frac{d}{d\tau}(U + \frac{dU}{dV}\delta V)\right),$$

and there is no expansion nor compression.

Fourthly, Let the body be compressed, without change of temperature, to its original volume V; then the heat given out is

$$+\delta V\frac{\tau - \kappa}{C n M}\left(\frac{1}{V} - \frac{dU}{dV}\right),$$

while the power absorbed in compression is $-\delta V\,.\,P$. The body being now restored in all respects to its primitive state, the sum of the two portions of power connected with change of volume, must, in virtue of the principle of *vis viva,* be equal to the sum of the four quantities of heat with their signs reversed. Those additions being made, and the sums divided by the common factor $\delta V\,\delta\tau$, the following equation is obtained:

$$\frac{dP}{d\tau} = \frac{1}{C n M}\left(\frac{1}{V} - \frac{dU}{dV}\right). \qquad (9.)$$

The integral of this partial differential equation is

$$U = \phi\,.\,\tau + \int dV\left(\frac{1}{V} - C n M\frac{dP}{d\tau}\right). \qquad (10.)$$

Now $\phi . \tau$ being the same for all densities, is the value of U for the perfectly gaseous state, or $\frac{\kappa}{\tau}$; for in that state, the integral $= 0$. The values of the partial differential coefficients are accordingly

$$\left.\begin{aligned} \frac{d\,\mathrm{U}}{d\,\mathrm{V}} &= \frac{1}{\mathrm{V}} - \mathrm{C}\,n\,\mathrm{M}\frac{d\,\mathrm{P}}{d\,\tau} \\ \frac{d\,\mathrm{U}}{d\,\tau} &= -\frac{\kappa}{\tau^2} - \mathrm{C}\,n\,\mathrm{M}\int d\,\mathrm{V}\,.\,\frac{d^2\,\mathrm{P}}{d\,\tau^2} \end{aligned}\right\} \quad (11.)$$

and they can, therefore, be determined in all cases in which the quantity $\kappa = \mathrm{C}\,n\,\mu\,b$, and the law of variation of the total elasticity with the volume and temperature are known, so as to complete the data required in order to apply equation 6 of this section to the calculation of the mechanical value of the variations of heat, due to changes of volume and molecular arrangement.

The total elasticity of an imperfect gas, according to equations VI. and XII. of the introduction, being

$$\mathrm{P} = \frac{\tau}{\mathrm{C}\,n\,\mathrm{M}\,\mathrm{V}}\left(1 - \mathrm{F}\left(\mathrm{D}, \frac{\tau}{\kappa}\right)\right) + f(\mathrm{D}),$$

its first and second partial differential coefficients with respect to the temperature are

$$\frac{d\,\mathrm{P}}{d\,\tau} = \frac{1}{\mathrm{C}\,n\,\mathrm{M}\,\mathrm{V}}\left(1 - \left(1 + \tau\frac{d}{d\,\tau}\right)\mathrm{F}\left(\mathrm{D}, \frac{\tau}{\kappa}\right)\right)$$

$$\frac{d^2\,\mathrm{P}}{d\,\tau^2} = -\frac{1}{\mathrm{C}\,n\,\mathrm{M}\,\mathrm{V}}\left(2\frac{d}{d\,\tau} + \tau\frac{d^2}{d\,\tau^2}\right)\mathrm{F}\left(\mathrm{D}, \frac{\tau}{\kappa}\right).$$

Consequently, for the imperfectly gaseous state,

$$\left.\begin{aligned} \mathrm{U} &= \frac{\kappa}{\tau} + \left(1 + \tau\frac{d}{d\,\tau}\right)\int d\,\mathrm{V}\,.\,\frac{\mathrm{F}\left(\mathrm{D}, \frac{\tau}{\kappa}\right)}{\mathrm{V}} \\ \frac{d\,\mathrm{U}}{d\,\mathrm{V}} &= \frac{1}{\mathrm{V}}\,.\left(1 + \tau\frac{d}{d\,\tau}\right)\mathrm{F}\left(\mathrm{D}, \frac{\tau}{\kappa}\right) \\ \frac{d\,\mathrm{U}}{d\,\tau} &= -\frac{\kappa}{\tau^2} + \left(2\frac{d}{d\,\tau} + \tau\frac{d^2}{d\,\tau^2}\right)\int d\,\mathrm{V}\,.\,\frac{\mathrm{F}\left(\mathrm{D}, \frac{\tau}{\kappa}\right)}{\mathrm{V}} \end{aligned}\right\} \quad (12.)$$

8. It is to be observed that the process followed in ascertaining the nature of the function U is analogous to that employed by M. Carnot in his theory of the motive power of heat, although founded on contrary

principles, and leading to different results. Carnot, in fact, considers heat to be something of a peculiar kind, whether a condition or a substance, the total amount of which in nature is incapable of increase or of diminution. It is not, therefore, according to his theory, convertible into mechanical power; but is capable, by its transmission through substances under particular circumstances, of causing mechanical power to be developed.

He supposes a body to go through certain changes of temperature and volume, and to return at last to its primitive volume and temperature, and conceives, in accordance with his view of the nature of heat, that it must have given out exactly the same quantity of heat that it has absorbed. The transmission of this heat he regards as the cause of the production of an amount of mechanical power depending on the quantity of heat transmitted, and on the temperature at which the transmission has taken place. According to these principles, a body, having received a certain quantity of heat, is capable of giving out, not only all the heat it has received, but also a quantity of mechanical power which did not before exist.

According to the theory of this essay, on the contrary, and to every conceivable theory which regards heat as a modification of motion, no mechanical power can be given out in the shape of expansion, unless the quantity of heat emitted by the body in returning to its primitive temperature and volume is *less* than the quantity of heat originally received; the excess of the latter quantity above the former disappearing as heat, to appear as expansive power, so that the sum of the *vis viva* in these two forms continues unchanged.

SECTION II.—OF REAL AND APPARENT SPECIFIC HEAT, ESPECIALLY IN THE STATE OF PERFECT GAS.

9. The apparent specific heat of a given substance is found by adding to the real specific heat (or the heat which retains its form in producing an elevation of one degree of temperature in unity of weight) that additional heat which disappears in producing changes of volume and of molecular arrangement, and which is determined by reversing the sign of Q^1 in equation 6 of Sect. I (so as to transform it from heat evolved to heat absorbed), and taking its *total* differential coefficient with respect to the temperature. Hence, denoting total apparent specific heat by K,

$$K = \frac{dQ}{d\tau} - \frac{d.Q^1}{d\tau} = \frac{dQ}{d\tau} - \frac{dQ^1}{d\tau} - \frac{dQ^1}{dV}\cdot\frac{dV}{d\tau}$$

$$= \frac{1}{CnM}\left\{\frac{3kM}{2\mu} + (\tau - \kappa)\left(\frac{dV}{d\tau}\left(\frac{1}{V} - \frac{dU}{dV}\right) - \frac{dU}{d\tau}\right)\right\} \quad (13.)$$

Another mode of expressing this coefficient is the following:

Denote the ratio $\frac{2\mu}{3kM}$ by N, and the real specific heat by

$$\mathfrak{k} = \frac{1}{C n M N}, \quad . \quad . \quad . \quad . \quad (14.)$$

then

$$K = \mathfrak{k}\left\{1 + N(\tau - \kappa)\left(\frac{dV}{d\tau}\left(\frac{1}{V} - \frac{dU}{dV}\right) - \frac{dU}{d\tau}\right)\right\}. \quad (15.)$$

The value of $\frac{dV}{d\tau}$ is to be determined from the conditions of each particular case, so that each substance may have a variety of apparent specific heats, according to the manner in which the volume varies with the temperature.

If the volume is not permitted to vary, so that $\frac{dV}{d\tau} = 0$, there is obtained the following result, being the apparent specific heat at constant volume:—

$$K_V = \frac{1}{C n M}\left(\frac{1}{N} - (\tau - \kappa)\frac{dU}{d\tau}\right) = \mathfrak{k}\left(1 - N(\tau - \kappa)\frac{dU}{d\tau}\right). \quad (16.)$$

10. Then the substance under consideration is a perfect gas, it has already been stated, equation (7), that

$$\frac{dU}{d\tau} = -\frac{\kappa}{\tau^2}, \qquad \frac{dU}{dV} = 0,$$

and because the volume of unity of weight is directly as the absolute temperature and inversely as the pressure,

$$\frac{1}{V}\frac{dV}{d\tau} = \frac{1}{\tau} - \frac{1}{P}\frac{dP}{d\tau}. \quad . \quad . \quad . \quad (17.)$$

Hence, the following are the values of the apparent specific heats of unity of weight of a theoretically perfect gas under different circumstances:—

General value of the total apparent specific heat:

$$
\left.
\begin{aligned}
K &= \frac{1}{CnM}\left\{\frac{1}{N} + (\tau - \kappa)\left(\frac{\kappa}{\tau^2} + \frac{dV}{Vd\tau}\right)\right\} \\
&= \frac{1}{CnM}\left\{\frac{1}{N} + (\tau - \kappa)\left(\frac{\kappa}{\tau^2} + \frac{1}{\tau} - \frac{dP}{Pd\tau}\right)\right\} \\
&\text{Apparent specific heat at constant volume:} \\
K_V &= \frac{1}{CnM}\left\{\frac{1}{N} + \frac{\kappa}{\tau} - \frac{\kappa^2}{\tau^2}\right\} = \mathfrak{k}\left(1 + N\left(\frac{\kappa}{\tau} - \frac{\kappa^2}{\tau^2}\right)\right) \\
&\text{Apparent specific heat under constant pressure:} \\
K_P &= \frac{1}{CnM}\left(\frac{1}{N} + 1 - \frac{\kappa^2}{\tau^2}\right) = \mathfrak{k}\left\{1 + N\left(1 - \frac{\kappa^2}{\tau^2}\right)\right\}
\end{aligned}
\right\} \quad (18.)
$$

The ratio of the apparent specific heat under constant pressure to the apparent specific heat at constant volume, is the following:—

$$
\frac{K_P}{K_V} = \frac{1 + N\left(1 - \frac{\kappa^2}{\tau^2}\right)}{1 + N\left(\frac{\kappa}{\tau} - \frac{\kappa^2}{\tau^2}\right)} = 1 + N\frac{1 - \frac{\kappa}{\tau}}{1 + N\left(\frac{\kappa}{\tau} - \frac{\kappa^2}{\tau^2}\right)} \quad (19.)
$$

The value of κ is unknown, and, as yet, no experimental data exist from which it can be determined. I have found, however, that practically, results of sufficient accuracy are obtained by regarding κ as so small in comparison with τ, that $\frac{\kappa}{\tau}$, and *à fortiori* $\frac{\kappa^2}{\tau^2}$, may be neglected in calculation.

Thus are obtained the following approximate results for perfect gases, and gases which may without material error be treated as perfect.

General value of the total apparent specific heat:—

$$
\left.
\begin{aligned}
K &= \frac{1}{CnM}\left(\frac{1}{N} + \frac{\tau}{V}\frac{dV}{d\tau}\right) = \mathfrak{k} + P\frac{dV}{d\tau} \\
&= \frac{1}{CnM}\left(\frac{1}{N} + 1 - \frac{\tau}{P}\frac{dP}{d\tau}\right) \\
&\text{Apparent specific heat at constant volume:—} \\
K_V &= \frac{1}{CnMN} = \mathfrak{k}, \\
&\text{being equal to the real specific heat.} \\
&\text{Apparent specific heat under constant pressure:—} \\
K_P &= \frac{1}{CnM}\left(\frac{1}{N} + 1\right) = \mathfrak{k}(1 + N).
\end{aligned}
\right\} \quad (20.)
$$

Ratio of those two specific heats :—

$$\frac{K_P}{K_V} = 1 + N. \quad . \quad . \quad . \quad . \qquad (21.)$$

This ratio is the quantity called by Poisson γ, in his researches on the propagation of sound.

11. It is unnecessary to do more than to refer to the researches of Poisson, and to those of Laplace, for the proof that the effect of the production of heat by the compression of air is the same as if the elasticity varied in proportion to that power of the density whose index is the ratio of the two specific heats; so that the actual velocity of sound is greater than that which it would have if there were no such development of heat, in the proportion of the square root of that ratio.

The following is the value of the velocity of sound in a gas, as given by Poisson in the second volume of his *Traité de Mécanique* :—

$$a = \sqrt{g \cdot \gamma \cdot (1 + E\,T) \frac{m\,h}{\Delta}}, \quad . \quad . \quad . \quad . \qquad (22.)$$

where a denotes the velocity of sound, g the velocity generated by gravity in unity of time, E the coefficient of increase of elasticity with temperature, at the freezing point of water, T the temperature measured from that point, m the specific gravity of mercury, Δ that of the gas at the temperature of melting ice, and pressure corresponding to a column of mercury of the height h. It follows that the ratio γ is given by the formula

$$\gamma = 1 + N \text{ nearly} = \frac{a^2 \Delta}{g\,m\,h\,(1 + E\,T)}. \quad . \quad . \qquad (23.)$$

Calculations have been made to determine the ratio γ from the velocity of sound; but as many of them involve erroneous values of the coefficient of elasticity E, the experiments have to be reduced anew.

The following calculation is founded on an experiment quoted by Poisson on the velocity of sound in atmospheric air, the values of E, m, and Δ being taken from the experiments of M. Regnault.

$$a = 340{\cdot}89 \text{ mètres per second.}$$

$$g = 9^{m}{\cdot}80896. \qquad h = 0^{m}{\cdot}76. \qquad T = 15^{\circ}{\cdot}9 \text{ Centigrade.}$$

$$E = 0{\cdot}003665; \quad \frac{m}{\Delta} = 10513.$$

Consequently, for atmospheric air,

$$\gamma = 1{\cdot}401.$$

The results of a reduction, according to correct data, of the experiments of Dulong upon the velocity of sound in atmospheric air, oxygen, and hydrogen, are as follows :—

Atmospheric air, ...	$\gamma = 1 \cdot 410$
Oxygen, ...	1·426
Hydrogen, ...	1·426

Thus it appears, that for the simple substances, oxygen and hydrogen, the ratio N is the same, while for atmospheric air it is somewhat smaller.*

12. The ordinary mode of expressing the specific heats of gases is to state their ratios to that of an *equal volume* of atmospheric air at the same pressure and temperature.

When $\frac{\kappa}{\tau}$ is a very small fraction, specific heats of *unity of volume* of a perfect gas are given by the equations—

$$\left. \begin{aligned} n \mathrm{M} \mathrm{K}_{\mathrm{v}} &= \frac{1}{\mathrm{C} \mathrm{N}} \\ n \mathrm{M} \mathrm{K}_{\mathrm{P}} &= \frac{1}{\mathrm{C}} \left(\frac{1}{\mathrm{N}} + 1 \right) \end{aligned} \right\} \quad . \quad . \quad . \quad (24.)$$

That is to say, the specific heat of unity of volume at constant volume is inversely proportional to the fraction by which the ratio of the two specific heats exceeds unity; a conclusion already deduced from experiment by Dulong.

The following is a comparison of the ratios of the apparent specific heats under constant pressure, of unity of volume of oxygen and hydrogen

* The following are some additional determinations of the value of γ for atmospheric air, founded upon experiments on the velocity of sound :—

Observers.	T Centigrade.	a Mètres per second.	γ
Bravais and Martins: mean of several experiments at temperatures varying from 5° to 11° Centigrade, reduced to 0° (*Comptes Rendus*, xix.)	0°	332·37	1·40955
Moll and Van Beek: reduced to	0°	332·25	1·40853
Stampfer and Myrbach: reduced to 0° (not corrected for moisture)	0°	332·96	1·41456
Académie des Sciences, 1738: (not corrected for moisture)	6°·1	337·10	1·418

A variation of one mètre per second in the velocity of sound at 0° corresponds to a variation of ·0085 in the value of γ.

respectively, to that of atmospheric air, as deduced from equation (24), with those determined experimentally by De la Roche and Bérard:—

Gas.	Ratio $\frac{n\,M\,K_p\,(\text{Gas})}{n\,M\,K_p\,(\text{Atmos. air})}$ By Theory.	By Experiment.
Oxygen, . . .	0·973	0·9765
Hydrogen, . . .	0·973	0·9033

This comparison exhibits a much more close agreement between theory and experiment than has been hitherto supposed to exist, the errors in the constants employed having had the effect of making the ratio $1+N$ seem greater for atmospheric air than for oxygen and hydrogen, while in fact it is smaller.

To treat the other substances on which both M. Dulong and MM. De la Roche and Bérard made experiments as perfect gases, would lead to sensible errors. I have, therefore, confined my calculations for the present to oxygen, hydrogen, and atmospheric air.

13. The heat produced by compressing so much of a perfect gas as would occupy unity of volume under the pressure unity, at the temperature 0° Centigrade, from its actual volume $n\,M\,V_1 = \frac{\tau}{P\,C}$, into a volume which is less in a given ratio s (when κ is neglected as compared with τ), is expressed by the following motion:—

$$n\,M\,Q' = -\frac{1}{C}\int_{V_1}^{s\,V_1} d\,V\,.\,\frac{\tau}{V} = -\,n\,M\,V_1\int_1^s P\,d\,s, \quad (25.)$$

being, in fact, equal to the mechanical power used in the compression. When the temperature is maintained constant, this becomes

$$n\,M\,Q'_{(\tau)} = \frac{\tau}{C}\log_e.\frac{1}{s}, \quad (26.)$$

which is obviously independent of the nature of the gas.

Hence, *equal volumes of all substances in the state of perfect gas, at the same pressure, and at equal and constant temperatures, being compressed by the same amount, disengage equal quantities of heat;* a law already deduced from experiment by Dulong.

14. The determination of the fraction N affords the means of calculating the mechanical or absolute value of specific heat, as defined by equation (1), section first. The data for atmospheric air being taken as follows:—

$$N = 0{\cdot}4, \qquad C = 274°{\cdot}6 \text{ Centigrade},$$

$\frac{1}{n\,\mathrm{M}}$ = height of an imaginary column of air of uniform density, at the temperature 0° Cent., whose pressure by weight on a given base is equal to its pressure by elasticity,

$$= 7990 \text{ mètres}, \quad = 26214 \text{ feet} :—$$

the real specific heat of atmospheric air, or the depth of fall equivalent to 1 Centigrade degree of temperature in that gas, is found to be

$$\mathfrak{k} = \frac{1}{\mathrm{C}\,n\,\mathrm{M}\,\mathrm{N}} = 72{\cdot}74 \text{ mètres} = 238{\cdot}66 \text{ feet.} \quad . \quad (27.)$$

The apparent specific heat of atmospheric air, under constant pressure, according to De la Roche and Bérard, is equal to that of liquid water at 0° Centigrade × 0·2669. The ratio of its real specific heat to the apparent specific heat of water at 0° Centigrade is, therefore,

$$\cdot 2669 \times \frac{1{\cdot}0}{1{\cdot}4} = \cdot 1906.$$

And, consequently, the mechanical value of the apparent specific heat of liquid water, at the temperature of melting ice, is

$$\left.\begin{array}{l} \frac{\mathfrak{k}\text{ (at. air)}}{\cdot 1906} = 381{\cdot}64 \text{ mètres} = 1252 \text{ feet per Centigrade degree,} \\ \qquad \text{or } 695{\cdot}6 \text{ feet per degree of Fahrenheit's scale.} \end{array}\right\} \quad (28.)$$

This quantity we shall denote by K_w. It is the mechanical equivalent of the ordinary *thermal unit.*

I have already pointed out (in article 2 of the first section) the causes which tend to make the apparent value of the mechanical equivalent of heat, in Mr. Joule's experiments, greater than the true value. The differences between the result I have just stated, and those at which he has arrived, do not seem greater than those causes are capable of producing when combined with the uncertainty of experiments, like those of Mr. Joule, on extremely small variations of temperature.

15. Besides the conditions of constant volume and constant pressure, there is a third condition in which it is of importance to know the apparent specific heat of an elastic fluid—namely, the condition of vapour at saturation, or in contact with its liquid.

The apparent specific heat of a vapour at saturation is the quantity of heat which unity of weight of that vapour receives or gives out, while its temperature is increased by one degree, its volume being at the same time compressed so as to bring it to the maximum pressure corresponding to the increased temperature.

It has been usually taken for granted, that this quantity is the same, with the variation for one degree of temperature, of what is called the *total heat* of evaporation. Such is, indeed, the case according to the theory of Carnôt; but I shall show that, according to the mechanical theory of heat, these two quantities are not only distinct, but in general of contrary signs.

I shall, for the present, consider such vapours only as may be treated in practice as perfect gases, so as to make the first of the equations (20) applicable.

It has been shown that the logarithm of the maximum elasticity of a vapour in contact with its liquid may be represented by the expression

$$\text{Log. P} = a - \frac{\beta}{\tau} - \frac{\gamma}{\tau^2}.$$

The coefficients a, β, γ, being those adapted for calculating the common logarithm of the pressure, I shall use the accented letters a', β', γ', to denote those suited to calculate the hyperbolic logarithm, being equal respectively to the former coefficients $\times$ 2·3025851.

Then for vapour at saturation,

$$\frac{d\,\text{P}}{\text{P}\,d\,\tau} = \frac{\beta'}{\tau^2} + \frac{2\,\gamma'}{\tau^3}. \quad . \quad . \quad . \quad (29.)$$

Making this substitution in the general equation (21.) we find the following value for the apparent specific heat of perfectly gaseous vapour at saturation:

$$\left.\begin{aligned} \text{K}_s &= \text{k} + \text{P}\frac{d\,\text{V}}{d\,\tau} = \text{k}\left(1 + \text{N}\,.\,\frac{\tau}{\text{V}}\frac{d\,\text{V}}{d\,\tau}\right) \\ &= \text{k}\left\{1 + \text{N}\left(1 - \frac{\tau}{\text{P}}\frac{d\,\text{P}}{d\,\tau}\right)\right\} \\ &= \frac{1}{\text{C}\,n\,\text{M}}\left(\frac{1}{\text{N}} + 1 - \frac{\beta'}{\tau} - \frac{2\,\gamma'}{\tau^2}\right) \end{aligned}\right\} \quad . \quad (30.)$$

16. For the vapours of which the properties are known, the negative terms of this expression exceed the positive at all ordinary temperatures, so that the kind of apparent specific heat now under consideration is a negative quantity:—that is to say, that if a given weight of vapour at saturation is increased in temperature, and at the same time maintained by compression at the maximum elasticity, the heat generated by the compression is greater than that which is required to produce the elevation of temperature, and a surplus of heat is given out; and on the other hand, if vapour at saturation is allowed to expand, and at the same time

maintained at the temperature of saturation, the heat which disappears in producing the expansion is greater than that set free by the fall of temperature; and the deficiency of heat must be supplied from without, *otherwise a portion of the vapour will be liquefied in order to supply the heat necessary for the expansion of the rest.*

This circumstance is obviously of great importance in meteorology, and in the theory of the steam-engine. There is as yet no experimental proof of it. It is true that, in the working of non-condensing engines, it has been found that the steam which escapes is always at the temperature of saturation corresponding to its pressure, and carries along with it a portion of water in the liquid state; but it is impossible to distinguish between the water which has been liquefied by the expansion of the steam, and that which has been carried over mechanically from the boiler.

The calculation of the proportion of vapour liquefied by a given expansion, requires the knowledge of the latent heat of evaporation, which forms the subject of the next section.

SECTION III.—OF THE LATENT AND TOTAL HEAT OF EVAPORATION, ESPECIALLY FOR WATER.

17. The latent heat of evaporation of a given substance at a given temperature, is the amount of heat which disappears in transforming unity of weight of the substance from the liquid state, to that of vapour of the maximum density for the given temperature, being consumed in producing an increase of volume, and an unknown change of molecular arrangement.

It is obvious, that if the vapour thus produced is reconverted into the liquid state *at the same temperature*, the heat given out during the liquefaction must be equal to that consumed during the evaporation; for as the sum of the expansive and compressive powers, and of those dependent on molecular arrangement during the whole process, is equal to zero, so must the sum of the quantities of heat absorbed and evolved.

The heat of liquefaction, at a given temperature, is therefore equal to that of evaporation, with the sign reversed.

18. If to the latent heat of evaporation at a given temperature, is added the quantity of heat necessary to raise unity of weight of the liquid from a certain fixed temperature, (usually that of melting ice), to the temperature at which the evaporation takes place, the result is called the total heat of evaporation from the fixed temperature chosen.

According to the theory of Carnôt, this quantity is called the constituent heat of vapour; and it is conceived, that if liquid at the temperature of melting ice be raised to any temperature and evaporated, and finally brought in the state of vapour to a certain given temperature, the whole heat expended will be equal to the constituent heat corresponding to that given temperature, and will be the same, whatsoever may have been the intermediate changes of volume, or the temperature of actual evaporation.

According to the mechanical theory of heat, on the other hand, the quantity of heat expended must vary with the intermediate circumstances; for otherwise no power could be gained by the alternate evaporation and liquefaction of a fluid at different temperatures.

19. The law of the latent and total heat of evaporation is immediately deducible from the principle of the constancy of the total *vis viva* in the two forms of heat and expansive power, when the body has returned to its primitive density and temperature, as already laid down in article 7.

That principle, when applied to evaporation and liquefaction, may be stated as follows:—

Let a portion of fluid in the liquid state be raised from a certain temperature to a higher temperature: let it be evaporated at the higher temperature: let the vapour then be allowed to expand, being maintained always at the temperature of saturation for its density, until it is restored to the original temperature, at which temperature let it be liquefied:—then, *the excess of the heat absorbed by the fluid above the heat given out, will be equal to the expansive power generated.*

To represent those operations algebraically,—let the lower absolute temperature be τ_0; the volume of unity of weight of liquid at that temperature v_0, and that of vapour at saturation V_0: let the pressure of that vapour be P_0; the latent heat of evaporation of unity of weight L_0; and let the corresponding quantities for the higher absolute temperature τ_1, be v_1, V_1, P_1, L_1. Let K_L represent the mean apparent specific heat of the substance in the liquid form between the temperatures τ_0 and τ_1. Then,—

First, Unity of weight of liquid being raised from the temperature τ_0 to the temperature τ_1, absorbs the heat,

$$K_L(\tau_1 - \tau_0),$$

and produces the expansive power,

$$\int_{v_0}^{v_1} dv \,.\, P.$$

Secondly, It is evaporated at the temperature τ_1, absorbing the heat

$$L_1,$$

and producing the expansive power,

$$P_1 (V_1 - v_1).$$

Thirdly, The vapour expands, at saturation, until it is restored to the original temperature τ. In this process it absorbs the heat,

$$-\int_{\tau_0}^{\tau_1} d\tau \,.\, K_s,$$

and produces the expansive power,

$$\int_{V_1}^{V_0} dV \,.\, P.$$

Fourthly, It is liquefied at the original temperature, giving out the heat

$$L_0,$$

and consuming the compressive power,

$$P_0 (V_0 - v_0).$$

The equation between the heat which has disappeared, and the expansive power which has been produced, is as follows :—

$$\left.\begin{array}{l} L_1 - L_0 + K_L (\tau_1 - \tau_0) - \int_{\tau_0}^{\tau_1} d\tau \,.\, K_s \\ = P_1 (V_1 - v_1) - P_0 (V_0 - v_0) + \int_{v_0}^{v_1} dv \,.\, P + \int_{V_1}^{V_0} dV \,.\, P. \end{array}\right\} \quad (31.)$$

If the vapour be such that it can be regarded as a perfect gas without sensible error, the substitution of $\mathfrak{k} + P \frac{dV}{d\tau}$ for K_s, and of $\frac{\tau}{C n M} = \mathfrak{k} N \tau$ for $P V$, transforms the above to

$$\left.\begin{array}{l} L_1 - L_0 + \{K_L - \mathfrak{k}(1 + N)\} (\tau_1 - \tau_0) \;. \\ = - P_1 v_1 + P_0 v_0 + \int_{v_0}^{v_1} dv \,.\, P = - \int_{P_1}^{P_0} dP \,.\, v \end{array}\right\} \quad (32.)$$

In almost all cases which occur in practice, v is so small as compared with V, that $-\int dP \,.\, v$ may be considered as sensibly $= 0$; and, therefore, (sensibly)

$$L_1 + K_L (\tau_1 - \tau_0) = L_0 + \mathfrak{k} (1 + N) (\tau_1 - \tau_0). \quad . \quad (33.)$$

Now this quantity, which I shall denote by H, is the total heat required to raise unity of weight of liquid from τ_0 to τ_1 of absolute temperature, and to evaporate it at the latter temperature. Therefore, *the total heat of evaporation, where the vapour may be treated as a perfect gas, increases sensibly at an uniform rate with the temperature of evaporation; and the coefficient of its increase with temperature is equal to the apparent specific heat of the vapour at constant pressure,* ꝁ$(1+\mathrm{N})$.

20. There have never been any experiments from which the apparent specific heat of steam under constant pressure can be deduced in the manner in which that of permanent gases has been ascertained.

The experiments of M. Regnault, however, prove that the total heat of evaporation of water increases uniformly with the temperature from 0° to 200° Centigrade, and thus far fully confirm the results of this theory.

The coefficient of increase is equal to

$$\mathrm{K_w} \times 0{\cdot}305.$$

Its mechanical value is, consequently,

$$116{\cdot}4 \text{ mètres} = 382 \text{ feet per Centigrade degree, or } 212 \text{ feet per degree of Fahrenheit.} \qquad (34.)$$

Although the principle of the conservation of *vis viva* has thus enabled us to ascertain the *law of increase* of the total heat of evaporation, it does not enable us to calculate *à priori* the constant $\mathrm{L_0}$ of the formula, being the latent heat of evaporation at the fixed temperature from which the total heat is measured; for the changes of molecular arrangement which constitute evaporation are unknown.

When the fixed temperature is that of melting ice, M. Regnault's experiments give 606·5 Centigrade degrees, applied to liquid water as the value of this constant; so that

$$\left.\begin{array}{l} \mathrm{H} = \mathrm{K_w}(606^\circ{\cdot}5 + {\cdot}305\,\mathrm{T}^\circ), \\ \qquad\text{for the Centigrade scale,} \\ \mathrm{H} = \mathrm{K_w}\Big(1091^\circ{\cdot}7 + {\cdot}305\,(\mathrm{T}^\circ - 32^\circ)\Big), \\ \qquad\text{for Fahrenheit's scale,} \end{array}\right\} \qquad (35.)$$

is the complete expression for the heat required to raise unity of weight of water from the temperature of melting ice to T° above the ordinary zero, and to evaporate it at the latter temperature. This formula has been given by M. Regnault as merely empirical; but we have seen that it closely represents the physical law when quantities depending on the expansion of water are neglected.

It must be remarked, that the unit of heat in M. Regnault's tables is not precisely the specific heat of water at 0° Centigrade, but its mean specific heat between the initial and final temperatures of the water in the calorimeter. The utmost error, however, which can arise from this circumstance, is less than $\frac{1}{1200}$ of the total heat of evaporation, so that it may safely be neglected.

The coefficient ·305 K_w = 382 feet per Centigrade degree, is the apparent specific heat of steam at constant pressure; that is to say, for steam,—

$$\left.\begin{array}{l} \mathfrak{k} + \dfrac{1}{C\,n\,M} = 382 \text{ feet per Centigrade degree, but } \dfrac{1}{C\,n\,M} = 153 \text{ ft.} \\ \quad\text{Therefore, the real specific heat of steam is } \mathfrak{k} = \dfrac{1}{C\,n\,M\,N} = 229 \\ \text{feet per Centigrade degree} = 127{\cdot}4 \text{ feet per deg. of Fahrenheit} \\ = K_w \times 183, \text{ and } N = \dfrac{153}{229} = \dfrac{2}{3}. \end{array}\right\} \quad (36.)$$

The quantity $-\int_{P_0}^{P_1} d\,P \,.\, v$ has been neglected, as already explained, in these calculations, on account of its smallness. When $\tau_0 = C$, or the fixed point is 0° Centigrade, this integral is nearly equal to

$$-\,v\,P_1 = -\frac{v}{V_1} \cdot \frac{\tau_1}{C\,n\,M} = -\,\mathfrak{k}\,N \,.\, \frac{v}{V_1}\,\tau_1, \qquad (37.)$$

which for steam, is equal to

$$-\,K_w \times {\cdot}122\,\frac{v}{V}\,\tau_1.$$

For a pressure of eight atmospheres, $\frac{v}{V} = \frac{1}{252}$ nearly, $\tau_1 = 445°{\cdot}5$ (T = 170°·9 Cent.); consequently, $-\,v\,P_1 = -\,K_w \times 0°{\cdot}22$ Cent., a quantity much less than the limit of errors of observation in experiments on latent heat. This shows that in practice we are justified in overlooking the influence of the volume of the liquid water on the heat of evaporation.

Section IV.—Of the Mechanical Action of Steam treated as a Perfect Gas, and the Power of the Steam Engine.

21. In the present limited state of our experimental knowledge of the density of steam at pressures differing much from that of the atmosphere,

it is desirable to ascertain whether any material error is likely to arise from treating it as a perfect gas. For this purpose the ratio of the volume of steam at 100° Centigrade under the pressure of one atmosphere, to that of the water which produces it at 4°·1 Centigrade, as calculated theoretically on the supposition of steam being a perfect gas, is to be compared with the actual ratio.

The weight of one volume of water at 4°·1 Centigrade being taken as unity, that of half a volume of oxygen at 0° Centigrade, under the pressure of one atmosphere, according to the experiments of M. Regnault, is . . . . 0·000714900
That of one volume of hydrogen, ... 0·000089578

The sum being, ... 0·000804478

The reciprocal of this sum being multiplied by $\frac{374 \cdot 6}{274 \cdot 6} = 1 \cdot 364166$, the ratio of dilatation of a perfect gas from 0° to 100° Centigrade, the result gives for the volume of steam of saturation at 100° Centigrade, as compared with that of water,

At 4°·1 . . . . 1695·72

And for its density, ... 0·00058972

The agreement of those results with the known volume and density of steam is sufficiently close to show, that at pressures less than one atmosphere, it may be regarded as a gas sensibly perfect; from which it may be concluded, that, in the absence of more precise data, the errors arising from treating it as a perfect gas at such higher pressures as occur in practice, will not be of much importance.

Representing, then, by v the volume of unity of weight of water at 4°·1 Centigrade, that of unity of weight of steam at any pressure and temperature will be given by the formula

$$V = \frac{1696\, v\, \omega}{(\tau)} \cdot \frac{\tau}{P}. \quad . \quad . \quad . \quad (38.)$$

ω representing the number of units of weight per unit of area in the pressure of one atmosphere, and (τ) the absolute temperature at which the pressure of saturation is one atmosphere; being for the Centigrade scale 374°·6, and for Fahrenheit's scale 674°·28.

The mechanical action of unity of weight of steam at the temperature τ and pressure P, during its entrance into a cylinder, before it is permitted to expand, is represented by the product of its pressure and volume, or by

$$P\,V = \frac{1696\,v\,\omega}{(\tau)} \,.\, \tau. \qquad . \quad . \quad . \quad . \quad (39.)$$

The coefficient $\frac{1696\,v\,\omega}{(\tau)}$ represents a certain depth of fall per degree of absolute temperature, and is the same with the coefficient $\frac{1}{C\,n\,M}$, already referred to.

By taking the following values of the factors:—$v = 0{\cdot}016$ cubic foot per pound avoirdupois, $\omega = 2117$ pounds avoirdupois per square foot, we find this coefficient to be

$$\left.\begin{array}{l} 153{\cdot}35 \text{ feet} = 46{\cdot}74 \text{ mètres per Centigrade degree,} \\ 85{\cdot}19 \text{ feet per degree of Fahrenheit.} \end{array}\right\} \quad (40.)$$

This determination may be considered correct to about $\frac{1}{1200}$ part. When French measures are used in the calculation, the following is the result:—

$v = 1$ cubic centimètre per gramme,

$\omega = 1033{\cdot}3$ grammes per square centimètre.

$$\left.\begin{array}{l} \frac{1}{C\,n\,M} = 46{\cdot}78 \text{ mètres per Centigrade degree,} \\ = 153{\cdot}48 \text{ feet, or } 85{\cdot}27 \text{ feet per degree of Fahrenheit.} \end{array}\right\} \quad (41.)$$

The difference, which is of no practical importance in calculating the power of the steam-engine, arises in the estimation of the density of liquid water.

22. Unit of weight of steam at saturation, of the elasticity P_1 and volume V_1, corresponding to the absolute temperature τ_1, being cut off from external sources of heat, it is now to be investigated what amount of power it will produce in expanding to a lower pressure P_2 and temperature τ_2.

It has already been shown, at the end of the second section, that if vapour at saturation is allowed to expand, it requires a supply of heat from without to maintain it at the temperature of saturation, otherwise a portion of it must be liquefied to supply the heat required to expand the rest. Hence, when unity of weight of steam at saturation, at the pressure P_1 and volume V_1, expands to a lower pressure P, being cut off from external sources of heat, it will not occupy the entire volume V corresponding to that pressure, according to equation (38), but a less volume, $S = m\,V$, where m represents the weight of water remaining in the gaseous state, the portion $1 - m$ having been liquefied during the expansion of the remainder. The expansive action of the steam will, therefore, be represented by

$$\int_{V_1}^{S_2} d\,S\,.\,P. \quad . \quad . \quad . \quad . \quad . \qquad (42.)$$

The law of variation of the fraction m flows from the following considerations:—Let δm represent the indefinitely small variation of m corresponding to the indefinitely small change of temperature $\delta \tau$; L, the latent heat of evaporation of unity of weight; K_s, as in equation (30), the specific heat of vapour at saturation, which is a negative coefficient varying with the temperature; then we must have

$$-\,L\,\delta m = m\,K_s\,\delta \tau, \text{ or } \frac{\delta m}{m} = -\,\frac{K_s}{L}\,\delta \tau,$$

in order that the heat produced by the liquefaction of δm may be equal to the heat required to expand m. Hence making, according to equation (30)—

$$K_s\,\delta \tau = k\left(\delta \tau + N\frac{\tau}{V}\,\delta V\right), \text{ and } \delta \tau = -\,\frac{\tau}{V}\,\delta V\,\frac{1}{\frac{\beta'}{\tau} + \frac{2\,\gamma'}{\tau^2} - 1}$$

we obtain

$$\frac{\delta m}{m} = -\,\frac{k\,\tau}{L}\left(N - \frac{1}{\frac{\beta'}{\tau} + \frac{2\,\gamma'}{\tau^2} - 1}\right)\frac{\delta V}{V}, \quad . \quad . \qquad (43.)$$

and denoting the coefficient of

$$\frac{\delta V}{V} \text{ by } -\,\nu,$$

$$\left.\begin{aligned}
&\frac{d \log. m}{d \log. V} = -\,\nu\,; \quad \frac{d \log. S}{d \log. V} = 1 - \nu, \\
&\text{and because} \\
&\frac{d \log. V}{d \log. P} = -\,1 + \frac{1}{\frac{\beta'}{\tau} + \frac{2\,\gamma'}{\tau^2}} \\
&\frac{d \log. m}{d \log. P} = +\,\nu\left(1 - \frac{1}{\frac{\beta'}{\tau} + \frac{2\,\gamma'}{\tau^2}}\right) \\
&\frac{d \log. S}{d \log. P} = -\,(1 - \nu)\left(1 - \frac{1}{\frac{\beta'}{\tau} + \frac{2\,\gamma'}{\tau^2}}\right) = -\,\sigma.
\end{aligned}\right\} \qquad (44.)$$

As the mean temperature of the liquid thus produced, more or less exceeds that of the remaining vapour, a small fraction of it will be reconverted into vapour, if the expansion is carried on slowly enough; but its amount is so small, that to take it into account would needlessly complicate the calculation, without making it to any material extent more accurate.

23. The extreme complexity of the exponent σ, considered as a function of the pressure P, would render a general formula for the expansive action $\int P\,dS$ very cumbrous in its application. For practical purposes, it is sufficient to consider the exponent σ as constant during the expansion which takes place in any given engine, assigning it an average value suitable to the part of the scale of pressures in which the expansion takes place. For engines in which the steam is introduced at pressures not exceeding four atmospheres, I conceive that it will be sufficiently accurate to make $\sigma = \frac{6}{7}$; while for engines in which the initial pressure lies between four and eight atmospheres, the suitable value is $\sigma = \frac{5}{6}$.

The utmost error which can arise from using these exponents is about $\frac{1}{100}$ of the whole power of the engine, and that only in extreme cases. Making, therefore,

$$P = P_1 \left(\frac{S}{V_1}\right)^{-\frac{1}{\sigma}}$$

we obtain for the value of the expansive action of unity of weight of steam,

$$\left.\begin{aligned} \int_{V_1}^{S_2} dS \, . \, P &= P_1 V_1 \frac{\sigma}{1-\sigma}\left(1 - \left(\frac{S_2}{V_1}\right)^{1-\frac{1}{\sigma}}\right) \\ &= P_1 V_1 \frac{\sigma}{1-\sigma}\left(1 - s^{1-\frac{1}{\sigma}}\right) \end{aligned}\right\} \quad (45.)$$

s being used to denote $\frac{S_2}{V_1}$, or the ratio of the volumes occupied by steam at the end and at the beginning of the expansion respectively.

A table to facilitate the computation is given in the Appendix.

The gross mechanical action of unity of weight of steam on one side of the piston is found by adding to the above quantity the action of the steam before it begins to expand, or $P_1 V_1$, and is therefore,

$$P_1 V_1 \left(\frac{1}{1-\sigma} - \frac{\sigma}{1-\sigma} s^{1-\frac{1}{\sigma}}\right). \quad . \quad . \quad (46.)$$

The values of the coefficients and exponent being

	$\frac{1}{1-\sigma}$	$\frac{\sigma}{1-\sigma}$	$1-\frac{1}{\sigma}$.
For initial pressures between			
1 and 4 atmospheres, .	7	6	$-\frac{1}{6}$,
4 and 8 atmospheres, .	6	5	$-\frac{1}{5}$.

24. The following deductions have to be made from the gross action, in order to obtain the action effective in overcoming resistance.

First, For loss of power owing to a portion of the steam being employed in filling steam-passages, and the space called the *clearance* of the cylinder at one end. Let the bulk of steam so employed be the fraction $c\,S_2$ of the space filled by steam at the end of the expansion; then the loss of power from this cause is

$$P_1\,c\,S_2 = c\,s\,P_1\,V_1.$$

Secondly, For the pressure on the opposite side of the piston, of the steam which escapes into the condenser, or into the atmosphere, as the case may be. Let P_3 be the pressure of this steam; the deduction to be made for its action is

$$P_3\,S_2\,(1-c) = P_3\,V_1\,(1-c)\,s.$$

These deductions having been made, there is obtained for the effect of unity of weight of water evaporated,

$$V_1\left\{P_1\left(\frac{1}{1-\sigma} - \frac{\sigma}{1-\sigma}s^{1-\frac{1}{\sigma}} - c\,s\right) - P_3\,(1-c)\,s\right\} \quad (47.)$$

25. The effect of the engine in unity of time is found by multiplying the above quantity by the number of units of weight of water evaporated in unity of time.

If this number be denoted by W,

$$W\,S_2\,(1-c) = W\,V_1\,(1-c)\,s = A\,u, \quad . \quad . \quad (48.)$$

will represent the cubical space traversed by the piston in unity of time, A denoting the area of the piston, and u its mean velocity.

Now, let the whole resistance to be overcome by the engine be reduced, by the principles of statics, to a certain equivalent pressure per unit of area of piston, and let this pressure be denoted by R. Then,

$$\mathrm{R\,A}\,u = \mathrm{R\,W\,V_1}(1 - c)\,s, \quad . \quad . \quad . \quad (49.)$$

expresses the effect of the engine in terms of the gross resistance.

We have now the means of calculating the circumstances attending the working of a steam-engine, according to the principle of the conservation of *vis viva*, or, in other words, of the equality of power and effect, which regulates the action of all machines that move with an uniform or periodical velocity.

This principle was first applied to the steam-engine by the Count de Pambour; and, accordingly, the formulæ which I am about to give only differ from those of his work in the expressions for the maximum pressure at a given temperature, and for the expansive action of the steam, which are results peculiar to the theory of this essay.

In the first place, the effect as expressed in terms of the pressure, is to be equated to the effect as expressed in terms of the resistance, as follows:—

$$\mathrm{R\,A}\,u = \mathrm{R\,W\,V_1}(1 - c)\,s = \mathrm{W\,V_1}\left\{\mathrm{P_1}\left(\frac{1}{1-\sigma} - \frac{\sigma}{1-\sigma}\,s^{1-\frac{1}{\sigma}} - c\,s\right) - \mathrm{P_3}(1 - c)\,s\right\} \quad . \quad . \quad . \quad . \quad (50.)$$

This is the fundamental equation of the action of the steam-engine, and corresponds with equation A. of M. de Pambour's theory.

26. Dividing both sides of equation (50) by the space traversed by the piston in unity of time, $\mathrm{W\,V_1}\,(1-c)\,s$, and transferring the pressure of the waste steam, $\mathrm{P_3}$, to the first side, we obtain this equation:—

$$\mathrm{R} + \mathrm{P_3} = \mathrm{P_1}\,\frac{\dfrac{1}{1-\sigma} - \dfrac{\sigma}{1-\sigma}\,s^{1-\frac{1}{\sigma}} - c\,s}{(1 - c)\,s}, \quad . \quad (51.)$$

which gives the means of determining the pressure $\mathrm{P_1}$ at which the steam must enter the cylinder, in order to overcome a given resistance and counter-pressure with a given expansion; or, supposing the expansion s to be variable at pleasure, and the initial pressure $\mathrm{P_1}$ fixed, the equation gives the means of finding, by approximation, the expansion best adapted to overcome a given resistance and counter-pressure.

The next step is to determine, from equations (XV.) of the introduction and (38) of this section, the volume $\mathrm{V_1}$ of unity of weight of steam corresponding to the maximum pressure $\mathrm{P_1}$. Then equation (48) gives the space traversed by the piston in unity of time, which, being multiplied by the resistance R per unit of area of piston, gives the gross effect of the engine.

27. If, on the other hand, the space traversed by the piston in unity of time is fixed, equation (48) gives the means of determining, from the evaporating power of the boiler W, either the volume V_1 of unity of weight of steam required to work the engine at a given velocity with a given expansion, or the expansion *s* proper to enable steam of a given initial density to work the engine at the given velocity. The initial pressure P_1 being then determined from the volume V_1, the resistance which the engine is capable of overcoming with the given velocity is to be calculated by means of equation (51).

28. This calculation involves the determination of the pressure P_1 from the volume V_1 of unity of weight of steam at saturation, which can only be done by approximation. The following formula will be found useful for this purpose:—

$$P_1 = \omega \left(\frac{V_0}{V_1}\right)^{\frac{12}{11}} \quad . \quad . \quad . \quad . \quad (52.)$$

where ω represents the pressure of one atmosphere, V_0 the volume of steam of saturation at that pressure (being 1696 times the volume of water at 4°·1 Cent., or 27·136 cubic feet per pound avoirdupois), and V_1 the volume of steam of saturation at the pressure P_1. This formula is only applicable between the pressures of one and eight atmospheres: that is to say, when the volume of steam is not greater than 27 cubic feet per pound, nor less than 4, and the temperature not lower than 100° Centigrade, nor higher than 171° Centigrade (which correspond to 212° and 340° Fahrenheit).

The greatest error in computing the pressure by means of this formula is about $\frac{1}{50}$ of an atmosphere, and occurs at the pressure of four atmospheres, so that it is $\frac{1}{200}$ of the whole pressure. This is sufficiently accurate for practice, in calculating the power of steam-engines; but should a more accurate result be required, the approximate value of the pressure may be used to calculate the temperature by means of equation (XV.); and the temperature thus determined, (which will be correct to $\frac{1}{5}$ of a Centigrade degree), may then be used in conjunction with the volume to compute a corrected value of the pressure, according to equation (38). The pressure, as thus ascertained, will be correct to $\frac{1}{2000}$ of its amount, which may be considered the greatest degree of accuracy attainable.

The most convenient and expeditious mode, however, of computing the pressure from the volume, or *vice versâ*, is by interpolation from the table given in the Appendix to this paper.

29. The resistance denoted by R may be divided into two parts; that which arises from the *useful work performed*, and that which is independent of it, being, in fact, the resistance of the engine when unloaded. Now it is evident, that the maximum *useful* effect of the steam has been attained,

as soon as it has expanded to a pressure which is in equilibrio with the pressure of the waste steam added to the resistance of the engine when unloaded; for any further expansion, though increasing the total effect, diminishes the useful effect. Therefore, if we make

$$R = R' + f,$$

R′ being the resistance arising from the useful work, and f the resistance of the engine when unloaded, both expressed in the form of pressure on the piston, the expansion corresponding to the maximum of useful effect will take place when

$$\left.\begin{array}{l} P_2 = P_3 + f, \\ \text{the corresponding ratio of expansion being} \\ s = \left(\frac{P_1}{P_3 + f}\right)^{\sigma} \end{array}\right\} \quad . \quad . \quad . \quad . \quad (53.)$$

The maximum useful effect with a given pressure on the safety-valve has been so fully discussed by M. de Pambour, that it is unnecessary to do more than to state that it takes place when the initial pressure in the cylinder is equal to that at the safety-valve: that is to say, when it and the useful resistance are the greatest that the safety-valve will permit.

30. Annexed is a table of the values of some of the quantities which enter into the preceding equations in the notation of the Count de Pambour's works :—

Expression in the Notation of this Paper.	Equivalent Expression in M. de Pambour's Notation.
$R = R' + f$	$(1 + \delta) r + f$
$A u$	$a v$
W	S × weight of one cubic foot of water.
P_3	p
s	$\frac{l + c}{l' + c}$
c	$\frac{c}{l' + c}$

31. As an illustration, I shall calculate the maximum useful effect of one pound, and of one cubic foot of water, in a Cornish double-acting

engine, in the circumstances taken by M. de Pambour as an example for that kind of engine, that is to say,—

Clearance one-twentieth of the stroke, or $c = \frac{1}{21}$

Resistance not depending on the useful load, $f =$ 72 lbs. per sq. ft.

Pressure of condensation, ... $P_3 =$ 576 lbs. " "

Consequently, to give the maximum useful effect,

$P_2 = P_3 + f$ = 648 lbs. " "

Total pressure of the steam when first admitted, $P_1 =$ 7200 lbs. " "

Volume of 1 lb. of steam $V_1 = 8{\cdot}7825$ cubic feet.

Therefore $P_1 V_1 = 63234$ lbs. raised one foot.

$$\frac{P_1}{P_2} = \frac{7200}{648}; \text{ and, consequently,}$$

Expansion to produce the maximum useful effect $s = \left(\frac{P_1}{P_2}\right)^{\frac{6}{7}} = 7{\cdot}877.$

Space traversed by the piston during the action of one pound of steam,

$$= V_1 (1 - c)\, s = 65{\cdot}886 \text{ cubic feet.}$$

Gross effect of one pound of steam, in pounds raised one foot high,

$$= P_1 V_1 \left(7 - 6\, s^{-\frac{1}{6}} - \frac{s}{21}\right) - P_3 V_1 (1 - c)\, s \quad = 112004$$

Deduct for resistance of engine when unloaded $f V_1 (1 - c)\, s$ = 4744

Effect of one pound of steam in overcoming resistance depending on useful load, ... 107260

This being multiplied by $62\frac{1}{2}$, gives for the effect of one cubic foot of water evaporated, in pounds raised one foot, . 6,703,750

It is here necessary to observe, that M. de Pambour distinguishes the useful resistance into two parts, the resistance of the useful load independently of the engine, and the increase in the resistance of the engine arising from the former resistance, and found by multiplying it by a constant fraction, which he calls δ. In calculating the net useful effect, he takes into account the former portion of the resistance only; consequently: Net useful effect as defined by M. de Pambour

$$= \frac{\text{Gross effect} - f V_1 (1 - c) s}{1 + \delta} \quad . \quad . \quad (54.)$$

The value of δ, for double-acting steam-engines generally, is considered by M. de Pambour to be $\frac{1}{7}$; consequently, to reduce the effect of one cubic foot of water, as calculated above, to that which corresponds with his definition, we must deduct $\frac{1}{8}$, which leaves,

5,865,781 lbs. raised one foot.

M. de Pambour's own calculation gives,

6,277,560,

being too large by about one-fifteenth.

32. In order to show the limit of the effect which may be expected from the expenditure of a given quantity of heat in evaporating water, and also to verify the approximate method employed in calculating the expansive action of the steam, I shall now investigate the *maximum gross effect*, including resistance of all kinds, producible by evaporating unity of weight of water at a higher temperature and liquefying it at a lower, and compare, in two examples, the power produced with the heat which disappears during the action of the steam, as calculated directly.

To obtain the maximum gross effect, the steam must continue to act expansively until it reaches the pressure of condensation, so that $P_2 = P_3$. The clearance must also be null, or $c = 0$. Making those substitutions in the formula (47), we find, for the maximum gross effect of unity of weight of water, evaporated under the pressure P_1, and liquefied under the pressure P_2,

$$P_1 V_1 \frac{1}{1 - \sigma} \left(1 - s^{1 - \frac{1}{\sigma}}\right) = P_1 V_1 \frac{1 - \left(\frac{P_2}{P_1}\right)^{1 - \sigma}}{1 - \sigma} \qquad (55.)$$

In order to calculate directly the heat which is converted into power in this operation, let τ_1, τ_2, respectively represent the absolute temperatures of evaporation and liquefaction, and L_2 the latent heat of evaporation at the lower temperature τ_2; then the total heat of evaporation at τ_1, starting from τ_2 as the fixed point, by equation (33), is

$$H_{2, 1} = L_2 + \cdot 305 \, K_w (\tau_1 - \tau_2).$$

This is the heat communicated to the water in raising it from τ_2, to τ_1 and evaporating it. Now a weight $1 - m$ of the steam is liquefied during the expansion at temperatures varying from τ_1 to τ_2, so that it may be looked upon as forming a mass of liquid water approximately at the mean temperature $\frac{\tau_1 + \tau_2}{2}$, and from which a quantity of heat, approximately represented by

$$K_w (1 - m) \frac{\tau_1 - \tau_2}{2},$$

must be abstracted, to reduce it to the primitive temperature τ_2.

Finally, the weight of steam remaining, m, has to be liquefied at the temperature τ_2, by the abstraction of the heat

$$m \, L_2.$$

The difference between the heat given to the water, and the heat abstracted from it, or

$$\left.\begin{array}{l} H_{2,1} - K_w (1 - m) \frac{\tau_1 - \tau_2}{2} - m \, L_2 \\ = (1 - m) L_2 + K_w \left(\cdot 305 - \frac{1 - m}{2} \right) (\tau_1 - \tau_2) \end{array}\right\} \quad (56.)$$

is the heat which has disappeared, and ought to agree with the expression (55) for the power produced, if the calculation has been conducted correctly.

As a first example, I shall suppose unity of weight of water to be evaporated under the pressure of four atmospheres, and liquefied under that of half an atmosphere; so that the proper values of the coefficients and exponent are

$$\frac{1}{1 - \sigma} = 7, \quad 1 - \sigma = \frac{1}{7}.$$

The data, in this case, for calculating the power are,

$$P_1 = 8468 \text{ lbs. per square foot.}$$

$$V_1 = 7 \cdot 584 \text{ cubic feet for 1 lb. of steam.}$$

$$P_1 V_1 = 64221 \text{ lbs. raised one foot.}$$

$$\frac{P_2}{P_1} = \frac{1}{8}, \text{ whence } s = 8^{\frac{6}{7}} = 5 \cdot 944.$$

Maximum possible effect of one pound of water,

$$= P_1 V_1 \times 7 \left(1 - \left(\frac{1}{8} \right)^{\frac{1}{7}} \right) = 115600 \text{ lbs. raised one foot.}$$

Being the mechanical equivalent of 92°·3 Centigrade applied to one pound of liquid water at 0° C.; or,

$$92^\circ{\cdot}3\ K_w.$$

Maximum possible effect of one cubic foot of water, 7,225,000 lbs. raised one foot.

In order to calculate directly the heat converted into power, we have,

$$\tau_1 = C + 144^\circ{\cdot}1 \text{ Cent. } \tau_2 = C + 81^\circ{\cdot}7.$$

$$L_2 = 549^\circ{\cdot}7\ K_w.$$

$$H_{21} = 568^\circ{\cdot}7\ K_w = \text{heat expended in the boiler.}$$

$$1 - m = {\cdot}14 \text{ nearly} = \text{proportion of steam liquefied during the expansion.}$$

The heat converted into mechanical power, as calculated from these data, is found to be,

$$91^\circ{\cdot}6\ K_w,$$

differing by only $0^\circ{\cdot}7$ from the amount as calculated from the power produced.

The direct method, however, is much less precise than the other, and is to be regarded as only a verification of the general principle of calculation.

The heat rendered effective, in the above example, is $\frac{92{\cdot}3}{568{\cdot}7}$, or less than *one-sixth* of that expended in the boiler.

As a second example, I shall suppose the steam to be produced at a pressure of eight atmospheres, and to expand to that of one atmosphere. In this case,

$$P_1 = 16936 \text{ lbs. per square foot.}$$

$$V_1 = 4{\cdot}03 \text{ cubic feet per lb. of steam.}$$

$$P_1 V_1 = 68252 \text{ lbs. raised one foot.}$$

$$\frac{P_2}{P_1} = \frac{1}{8} \therefore s = 5{\cdot}657 = 8^{\frac{5}{6}}.$$

Maximum possible effect of one pound of water,

$$= P_1 V_1 \times 6\left(1 - \left(\frac{1}{8}\right)^{\frac{1}{6}}\right) = 119{,}942 \text{ lbs. raised one foot.}$$

Being the equivalent of $95^\circ{\cdot}8\ K_w$ (Centigrade).

Maximum possible effect of one cubic foot of water = 7,496,375 lbs. raised one foot.

The data for calculating directly the heat rendered effective are,

$$\tau_1 = \text{C} + 170^\circ\cdot 9 \text{ Cent. } \tau_2 = \text{C} + 100^\circ.$$

$$\text{L}_2 = 537^\circ \text{ K}_w.$$

$$\text{H}_{2,1} = 558^\circ\cdot 6 \text{ K}_w = \text{heat expended in the boiler.}$$

$$1 - m = \cdot 148 \text{ nearly} = \text{steam liquefied during the expansion.}$$

Whence, the heat converted in power, as calculated directly, is

$$95^\circ\cdot 8 \text{ K}_w,$$

agreeing with the calculation from the power produced.

In this example, the heat rendered effective is $\frac{95\cdot 8}{558\cdot 6}$, or somewhat more than one-sixth of that expended in the boiler.

33. The results of the calculations of maximum possible effect, of which examples have just been given, are *limits* which may be approached in practice by Cornish and similar engines, but which cannot be fully realised; and yet it has been shown, that in those theoretical cases only about *one-sixth* of the heat expended in the boiler is rendered effective. In practice, of course, the proportion of heat rendered effective must be still smaller; and, in fact, in some unexpansive engines, it amounts to only *one-twenty-fourth*, or even less.

Dr. Lyon Playfair, in a memoir on the Evaporating Power of Fuel, has taken notice of the great disproportion between the heat expended in the steam-engine and the work performed. It has now been shown that this waste of heat is, to a great extent, a necessary consequence of the nature of the machine. It can only be reduced by increasing the initial pressure of the steam, and the extent of the expansive action; and to both of those resources there are practical limits, which have already, in some instances, been nearly attained.

APPENDIX TO THE FOURTH SECTION.

Containing Tables to be used in Calculating the Pressure, Volume, and Mechanical Action of Steam, Treated as a Perfect Gas.

The object of the First of the annexed Tables is to facilitate the calculation of the volume of steam of saturation at a given pressure, of the pressure of steam of saturation at a given volume, and of its mechanical action at full pressure.

The pressures are expressed in pounds avoirdupois per square foot, and the volumes by the number of cubic feet occupied by one pound avoirdupois of steam, when considered as a perfect gas; those denominations being the most convenient for mechanical calculations in this country.

The columns to be used in determining the pressure from the volume, and *vice versâ*, are the third, fourth, sixth, and seventh.

The third column contains the common logarithms of the pressures of steam of saturation for every fifth degree of the Centigrade thermometer, from $-30°$ to $+260°$: that is to say, for every ninth degree of Fahrenheit's thermometer, from $-22°$ to $+500°$.

The fourth column gives the differences of the successive terms of the third column.

The sixth column contains the common logarithms of the volume of one pound of steam of saturation corresponding to the same temperatures.

The seventh column contains the differences of the successive terms of the sixth column, which are negative; for the volumes diminish as the pressures increase.

By the ordinary method of taking proportional parts of the differences, the logarithms of the volumes corresponding to intermediate pressures, or the logarithms of the pressures corresponding to intermediate volumes, can be calculated with great precision. Thus, let $X + h$ be the logarithm of a pressure not found in the table, X being the next less logarithm which is found in the table; let Y be the logarithm of the volume corresponding to X, and $Y - k$ the logarithm of the volume corresponding to $X + h$; let H be the difference between X and the next greater logarithm in the table, as given in the fourth column, and K the corresponding difference in the seventh column; then by the proportion

$$H : K :: h : k$$

either $Y - k$ may be found from $X + h$, or $X + h$ from $Y - k$.

In the fifth and eighth columns respectively, are given the actual pressures and volumes corresponding to the logarithms in the third and sixth columns, to five places of figures.

In the ninth column are given the values of the quantity denoted by $P_1 V_1$ in the formulæ, which represents the mechanical action of unity of weight of steam at full pressure, or before it has begun to expand, in raising an equal weight. Those values are expressed in feet, being the products of the pressures in the fifth column by the volumes in the eighth, and have been found by multiplying the absolute temperature in Centigrade degrees by 153·48 feet. Intermediate terms in this column, for a given pressure or a given volume, may be approximated to by the method of differences, the constant difference for 5° Centigrade being 767·4 feet; but it is more accurate to calculate them by taking the product of the pressure and volume.

When the pressure is given in other denominations, the following logarithms are to be added to its logarithm, in order to reduce it to pounds avoirdupois per square foot:—

For Millimètres of mercury, ...	0·44477
„ Inches of mercury, ...	1·84960
„ Atmospheres of 760 millimètres, . .	3·32559
„ Atmospheres of 30 inches, . . .	3·32672
„ Kilogrammes on the square centimètre, .	3·31136
„ Kilogrammes on the circular centimètre,	3·41627
„ Kilogrammes on the square mètre, .	$\bar{1}$·31136
„ Pounds avoirdupois on the square inch, .	2·15836
„ Pounds avoirdupois on the circular inch,	2·26327

To reduce the logarithm of the number of cubic mètres occupied by one kilogramme to that of the number of cubic feet occupied by one pound avoirdupois, add 1·20463.

The logarithms are given to five places of decimals only, as a greater degree of precision is not attainable in calculations of this kind.

The Second Table is for the purpose of calculating the mechanical action of steam in expansive engines.

The first column contains values of the fraction of the entire capacity of the cylinder which is filled with steam before the expansion commences (being the quantity $\frac{1}{s}$ of the formulæ), for every hundredth part, from 1·00, or the whole cylinder, down to 0·10, or one-tenth.

If l be the entire length of stroke, l' the portion performed at full pressure, and c the fraction of the entire capacity of the cylinder allowed for clearance, then

$$\frac{l'}{l} = \frac{\frac{1}{s} - c}{1 - c}, \quad \text{and} \ \frac{1}{s} = (1 - c)\frac{l'}{l} + c.$$

The *entire capacity of the cylinder* is to be understood to include clearance at one end only.

The second column gives the reciprocals of the quantities in the first, or the values of the ratio of expansion s.

The third and fourth columns, headed Z, give the values of the quantity

$$\frac{1}{1-\sigma} - \frac{\sigma}{1-\sigma} s^{1-\frac{1}{\sigma}}$$

of article 23, which represents the ratio of the entire gross action of the steam to its action at full pressure, without allowing for clearance. The third column is to be used for initial pressures of from one to four atmospheres; and the fourth for initial pressures of from four to eight atmospheres.

The deduction to be made from the quantity Z for clearance is $c s$, or the product of the fraction of the cylinder allowed for clearance by the ratio of expansion. Hence, to calculate from the tables the net mechanical action of unity of weight of steam, allowing for the counter-pressure of the waste steam P_3, as well as for clearance, we have the formula

$$P_1 V_1 (Z - c s) - P_3 V_1 (1 - c) s,$$

being equivalent to the formula (47) of this paper.

TABLE I.—Pressure and Volume of Steam, and its Action at Full Pressure.

(1.) Temperature Fahrenheit.	(2.) Temperature Centigrade.	(3.) Logarithm of Pressure in lbs. per Square Foot.	(4.) Differences.	(5.) Pressure in lbs. per Square Foot.	(6.) Logarithm of Volume of one lb. of Steam in Cubic Feet.	(7.) Differences.	(8.) Volume of one lb. of Steam in Cubic Feet.	(9.) Action of a given Weight of Steam in raising an equal Weight in Feet, at full Pressure $= P_1 V_1$
−22°	−30°	$\bar{1}$·99278	0·20563	0·9835	4·58173	0·19684	38171	37541
−13	−25	0·19841	0·19602	1·5791	4·38489	0·18741	24260	38309
− 4	−20	0·39443	0·18710	2·4799	4·19748	0·17865	15757	39076
+ 5	−15	0·58153	0·17864	3·8153	4·01883	0·17036	10443	39843
14	−10	0·76017	0·17085	5·7567	3·84847	0·16272	7054·6	40611
23	− 5	0·93102	0·16348	8·5314	3·68575	0·15550	4850·1	41378
32	0	1·09450	0·15661	12·431	3·53025	0·14877	3390·4	42146
41	+ 5	1·25111	0·15012	17·828	3·38148	0·14242	2407·0	42913
50	10	1·40123	0·14404	25·190	3·23906	0·13648	1734·0	43680
59	15	1·54527	0·13836	35·097	3·10258	0·13093	1266·4	44448
68	20	1·68363	0·13284	48·265	2·97165	0·12553	936·81	45215
77	25	1·81647	0·12780	65·535	2·84612	0·12061	701·65	45983
86	30	1·94427	0·12297	87·957	2·72551	0·11590	531·51	46750
95	35	2·06724	0·11842	116·75	2·60961	0·11146	407·01	47517
104	40	2·18566	0·11410	153·34	2·49815	0·10725	314·88	48285
113	45	2·29976	0·11002	199·42	2·39090	0·10328	245·98	49052
122	50	2·40978	0·10614	256·91	2·28762	0·09950	193·92	49820
131	55	2·51592	0·10247	328·04	2·18812	0·09593	154·21	50587
140	60	2·61839	0·09897	415·33	2·09219	0·09253	123·65	51354
149	65	2·71736	0·09566	521·63	1·99966	0·08931	99·922	52122
158	70	2·81302	0·09250	650·16	1·91035	0·08625	81·349	52889
167	75	2·90552	0·08953	804·49	1·82410	0·08336	66·696	53657
176	80	2·99505	0·08658	988·67	1·74074	0·08050	55·048	54424
185	85	3·08163		1206·8	1·66024		45·734	55191

194	90	3·16551	0·08388	1463·9	1·58236	0·07788	38·226	55959
203	95	3·24680	0·08129	1765·2	1·50698	0·07538	32·135	56726
212	100	3·32559	0·07879	2116·4	1·43403	0·07295	27·166	57494
221	105	3·40199	0·07640	2523·4	1·36339	0·07064	23·088	58261
230	110	3·47614	0·07415	2993·2	1·29492	0·06847	19·721	59028
239	115	3·54810	0·07196	3532·6	1·22857	0·06635	16·927	59796
248	120	3·61798	0·06988	4149·3	1·16423	0·06434	14·596	60563
257	125	3·68586	0·06788	4851·3	1·10182	0·06241	12·642	61331
266	130	3·75183	0·06597	5647·2	1·04125	0·06057	10·996	62098
275	135	3·81597	0·06414	6545·9	0·98244	0·05881	9·6037	62865
284	140	3·87835	0·06238	7557·0	0·92533	0·05711	8·4204	63633
293	145	3·93904	0·06069	8690·4	0·86985	0·05548	7·4105	64400
302	150	3·99811	0·05907	9956·6	0·81592	0·05393	6·5452	65168
311	155	4·05562	0·05751	11366	0·76350	0·05242	5·8010	65935
320	160	4·11163	0·05601	12931	0·71251	0·05099	5·1583	66702
329	165	4·16619	0·05456	14662	0·66292	0·04959	4·6017	67470
338	170	4·21938	0·05319	16572	0·61464	0·04828	4·1176	68237
347	175	4·27122	0·05184	18673	0·56766	0·04698	3·6954	69005
356	180	4·32178	0·05056	20979	0·52190	0·04576	3·3258	69772
365	185	4·37110	0·04932	23502	0·47733	0·04457	3·0014	70539
374	190	4·41922	0·04812	26256	0·43391	0·04342	2·7159	71307
383	195	4·46618	0·04696	29254	0·39160	0·04231	2·4638	72074
392	200	4·51204	0·04586	32512	0·35034	0·04126	2·2405	72842
401	205	4·55682	0·04478	36043	0·31011	0·04023	2·0423	73609
410	210	4·60057	0·04375	39863	0·27087	0·03924	1·8663	74376
419	215	4·64331	0·04274	43986	0·23258	0·03829	1·7084	75144
428	220	4·68507	0·04176	48425	0·19524	0·03734	1·5676	75911
437	225	4·72592	0·04085	53201	0·15875	0·03649	1·4413	76679
446	230	4·76586	0·03994	58326	0·12314	0·03561	1·3278	77446
455	235	4·80492	0·03906	63815	0·08836	0·03478	1·2256	78213
464	240	4·84311	0·03819	69680	0·05441	0·03395	1·1335	78981
473	245	4·88051	0·03740	75947	0·02121	0·03320	1·0501	79748
482	250	4·91711	0·03660	82625	$\bar{1}$·98877	0·03244	0·97447	80516
491	255	4·95293	0·03582	89728	$\bar{1}$·95707	0·03170	0·90588	81283
500	260	4·98800	0·03507	97275	$\bar{1}$·92608	0·03099	0·84349	82050

TABLE II.—Expansive Action of Steam.

(1.)	(2.)	(3.)	(4.)	(1.)	(2.)	(3.)	(4.)
		Coefficient of Gross Action = Z.				Coefficient of Gross Action = Z.	
Fraction of Cylinder filled with Steam at full Pressure $=\frac{1}{s}$.	Ratio of Expansion $= s$.	Initial Pressure One to Four Atmospheres.	Initial Pressure Four to Eight Atmospheres.	Fraction of Cylinder filled with Steam at full Pressure $=\frac{1}{s}$.	Ratio of Expansion $= s$.	Initial Pressure One to Four Atmospheres.	Initial Pressure Four to Eight Atmospheres.
1·00	1·000	1·000	1·000	·54	1·852	1·586	1·580
·99	1·010	1·010	1·010	·53	1·887	1·602	1·596
·98	1·020	1·020	1·020	·52	1·923	1·620	1·613
·97	1·031	1·030	1·030	·51	1·961	1·637	1·630
·96	1·042	1·041	1·041	·50	2·000	1·655	1·647
·95	1·053	1·051	1·051	·49	2·041	1·673	1·665
·94	1·064	1·062	1·062	·48	2·083	1·691	1·683
·93	1·075	1·072	1·072	·47	2·128	1·709	1·701
·92	1·087	1·083	1·083	·46	2·174	1·728	1·719
·91	1·099	1·094	1·093	·45	2·222	1·748	1·738
·90	1·111	1·104	1·104	·44	2·273	1·767	1·757
·89	1·124	1·115	1·115	·43	2·326	1·787	1·777
·88	1·136	1·126	1·126	·42	2·381	1·808	1·796
·87	1·149	1·138	1·137	·41	2·439	1·829	1·817
·86	1·163	1·149	1·149	·40	2·500	1·850	1·837
·85	1·176	1·160	1·160	·39	2·564	1·871	1·858
·84	1·190	1·172	1·171	·38	2·632	1·894	1·880
·83	1·205	1·183	1·183	·37	2·703	1·916	1·902
·82	1·220	1·195	1·195	·36	2·778	1·939	1·924
·81	1·235	1·207	1·206	·35	2·857	1·963	1·947
·80	1·250	1·219	1·218	·34	2·941	1·987	1·970
·79	1·266	1·231	1·230	·33	3·030	2·012	1·994
·78	1·282	1·243	1·242	·32	3·125	2·038	2·019
·77	1·299	1·256	1·255	·31	3·225	2·064	2·044
·76	1·316	1·268	1·267	·30	3·333	2·091	2·070
·75	1·333	1·281	1·280	·29	3·448	2·119	2·097
·74	1·351	1·294	1·292	·28	3·571	2·147	2·124
·73	1·370	1·307	1·305	·27	3·704	2·176	2·152
·72	1·389	1·320	1·318	·26	3·846	2·207	2·181
·71	1·408	1·333	1·331	·25	4·000	2·238	2·211
·70	1·429	1·346	1·344	·24	4·167	2·270	2·242
·69	1·449	1·360	1·358	·23	4·348	2·304	2·273
·68	1·471	1·374	1·371	·22	4·545	2·338	2·306
·67	1·493	1·387	1·385	·21	4·762	2·374	2·341
·66	1·515	1·401	1·399	·20	5·000	2·412	2·376
·65	1·538	1·416	1·413	·19	5·263	2·451	2·413
·64	1·563	1·430	1·427	·18	5·556	2·492	2·452
·63	1·587	1·445	1·441	·17	5·882	2·534	2·492
·62	1·613	1·459	1·456	·16	6·250	2·579	2·434
·61	1·640	1·474	1·471	·15	6·667	2·626	2·579
·60	1·667	1·490	1·486	·14	7·143	2·676	2·626
·59	1·695	1·505	1·501	·13	7·692	2·730	2·675
·58	1·724	1·521	1·516	·12	8·333	2·786	2·728
·57	1·754	1·537	1·532	·11	9·091	2·847	2·784
·56	1·786	1·553	1·547	·10	10·000	2·912	2·845
·55	1·818	1·569	1·563				

XV.—NOTE AS TO THE DYNAMICAL EQUIVALENT OF TEMPERATURE IN LIQUID WATER, AND THE SPECIFIC HEAT OF ATMOSPHERIC AIR AND STEAM:

Being a Supplement to a Paper On the Mechanical Action of Heat.*

33. In my paper on the Mechanical Action of Heat (*see p. 234*), published in the first part of the twentieth volume of the *Transactions of the Royal Society of Edinburgh*, some of the numerical results depend upon the dynamical equivalent of a degree of temperature in liquid water. The value of that quantity which I then used, was calculated from the experiments of De la Roche and Bérard on the apparent specific heat of atmospheric air under constant pressure, as compared with liquid water.

The experiments of Mr. Joule on the production of heat by friction give, for the specific heat of liquid water, an equivalent about one-ninth part greater than that which is determined from those of De la Roche and Bérard. I was formerly disposed to ascribe this discrepancy, in a great measure, to the smallness of the differences of temperature measured by Mr. Joule, and to unknown causes of loss of power in his apparatus, such as the production of sound and of electricity; but, subsequently to the publication of my paper, I have seen the detailed account of Mr. Joule's last experiments in the *Philosophical Transactions* for 1850, which has convinced me, that the uncertainty arising from the smallness of the elevations of temperature, is removed by the multitude of experiments (being forty on water, fifty on mercury, and twenty on cast iron); that the agreement amongst the results from substances so different, shows that the error by unknown losses of power is insensible, or nearly so; and that the necessary conclusion is, that the dynamical value assigned by Mr. Joule to the specific heat of liquid water—viz., 772 feet per degree of Fahrenheit—does not err by more than two, or, at the utmost, three feet; and therefore, that the discrepancy originates chiefly in the experiments of De la Roche and Bérard.

I therefore take the earliest opportunity of correcting such of my calculations as require it, so as to correspond with Mr. Joule's equivalent.

* Read before the Royal Society of Edinburgh on December 2, 1850, and published in the *Transactions* of that Society, Vol. XX., Part II.

They relate to the specific heat of atmospheric air as compared with liquid water, and to that of steam, and are contained in the second and third sections of my paper, Articles 14 and 20; equations 28, 34, and 36.

Specific Heat of Atmospheric Air as Compared with liquid Water.—(Section II., Article 14.)

The dynamical values of the specific heat of atmospheric air are calculated independently from the velocity of sound, without reference to the specific heat of liquid water; and from the closeness of the agreement of the experiments of MM. Bravais and Martins, Moll and Van Beek, Stampfer and Myrbach, Wertheim and others, it is clear that the limits of error are about $\frac{1}{300}$ for the velocity of sound, $\frac{1}{150}$ for the ratio, and from $\frac{1}{40}$ to $\frac{1}{50}$ for the dynamical values of the specific heat of air, at constant volume and constant pressure. Those values, as given by equation (27), are—

Real specific heat,—

$$\mathfrak{k} = 238{\cdot}66 \text{ feet} = 72{\cdot}74 \text{ mètres per Centigrade degree,}$$
$$= 132{\cdot}6 \text{ feet per degree of Fahrenheit.}$$

Apparent specific heat under constant pressure,—

$$K_P = 334 \text{ feet} = 101{\cdot}8 \text{ mètres per Centigrade degree,}$$
$$= 185{\cdot}6 \text{ feet per degree of Fahrenheit.}$$

The ratio of these two quantities being taken as

$$\frac{K_P}{\mathfrak{k}} = 1 + N = 1{\cdot}4.$$

The dynamical equivalent of the specific heat of liquid water, as determined by Mr. Joule, is

$$K_W = 1389{\cdot}6 \text{ feet} = 423{\cdot}54 \text{ mètres per Centigrade degree,}$$
$$= 772 \text{ feet per degree of Fahrenheit.}$$

The specific heat of air, that of liquid water being taken as unity, has therefore the following values:—

Real specific heat,—

$$\frac{\mathfrak{k}}{K_W} = \frac{132{\cdot}6}{772} = 0{\cdot}1717.$$

Apparent specific heat under constant pressure,—

$$\frac{K_P}{K_W} = \frac{185{\cdot}6}{772} = 0{\cdot}2404.$$

This last quantity, according to De la Roche and Bérard, is

	0·2669
The discrepancy being	0·0265

or *one-ninth* of the value, according to Mr. Joule's equivalent.

Specific Heat of Steam.—(Section III., Art. 20.)

The apparent specific heat of steam (equations 34 and 36) as a gas under constant pressure, is equal to that of liquid water × 0·305. Its dynamical value is, therefore,

$$K_P = \mathfrak{k} + \frac{1}{C n M} = 1389{\cdot}6 \times 0{\cdot}305$$

$$= 422{\cdot}83 \text{ feet} = 129{\cdot}18 \text{ mètres per Cent. degree.}$$

But

$$\frac{1}{C n M} = 153{\cdot}48 \text{ feet} = 46{\cdot}78 \text{ mètres per Cent. degree.}$$

Therefore, the real specific heat is

$$\mathfrak{k} = 269{\cdot}35 \text{ feet} = 82{\cdot}40 \text{ mètres per Cent. degree.}$$

Or, that of liquid water being taken as unity,

$$\frac{\mathfrak{k}}{K_W} = \frac{269{\cdot}35}{1389{\cdot}6} = 0{\cdot}194.$$

The ratio of these two values of the specific heat of steam is

$$1 + N = 1{\cdot}57.$$

Their dynamical equivalents for Fahrenheit's scale are,

$$\mathfrak{k} = 149{\cdot}64 \text{ feet}, \quad . \quad . \quad . \quad K_P = 235{\cdot}46 \text{ feet}.$$

Neither the formulæ in the fourth Section, respecting the working of the steam-engine, nor the tables at the end of the paper, require any alteration; for the action of steam at full pressure being calculated from data independent of its specific heat, is not at all affected by the discrepancy I have mentioned; and the expansive action is not affected to an extent appreciable in practice.

XVI.—ON THE POWER AND ECONOMY OF SINGLE-ACTING EXPANSIVE STEAM-ENGINES:

Being a Supplement to the Fourth Section of a Paper On the Mechanical Action of Heat.*

34. The objects of this paper are twofold: *First,* To compare the results of the formulæ and tables relative to the power of the steam-engine, which have been deduced from the dynamical theory of heat, with those of experiments on the actual duty of a large Cornish engine at various rates of expansion; and, *Secondly,* To investigate and explain the method of determining the rate of expansion, and, consequently, the dimensions and proportions of a Cornish engine, which, with a given maximum pressure of steam in the cylinder, at a given velocity, shall perform a given amount of work at the least possible pecuniary cost, taking into account the expense of fuel, and the interest of the capital required for the construction of the engine.

This problem is solved with the aid of the tables already printed, by drawing two straight lines on a diagram annexed to this paper.

The merit of first proposing the question of the economy of expansive engines in this definite shape belongs, I believe, to the Artizan Club, who have offered premiums for its solution; having done so (to use their own words) "with a view to enable those who, from their position, cannot take part in the discussions of the various scientific societies, to give the profession the benefit of their studies and experience." The 5th of April is the latest day fixed by them for receiving papers; and as this communication cannot possibly be read to a meeting before the 7th April, nor published until some months afterwards, I trust I may feel confident that it will not be considered as interfering with their design.

Formulæ Applicable to the Cornish Engine.

35. The equations of motion of the steam-engine, in this and the original paper, are the same in their general form with those of M. de

* Read before the Royal Society of Edinburgh on April 21, 1851, and published in the *Transactions* of that Society, Vol. XX., Part II.

Pambour. The differences consist in the expressions for the pressure and volume of steam, and for the mechanical effect of its expansion; the former of which were deduced from a formula suggested by peculiar hypothetical views, and the latter from the dynamical theory of heat.

Those equations are Nos. (50) and (51) of the original paper. *(See p. 271.)* I shall now express them in a form more convenient for practical use, the notation being as follows:—

Let A be the area of the piston;

l, the length of stroke;

n, the number of double strokes in unity of time;

c, the fraction of the total bulk of steam above the piston when down, allowed for clearance, and for filling steam-passages; so that the total bulk of steam at the end of the effective stroke is

$$\frac{lA}{1-c} \quad . \quad . \quad . \quad . \quad . \quad (a.)$$

l', the length of the portion of the stroke performed when the steam is cut off.

s, the ratio of expansion of the steam, so that

$$\left.\begin{aligned} \frac{1}{s} &= (1-c)\frac{l'}{l} + c \\ \frac{l'}{l} &= \frac{\frac{1}{s} - c}{1-c} \end{aligned}\right\} \quad . \quad . \quad . \quad (b.)$$

Let W be the weight of steam expended in unity of time.

P_1, the pressure at which it enters the cylinder.

V_1, the volume of unity of weight of steam at saturation at the pressure P_1; which may be found from Table I. of the Appendix to the original paper. *(See p. 282.)*

F, the sum of all the resistances not depending on the useful load, reduced to a pressure per unit of area of piston; whether arising from imperfect vacuum in the condenser, resistance of the air-pump, feed-pump, and cold-water pump, friction, or any other cause.

R, the resistance arising from the *useful load,* reduced to a pressure per unit of area of piston.

Z, the ratio of the total action of steam working at the expansion s, to its action without expansion. Values of this ratio are given in the second table of the Appendix to the original paper.

Then the following are the two fundamental equations of the motion of the steam-engine as comprehended in equation (50) of the original paper.

First, Equality of power and effect,—

$$\mathrm{R\,A}\,l\,n = \mathrm{W\,V_1}\,\{\mathrm{P_1}(\mathrm{Z} - c\,s) - \mathrm{F}(1 - c)\,s\}. \qquad (c.)$$

Secondly, Equality of two expressions for the weight of steam expended in unity of time,—

$$\mathrm{W} = \frac{\mathrm{A}\,l\,n}{\mathrm{V_1}(1 - c)\,s}. \qquad (d.)$$

From these two equations is deduced the following, expressing the ratio of the mean load on the piston to the initial pressure of the steam:—

$$\frac{\mathrm{R} + \mathrm{F}}{\mathrm{P_1}} = \frac{\mathrm{Z} - c\,s}{(1 - c)\,s}, \qquad (e.)$$

being equivalent to equation (51).

In computing the effect of Cornish engines these formulæ require to be modified, owing to the following circumstances:—

The terms depending on the clearance c have been introduced into equations (c), (d), on the supposition that the steam employed in filling the space above the piston at the top of its stroke is lost, being allowed to escape into the condenser, without having effected any work; so that a weight of steam $\mathrm{W}\,c\,s$ is wasted, and an amount of power $\mathrm{W\,V_1}\,(\mathrm{P_1} - \mathrm{F})\,c\,s$ lost, in unity of time. But in Cornish engines this is not the case; for by closing the equilibrium-valve at the proper point of the up or out-door stroke, nearly the whole quantity of steam necessary to fill the clearance and valve-boxes may be kept imprisoned above the piston, so as to make the loss of power depending on it insensible in practice. This portion of steam is called a cushion, from its preventing a shock at the end of the upstroke; and, as Mr. Pole in his valuable work on the Cornish engine has observed, its alternate compression and expansion compensate each other, and have no effect on the duty of the engine. The proper moment of closing the equilibrium-valve is fixed by trial, which is, perhaps, the best way; but if it is to be fixed by theory, the following is the proper formula: let l'' be the length of the portion of the upstroke remaining to be performed after the equilibrium-valve has been closed: then—

$$\frac{l''}{l} = \frac{c\,(s - 1)}{1 - c}. \qquad (f.)$$

A slight deviation from this adjustment will produce little effect in practice, if the fraction c is small.

In forming the equations of motion, therefore, of the Cornish engine, we may, without material error, in practice omit the terms denoting a waste of steam and loss of power due to clearance and filling of steam-passages; and the results are the following:—

Equation of effect and power in unity of time:—

Useful effect $E = R A l n = W V_1 \{P_1 Z - F\}$. . (57.)

Weight of steam expended in unity of time:—

$$W = \frac{A l n}{V_1 s}. \quad . \quad . \quad . \quad . \qquad (58.)$$

From those two fundamental equations the following are deduced:—

Ratio of mean load on piston to maximum pressure,—

$$\frac{R + F}{P_1} = \frac{Z}{s}. \quad . \quad . \quad . \quad . \qquad (59.)$$

Duty of unity of weight of steam,—

$$\frac{E}{W} = V_1 (P_1 Z - F), \quad . \quad . \quad . \qquad (60.)$$

which, being multiplied by the number of units of weight of steam produced by a given weight of fuel, gives the duty of that weight of fuel.

Weight of steam expended per stroke,—

$$\frac{W}{n} = \frac{A l}{V_1 s}. \quad . \quad . \quad . \quad . \qquad (61.)$$

In fact, it is clear that if any five quantities out of the following seven be given, the other two may be determined by means of the equations:

$R + F$, the mean load on unit of area of piston.

P_1, the maximum pressure of steam in the cylinder.

s, the ratio of expansion.

W, the weight of steam produced in unity of time.

A, the area of the piston.

l, the length of stroke.

n, the number of strokes in unity of time.

The other quantities, E, V_1, Z, are functions of those seven.

Comparison of the Theory with Mr. Wicksteed's Experiments.

36. In order to test the practical value of this theory, I shall compare its results with those of the experiments which were made by Mr. Wicksteed on the large Cornish pumping engine, built under the direction of that eminent engineer, by Messrs. Harvey and West, for the East London

Water-Works at Old Ford, and which were published in 1841. The dimensions and structure of the engine, and the details of the experiments, are stated with such minuteness and precision, that there is none of that uncertainty respecting the circumstances of particular cases, which is the most frequent cause of failure in the attempt to apply theoretical principles to practice.

The engine was worked under a uniform load at five different rates of expansion successively. The number of strokes, and the consumption of steam during each trial, having been accurately registered, Mr. Wicksteed gives a table showing the weight of steam consumed per stroke for each of the five rates of expansion. I shall now compute the weight of steam per stroke theoretically, and compare the results.

Throughout these calculations I shall uniformly use the foot as the unit of length, the avoirdupois pound as that of weight, and the hour as that of time. Pressures are consequently expressed in pounds per square foot for the purpose of calculation; although in the table of experiments I have reduced them to pounds per square inch, as being the more familiar denomination.

The data respecting the dimensions and load of the engine, which are constant throughout the experiments, are the following:—

Area of piston, ...	$A = 34{\cdot}854$ square feet.
Stroke, ...	$l = 10$ feet.
Cubic space traversed by piston during one down stroke, ...	$= Al = 348{\cdot}54$ cubic feet.
Clearance and valve-boxes, ...	18·00 „
Sum, .	366·54 „

Therefore, $c = 0{\cdot}05$

R = useful load of piston, ...	$= 1597{\cdot}0$ lbs. per sq. ft.
F = additional resistance, ...	$= 266{\cdot}6$ „
$R + F$ = total mean pressure on piston, . .	$= 1863{\cdot}6$ „

The mode of calculation is the following:—

Mr. Wicksteed states the fraction $\frac{l'}{l}$ of the stroke performed at full pressure in each experiment. From this the ratio of expansion s is computed by equation (b), giving in this case

$$\frac{1}{s} = 0{\cdot}95\,\frac{l'}{l} + 0{\cdot}05.$$

The value of Z corresponding to s is then found by means of the third column of table second; that column being selected because the initial pressures were all below four atmospheres. This affords the means of determining the initial pressure of the steam by equation (59), viz. —

$$P_1 = \frac{s}{Z} (R + F) = 1863 \cdot 6 \frac{s}{Z}.$$

By using table first according to the directions prefixed to it, the volume of one pound of steam at the pressure P_1, in cubic feet, is calculated, and thence, by equation (60), the weight of steam per stroke, according to theory, which is compared with the weight as ascertained by experiment.

Further, to illustrate the subject, the useful effect, or duty of a pound of steam, is computed according to the theory and the experiments respectively, and the results compared.

The following table (*See p. 294*) exhibits the results.

This comparison sufficiently proves that the results of the theory are practically correct.

It is remarkable, that in every instance except one (experiment E), the experimental results show a somewhat less expenditure of steam per stroke, and a greater duty per pound of steam, than theory indicates. This is to be ascribed to the fact, that although the action of the steam is computed theoretically, on the assumption that during the expansion it is cut off from external sources of heat, yet it is not exactly so in practice; for the cylinder is surrounded with a jacket or casing communicating with the boiler, in which the temperature is much higher than the highest temperature in the cylinder, the pressure in the boiler being more than double the maximum pressure of the steam when working, as columns (2) and (5) show. There is, therefore, a portion of steam, of whose amount no computation can be made, which circulates between the boiler and the jacket, serving to convey heat to the cylinder, and thus augment by a small quantity the action of the steam expended; and hence the formulæ almost always err on the safe side.

Supposing one pound of the best Welsh coals to be capable (as found by Mr. Wicksteed) of evaporating 9·493 lbs. of water at the pressure in the boiler during the experiment F, then the duty of a Cornish bushel, or 94 lbs. of such coals, in the circumstances of that experiment would be—

By theory,		88,288,000 ft. lbs.
By experiment,		90,801,000 „
	Difference,	+ 2,513,000 „

COMPARISON OF THE THEORY WITH MR. WICKSTEED'S EXPERIMENTS ON THE CORNISH PUMPING ENGINE AT OLD FORD.

Number of Experiments.	Pressure in the Boiler, Pounds per Square Inch.	Steam cut off at $\frac{l'}{l}$ of the Stroke.	Ratio of Expansion, *s*.	Maximum Pressure in the Cylinder, Pounds per Square Inch.	Pounds of Steam expended per Stroke.		Difference.	Duty of one Pound of Steam.		Difference.
					By Theory.	By Experiment.		By Theory.	By Experiment.	
								Foot Pounds.	Foot Pounds.	
B.	30·45	0·603	1·605	14·27	7·781	7·536	0·245	71530	73860	+2330
C.	33·20	0·477	1·988	15 59	6·963	6·463	−0·500	79936	86123	+6187
D.	39·2	0·397	2·342	16·9	6·236	6·200	−0·036	89275	89776	+ 501
E.	41·2	0 352	2·605	17·89	5·905	5·985	+0·085	94258	93002	−1256
F.	45·7	0·313	2·882	18·93	5·626	5·470	−0·156	98940	101756	+2816
(1)	(2)	(3)	(4)	(5)	(6)	(7)	(8)	(9)	(10)	(11)

Economy of Single-Acting Expansive Engines.

37. By increasing the ratio of expansion in a Cornish engine, the quantity of steam required to perform a given duty is diminished; and the cost of fuel, and of the boilers, is lowered. But at the same time, as the cylinders and every part of the engine must be made larger, to admit of a greater expansion, the cost of the engine is increased. It thus becomes a problem of maxima and minima to determine what ratio of expansion ought to be adopted under given circumstances, in order that the sum of the annual cost of fuel, and the interest of the capital employed in construction, may be the least possible, as compared with the work done.

That this problem may admit of a definite solution, the following five quantities must be given:—

P_1, the initial pressure in the cylinder.

F, the resistance not depending on the useful load.

$l\,n$, the amount of the length of the effective strokes made in unity of time.

h, the annual cost of producing unity of weight of steam in unity of time, which consists of two parts—the price of fuel, and the interest of the cost of the boilers.

k, the interest of the cost of the engine, per unit of area of piston.

Hence the annual expenditure to be taken into consideration, reduced to unity of weight of steam, is

$$h + k\frac{A}{W} = h + k\frac{V_1 s}{l\,n}.$$

And the useful effect of unity of weight of steam being

$$V_1 (P_1 Z - F s),$$

the problem is to determine the ratio of expansion s, so that

$$\frac{V_1 (P_1 Z - F s)}{h + k\dfrac{V_1 s}{l\,n}}$$

shall be a maximum.

Dividing the numerator of this fraction by $V_1 P_1$, and the denominator by $\dfrac{k V_1}{l\,n}$, both of which are constants in this problem, we find that it will be solved by making the ratio

$$\frac{Z - \frac{F}{P_1} s}{\frac{h l n}{k V_1} + s} \quad . \quad . \quad . \quad . \quad . \quad (62.)$$

a maximum.

The algebraical solution would be extremely complicated and tedious. The graphic solution, on the other hand, is very simple and rapid, and sufficiently accurate for all practical purposes; and I have therefore adopted it.

In the diagram (*See Plate II., Fig. 1*), the axis of abscissæ $-$X O $+$X, is graduated from O towards $+$X into divisions representing ratios of expansion, or values of s. The divisions of the axis of ordinates O Y represent values of Z. The curve marked "locus of Z," is laid down from the third column of Table II. of the Appendix to the original paper, being applicable to initial pressures not exceeding four atmospheres.

Through the origin O draw a straight line B O A, at such an inclination to $-$X O $+$X that its ordinates are represented by $\frac{F}{P_1} s$. Then the ordinates measured from this inclined line to the locus of Z represent the value of the numerator $Z - \frac{F}{P_1} s$, of the ratio (62), corresponding to the various values of s.

Take a point at C on the line B O A, whose abscissa, measured along O $-$ X, represents $-\frac{h n l}{k V_1}$. Then the ordinates, measured from B O A, of any straight line drawn through C, vary proportionally to the denominator $\frac{h n l}{k V_1} + s$ of the ratio (62).

Through the point C, therefore, draw a straight line C T, touching the locus of Z: then the ratio (62) is a maximum at the point of contact T, and the abscissa at that point represents the ratio of expansion required.

Example.

38. To exemplify this method, let us take the following data:—

Greatest pressure in the cylinder $P_1 = 20$ lbs. per square inch, $= 2880$ lbs. per square foot.

The corresponding value of V_1 is 20·248 cubic feet per pound of steam.

To obtain this initial pressure in the cylinder, it will be necessary to have a pressure of about 50 lbs. per square inch in the boiler.

F, resistance not depending on the useful load $=$ 2 lbs. per square inch, $=$288 lbs. per square foot, $= \frac{1}{10} P_1$.

$l\,n$, amount of down strokes, = 4800 feet per hour; being the average speed found to answer best in practice.

To estimate h, the annual cost of producing one pound of steam per hour, I shall suppose that the engine works 6000 hours per annum; that the cost of fuel is one penny per 100 lbs. of steam;* that the cost of boiler for each pound of steam per hour is 0·016 ton, at £27, = £0·432; and that the interest of capital is five per cent. per annum. Hence h is thus made up—

Fuel for 6000 lbs. of steam at 0·01d., . . .	£0·2500
Interest on £0·432, at 5 per cent., . . .	0·0216
	h = £0·2716

Estimating the cost of the engine at £250 per square foot of piston, we find k = 5 per cent. per annum on £250 = £12·5.

$$\text{and } \frac{h}{k} = 0\text{·}0217;\ \frac{h\,l\,n}{k\,V_1} = 5\text{·}144.$$

The line B O A, then, is to be drawn so that its ordinates are

$$\frac{F}{P_1}s = \frac{1}{10}s.$$

The point C is taken on this line, at $\frac{h\,l\,n}{k\,V_1} = 5\text{·}144$ divisions of the axis of abscissæ to the left of O Y.

The tangent C T being drawn, is found to touch the locus of Z at 2·800 divisions to the right of O Y.

Then s = 2·800 is the ratio of expansion sought, corresponding to the greatest economy.

If we make c=0·05, as in Mr. Wicksteed's engine, then the fraction of the stroke to be performed at full pressure is

$$\frac{l'}{l} = 0\text{·}323,$$

being nearly the same as in experiment F.

The mean resistance of the useful load per square foot of piston is

$$R = \frac{Z}{s}P_1 - F = 1713\text{·}6 \text{ lbs.}$$

* This estimate is made on the supposition that coals capable of producing nine times their weight of steam are worth about 16s. 9d. per ton.

The duty of one square foot of piston per hour—

$$R\, l\, n, = 8{,}225{,}300 \text{ foot-lbs.}$$

And one horse-power being 1,980,000 foot-lbs. per hour, the real horse-power of the engine is

4·154 per square foot of piston.

The duty of one pound of steam is

$$R\, V_1\, s = 97{,}154 \text{ foot-lbs.}$$

To give an example of a special case, let the duty to be performed be 198,000,000 foot-pounds per hour, being equal to 100 real horse-power, for 6,000 hours per annum. This being called E, we find from the above data that the area of piston required is

$$A = \frac{E}{R\, l\, n} = 24{\cdot}072 \text{ square feet.}$$

The consumption of steam per hour is

$$W = \frac{E}{R\, V_1\, s} = 2038 \text{ lbs.,}$$

which requires $2038 \times 0{\cdot}016 = 32{\cdot}608$ tons of boilers.

The expenditure of steam per annum is

$$2038 \times 6000 = 12{,}228{,}000 \text{ lbs.}$$

Hence we have the following estimate:—

Cost of engine, 24·072 square feet of piston at £250, ...	£6018·000
Cost of boilers, 32·608 tons at £27, ...	880·416
Total capital expended, ...	£6898·416
Interest at five per cent. per annum, ...	344·921
Cost of fuel per annum, 12,228,000 lbs. of steam at 0·01d., ...	509·500
Annual cost for interest and fuel, . .	£854·421

I wish it to be understood that the rates I have adopted in the foregoing calculations, for interest, cost of fuel, and cost of construction, are not intended as estimates of their average amount, nor of their amount in any particular case; but are merely assumed in order to illustrate, by a numerical example, the rules laid down in the preceding article. It is, of course, the business of the engineer to ascertain those data with reference to the special situation and circumstances of the proposed work; and having done so, the method explained in this paper will enable him to determine the dimensions and ratio of expansion which ought to be adopted for the engine, in order that it may effect its duty with the greatest possible economy.

XVII.—ON THE ECONOMY OF HEAT IN EXPANSIVE MACHINES:*

Forming the Fifth Section of a Paper ON THE MECHANICAL ACTION OF HEAT.

39. A MACHINE working by expansive power consists essentially of a portion of some substance to which heat is communicated, so as to expand it, at a higher temperature, being abstracted from it, so as to condense it to its original volume, at a lower temperature. The quantity of heat given out by the substance is less than the quantity received; the difference disappearing as heat, to appear in the form of expansive power.

The heat originally received by the working body may act in two ways: to raise its temperature, and to expand it. The heat given out may also act in two ways: to lower the temperature, and to contract the body. Now, as the conversion of heat into expansive power arises from changes of volume only, and not from changes of temperature, it is obvious, that the proportion of the heat received which is converted into expansive power will be the greatest possible, when the reception of heat, and its emission, each take place at a constant temperature.

40. Carnôt was the first to assert the law, *that the ratio of the maximum mechanical effect to the whole heat expended in an expansive machine, is a function solely of the two temperatures at which the heat is respectively received and emitted, and is independent of the nature of the working substance.* But his investigations not being based on the principle of the dynamical convertibility of heat, involve the fallacy that power can be produced out of nothing.

41. The merit of combining Carnôt's *Law*, as it is termed, with that of the convertibility of heat and power, belongs to Mr. Clausius and Professor William Thomson; and in the shape into which they have brought it, it may be stated thus:—

The maximum proportion of heat converted into expansive power by any machine, is a function solely of the temperatures at which heat is received and emitted by the working substance; which function for each pair of temperatures is the same for all substances in nature.

* Read before the Royal Society of Edinburgh on April 21, 1851, and published in the *Transactions* of that Society, Vol. XX., Part II.

Plate II.

Fig. 1.

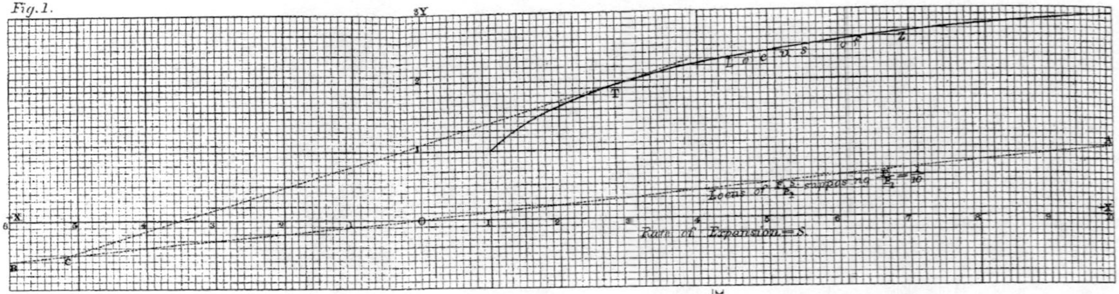

Fig. 2.

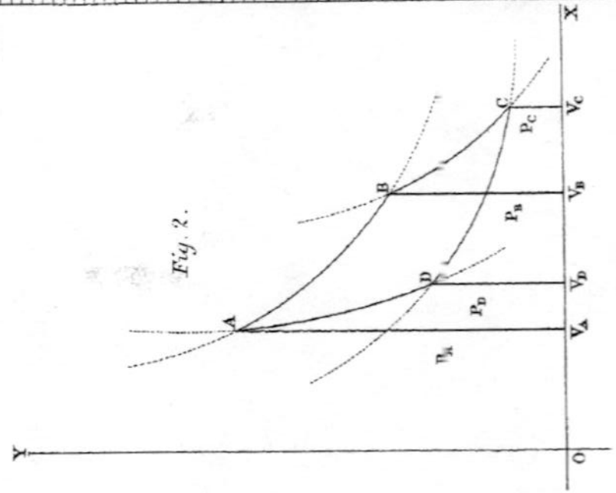

This law is laid down by Mr. Clausius, as it originally had been by Carnôt, as an independent axiom; and I had at first doubts as to the soundness of the reasoning by which he maintained it. Having stated those doubts to Professor Thomson, I am indebted to him for having induced me to investigate the subject thoroughly; for although I have not yet seen his paper, nor become acquainted with the method by which he proves Carnôt's law, I have received from him a statement of some of his more important results.

42. I have now come to the conclusions,—First, *That Carnôt's Law is not an independent principle in the theory of heat; but is deducible, as a consequence, from the equations of the mutual conversion of heat and expansive power, as given in the first section of this paper.*

Secondly, *That the function of the temperatures of reception and emission, which expresses the maximum ratio of the heat converted into power to the total heat received by the working body, is the ratio of the difference of those temperatures to the absolute temperature of reception diminished by the constant, which I have called* $\kappa = C\,n\,\mu\,b$; and which must, as I have shown in the introduction, be the same for all substances, in order that molecular equilibrium may be possible.

43. Let abscissæ, parallel to O X in the diagram, Plate II., Fig. 2, denote the volumes successively assumed by the working body, and ordinates, parallel to O Y, the corresponding pressures. Let τ_1 be the constant absolute temperature at which the reception of heat by the body takes place: τ_0, the constant absolute temperature at which the emission of heat takes place. Let A B be a curve such that its ordinates denote the pressures, at the temperature of reception τ_1, corresponding to the volumes denoted by abscissæ. Let D C be a similar curve for the temperature of emission τ_0. Let A D and B C be two curves, expressing by their co-ordinates how the pressure and volume must vary, in order that the body may change its temperature, without receiving or emitting heat; the former corresponding to the most condensed and the latter to the most expanded state of the body, during the working of the machine.

The quantity of heat received or emitted during an operation on the body involving indefinitely small variations of volume and temperature, is expressed by adding to equation (6) of section fourth the heat due to change of temperature only, in virtue of the real specific heat. We thus obtain the differential equation

$$\delta Q' - \delta Q = - \frac{\tau - \kappa}{C\,n\,M} \left\{ \delta V . \left(\frac{1}{V} - \frac{d\,U}{d\,V} \right) - \delta \tau \, . \, \frac{d\,U}{d\,\tau} \right\}$$

$$- \mathfrak{k}\,\delta\tau,$$

in which the negative sign denotes absorption, and the positive emission.

If we now put for $\frac{d\,U}{d\,V}, \frac{d\,U}{d\,\tau}$, their values according to equation (11), we find

$$\delta\,Q' - \delta\,Q = -(\tau - \kappa)\frac{d\,P}{d\,\tau}\,.\,\delta\,V$$

$$-\left\{\mathfrak{k} + \frac{1}{C\,n\,M}\left(\frac{\kappa}{\tau} - \frac{\kappa^2}{\tau^2}\right) + (\tau - \kappa)\frac{d}{d\,\tau}\int\frac{d\,P}{d\,\tau}\,d\,V\right\}\delta\,\tau. \quad (63.)$$

The first term represents the variation of heat due to variation of volume only; the second, that due to variation of temperature. Let us now apply this equation to the cycle of operations undergone by the working body in an expansive machine, as denoted by the diagram.

First operation.—The body, being at first at the volume V_A and pressure P_A, is made to expand, by the communication of heat at the constant temperature τ_1, until it reaches the volume V_B and pressure P_B, A B being the locus of the pressures.

Here $\delta\,\tau = 0$; therefore, the total heat received is

$$\left.\begin{aligned} H_1 = -\,Q'_1 &= (\tau_1 - \kappa)\int_{V_A}^{V_B}\frac{d\,P}{d\,\tau}\,d\,V \\ &= (\tau_1 - \kappa)\{\phi\,(V_B, \tau_1) - \phi\,(V_A, \tau_1)\}. \end{aligned}\right\} \quad . \quad (a.)$$

Second operation.—The body, being prevented from receiving or emitting heat, expands until it falls to the temperature τ_0, the locus of the pressures being the curve B C. During this operation the following condition must be fulfilled,—

$$0 = \delta\,Q' - \delta\,Q,$$

which, attending to the fact that V is now a function of τ, and transforming the integrals as before, gives the equation

$$0 = \mathfrak{k} + \frac{1}{C\,n\,M}\left(\frac{\kappa}{\tau} - \frac{\kappa^2}{\tau^2}\right) + (\tau - \kappa)\left(\frac{d}{d\,\tau} + \frac{d\,V}{d\,\tau}\cdot\frac{d}{d\,V}\right)\phi\,(V, \tau).$$

This equation shows that

$$\phi\,(V_B, \tau_1) - \phi\,(V_C, \tau_0) = \psi\,(\tau_1, \tau_0). \quad . \quad . \quad (b.)$$

Third operation.—The body, by the abstraction of heat, is made to contract, at the constant temperature τ_0, to the volume V_D and pressure P_D, which are such as to satisfy conditions depending on the fourth operation. C D is the locus of the pressures. The heat emitted is evidently

$$H_0 = Q'_0 = (\tau_0 - \kappa)\{\phi\,(V_C, \tau_0) - \phi\,(V_D, \tau_0)\}. \quad . \quad (c.)$$

Fourth operation.—The body, being prevented from receiving or emitting heat, is compressed until it recovers its original temperature τ_1, volume V_A, and pressure P_A; the locus of the pressures being D A. During this operation, the same conditions must be fulfilled as in the second operation; therefore,

$$\phi(V_A, \tau_1) - \phi(V_D, \tau_0) = \psi(\tau_1, \tau_0). \quad . \quad . \quad (d.)$$

ψ being the same function as in equation (*b*).

By comparing equations (*b*) and (*d*), we obtain the relation which must subsist between the four volumes to which the body is successively brought, in order that the maximum effect may be obtained from the heat. It is expressed by the equation

$$\phi(V_B, \tau_1) - \phi(V_A, \tau_1) = \phi(V_C, \tau_0) - \phi(V_D, \tau_0). \quad . \quad (64.)$$

From this and equations (*a*) and (*c*), it appears that

$$\frac{H_0}{H_1} = \frac{\tau_0 - \kappa}{\tau_1 - \kappa}. \quad . \quad . \quad . \quad . \quad (65.)$$

That is to say: *when no heat is employed in producing variations of temperature, the ratio of the heat received to the heat emitted by the working body of an expansive machine, is equal to that of the absolute temperatures of reception and emission, each diminished by the constant* κ, *which is the same for all substances.*

Hence, let

$$\Pi = -Q'_1 - Q'_0 = H_1 - H_0$$

denote the maximum amount of power which can be obtained out of the total heat H_1, in an expansive machine working between the temperatures τ_1 and τ_0. Then

$$\frac{\Pi}{H_1} = \frac{\tau_1 - \tau_0}{\tau_1 - \kappa}, \quad . \quad . \quad . \quad . \quad (66.)$$

being the law which has been enunciated in Article 42, and which is deduced entirely from the principles already laid down in the introduction and first section of this paper.

The value of the constant κ is unknown; and the nearest approximation to accuracy which we can at present make, is to neglect it in calculation, as being very small as compared with τ.

44. This approximation having been adopted, I believe it will be found that the formula (66), although very different in appearance from that arrived at by Professor Thomson, gives nearly the same numerical results. For example: let the machine work between the temperatures 140° and 30° Centigrade: then

$$\tau_1 = 414^\circ{\cdot}6,\ \tau_0 = 304^\circ{\cdot}6,$$

and

$$\frac{\Pi}{\text{H}_1} = 0{\cdot}2653.$$

Professor Thomson has informed me, that for the same temperatures he finds this ratio to be 0·2713.*

45. To make a steam-engine work according to the conditions of maximum effect here laid down, the steam must enter the cylinder from the boiler without diminishing in pressure, and must be worked expansively down to the pressure and temperature of condensation. It must then be so far liquefied by conduction alone, that on the liquefaction being completed by compression, it may be restored to the temperature of the boiler by means of that compression alone. These conditions are unattainable in steam-engines as at present constructed, and different from those which form the basis of the formulæ and tables in the fourth section of this paper; hence it is found, both by experiment and by calculation from those formulæ, that the proportion of the total heat converted into power in any possible steam-engine is less than that indicated by equation (66).

The annexed table illustrates this.

The heat transformed into power, as given in the fifth column, has been reduced to Centigrade degrees in liquid water, by dividing the duty of a pound of steam by Mr. Joule's equivalent, 1389·6 feet per Centigrade degree. Hence, the first two numbers in that column are less than those given in Art. 32, which were computed from too small an equivalent.

The first two cases fulfil the conditions required by Carnôt's law in every respect except one:—viz., that the steam remaining at the end of the stroke, instead of being partially liquefied by refrigeration, and then reduced to water at the temperature of the boiler by compression, is supposed to be entirely liquefied by refrigeration. This occasions the loss of the heat necessary to raise the water from the temperature of the condenser to that of the boiler; but at the same time, there is a gain of the power which would be required to liquefy part of the steam by compression, and those two quantities partially compensate for each other's effects on the ratio of the power to the heat expended, so that although it is below the maximum, the difference is small.

* From information which I have received from Professor Thomson subsequently to the completion of this paper, it appears that his formula becomes identical with the approximate formula here proposed, on making the function called by him $\mu = \frac{\text{J}}{\tau}$, J being Joule's equivalent.

Mr. Joule also, some time since, arrived at this approximate formula in the particular case of a perfect gas.

CASE.	Absolute Temperature in the Boiler = τ_1 Centigrade.	Absolute Temperature in the Condenser = τ_0 Centigrade.	Total heat expended in Centigrade Degrees applied to Liquid Water.	Heat Transformed into Expansive Power, in Centigrade Degrees, applied to Liquid Water.	Proportion of Heat rendered Effective.	Maximum Proportion according to Carnôt's Law.
First Ideal Example in Section IV., Art. 32, .	144°·1 + 274°·6 = 418°·7	81°·7 + 274°·6 = 356°·3	568°·7	83°·2	0·1463	0·1490
Second Ideal Example, .	170°·9 + 274°·6 = 445°·5	100° + 274°·6 = 374°·6	558°·6	86°·3	0·1545	0·1592
Mr. Wicksteed's Engine, Experiment F, by calculation, ...	135°·2 + 274°·6 = 409°·8	30° + 274°·6 = 304°·6	617°·7	71°·2	0·1153	0·2567
Do., by observation, . .	Ditto.	Ditto.	Ditto.	73°·23	0·1185	Ditto.
(1)	(2)	(3)	(4)	(5)	(6)	(7)

In the third and fourth examples, founded on the calculated and observed duty of Mr. Wicksteed's engine during experiment F, the actual ratio is less than half the maximum. This waste of heat is to be ascribed to the following causes :—

First, The mode of liquefaction, which has already been referred to.

Secondly, The initial pressure in the cylinder is but 18·93 lbs. on the square inch, while that in the boiler is 45·7; so that although the steam is produced at 135°·2 Centigrade, it only begins to work at 107°·26. This great fall of pressure is accounted for by the fact, that the steam for each stroke, which is produced in the boiler in about seven or eight seconds, escapes suddenly into the cylinder in a fraction of a second.

Thirdly, The expansive working of the steam, instead of being continued down to 30° Centigrade, the temperature of the condenser, stops at a much higher temperature, 74°·66. This is the most important cause of loss of power.

If we now take for τ_1 and τ_0 the absolute temperatures at the beginning and end of the expansive working, and calculate the maximum duty of one pound of steam by Carnôt's Law between those temperatures, we find,—

$$\tau_1 = 107^\circ{\cdot}26 + 274^\circ{\cdot}6 = 381^\circ{\cdot}86$$

$$\tau_0 = 74^\circ{\cdot}66 + 274^\circ{\cdot}6 = 349^\circ{\cdot}26$$

$$\frac{\Pi}{H_1} = 0{\cdot}08542$$

$H_1 = 564^\circ{\cdot}5$; $\therefore \Pi =$ . . . . $48^\circ{\cdot}22$

To this has to be added the duty, at full pressure, of steam at τ_0, diminished by one-third for back-pressure and friction, and by one-fifteenth for liquefaction in the cylinder, = ... $23^\circ{\cdot}14$

The whole amounting to . . . $71^\circ{\cdot}36$

Which agrees very nearly with 73°·23, the observed duty, and almost exactly with 71°·2, the duty as calculated by the formulæ and tables of section fourth.

These examples show clearly the nature and causes of the waste of heat in the steam-engine.

XVIII.—ON THE ABSOLUTE ZERO OF THE PERFECT GAS THERMOMETER:*

Being a Note to a Paper ON THE MECHANICAL ACTION OF HEAT.
(*See p. 234.*)

TEMPERATURE being measured by the pressure of a perfect gas at constant density, the absolute zero of temperature is that point on the thermometric scale at which, if it were possible to maintain a perfect gas at so low a temperature, the pressure would be null.

The position of this point is of great importance, both theoretically and practically, for by reckoning temperatures from it, the laws of phenomena depending on heat are reduced to a more simple form than they are when any other zero is adopted.

As we cannot obtain any substance in the perfectly gaseous condition (that is to say, entirely devoid of cohesion), we cannot determine the position of the absolute thermometric zero by direct experiment, which furnishes us with approximate positions only. Those approximate positions are always too high; because the effect of cohesion is to make the pressure of a gas diminish more rapidly with a diminution of temperature, than if it were devoid of cohesion.

As a gas is rarefied, the cohesion of its particles diminishes, not only in absolute amount, but also in the proportion which it bears to the pressure due to heat. The gas, therefore, approaches more and more nearly to the state of a perfect gas as its density diminishes; and from a series of experiments on the rate of increase of its elasticity with temperature, at progressively diminishing densities, may be calculated the positions of a series of points on the thermometric scale, approaching more and more nearly to the true absolute zero.

By observing the law which those successive approximations follow, the true position of the absolute zero can be determined.

Having performed this operation by means of a graphic process, soon after the publication of the experiments of M. Regnault on the elasticity and expansion of gases, I stated the result in a paper on the Elasticity of Vapours (*Edinburgh New Philosophical Journal*, July, 1849), (*See p. 1*), and

* Read before the Royal Society of Edinburgh on January 4, 1853, and published in the *Transactions* of that Society, Vol. XX. Part IV.

also in a paper on the Mechanical Action of Heat (*Trans. Royal Soc. Edin.*, Vol. XX., Part I.), (*See p. 234*)—viz., that the absolute zero is—

274°·6 Centigrade, or 494°·28 Fahrenheit, } below the temperature of melting ice;

or 462°·28 below the ordinary zero of Fahrenheit's scale.

To enable others to judge of the accuracy of this result, I shall now explain the method by which it was obtained.

Let E denote the mean rate of increase, per degree, between the freezing and boiling points, of the pressure of a gas whose volume is maintained constant. Then the reciprocal of this coefficient, $\frac{1}{E}$, is an approximation to the number of degrees below the freezing point, at which the absolute zero is situated.

The experimental data in the following table were copied from the memoirs of M. Regnault on the Expansion of Gases. The numbers in the first column designate the series of experiments. The second column contains the pressures of the gases at the freezing point. The third column contains the mean coefficients of increase of pressure per Centigrade degree, between 0° and 100° Centigrade. The fourth column contains the reciprocals of those coefficients, with the negative sign, being approximate positions of the absolute zero, in Centigrade degrees, below the temperature of melting ice. The gases employed were atmospheric air and carbonic acid.

The approximate positions of the absolute zero contained in this table were laid down on a diagram, in which they were marked by crosses. The longitudinal divisions representing Centigrade degrees divided into tenths; the transverse divisions, atmospheres of pressure at 0° Centigrade, also divided into tenths. The positions of the crosses indicating at once the pressures in the second column of the table, and the approximate zeros in the fourth; and the numbers affixed to them corresponding with those in the first column.

As the effect of cohesion is greater and more easily eliminated in carbonic acid gas than in atmospheric air, the determination of the true absolute zero was made from the experiments on the former gas. The approximate positions of the absolute zero for carbonic acid lie nearly in a straight line. A straight line having been drawn so that it should, as nearly as possible, traverse them, was found to intersect the line corresponding to the zero of pressure, that is, to the state of perfect gas, at a point on the scale of temperatures 274·6 Centigrade degrees below the temperature of melting ice, which point was accordingly taken as the true absolute zero of the perfect gas thermometer.

No.	Pressure at 0° Centigrade in Atmospheres.	Coefficient of Increase of Elasticity with Temperature = E.	Approximate Positions of the Absolute Zero in Centigrade Degrees $= \frac{1}{E}$.
		CARBONIC ACID.	
1.	0·9980	0·0036856	° −271·33
2.	1·1857	0·0036943	−270·63
3.	2·2931	0·0037523	−266·50
4.	4·7225	0·0038598	−259·08
		ATMOSPHERIC AIR.	
1.	0·1444	0·0036482	−274·11
2.	0·2294	0·0036513	−273·88
3.	0·3501	0·0036542	−273·66
4.	0·4930	0·0036587	−273·32
5.	0·4937	0·0036572	−273·43
6.	1·0000	0·0036650	−272·85
7.	2·2084	0·0036760	−272·03
8.	2·2270	0·0036800	−271·74
9.	2·8213	0·0036894	−271·05
10.	4·8100	0·0037091	−269·61

So far as their irregularity permits, the experiments on atmospheric air confirm this result, for the approximate positions of the absolute zero deduced from them evidently tend towards the very same point on the diagram with those deduced from the experiments on carbonic acid.

The values of the coefficient of dilatation and of increase of pressure of a perfect gas, per degree, in fractions of its volume and pressure, at the temperature of melting ice, are, accordingly,—

$$\text{For the Centigrade Scale} \quad \frac{1}{274{\cdot}6} = 0{\cdot}00364166$$

$$\text{For Fahrenheit's Scale} \quad \frac{1}{494{\cdot}28} = 0{\cdot}00202314.$$

XIX.—ON THE MECHANICAL ACTION OF HEAT.*

SECTION VI.—A REVIEW OF THE FUNDAMENTAL PRINCIPLES OF THE MECHANICAL THEORY OF HEAT; WITH REMARKS ON THE THERMIC PHENOMENA OF CURRENTS OF ELASTIC FLUIDS, AS ILLUSTRATING THOSE PRINCIPLES.

46. I HAVE been induced to write this section in continuation of a paper on the Mechanical Action of Heat, by the publication (in the *Philosophical Magazine* for December, 1852, Supplementary Number) of a series of experiments by Mr. Joule and Professor William Thomson, On the Thermal Effects experienced by Air in rushing through Small Apertures. Although those authors express an intention to continue the experiments on a large scale, so as to obtain more precise results; yet the results already obtained are sufficient to constitute the first step towards the experimental determination of that most important function in the theory of the mechanical action of heat, which has received the name of Carnôt's function.

By the theoretical investigations of Messrs. Clausius and Thomson,—which are based simply on the fact of the convertibility of heat and mechanical power, the determination of their relative value by Mr. Joule, and the properties of the function called *temperature*, without any definite supposition as to the nature of heat,—Carnôt's function is left wholly indeterminate.

By the investigations contained in the previous sections of this paper, and in a paper on the Centrifugal Theory of Elasticity (*See p. 49*),—in which the supposition is made, that heat consists in the revolutions of what are called molecular vortices, so that the elasticity arising from heat is in fact centrifugal force,—a *form* is assigned to Carnôt's function; but its numerical values are left to be ascertained by experiment.

The recent experiments of Messrs. Joule and Thomson serve (so far as the degree of precision of their results permits) at once to determine numerical values of Carnôt's function for use in practice, and to test the

* Read before the Royal Society of Edinburgh on January 17, 1853, and published in the *Transactions* of that Society, Vol. XX., Part IV.

accuracy with which the phenomena of heat are represented by the consequences of the hypothesis of molecular vortices, from which the investigation in this paper sets out.

SUB-SECTION 1.—PROPERTIES OF EXPANSIVE HEAT.

47. To show more clearly the nature of the questions, towards the decision of which these experiments are a step, I shall now briefly review the fundamental principles of the theory of heat, and the reasoning on which they are based; and the object of this being illustration rather than research, I shall use algebraical symbols no further than is absolutely necessary to brevity and clearness, and shall follow an order of investigation which, though the same in its results with that pursued in the previous sections of this paper, is different in arrangement.

By a mind which admits as an axiom that, in the present order of things, physical power cannot be annihilated, nor produced out of nothing, the law of the mutual convertibility of heat and motive power must be viewed as a necessary corollary from this axiom, and Mr. Joule's experiments, as the means of determining the relative numerical value of those two forms of power. By a mind which does not admit the necessity of the axiom, these experiments must be viewed also as the proof of the law.

This law was virtually, though not expressly, admitted by those who introduced the term latent heat into scientific language; for when divested of ideas connected with the hypothesis of a subtle fluid of caloric, and regarded simply as the expression of a fact, this term denotes heat which has disappeared during the appearance of expansive power in a mass of matter, and which may be made to reappear by the expenditure of an equal amount of compressive power.

48. Without for the present framing any mechanical hypothesis as to the nature of heat, let us conceive that unity of weight of any substance occupying the bulk V under the pressure P, and possessing the absolute quantity of thermometric heat whose mechanical equivalent is Q, undergoes the indefinitely small increase of volume $d\,V$; and let us investigate how much heat becomes latent, or is converted into expansive power, during this process; the thermometric heat being maintained constant, so that the heat which disappears must be supplied from some external source.

During the expansion $d\,V$, the body, by its elastic pressure P, exerts the mechanical power $P\,d\,V$. Part of this power is produced by molecular attractions and repulsions; and although this part may be modified by the influence of heat upon the distribution of the particles of the body, it is not the direct effect of heat. The remainder must be considered as directly caused by the heat possessed by the body, of which the pressure

P is a function; and to this portion of the power developed, the heat which disappears during the expansion must be equivalent.

To determine the portion of the mechanical power $P\,dV$ which is the effect of heat, let the total heat of the body, Q, be now supposed to vary by an indefinitely small quantity dQ. Then the mechanical power of expansion $P\,dV$ will vary by the indefinitely small quantity

$$dQ \times \frac{dP}{dQ}\,dV.$$

This is the development of power for the expansion dV, caused by each indefinitely small portion dQ of the total heat possessed by the body; and, consequently, the whole mechanical power for the expansion dV due to the whole heat possessed by the body Q, is expressed as follows:—

$$Q\frac{dP}{dQ}\cdot dV, \quad . \quad . \quad . \quad . \quad (67.)$$

and this is the equivalent of the heat transformed into mechanical power, or the latent heat of expansion of unity of weight, for the small increment of volume dV, at the volume V and total heat Q.

Now, a part only of this power, viz.—

$$P\,dV,$$

is visible mechanical energy, expended in producing velocity in the expanding body itself, or in overcoming the resistance of the bodies which inclose it. The remainder

$$\left(Q\frac{dP}{dQ} - P\right)dV, \quad . \quad . \quad . \quad (68.)$$

is, therefore, expended in overcoming molecular attraction.

Molecular attraction depends on the density and distribution of the particles of the body; and is, consequently, a function of the volume and total heat of unity of weight. It is, therefore, possible to find a potential S, being a function of V and Q, of such a nature, that the difference between its two values

$$S_2 - S_1,$$

corresponding respectively to two sets of values of the volume and total heat (V_1, Q_1 and V_2, Q_2), shall represent the power which is the equivalent of the heat consumed in overcoming molecular attraction, during the passage of the body from the volume V_1 and heat Q_1 to the volume V_2

and heat Q_2. The form of the expression (68) shows that this potential has the following property:—

$$\frac{d\,S}{d\,V} = Q\frac{d\,P}{d\,Q} - P. \quad . \quad . \quad . \quad (69.)$$

The integration of which partial differential equation gives the following value for the potential of molecular action:—

$$S = \int \left(Q\frac{d\,P}{d\,Q} - P\right) d\,V + \phi\,(Q), \quad . \quad . \quad (70.)$$

$\phi\,(Q)$ being some unknown function of the heat only, and the integral being taken as if the heat Q were constant.

The heat which disappears in overcoming molecular action, during a small increase of total heat $d\,Q$, while the volume remains constant, is expressed as follows:—

$$\frac{d\,S}{d\,Q}\,d\,Q = \left\{Q\int\frac{d^2\,P}{d\,Q^2}\,d\,V + \psi'\,(Q)\right\} d\,Q, \quad . \quad (71.)$$

the heat Q being treated as a constant in the integration.

If we now investigate the entire quantity of heat, both sensible and latent, which is consumed by a body during a simultaneous small change of total heat $d\,Q$ and volume $d\,V$, we find the following results:—

Sensible heat (which retains its condition), .	$= d\,Q$
Latent heat, or heat which disappears in overcoming molecular action, . .	$\frac{d\,S}{d\,Q}\,d\,Q + \frac{d\,S}{d\,V}\,d\,V.$
Latent heat equivalent to the visible mechanical effect, ...	$P\,d\,V$

The amount being

$$\left.\begin{array}{l} d\,Q + d\,.\,S + P\,d\,V = \left(1 + \frac{d\,S}{d\,Q}\right) d\,Q + \left(\frac{d\,S}{d\,V} + P\right) d\,V = \\ \left(1 + Q\int\frac{d^2\,P}{d\,Q^2}\,d\,V + \phi'\,Q\right) d\,Q + Q\,.\,\frac{d\,P}{d\,Q}\,.\,d\,V. \end{array}\right\} (72.)$$

This formula expresses completely the relations between heat, molecular action, and expansion, in all those cases in which the expansive power developed, $P\,d\,V$, is entirely communicated to the bodies inclosing the substance which expands.

49. The following coefficients are contained in, or deducible from it.

The ratio of the specific heat at constant volume to the real specific heat :—

$$\frac{K_v}{k} = 1 + \frac{dS}{dQ} = 1 + Q \int \frac{d^2 P}{d Q^2} \cdot dV + \phi'(Q). \quad . \quad (73.)$$

The coefficient of latent heat of expansion at constant heat :—

$$\frac{dS}{dV} + P = Q \frac{dP}{dQ}. \quad . \quad . \quad . \quad (74.)$$

The ratio of the specific heat at constant pressure to the real specific heat is found as follows. To have the pressure constant we must have

$$\frac{dP}{dQ} dQ + \frac{dP}{dV} dV = 0; \text{ or, } \frac{dV}{dQ} = -\frac{\frac{dP}{dQ}}{\frac{dP}{dV}}$$

consequently, the ratio in question is,

$$\left.\begin{aligned} \frac{K_p}{k} = 1 + \frac{dS}{dQ} - \left(\frac{dS}{dV} + P\right)\frac{\frac{dP}{dQ}}{\frac{dP}{dV}} = 1 + Q \int \frac{d^2 P}{d Q^2} \cdot dV \\ + \phi'(Q) - Q \frac{\left(\frac{dP}{dQ}\right)^2}{\frac{dP}{dV}} \end{aligned}\right\} (75.)$$

50. In order to investigate the laws according to which heat is converted into mechanical power, in a machine working by the expansion of an elastic body, it will be convenient to use a function,

$$F = \int \frac{dP}{dQ} dV \; (Q = \text{const.}),$$

of such a nature that the difference between two of its values, corresponding to different volumes of the body at the same total heat, represents the ratio of the heat converted into power by expansion between those volumes, to the given constant total heat. I shall call this function a *heat-potential.*

Introducing this function into equation (72), we find for the total heat consumed by a body during the increase of total heat $d\,Q$, and the expansion $d\,V$,

$$dQ + d.S + P\,dV = \Big(1 + \phi'.(Q)\Big)\,dQ + Q\,d.F \qquad (76.)$$

$$\left(\text{observing that } d\,.\,\mathrm{F} = \frac{d\,\mathrm{F}}{d\,\mathrm{Q}}\,d\,\mathrm{Q} + \frac{d\,\mathrm{F}}{d\,\mathrm{V}}\,d\,\mathrm{V}\right.$$

$$\left. = \left(\int \frac{d^2\,\mathrm{P}}{d\,\mathrm{Q}^2}\,d\,\mathrm{V}\,.\right) d\,\mathrm{Q} + \frac{d\,\mathrm{P}}{d\,\mathrm{Q}}\,d\,\mathrm{V}\,.\right).$$

Let us now suppose that the body changes its volume without either losing or gaining heat by conduction. This condition is expressed by the equation

$$0 = (1 + \phi'\,.\,\mathrm{Q})\,d\,\mathrm{Q} + \mathrm{Q}\,d\,.\,\mathrm{F},$$

from which we deduce the following,

$$-\,d\,.\,\mathrm{F} = \frac{1 + \phi'\,.\,(\mathrm{Q})}{\mathrm{Q}}\,.\,d\,\mathrm{Q}, \qquad (77.)$$

which expresses the following theorem:—

When the quantity of heat in a body is varied by variation of volume only, the variation of the heat-potential depends on the heat only, and is independent of the volume.

In order that a machine working by the expansive power of heat may produce its greatest effect, all the heat communicated from external bodies should be employed in producing expansive power, and none in producing variations of the quantity of heat in the body; for heat employed for the latter purpose would be wasted, so far as the production of visible motion is concerned. To effect this the body must receive heat by conduction, and convert it into expansive power, while containing a certain constant quantity of heat Q_1; give out by conduction heat produced by compression, while containing a smaller constant quantity of heat Q_2; and change between those two quantities of thermometric heat by means of changes of volume only, without conduction. For this purpose a cycle of operations must be performed similar to that described by Carnot, as follows:—

(I.) Let F_A be the initial value of the heat-potential; let the body expand at the constant heat Q_1 till the heat-potential becomes F_B. Then the heat received and converted into expansive power is

$$H_1 = Q_1\,(F_B - F_A).$$

(II.) Let the body further expand without receiving or emitting heat till the quantity of heat in it falls to Q_2; the heat-potential varying according to equation (77), and becoming at length F_C. The heat converted into expansive power in this operation is

$$Q_1 - Q_2.$$

(III.) Let the body be compressed, at the constant heat Q_2, till the heat-potential becomes F_D; a quantity differing from the initial heat-potential F_A by as much as F_C differs from F_B. In this operation the following amount of power is reconverted into heat, and given out by conduction:—

$$H_2 = Q_2 (F_C - F_D).$$

(IV.) Let the body be further compressed, till the heat-potential returns to F_A, its original value. Then, by the power expended in this compression alone, without the aid of conduction, the total heat of the body will be restored to its original amount, exactly reversing the operation II.

At the end of this cycle of operations, the following quantity of heat will have been converted into mechanical power:—

$$H_1 - H_2 = Q_1 (F_B - F_A) - Q_2 (F_C - F_D),$$

but it is obvious that the difference between the heat-potentials is the same in the first and third operations; therefore, the useful effect is simply

$$H_1 - H_2 = (Q_1 - Q_2)(F_B - F_A),$$

while the whole heat expended is,

$$H_1 = Q_1 (F_B - F_A). \quad \ldots \quad (78.)$$

Hence, *the ratio of the heat converted into mechanical effect, in an expansive machine, working to the greatest advantage, to the whole heat expended, is the same with that which the difference between the quantities of heat possessed by the expansive body during the operations of receiving and emitting heat respectively, bears to the quantity of heat possessed by it during the operation of receiving heat; and is independent of the nature and condition of the body.*

This theorem is thus expressed symbolically,—

$$\frac{H_1 - H_2}{H_1} = \frac{\text{Effect}}{\text{Heat expended}} = \frac{Q_1 - Q_2}{Q_1}. \quad \ldots \quad (79.)$$

51. When a body expands without meeting with resistance, so that all its expansive power is expended in giving velocity to its own particles, and when that velocity is ultimately extinguished by friction, then a quantity of heat equivalent to the expansive power is reproduced.

The heat consumed is expressed by taking away the term representing the expansive power, $P\,d\,V$, from the expression (72), the remainder of which consists merely of the variation of actual heat, and the heat expended in overcoming molecular attraction, viz.:—

$$d\,Q + d\,.\,S = \left(1 + \frac{d\,S}{d\,Q}\right) d\,Q + \frac{d\,S}{d\,V}\,d\,V$$

$$= \left(1 + Q\int \frac{d^2 P}{d Q^2} d V + \phi' . (Q)\right) d Q + \left(Q \frac{d P}{d Q} - P\right) d V.$$

This expression is a complete differential, and may be written thus :—

$$d (Q + S) = d \left\{ Q + \phi (Q) + \left(Q \frac{d}{d Q} - 1\right) \int P d V \right\} \quad (80.)$$

(Q being treated as a constant in performing the integration $\int P \, d V$).

Its integral, Q + S, the sum of the heat of the body, and of the potential of its molecular actions, is the same quantity which I have denoted by the symbol Ψ in the tenth article of a paper on the Centrifugal Theory of Elasticity (*See p. 62*), and whose differences are there stated to represent the total amount of power which must be exercised on a body, whether in the form of expansive or compressive power, or in that of heat, to make it pass from one volume and temperature to another. This integral corresponds also to the function treated of by Professor William Thomson in the fifth part of his paper on the Dynamical Theory of Heat, under the name of "Total Mechanical Energy."

52. We have now obtained a system of formulæ, expressing all the relations between heat and expansive power, analogous to those deduced from a consideration of the properties of temperature, by Messrs. Clausius and Thomson, and from the hypothesis of molecular vortices in the previous sections of this paper; but, in the present section, both the theorems and the investigations are distinguished from former researches by this circumstance—that they are independent, not only of any hypothesis respecting the constitution of matter, but of the properties, and even of the existence, of such a function as temperature; being, in fact, simply the necessary consequences of the following

Definition of Expansive Heat.

Let the term Expansive Heat *be used to denote a kind of physical energy convertible with, and measurable by, equivalent quantities of mechanical power, and augmenting the expansive elasticity of matter in which it is present.*

52A. It is further to be remarked, that the theorems and formulæ in the preceding articles of this section are applicable, not only to heat and expansive power, but to any two directly convertible forms of physical energy, one of which is actual, and the other potential. They are, in fact, the principles of the conversion of energy in the abstract, when interpreted according to the following definitions of the symbols :—

Let Q denote the quantity of a form of actual physical energy present in a given body;

V, a measurable state, condition, or mode of existence of the body, whose tendency to increase is represented by

P, a force, depending on the condition V, the energy Q, and permanent properties of the body, so that

$P\,dV$ is the increment of a form of potential energy, corresponding to a small increment dV of the condition V.

Let dS be the quantity whereby the increment of potential energy $P\,dV$ falls short of the quantity of actual energy of the form Q, which is converted into the potential form by the change of condition dV.

Then, as in equation (69),

$$\frac{dS}{dV} = Q\frac{dP}{dQ} - P,$$

an equation from which all those in the previous articles are deducible, and which comprehends the whole theory of the mutual conversion of the actual form of energy Q, and the potential form $\int P\,dV$, whatsoever those forms may be, when no other form of energy interferes. The application of these principles to any form, or any number of forms, of actual and potential energy, is the subject of a paper read before the Philosophical Society of Glasgow, on the 5th January, 1853, and published in the *Philosophical Magazine* for February, 1853. (*See p. 203.*)

Sub-Section 2.—Properties of Temperature.

53. Still abstaining from the assumption of any mechanical hypothesis, let us proceed a step beyond the investigation of the foregoing articles, and introduce the consideration of *temperature*—that is to say, of an arbitrary function increasing with heat, and having the following properties:—

Definition of Equal Temperatures.

Two portions of matter are said to have equal temperatures when neither tends to communicate heat to the other.

Corollary.

All bodies absolutely destitute of heat have equal temperatures.

The ratio of the real specific heats of two substances is that of the quantities of heat which equal weights of them possess at the same temperature.

Theorem.

The ratio of the real specific heats of any pair of substances is the same at all temperatures.

For, suppose equal weights of a pair of homogeneous substances to be in contact, containing heat in such proportions as to be in equilibrio. Then, let additional portions of each substance, of equal weight, and destitute of heat, be added to the original masses; so that the quantities of heat in unity of weight may be diminished in each substance, but may continue to be in the same ratio. Then, if the equality of temperature do not continue, portions of heat which were in equilibrio must have lost that equilibrium, merely by being transferred to other particles of a pair of homogeneous substances, which is absurd. Therefore, the temperatures continue equal.

It follows, that the quantity of heat in unity of weight of a substance at a given temperature, may be expressed by the product of a quantity depending on the nature of the substance, and independent of the temperature, multiplied by a function of the temperature, which is the same for all substances.

Let τ denote the temperature of a body according to the scale adopted; κ, the position, on the same scale, of the temperature corresponding to absolute privation of heat; $\mathfrak{k}$, a quantity depending on the nature of the substance, and independent of temperature. Then the quantity of heat in unity of weight may be expressed as follows:—

$$Q = \mathfrak{k}\,(\psi \cdot \tau - \psi \cdot \kappa). \qquad (81.)$$

54. If we introduce this notation into the formula (79), which expresses the proportion of the total heat expended which is converted into useful power by an expansive machine working to the best advantage, the quantity $\mathfrak{k}$, peculiar to the substance employed, disappears, and we obtain Carnot's theorem, as modified by Messrs. Clausius and Thomson—viz., that this ratio is a function solely of the temperatures at which heat is received and emitted respectively, and is independent of the nature of the substance; or symbolically,

$$\frac{\text{Effect}}{\text{Heat expended}} = \frac{Q_1 - Q_2}{Q_1} = \frac{\psi \cdot \tau_1 - \psi \cdot \tau_2}{\psi \cdot \tau_1 - \psi \cdot \kappa}. \qquad (82.)$$

55. Let us now apply the same notation to the formula (67) for the latent heat of a small expansion, $d\,V$, at constant heat, viz:—

$$Q \frac{d\,P}{d\,Q} \cdot d\,V,$$

we have evidently

$$\frac{d\,\mathrm{P}}{d\,\mathrm{Q}} = \frac{1}{\frac{d\,\mathrm{Q}}{d\,\tau}} \cdot \frac{d\,\mathrm{P}}{d\,\tau} = \frac{1}{\mathfrak{k}\,\psi'\,.\,\tau} \cdot \frac{d\,\mathrm{P}}{d\,\tau},$$

and, consequently, the heat which disappears by the expansion $d\,\mathrm{V}$ is

$$\mathrm{Q}\frac{d\,\mathrm{P}}{d\,\mathrm{Q}} \cdot d\,\mathrm{V} = \frac{\psi\,.\,\tau - \psi\,.\,\kappa}{\psi'\,.\,\tau} \cdot \frac{d\,\mathrm{P}}{d\,\tau} \cdot d\,\mathrm{V}, \quad . \quad (83.)$$

from which formula the specific quantity $\mathfrak{k}$ has disappeared.

Now, in the notation of Professor Thomson we have

$$\frac{\psi\,.\,\tau - \psi\,.\,\kappa}{\psi'\,.\,\tau} = \frac{\mathrm{J}}{\mu},$$

where J is Joule's equivalent, and μ a function of the temperature, the same for all substances, to be determined empirically; and, consequently,

$$\text{hyp. log. } (\psi\,.\,\tau - \psi\,.\,\kappa) = \frac{1}{\mathrm{J}}\int^{\tau} \mu\,d\,\tau,$$

or,

$$\left.\begin{aligned} \psi\,.\,\tau - \psi\,.\,\kappa &= \varepsilon^{\frac{1}{\mathrm{J}}\int^{\tau} \mu\,d\,\tau} \\ \text{and} \quad \mathrm{Q} = \mathfrak{k}\,(\psi\,.\,\tau - \psi\,.\,\kappa) &= \mathfrak{k}\,.\,\varepsilon^{\frac{1}{\mathrm{J}}\int^{\tau} \mu\,d\,\tau} \end{aligned}\right\} \quad . \quad (84.)$$

These expressions will be recognised by those who have studied Professor Thomson's papers on the dynamical theory of heat. By introducing the value given above of the quantity of heat in unity of weight, into the formulæ of the preceding articles of this section, they are at once transformed to those of Professor Thomson, and in particular, the formulæ (79) and (82) become the following:—

$$\left.\begin{aligned} {}^{*}\frac{\text{Effect of Machine}}{\text{Heat expended}} &= \frac{\varepsilon^{\frac{1}{\mathrm{J}}\int^{\tau_1} \mu\,d\,\tau} - \varepsilon^{\frac{1}{\mathrm{J}}\int^{\tau_2} \mu\,d\,\tau}}{\varepsilon^{\frac{1}{\mathrm{J}}\int^{\tau_1} \mu\,d\,\tau}} \\ &= 1 - \varepsilon^{-\frac{1}{\mathrm{J}}\int_{\tau_2}^{\tau_1} \mu\,d\,\tau} \end{aligned}\right\} \quad (85.)$$

*** It is to be observed, that in Professor Thomson's notation, heat is supposed to be measured by an arbitrary unit, whose ratio to a unit of mechanical power is denoted by J; while in this paper, the same unit is employed in expressing quantities of heat and of mechanical power.**

SUB-SECTION 3.—ON THE HYPOTHESIS OF MOLECULAR VORTICES.

56. The use of a mechanical hypothesis in the theory of heat, as in other branches of physics, is to render it a branch of mechanics, the only complete physical science; and to deduce its principles from the laws of force and motion, which are better understood than those of any other phenomena.

The results of the investigations in the preceding part of this section are consistent alike with all conceivable hypotheses which ascribe the phenomena of heat to invisible motions amongst the particles of bodies.

Those investigations, however, leave undetermined the relation between temperature and quantity of heat, except in so far as they show that it must follow the same law of variation in all substances.

By adopting a definite hypothesis, we are conducted to a definite relation between temperature and quantity of heat; which, being introduced into the formulæ, leads to specific results respecting the phenomena of the mutual transformation of heat and visible mechanical power; and those results, being compared with experiment, furnish a test of the soundness of the hypothesis.

Thus, the hypothesis of molecular vortices, which forms the basis of the investigations in the first five sections of this paper, and in a paper on the centrifugal theory of elasticity, leads to the conclusion, that, if temperature be measured by the expansion of a perfect gas, the total quantity of heat in a body is simply proportional to the elevation of its temperature above the temperature of absolute privation of heat; or, in the notation of the preceding article,

$$\psi \cdot \tau = \tau, \psi' \cdot \tau = 1,$$

and

$$Q = \mathfrak{k}\,(\tau - \kappa), \quad . \quad . \quad . \quad . \quad (86.)$$

$\mathfrak{k}$ being the real specific heat of the body.

If this value be substituted for the quantity of heat Q, in all the formulæ, from (67) to (80) inclusive, which are founded simply on the definition of expansive heat, it reproduces all the formulæ which, in this and the other paper referred to, have been deduced directly from the hypothesis. In the sequel, I shall apply one of these formulæ to the calculation, from the experiments of Professor Thomson and Mr. Joule on the heating of currents of air by friction, of approximate values of the absolute temperature corresponding to total privation of heat, that the mutual consistency of those values may serve as a test of the soundness of the hypothesis, and the accuracy of the formulæ deduced from it.

57. Before proceeding further, it may be desirable to point out how far

this hypothesis agrees with, and how far it differs from, that proposed by Mr. Herapath and Mr. Waterston, which supposes bodies to consist of extremely small and perfectly elastic particles, which fly about in all directions with a velocity whose half-square is the mechanical equivalent of the heat possessed by unity of weight, and are prevented from dispersing by their collisions with each other and with the particles of surrounding bodies. Let v be the velocity of motion, then

$$\frac{v^2}{2g} = Q,$$

represents the heat possessed by unity of weight, expressed in terms of the force of gravity.

The expansive pressure due to such motions is found by conceiving a hard, perfectly elastic plane of the area unity to be opposed to the collision of the particles, and calculating the pressure which would be required to maintain its position against them. If all the particles were to strike and rebound from such a plane at right angles, the pressure would be represented thus:

$$\frac{v^2}{g} \cdot \frac{1}{V},$$

where V is the volume which contains so many particles as amount to unity of weight. But the particles are supposed to fly in equal numbers in all directions. Then, if θ denote the angle of incidence on the plane

$$\frac{\sin\theta\, d\,\theta}{\int_0^{\frac{\pi}{2}} \sin\theta\, d\,\theta} = \sin\theta\, d\,\theta$$

represents the proportion of the whole particles which fly in those directions which make the angle θ with the normal to the plane. Of this proportion, again, the fraction $\cos\theta$ only strikes the plane; while the force of the blow also is less than that of a normal blow in the ratio $\cos\theta : 1$. Hence, the mean force of collision is

$$\int_0^{\frac{\pi}{2}} \cos^2\theta \sin\theta\, d\,\theta = \frac{1}{3}$$

of the force of a perpendicular collision; so that the expansive pressure is represented by

$$\frac{1}{3} \cdot \frac{v^2}{g} \cdot \frac{1}{V} = \frac{2}{3} \cdot \frac{Q}{V} = P.$$

Hence, according to this hypothesis, we should have for a perfect gas

$$PV = \frac{2}{3} Q,$$

or the product of the pressure and volume of a mass of a perfect gas equal to two-thirds of the mechanical equivalent of its total heat.

It is known, however, that the product of the pressure and volume of a mass of sensibly perfect gas is only about four-tenths of the equivalent of its total heat. The hypothesis, therefore, requires modification.

By supposing the particles to attract each other, or to be of appreciable bulk compared with the distances between them, the ratio in question is diminished; but either of these suppositions is inconsistent with the perfectly gaseous condition.

It appears to me, that, besides this difficulty connected with the gaseous condition, there exists also great difficulty in conceiving how the hypothesis can be applied to the solid condition, in which the particles preserve definite arrangements. The limited amount of time and attention, however, which I have hitherto bestowed on this hypothesis, is not sufficient to entitle me to pronounce whether these difficulties admit of a solution.

58. The idea of ascribing expansive elasticity to the centrifugal force of vortices or eddies in elastic atmospheres surrounding nuclei of atoms, originated with Sir Humphry Davy. The peculiarity of the view of the hypothesis taken in this paper consists in the function ascribed to the nuclei or central physical points of the atoms, which, besides retaining the atmospheres round them by their attraction, are supposed, by their actions on each other, to constitute the medium which transmits radiant heat and light; so that heat is radiant or thermometric, according as it affects the nuclei or their atmospheres.

In this form the hypothesis of molecular vortices is not a mere special supposition to elucidate the theory of expansive heat, but becomes connected with the theory of the elasticity of matter in all conditions, from solid to gaseous, and with that of the transmission of radiations.

I have already investigated mathematically the consequences of this hypothesis by two different processes, which are necessarily somewhat complicated.

When the question, however, is confined to the relations between temperatures and quantities of heat, a more simple process may be followed, analogous to that which has been applied in the preceding article to the hypothesis of molecular collisions.

If a mass of elastic fluid, so much rarefied that the effect of molecular attraction is insensible, be entirely filled with vortices, eddies, or circulating currents of any size and figure, so that every particle moves with the common velocity w, then, if the planes of revolution of these eddies be uniformly distributed in all possible positions, it follows, from reasoning precisely similar to that employed in the preceding article, that the pressure exerted by the fluid against a plane, in consequence of the centrifugal force of the eddies, has the following value in terms of gravity :—

$$\frac{1}{3} \cdot \frac{w^2}{g} \cdot \frac{1}{V}, \quad . \quad . \quad . \quad . \quad (87.)$$

or two-thirds of the hydrostatic pressure due to the velocity of the eddies w; V being, as before, the volume occupied by unity of weight.

It is, however, reasonable to suppose, that the motion of the particles of atomic atmospheres does not consist merely in circulating currents; but that those currents are accompanied with a certain proportionate amount of vibration,—a kind of motion which does not produce centrifugal force. To these we have to add the oscillations of the atomic nuclei, in order to obtain the mechanical equivalent of the whole molecular motions; which is thus found to be expressed for unity of weight by

$$k \frac{w^2}{2 g} \cdot = Q, \quad . \quad . \quad . \quad . \quad (88.)$$

k being a specific coefficient. Hence it follows (denoting $\frac{2 k}{3}$ by N), that the expansive pressure due to molecular motions in a perfect gas is equal to the mechanical equivalent of those motions in unity of volume multiplied by a specific constant

$$N \cdot \frac{Q}{V} \quad . \quad . \quad . \quad . \quad (89.)$$

The coefficient N has to be determined by experiment; its value for atmospheric air is known to be between 0·4 and 0·41.

In order to account for the transmission of pressure throughout the molecular atmospheres, it is necessary to suppose them possessed of a certain amount of inherent elasticity, however small, varying proportionally to density, and independent of heat. Let this be represented by

$$\frac{h}{V},$$

then

$$P = (N Q + h) \frac{1}{V}, \quad . \quad . \quad . \quad (90.)$$

is the total pressure of a perfect gas.

Equilibrium of heat and pressure between portions of two different perfect gases in contact requires that the pressures independent of heat and the pressures caused by heat, shall separately be in equilibrio. Let the suffixes a and b be used to distinguish quantities relative to two different substances in the perfectly gaseous condition. Then the first condition of equilibrium is expressed as follows :—

$$\left(\frac{h}{V}\right)(a) = \left(\frac{h}{V}\right)(b), \quad . \quad . \quad . \quad (91.)$$

that is to say, *the densities of two perfect gases in equilibrio are inversely proportional to the coefficients of elasticity of their atomic atmospheres.*

The second condition is expressed as follows :—

$$\left(\frac{N\,Q}{V}\right)(a) = \left(\frac{N\,Q}{V}\right)(b),$$

which, being taken in connection with the first condition, gives

$$\left(\frac{N}{h}\,Q\right)(a) = \left(\frac{N}{h}\,Q\right)(b). \quad . \quad . \quad . \quad (92.)$$

Now, by equation (90), we have

$$\frac{N}{h}\,Q = \frac{P\,V}{h} - 1.$$

Hence the condition of equilibrium of heat between two perfect gases is

$$\left(\frac{P\,V}{h}\right)(a) = \left(\frac{P\,V}{h}\right)(b), \quad . \quad . \quad . \quad (93.)$$

consequently, *temperature may be measured by the product of the pressure and volume of a perfect gas, divided by a coefficient, which is proportional to the volume of the gas at a standard pressure and temperature.*

Temperatures thus measured are reckoned from the point known as the *zero of gaseous tension*, or *absolute zero of a perfect gas thermometer*, 274°·6 Centigrade below the temperature of melting ice.

Let V_0 denote the volume of unity of weight of a perfect gas, at a standard pressure P_0, and absolute temperature τ_0; then any other absolute temperature has the following value :—

$$\tau = \tau_0 \frac{P\,V}{P_0\,V_0} = \frac{\tau_0}{P_0\,V_0}(N\,Q + h), \quad . \quad . \quad (94.)$$

while the absolute temperature of total privation of heat is

$$\kappa = \tau_0 \frac{h}{P_0\,V_0}. \quad . \quad . \quad . \quad . \quad (94\text{A}.)$$

Hence it appears that quantity of heat in unity of weight bears the following relation to temperature,—

$$Q = \frac{1}{N}(P V - h) = \frac{P_0 V_0}{N \tau_0}(\tau - \kappa), \quad . \quad . \quad (95.)$$

in which, if we substitute the symbol of real specific heat,

$$\mathfrak{k} = \frac{P_0 V_0}{N \tau_0}, \quad . \quad . \quad . \quad . \quad . \quad (96.)$$

we obtain the formula already given (86) for the relation between heat and temperature.*

59. The introduction of the value given above of the quantity of heat in terms of temperature, into the formula (67), gives for the latent heat of a small expansion d V at constant temperature

$$(\tau - \kappa)\frac{d P}{d \tau} \cdot d V. \quad . \quad . \quad . \quad . \quad (97.)$$

The formulæ (79) and (82), for the proportion of heat rendered available by an expansive engine working to the greatest advantage, becomes

$$\frac{\tau_1 - \tau_2}{\tau_1 - \kappa}, \quad . \quad . \quad . \quad . \quad . \quad (98.)$$

or the ratio of the difference between the temperatures of receiving and emitting heat, to the elevation of the former temperature above that of total privation of heat. This is the law already arrived at by a different process in Section V of this paper.

When the same substitution is made in equation (80), which represents the total energy, whether as heat or as compressive power, which must be applied to unity of weight of a substance to produce given changes of heat and volume, the following result is obtained :—

$$\begin{aligned} d \,.\, \Psi = d Q + d \,.\, S &= \left\{ \mathfrak{k} + f'(\tau) + (\tau - \kappa)\int \frac{d^2 P}{d \tau_2} d V \right\} d \tau \\ &+ \left\{ (\tau - \kappa)\frac{d P}{d \tau} - P \right\} d V \\ &= d \,.\, \left\{ \mathfrak{k}\, \tau + f(\tau) + \left((\tau - \kappa)\frac{d}{d \tau} - 1 \right) \int P\, d V \right\} \quad . \quad (99.) \end{aligned}$$

As it cannot be simplified, it is unnecessary here to recapitulate the investigation, which leads to the conclusion that the functions f (τ) and f' (τ) have the following values :—

* See Appendix, Note A, p. 336.

$$f(\tau) = \text{k} \, \text{N} \left(\kappa \text{ hyp. log. } \tau + \frac{\kappa^2}{\tau} \right); \; f'(\tau) = \text{k} \, \text{N} \left(\frac{\kappa}{\tau} - \frac{\kappa^2}{\tau^2} \right). \quad (99 \text{ A.})$$

We have thus reproduced equation (26) of the paper formerly referred to, on the Centrifugal Theory of Elasticity. (*See p. 49.*)

The coefficient of the variation of temperature in the first form of equation (99) is the specific heat of the substance at constant volume. Denoting this by K_v, the formula becomes

$$d \, . \, \Psi = K_v \, . \, d\tau + \left\{ (\tau - \kappa) \frac{d \, P}{d\tau} - P \right\} d \, V. \quad . \quad (100.)$$

SUB-SECTION 4.—THERMIC PHENOMENA OF CURRENTS OF ELASTIC FLUIDS.

60. When a gas previously compressed is allowed to escape through small apertures, as in the experiments of Mr. Joule and Professor Thomson, and has its velocity destroyed entirely by the mutual friction of its particles, without impediment from any other substance, and without conduction of heat to or from any other substance; then its condition is expressed by making

$$d \, \Psi = 0,$$

that is to say,

$$- \, d\tau = \frac{1}{K_v} \, . \left\{ \tau \, . \left(\frac{d \, P}{d\tau} - \frac{P}{\tau} \right) - \kappa \frac{d \, P}{d\tau} \right\} d \, V. \; . \quad (101.)$$

If we assume (as is really the case in the experiments) that the specific heat of the gas at constant volume does not sensibly vary within the limits of the experiments as to temperature and volume, so that K_v is sensibly constant, and also that the variation of temperature is very small as compared with the absolute temperatures, then we have the following approximate integral:

$$- \, \Delta\tau = \frac{1}{K_v} \, . \left\{ \tau \int_{V_1}^{V_2} \left(\frac{d \, P}{d\tau} - \frac{P}{\tau} \right) d \, V - \kappa \int_{V_1}^{V_2} \frac{d \, P}{d\tau} \, . \, d \, V \right\}, \quad (102.)$$

which represents the cooling effect of an expansion from the volume V_1 to the volume V_2.

If it were possible to obtain any substance in the state of perfect gas to be used in experiments of this kind, the first integral in the above expression would disappear, because for a perfect gas

$$\frac{d \, P}{d\tau} = \frac{P}{\tau},$$

and as the other term is negative, the result would be a slight heating effect. As no gas, however, is perfect, and as $\frac{d\,P}{d\,\tau}$ always exceeds $\frac{P}{\tau}$, the mode of reducing the experimental data is to calculate the value of the first term, which represents the effect of cohesion, from the known properties of the gas, to subtract from it the actual cooling, and from the remainder to compute values of κ, the temperature of absolute privation of heat, according to the following formula :—

$$\kappa = \frac{\frac{1}{K_v}\,\tau \int_{V_1}^{V_2} \left(\frac{d\,P}{d\,\tau} - \frac{P}{\tau}\right) d\,V - (-\Delta\,\tau)}{\frac{1}{K_v}\int_{V_1}^{V_2} \frac{d\,P}{d\,\tau}\,d\,V} \quad . \quad (103.)$$

When the gas is nearly perfect, as in the case of atmospheric air, it is unnecessary to take into consideration its deviation from the perfect condition in computing the integral in the denominator, whose approximate value is found to be

$\frac{P_0\,V_0}{K_v\,\tau_0}$. hyp. log. $\frac{V_2}{V_1} = N$. hyp. log. $\frac{P_1}{P_2}$ nearly (τ being nearly constant),

and K_v nearly $= \text{k}$.

The value of the integral in the numerator is found as follows :—

The centrifugal theory of elasticity indicates that the pressure of an imperfect gas may be represented by the following formula :—

$$P = P_0 \frac{V_0}{V}\left\{\frac{\tau}{\tau_0} + A_0 - \frac{A_1}{\tau} - \frac{A_2}{\tau^2} - \&c.\right\}, \quad . \quad (104.)$$

where V_0 is the volume in the perfectly gaseous state, at a standard pressure P_0, and absolute temperature τ_0, and A_0, A_1, &c., are a series of functions of the density, to be determined empirically. From this formula it is easily seen that

$$\frac{d\,P}{d\,\tau} - \frac{P}{\tau} = P_0 \frac{V_0}{V}\left\{-\frac{A_0}{\tau} + \frac{2\,A_1}{\tau^2} + \&c.\right\}, \quad . \quad (105.)$$

so that the first term in the numerator of the expression (103) has the following value :—

$$\frac{\tau}{K_v}\int_{V_1}^{V_2}\left(\frac{d\,P}{d\,\tau} - \frac{P}{\tau}\right) d\,V = \frac{P_0 V_0}{K_v}\left\{-\int_{V_1}^{V_2}\frac{A_0}{V}\,d\,V + \frac{2}{\tau}\int_{V_1}^{V_2}\frac{A}{\tau}\,d\,V + \&c.\right\}, \ (106.)$$

in which $\frac{P_0\,V_0}{K_v} = N\,\tau_0$ nearly.

In order to represent correctly the result of M. Regnault's experiments on the elasticity and expansion of gases, it was found sufficient to use, in the formula for the pressure (104), the first three terms; and the functions of the density which occur in these terms, as determined empirically from the experiments, were found to have the following values, in which the unit of volume is the theoretical volume of unity of weight of air under the pressure of one atmosphere, at the temperature of melting ice,* and the values of the constants are given for the Centigrade scale.

$$\frac{A_0}{V} = b \left(\frac{1}{V}\right)^{\frac{3}{2}}; \quad \frac{A_1}{V} = a \left(\frac{1}{V}\right)^{\frac{8}{5}}. \qquad . \qquad . \qquad (107.)$$

Com. log. $b = \bar{3}{\cdot}8181545$; Com. log. $a = 0{\cdot}3176168$.

Hence it appears that the integrals in the formula (106) have the following values :—

$$\int_{V_1}^{V_2} \frac{A_0}{V} dV = 2b \,.\, \Delta \,.\, \left(\frac{1}{V}\right)^{\frac{1}{2}}; \quad \frac{2}{\tau}\int_{V_1}^{V_2} \frac{A_1}{V} dV = \frac{10}{3} \,.\, \frac{a}{\tau_0} \,.\, \frac{\tau_0}{\tau} \,.\, \Delta \,.\, \left(\frac{1}{V}\right)^{\frac{6}{10}} \quad (107\,A.)$$

in which the common logarithms of the constants are

$$\text{Com. log. } 2\,b = 2{\cdot}1101845; \quad \text{log. } \frac{10}{3} \,.\, \frac{a}{\tau_0} = \bar{2}{\cdot}4017950;$$

and these values suit any scale of temperatures.

In calculating, for use in these formulæ, the densities $\frac{1}{V}$ from the observed pressures, it is sufficiently near the truth, in the case of air, to use the approximate equation

$$\frac{1}{V} = \frac{\tau_0}{\tau} \,.\, \text{P (in atmospheres)}.$$

The common logarithm of τ_0, the absolute temperature of melting ice, for the Centigrade scale, is 2·4387005.

The constant N for atmospheric air is 0·4 nearly; therefore

$$\text{Com. log. (N} \times \text{hyp. log. 10)} = \bar{1}{\cdot}9642757.$$

The following, therefore, is the approximate value of the formula (103) to be used (with the numerical constants already given) in reducing the experiments of Mr. Joule and Professor Thomson on atmospheric air, so as to obtain approximate values of the absolute temperature of total privation of heat :—

* This unit of volume is greater than the actual volume of air, under the circumstances described, in the ratio of 1·00085 to 1.

$$\kappa = \left\{ \mathrm{N}\,\tau_0 \left(\frac{10\,a}{3\,\tau_0} \cdot \left(\frac{\tau_0}{\tau}\right)^{\frac{16}{10}} \Delta \,.\, (\mathrm{P}^{\frac{6}{10}}) - 2\,b \left(\frac{\tau_0}{\tau}\right)^{\frac{1}{2}} \Delta \,.\, (\mathrm{P}^{\frac{1}{2}}) \right) - (-\Delta\,\tau) \right\}$$

$$\div\ \mathrm{N}\ \text{hyp. log. } 10 \times \Delta\,.\ \text{com. log. } \frac{1}{\mathrm{P}} \quad . \quad (108.)$$

In using this formula, the mean absolute temperature should be taken as the value of τ.

The following table shows the values of the quantity κ, computed from ten mean experimental data, taken respectively from the first ten series of experiments described in the recent paper of Messrs. Joule and Thomson, in the supplementary number of the *Philosophical Magazine* for December, 1852. The temperatures in the table, for the sake of convenience, are reduced to the Centigrade scale, because that scale has been used throughout the previous sections of this paper.

The final pressure in each case was that of the atmosphere.

Professor Thomson and Mr. Joule have expressed the opinion, which is undoubtedly correct, that those experiments in which the largest quantities of air were used were the least liable to error from disturbing causes, such as the conduction of heat.

Now, it may be observed in the table, that the calculated values of κ are generally greatest, and the discrepancies amongst them least, for the experiments in which most air was used. To illustrate this, the results of the last eight series are arranged below in the order of the quantities of air employed.

Cubic inches per second,	1·4	2·8	5·6	5·6	6·4	8·4	11·2	11·2
Values of κ,	1·683	1·762	2·09	2·228	1·51	2·087	2·345	2·14

It is further to be remarked, that the discrepancy between the highest and the lowest of the values of κ is

$$2^\circ{\cdot}345 - 1^\circ{\cdot}08 = 1^\circ{\cdot}265 \text{ Centigrade}:$$

a quantity which corresponds to a difference of less than *one three-hundredth part* in computing the proportion of heat converted into mechanical power by any ordinary expansive engine, according to the formula (98), which has been deduced from the hypothesis of molecular vortices.

The experiments, therefore, may be considered as tending to prove, that the formulæ deduced from this hypothesis are sufficiently correct for practical purposes; and also as affording a strong probability that the principles to which it leads are theoretically exact, and that the temperature of absolute privation of heat is a real fixed point on the scale,

SERIES AND NUMBER OF EXPERIMENTS.	Quantity of Air used in Cubic Inches per second.	Initial Temperature. Centigrade.	Mean Absolute Temperature τ.	Initial Pressure in Atmospheres.	Actual Cooling. Centigrade.	Calculated Cooling by Cohesion. Centigrade.	Difference.	Approximate Values of κ, the Absolute Temperature of Total Privation of Heat. Centigrade.
Series I. Mean of 3 Experiments,	Not noted.	16°·11	290°·4	3·61	0°·76	1°·70	0°·94	1°·83
Series II. Mean of 5 Experiments,	Not noted.	71·11	345·6	2·64	0·28	0·70	0·42	1·08
Series III. Mean of 7 Experiments,	6·4	77·11	351·5	4·28	0·37	1·25	0·88	1·51
Series IV. Mean of 3 Experiments,	5·6	17·39	281·7	3·04	0·56	1·49	0·93	2·09
Series V. Mean of 2 Experiments,	8·4	3·89	278·0	4·11	1·00	2·18	1·18	2·087
Series VI. Mean of 4 Experiments,	11·2	5·21	279·4	4·40	0·92	2·31	1·39	2·345
Series VII. Mean of 5 Experiments,	1·4	17·87	292·3	1·894	0·23	0·66	0·43	1·683
Series VIII. Mean of 5 Experiments,	2·8	18·04	292·5	2·41	0·31	0·93	0·62	1·762
Series IX. Mean of 4 Experiments,	5·6	18·21	292·6	2·92	0·43	1·385	0·955	2·228
Series X. Mean of 3 Experiments,	11·2	18·37	292·6	4·11	0·72	1·93	1·21	2·14

somewhat more than two Centigrade degrees above the absolute zero of a perfect gas-thermometer (which is, of course, an imaginary point); that is to say, about $272\frac{1}{2}$ Centigrade degrees, or $490\frac{1}{2}$ degrees of Fahrenheit, below the temperature of melting ice.

If these conclusions be correct, it follows, that when the temperatures T_1 and T_2, between which an expansive engine works, are measured from the ordinary zero points of the Centigrade and of Fahrenheit's scales respectively, the following are the utmost proportions of the total heat expended which it can be made to convert into mechanical power:—

$$\left.\begin{array}{ll}\text{For the Centigrade scale,} & \dfrac{T_1 - T_2}{T_1 + 272\frac{1}{2}} \\[2ex] \text{For Fahrenheit's scale,} & \dfrac{T_1 - T_2}{T_1 + 458\frac{1}{2}}\end{array}\right\} \quad . \quad (109.)$$

In the fifth section of this paper, where a comparison is made between the actual duty of the Cornish engine at Old Ford, as determined by Mr. Wicksteed, and the greatest possible duty which could be obtained from a given quantity of heat by a theoretically perfect engine working between the same temperatures, the constant κ is treated as being so small that it may be neglected in practice. If the value of κ is really $2^{\circ}{\cdot}1$ Centigrade, as computed above, the calculated maximum theoretical duty in Section V is too small by about one one-hundred-and-ninetieth part of its amount,—a quantity of no practical importance in such calculations.

61. It may be anticipated, that when Mr. Joule and Professor Thomson shall have performed experiments on the thermic phenomena exhibited by air in more copious currents, and by gases of more definite composition, and more simple laws of elasticity, much more precise results will be obtained.

When a gas deviating considerably from the perfectly gaseous condition, or a vapour near the point of saturation, is employed, it will no longer be sufficiently accurate to treat the specific heat at constant volume as a constant quantity, nor the cooling effect as very small. It will, therefore, be necessary to employ, for the reduction of the experiments, the integral form of equation (99)—that is to say,

$$0 = \Delta \Psi = \Delta \left\{ \mathfrak{k}\, \tau + \mathfrak{k}\, N\, \kappa \left(\text{hyp. log. } \tau + \frac{\kappa}{\tau}\right) \right.$$

$$\left. + \left((\tau - \kappa)\frac{d}{d\,\tau} - 1\right)\int P\, d\, V \right\}$$

$$= \text{k}(\tau_2 - \tau_1) + \Delta \int \left(\tau \frac{d\,P}{d\,\tau} - P \right) d\,V$$

$$- \kappa \left\{ \Delta \int \frac{d\,P}{d\,\tau} d\,V - \text{k}\,N \left(\Delta \,.\, \frac{\kappa}{\tau} + \Delta \text{ hyp. log. } \tau \right) \right\}. \quad (110.)$$

62. Preliminary to the application of this equation, it is necessary to determine the mechanical value of the real specific heat k. Supposing the law which connects the pressure, density, and temperature of the gas to be known, it is sufficient for this purpose to have an accurate experimental determination, either of the apparent specific heat at constant pressure for a given temperature, or the velocity of sound in the gas under given circumstances.

First, let us suppose that the apparent specific heat at constant pressure is known.

The value of this coefficient (Centrifugal Theory of Elasticity, Art. 12) is

$$K_p = \text{k} + (\tau - \kappa) \left\{ \frac{P_0 V_0}{\tau_0} \frac{\kappa}{\tau^2} + \int \frac{d^2 P}{d\,\tau^2} d\,V + \frac{\left(\frac{d\,P}{d\,\tau} \right)^2}{- \frac{d\,P}{d\,V}} \right\} \quad (111.)$$

In order that the lower limit of the integral may correspond with the condition of perfect gas, it is convenient to transform it into one in terms of the density. Let D be the weight of unity of volume, then

$$\int \frac{d^2 P}{d\,\tau^2} d\,V = - \int_0^D \frac{1}{D^2} \cdot \frac{d^2 P}{d\,\tau^2} d\,D. \quad . \quad . \quad (111\text{ A.})$$

If, then, we have the pressure of the gas under consideration expressed by the following approximate formula :—

$$P = \frac{P_0 V_0}{V} \left\{ \frac{\tau}{\tau_0} + A_0 - \frac{A_1}{\tau} \right\}.$$

The following will be the values of the functions of the pressure which enter into the above equation :—

$$\frac{dP}{d\,\tau} = \frac{P_0 V_0}{V} \left\{ \frac{1}{\tau_0} + \frac{A_1}{\tau^2} \right\}; \; \frac{d^2 P}{d\,\tau^2} = -2 \frac{P_0 V_0}{V} \cdot \frac{A_1}{\tau^3} = -2 P_0 V_0 \frac{A_1 D}{\tau^3}$$

$$\left.\begin{aligned}
&\int \frac{d^2 P}{d\tau^2} dV = -\int_0^D \frac{1}{D^2} \cdot \frac{d^2 P}{d\tau^2} dD = +2\frac{P_0 V_0}{\tau^3} \cdot \int_0^D \frac{A_1}{D} dD; \\
&-\frac{dP}{dV} = D^2 \frac{dP}{dD} = P_0 V_0 D^2 \left\{ \frac{\tau}{\tau_0} + \frac{d.A_0 D}{dD} - \frac{1}{\tau} \cdot \frac{d.A_1 D}{dD} \right\} \\
&\therefore \frac{\left(\frac{dP}{d\tau}\right)^2}{-\frac{dP}{dV}} = P_0 V_0 \cdot \frac{\left(\frac{1}{\tau_0} + \frac{A_1}{\tau^2}\right)^2}{\frac{\tau}{\tau_0} + \frac{d.A_0 D}{dD} - \frac{1}{\tau} \cdot \frac{d.A_1 D}{dD}}
\end{aligned}\right\} . \quad (111\text{ B.})$$

To illustrate the application of these formulæ, let us calculate the difference between the real specific heat and the apparent specific heat, at constant pressure, of carbonic acid gas, at the temperature of melting ice, and at the density which, if the gas were perfect, would correspond to a pressure of one atmosphere at the temperature of melting ice. Let this density be denoted by D_0, and its reciprocal by V_0. As the constants have been deduced from M. Regnault's experiments, the calculations will be made in French measures and for the latitude of Paris.

The actual density of carbonic acid at 0° Centigrade, and under one atmosphere of pressure, exceeds the theoretical density, in the perfectly gaseous state, in the ratio of 1·0065 to 1 nearly. Hence, the height of a homogeneous atmosphere of actual carbonic acid at 0° Centigrade being . . . . 5225·5 metres,
the corresponding height in the state of perfect gas is $P_0 V_0 = 5259{\cdot}5$ „
and $\frac{P_0 V_0}{\tau} = 19{\cdot}53$ metres per Centigrade degree $= 62{\cdot}84$ feet.

The functions which express the influence of density on the deviation of carbonic acid gas from the perfectly gaseous state, have the following values:—

$$\left.\begin{aligned}
&A_0 = b \cdot \frac{D}{D_0};\ A_1 = a \cdot \frac{D}{D_0};\ \text{when} \\
&\text{Com. log. } b = \bar{3}{\cdot}1083932;\ \text{Com. log. } a = 0{\cdot}3344538 \\
&\qquad b = 0{\cdot}00128349 \qquad\qquad a = 2{\cdot}16: \\
&\int_0^D \frac{A_1}{D} dD = \int_0^{\frac{D}{D_0}} A_1 \frac{D_0}{D} \cdot \frac{d.D}{D_0} = a \cdot \frac{D}{D_0};\ \frac{d}{dD} \cdot A_0 D \\
&\qquad = 2b \cdot \frac{D}{D_0};\ \frac{d}{dD} \cdot A_1 D = 2a \frac{D}{D_0}
\end{aligned}\right\} . \quad (111\text{ C.})$$

For the purposes of a first approximation, we may assume that the value of κ already found is sufficiently near the truth—viz., 2°·1 Centigrade, so that, in the present instance, $\tau - \kappa = 272°\cdot 5$ Centigrade.

Then we find the following results when $\tau = \tau_0$, and $\mathrm{D} = \mathrm{D}_0$:

	Metres.	Feet.
$(\tau - \kappa) \dfrac{\mathrm{P}_0 \mathrm{V}_0}{\tau_0} \cdot \dfrac{\kappa}{\tau^2} =$ per Centigrade degree, . . .	0·145	0·48
$(\tau - \kappa) \displaystyle\int \frac{d^2 \mathrm{P}}{d \tau^2} d \mathrm{V} =$ per Centigrade degree, . . .	0·150	0·49
Sum $= \mathrm{K}_v - \mathfrak{k} =$ excess of apparent specific heat at constant volume above real specific heat, ...	0·295	0·97
$(\tau - \kappa) \dfrac{\left(\dfrac{d \mathrm{P}}{d \tau}\right)^2}{-\dfrac{d \mathrm{P}}{d \mathrm{V}}} =$ difference between apparent specific heats at constant volume and at constant pressure, . . .	19·565	64·19
$\mathrm{K}_p - \mathfrak{k} =$ excess of apparent specific heat at constant pressure above real specific heat, ...	19·860	65·16

$\frac{5}{9}$ of the above quantities are of course the corresponding quantities for Fahrenheit's scale.

Secondly, If the velocity of sound in the gas is given, let this $= u$. Then we know that

$$u^2 = g \cdot \frac{d \mathrm{P}}{d \mathrm{D}} \cdot \frac{\mathrm{K}_p}{\mathrm{K}_v} \quad . \quad . \quad . \quad (112.)$$

in which

$$\frac{d \mathrm{P}}{d \mathrm{D}} = \mathrm{P}_0 \mathrm{V}_0 \left\{ \frac{\tau}{\tau_0} + \frac{d \cdot \mathrm{A}_0 \mathrm{D}}{d \mathrm{D}} - \frac{1}{\tau} \frac{d \cdot \mathrm{A}_1 \mathrm{D}}{d \mathrm{D}} \right\}. \quad (112\ \mathrm{A.})$$

So that from the velocity of sound we can calculate the ratio of the specific heats at constant pressure and at constant volume. Let this ratio be denoted by γ, and let

$$\mathrm{K}_v = \mathfrak{k} + c; \ \mathrm{K}_p = \mathfrak{k} + c';$$

then

$$\gamma = \frac{\mathfrak{k} + c'}{\mathfrak{k} + c}; \text{ and } \mathfrak{k} = \frac{c' - \gamma c}{\gamma - 1} \quad . \quad . \quad (112\ \mathrm{B.})$$

in which c' and c are to be calculated as above.*

63. In using the formula (110) for a gas whose pressure is represented by the formula

$$P = \frac{P_0 V_0}{V} \left\{ \frac{\tau}{\tau_0} + A_0 - \frac{A_1}{\tau} \right\}$$

the integrals may be transformed so as to be taken, with respect to the density, as in the preceding article. Thus we obtain

$$\Delta \int \left(\tau \frac{d P}{d \tau} - P \right) d V = - \Delta \int \frac{1}{D^2} \left(\tau \frac{d P}{d \tau} - P \right) d D -$$

$$- P_0 V_0 \Delta \left\{ \frac{2}{\tau} \int_0^D \frac{A_1}{D} d D - \int \frac{A_0}{D} d D \right\} \int \frac{d P}{d \tau} d V =$$

$$- \int \frac{1}{D^2} \cdot \frac{d P}{d \tau} d D = - P_0 V_0 \Delta \left(\frac{1}{\tau_0} \text{hyp. log. } D + \frac{1}{\tau^2} \int \frac{A_1}{D} d D \right) \quad (113.)$$

For carbonic acid, the first of these formulæ becomes simply

$$+ P_0 V_0 \left\{ \frac{2 a}{D_0} \left(\frac{D_1}{\tau_1} - \frac{D_2}{\tau_2} \right) - \frac{b}{D_0} \left(D_1 - D_2 \right) \right\}$$

and the second, (113 A.)

$$+ P_0 V_0 \left\{ \frac{1}{\tau_0} \text{hyp. log.} \frac{D_1}{D_2} - \frac{a}{D_0} \left(\frac{D_1}{\tau_1^2} - \frac{D_2}{\tau_2^2} \right) \right\}$$

APPENDIX.

NOTE A. (to Article 58).—Since this section was read, the theoretical views relative to the relation between heat and temperature contained in it and the previous sections of this paper, have received a strong confirmation by the publication by M. Regnault of the fact, that he has found the specific heat of air to be sensibly constant at all temperatures from − 30° Centigrade to + 225°, and at all pressures from one to ten atmospheres (*Comptes Rendus*, April 18, 1853); so that equal lengths on the scale of the air thermometer represent equal quantities of heat.

NOTE B. (to Article 62).—Until very recently, there existed no exact

* See Appendix, Note B.

TABLE SHOWING THE RELATIONS BETWEEN THE PRECEDING FORMULÆ AND EXPERIMENTAL RESULTS.

Gases.	Experimental Data.			Theoretical Results.			Velocity of Sound at 0° Centigrade.		
	P V at 0° C. Regnault.	$K_P' \div K_W$ Regnault.	K_P	$K_P - K_V$	K_V	$K_P : K_V$	By Theory.	By Observation.	Observers.
	Feet.		Feet.	Feet.	Feet.		Ft. per sec.	Ft. per sec.	
Air, ...	26214·4	0·2379	330·6	96·0	234·6	1·4094	1090·4	1090·5	Bravais and Martins.
								1090·1	Moll and Van Beck.
Oxygen, . . .	23710·4	0·2182	303·2	86·8	216·4	1·4014	1036·4	1042·3	Dulong.
Hydrogen, . . .	378819·0	3·4046	4731·0	1388·0	3343·0	1·4150	4153·3	4165·1	Dulong.
Carbonic Oxide, . .	27097·8	0·2479	344·5	99·25	245·25	1·4047	1106·8	1107·0	Dulong.
Carbonic Acid, . .	17144·7	0·2169	300·7	64·2	236·5	1·2714	837·55	858·28	Dulong.

experimental determination of the specific heat of any gas. The specific heat of air at constant pressure, as compared with that of water, was calculated theoretically in the previous part of this paper, from Joule's equivalent and the velocity of sound, and found to be 0·24. This value has since been confirmed very closely by Mr. Joule's experiments, whose mean result was 0·23, and still more exactly by M. Regnault's experiments, already referred to, which give the value 0·2379. The table (*See p. 337*), shows the results of the application of the formulæ of this paper to the specific heats of five different gases at constant pressure, selected from M. Regnault's table (*Comptes Rendus*, April 18), as being those in which the velocity of sound can be computed, and has been determined experimentally. The table shows also a comparison of the calculated and observed velocities of sound. This table appeared originally, in French measures, in the *Philosophical Magazine* for June, 1853; the metres are here reduced to feet. K_P, K_V, and K_W, are expressed in feet of fall per Centigrade degree. K_W (Joule's equivalent) = 1389·6

The real specific heat of carbonic acid gas is 235·5 feet of fall per Centigrade degree. That of the other gases does not differ from the apparent specific heat at constant volume by an amount appreciable in practice.

XX.—ON THE GEOMETRICAL REPRESENTATION OF THE EXPANSIVE ACTION OF HEAT, AND THE THEORY OF THERMODYNAMIC ENGINES.*

SECTION I.—INTRODUCTION AND GENERAL THEOREMS.

1. THE first application of a geometrical diagram to represent the expansive action of heat was made by James Watt, when he contrived the well-known steam-engine indicator, subsequently altered and improved by others in various ways. As the diagram described by Watt's Indicator is the type of all diagrams representing the expansive action of heat, its general nature is exhibited in Fig. 1.

Let abscissæ, measured along, or parallel to, the axis O X represent the volumes successively assumed by a given mass of an elastic substance, by whose alternate expansion and contraction heat is made to produce motive power; O V_A and O V_B being the least and greatest volumes which the substance is made to assume, and O V any intermediate volume. For brevity's sake, these quantities will be denoted by V_A, V_B, and V, respectively. Then $V_B - V_A$ may represent the space traversed by the piston of an engine during a single stroke.

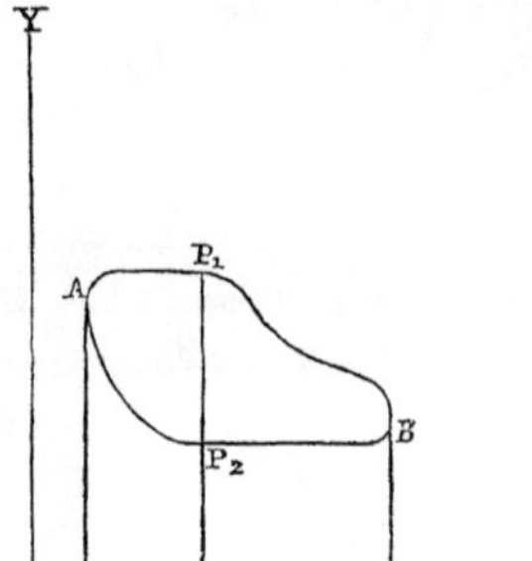

Fig. 1.

Let ordinates, measured parallel to the axis O Y, and at right angles to O X, denote the expansive pressures successively exerted by the substance at the volumes denoted by the abscissæ. During the increase of volume from V_A to V_B, the pressure, in order that motive power may be produced, must be, on the whole, greater than during the diminution of volume from V_B to V_A; so that, for instance, the ordinates V P_1 and V P_2, or the symbols

* Read before the Royal Society of London on January 19, 1854, and published in the *Philosophical Transactions* for 1854.

P_1 and P_2, may represent the pressures corresponding to a given volume V during the expansion and contraction of the substance respectively.

Then the area of the curvilinear figure, or *Indicator-diagram*, A P_1 B P_2 A, will represent the motive power, or "potential energy," *developed* or *given out* during a complete stroke, or cycle of changes of volume of the elastic substance. The algebraical expression for this area is

$$\int_{V_A}^{V_B} (P_1 - P_2)\, d\,V. . \qquad . \quad . \quad . \quad (1.)$$

The practical use of such diagrams, in ascertaining the power and the mode of action of the steam in steam-engines, where the curve A P_1 B P_2 A is described by a pencil attached to a pressure-gauge on a card whose motion corresponds with that of the piston, is sufficiently well known.

2. It appears that the earliest application of *diagrams of energy* (as they may be called) to prove and illustrate the theoretical principles of the mechanical action of heat, was made either by Carnôt, or by M. Clapeyron in his account of Carnôt's theory; but the conclusions of those authors were in a great measure vitiated by the assumption of the substantiality of heat.

In the fifth section of a paper on the Mechanical Action of Heat, published in the *Transactions of the Royal Society of Edinburgh* (*See p. 300*), a diagram of energy is employed to demonstrate the general law of the economy of heat in thermodynamic engines according to the correct principle of the action of such machines—viz., that the area of the diagram represents at once the potential energy or motive power which is developed at each stroke and the mechanical equivalent of the actual energy, or heat, which permanently disappears.

As the principles of the expansive action of heat are capable of being presented to the mind more clearly by the aid of diagrams of energy than by means of words and algebraical symbols alone, I purpose, in the present paper, to apply those diagrams partly to the illustration and demonstration of propositions already proved by other means, but chiefly to the solution of new questions, especially those relating to the action of heat in all classes of engines, whether worked by air, or by steam, or by any other material; so as to present, in a systematic form, those theoretical principles which are applicable to all methods of transforming heat to motive power by means of the changes of volume of an elastic substance.

Throughout the whole of this investigation, quantities of heat, and coefficients of specific heat, are expressed, not by units of temperature in a unit of weight of water, but by equivalent quantities of mechanical power, stated in foot-pounds, according to the ratio established by Mr. Joule's experiments on friction (*Phil. Trans.*, 1850); that is to say,

772 foot-pounds per degree of Fahr., or

1389·6 foot-pounds per Centigrade degree,

applied to one pound of liquid water at atmospheric temperatures.

3. *Of Isothermal Curves, and Curves of No Transmission of Heat.*

A curve described on a diagram of energy, such that its ordinates represent the pressures of a homogeneous substance corresponding to various volumes, while the total *sensible* or *actual heat* present in the body is maintained at a constant value, denoted, for example, by Q, may be called the *isothermal curve of* Q for the given substance (See Fig. 2). Suppose, for instance, that the co-ordinates of the point A, V_A and P_A, represent respectively a volume and a pressure of a given substance, at which the actual heat is Q; and the co-ordinates of the point B—viz., V_B and P_B, another volume and pressure at which the actual heat is the same; then are the points A and B situated on the same isothermal curve Q Q.

On the other hand, let the substance be allowed to expand from the volume and pressure V_A, P_A, without receiving or emitting heat; and when it reaches a certain volume, V_C, let the pressure be represented by P_C, which is less than the pressure would have been had the actual heat been maintained constant, because by expansion heat is made to disappear. Then C will be a point on a certain curve N N passing through A, which may be called a *Curve of No Transmission.*

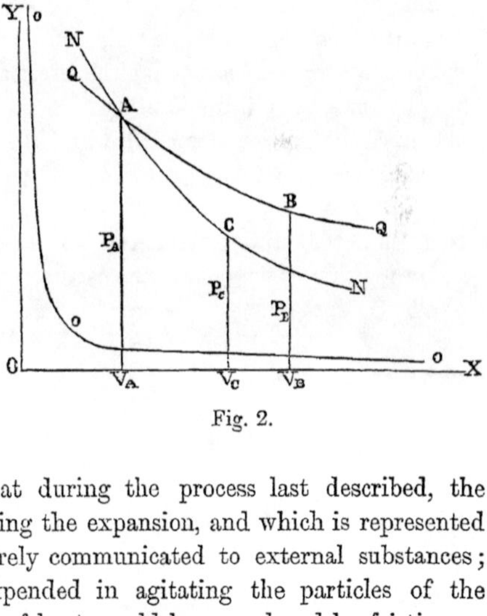

Fig. 2.

It is to be understood, that during the process last described, the potential energy developed during the expansion, and which is represented by the area A C V_C V_A, is entirely communicated to external substances; for if any part of it were expended in agitating the particles of the expanding substance, a portion of heat would be reproduced by friction.

If *o o o* be a curve whose ordinates represent the pressures corresponding to various volumes when the substance is absolutely destitute of heat, then this curve, which may be called the *Curve of Absolute Cold*, is at once an isothermal curve and a curve of no transmission.

So far as we yet know, the curve of absolute cold is, for all substances, an asymptote to all the other isothermal curves and curves of no transmission, which approach it and each other indefinitely as the volume of the substance increases without limit.

NOTE.—The following remarks are intended to render more clear the precise meaning of the term *Total Actual Heat.*

The Total Actual Heat of a given mass of a given substance at a given temperature, is the quantity of physical energy present in the mass *in the form of heat* under the given circumstances.

If, for the purpose of illustrating this definition, we assume the hypothesis that heat consists in molecular revolutions of a particular kind, then the Total Actual Heat of a mass is measured by the mechanical power corresponding to the *vis viva* of those revolutions, and is represented by

$$\frac{1}{2} \Sigma \, . \, m \, v^2,$$

m being the mass of any circulating molecule, and v^2 the mean square of its velocity.

But the meaning of the term Total Actual Heat may also be illustrated without the aid of any hypothesis.

For this purpose, let us take the ascertained fact of the production of heat by the expenditure of mechanical power in friction, according to the numerical proportion determined by Mr. Joule; and let E denote the quantity of mechanical power which must be expended in friction, in order to raise the temperature of unity of weight of a given substance from that of absolute privation of heat to a given temperature τ.

During this operation, let the several elements of the external surface of the mass undergo changes of relative position expressed by the variations of quantities denoted generally by p, and let the increase of each such quantity as p be resisted by an externally-applied force such as P.

Then, during the elevation of temperature from absolute cold to τ, the energy converted to the potential form in overcoming the external pressures P will be

$$\Sigma \, . \int \mathrm{P} \, d \, p.$$

Also, let the internal particles of the mass undergo changes of relative position, expressed by the variations of quantities denoted generally by r, and let the increase of each such quantity as r be resisted by an internal molecular force such as R:

Then the energy converted to the potential form in overcoming internal molecular forces will be

$$\Sigma \, . \int \mathrm{R} \, d \, r.$$

Subtracting these quantities of energy converted to the potential form by means of external pressures and internal forces, from the whole power converted into heat by friction in order to raise the temperature of the mass from that of absolute privation of heat to the given temperature τ, we find the following result:—

$$Q = E - \Sigma . \int P \, dp - \Sigma . \int R \, dr;$$

and this remainder is the quantity of energy which *retains the form of heat* in unity of weight of the given substance at the given temperature; that is to say, the Total Actual Heat.

It is obvious that Total Actual Heat cannot be ascertained directly; first, because the temperature of total privation of heat is unattainable; and, secondly, because the molecular forces R are unknown.

It can, however, be determined indirectly from the latent heat of expansion of the substance. For the heat which disappears during the expansion of unity of weight of an elastic substance at constant actual heat from the volume V_A to the volume V_B, under the constant or variable pressure P, is expressed (as will be shown in the sequel) by

$$Q . \frac{d}{dQ} \int_{V_A}^{V_B} P \, dV;$$

so that from a sufficient number of experiments on the amount of heat transformed to potential energy by the expansion of a given substance, the relations, for that substance, between pressure, volume, and total actual heat, may be determined.

4. Proposition I.—Theorem. *The mechanical equivalent of the heat absorbed or given out by a substance in passing from one given state as to pressure and volume to another given state, through a series of states represented by the co-ordinates of a given curve on a diagram of energy, is represented by the area included between the given curve and two curves of no transmission of heat drawn from its extremities, and indefinitely prolonged in the direction representing increase of volume.*

(Demonstration: see Fig. 3.) Let the co-ordinates of any two points, A and B, represent respectively the volumes and pressures of the substance in any two conditions; and let a curve of any figure, A C B, represent, by the co-ordinates of its points, an arbitrary succession of volumes and pressures through which the substance is made to pass, in changing from the condition A to the condition B. From the points A and B respectively let two curves of no transmission A M, B N, extend indefinitely towards X; then the area referred to in the enunciation is that contained between the given arbitrary curve A C B and the two indefinitely prolonged curves of no transmission; areas above the curve A M being considered as representing heat absorbed by the substance, and those below heat given out.

To fix the ideas, let us, in the first place, suppose the area M A C B N

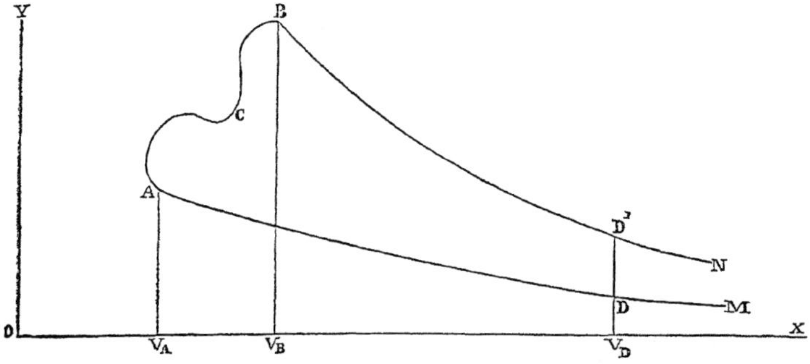

Fig. 3.

to be situated above A M. After the substance has reached the state B, let it be expanded according to the curve of no transmission B N, until its volume and pressure are represented by the co-ordinates of the point D′. Next, let the volume V_D be maintained constant, while heat is abstracted, until the pressure falls so as to be represented by the ordinate of the point D, situated on the curve of no transmission A M. Finally, let the substance be compressed, according to this curve of no transmission, until it recovers its primitive condition A. Then the area ACBD′DA, which represents the whole potential energy developed by the substance during one cycle of operations, represents also the heat which disappears, that is, the difference between the heat absorbed by the substance during the change from A to B, and emitted during the change from D′ to D; for if this were not so, the cycle of operations would alter the amount of energy in the universe, which is impossible.

The farther the ordinate V_D D D′ is removed in the direction of X, the smaller does the heat emitted during the change from D′ to D become; and, consequently, the more nearly does the area ACBD′DA approximate to the equivalent of the heat absorbed during the change from A to B; to which, therefore, the area of the indefinitely prolonged diagram M A C B N is exactly equal. Q. E. D.

It is easy to see how a similar demonstration could have been applied, *mutatis mutandis*, had the area lain below the curve A M. It is evident also, that when this area lies, part above and part below the line A M, the difference between these two parts represents the difference between the heat absorbed and the heat emitted during different parts of the operation.

5. *First Corollary.*—THEOREM. *The difference between the whole heat absorbed, and the whole expansive power developed, during the operation represented by any curve, such as* A C B, *on a diagram of energy, depends on the initial and final conditions of the substance alone, and not on the intermediate process.*

(Demonstration.) In Fig. 3, draw the ordinates $A\,V_A$, $B\,V_B$ parallel to O Y. Then the area $V_A\,A\,C\,B\,V_B$ represents the expansive power developed during the operation ACB; and it is evident that the difference between this area and the indefinitely-prolonged area M A C B N, which represents the heat received by the substance, depends simply on the positions of the points A and B, which denote the initial and final conditions of the substance as to volume and pressure, and not on the form of the curve ACB, which represents the intermediate process. Q.E.D.

To express this result symbolically, it is to be considered, that the excess of the heat or actual energy *received* by the substance above the expansive power or potential energy *given out* and exerted on external bodies, in passing from the condition A to the condition B, is equal to the whole energy *stored up* in the substance during this operation, which consists of two parts, viz.—

Actual energy; being the increase of the actual or sensible heat of the substance in passing from the condition A to the condition B, which is to be represented by this expression,

$$\Delta . Q = Q_B - Q_A;$$

Potential energy; being the power which is stored up in producing changes of molecular arrangement during this process; and which it appears from the theorem just proved, must be represented, like the actual energy, by the difference between a function of the volume and pressure corresponding to A, and the analogous function of the volume and pressure corresponding to B; that is to say, by an expression of the form,

$$\Delta S = S_B - S_A.$$

Let

$$H_{A,B} = \text{area } M\,A\,C\,B\,N$$

represent the heat received by the substance during the operation A C B, and

$$\int_{V_A}^{V_B} P\,d\,V = \text{area } V_A\,A\,C\,B\,V_B$$

the power or potential energy given out.

Then, the theorem of this article is expressed as follows:—

$$H_{A,B} - \int_{V_A}^{V_B} P\,d\,V = Q_B - Q_A + S_B - S_A = \Delta\,Q + \Delta\,.\,S, \quad (2.)$$

being a form of the general equation of the expansive action of heat, in which the *potential of molecular action*, S, remains to be determined.

6. *Second Corollary* (see Fig. 4).—The *Latent Heat of Expansion* of a substance from one given volume V_A to another V_B, for a given amount of actual heat Q;

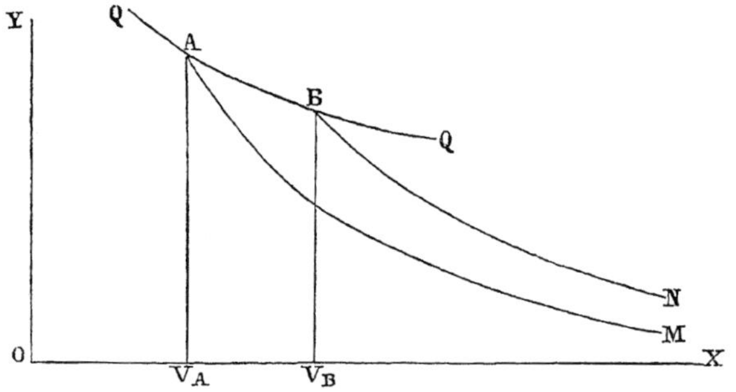

Fig. 4.

that is to say, the heat which must be absorbed by the substance in expanding from the volume V_A to the volume V_B, in order that the actual heat Q may be maintained constant, is represented geometrically as follows: Let Q Q be the isothermal curve of the given actual heat Q on the diagram of energy; A, B two points on this curve, whose co-ordinates represent the two given volumes and the corresponding pressures. Through A and B draw the two curves of no transmission A M, B N, produced indefinitely in the direction of X. Then the area contained between the portion of isothermal curve A B, and the indefinitely-produced curves A M, B N, represents the mechanical equivalent of the latent heat sought, whose symbolical expression is formed from equation (2) by making $Q_B - Q_A = 0$, and is as follows:—

$$H_{A,B}\,(\text{for } Q = \text{const.}) = \int_{V_A}^{V_B} P\,d\,V + S_B - S_A. \quad . \quad (3.)$$

Section II.—Propositions Relative to Homogeneous Substances.

7. Proposition II.—Theorem. *In Fig.* 5, *let* $A_1\,A_2\,M$, $B_1\,B_2\,N$ *be any two curves of no transmission, indefinitely extended in the direction of* X,

intersected in the points A_1, B_1, A_2, B_2, *by two isothermal curves* $Q_1 A_1 B_1 Q_1$, $Q_2 A_2 B_2 Q_2$, *which are indefinitely near to each other; that is to say, which correspond to two quantities of actual heat,* Q_1 *and* Q_2, *differing by an indefinitely small quantity* $Q_1 - Q_2 = \delta Q$.

Then the elementary quadrilateral area, $A_1 B_1 B_2 A_2$, *bears to the whole*

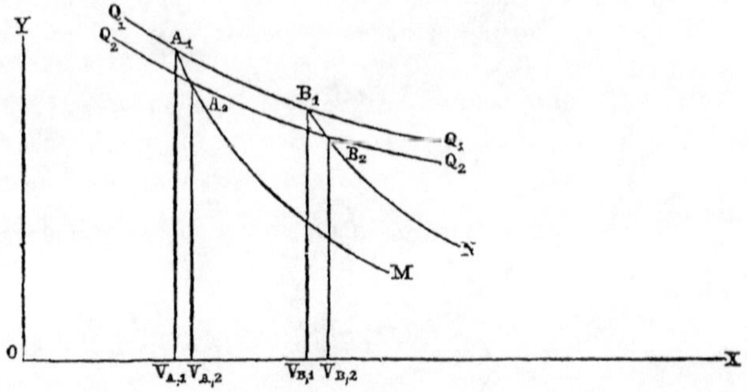

Fig 5.

indefinitely-prolonged area $M A_1 B_1 N$, *the same proportion which the indefinitely small difference of actual heat* δQ *bears to the whole actual heat* Q_1; *or*

$$\frac{\text{area } A_1 B_1 B_2 A_2}{\text{area } M A_1 B_1 N} = \frac{\delta Q}{Q_1}$$

(Demonstration.) Draw the ordinates $A_1 V_{A1}$, $A_2 V_{A2}$, $B_1 V_{B1}$, $B_2 V_{B2}$. Suppose, in the first place, that δQ is an aliquot part of Q_1, obtained by dividing the latter quantity by a very large integer n, which we are at liberty to increase without limit.

The entire indefinitely-prolonged area $M A_1 B_1 N$ represents a quantity of heat which is converted into potential energy during the expansion of the substance from V_{A1} to V_{B1}, in consequence of the continued presence of the total actual heat Q_1; for if no heat were present no such conversion would take place. *Mutatis mutandis*, a similar statement may be made respecting the area $M A_2 B_2 N$. By increasing without limit the number n and diminishing δQ, we may make the expansion from V_{A2} to V_{B2} as nearly as we please an identical phenomenon with the expansion from V_{A1} to V_{B1}. The quadrilateral $A_1 B_1 B_2 A_2$ represents the diminution of conversion of heat to potential energy, which results from the abstraction of any one whatsoever of the n small equal parts δQ into which the actual heat Q_1 is supposed to be divided, and it therefore represents the effect, in conversion of heat to potential energy, of the presence of any one of those small portions of actual heat. And as all those portions δQ are similar and similarly circumstanced, the effect of the presence of the whole actual heat Q_1 in causing conversion of heat to potential energy, will be simply the sum of the effects of all its small portions, and will bear the same ratio to the effect of one of those small portions which the whole actual heat bears to the small portion. Thus, by virtue of the general law enunciated below and assumed as an axiom, the theorem is proved when δQ is an aliquot part of Q_1; but δQ is either an aliquot part, or a sum of aliquot parts, or may be indefinitely approximated to by a series of aliquot parts; so that the theorem is universally true. Q. E. D.

The symbolical expression of this theorem is as follows: when the actual heat Q_1, at any given volume, is varied by the indefinitely small quantity δQ, let the pressure vary by the indefinitely small quantity $\frac{dP}{dQ}\delta Q$; then the area of the quadrilateral $A_1 B_1 B_2 A_2$ will be represented by

$$\delta Q \cdot \int_{V_{A,1}}^{V_{B,1}} \frac{dP}{dQ} dV,$$

and, consequently, that of the whole figure $M A_1 B_1 N$, or the latent heat of expansion from $V_{A,1}$ to $V_{B,1}$, at Q_1, by

$$H_1 = Q_1 \int_{V_{A1}}^{V_{B1}} \frac{dP}{dQ} dV; \quad . \quad . \quad . \quad (4.)$$

a result identical with that expressed in the sixth section of a paper published in the *Transactions of the Royal Society of Edinburgh.* (*See p. 310.*)

The demonstration of this theorem is an example of a special application of the following

GENERAL LAW OF THE TRANSFORMATION OF ENERGY.

The effect of the presence, in a substance, of a quantity of actual energy, in causing transformation of energy, is the sum of the effects of all its parts:— a law first enunciated in a paper read by me to the Philosophical Society of Glasgow on the 5th of January, 1853. (*See p. 203.*)

8. GENERAL EQUATION OF THE EXPANSIVE ACTION OF HEAT.

The two expressions for the latent heat of expansion at constant actual heat, given in equations (3) and (4) respectively, being equated, furnish the means of determining the potential energy of molecular action S, so far as it depends on volume, and thus of giving a definite form to the general equation (2).

The two expressions referred to may be thus stated in words:—

I. The heat which disappears in producing a given expansion, while the actual heat present in the substance is maintained constant, is equivalent to the sum of the potential energy given out in the form of expansive power, and the potential energy stored up by means of molecular attractions.

II. It is also equivalent to the potential energy due to the action during the expansion of a pressure $Q\frac{dP}{dQ}$, at each instant equal to what the pressure would be, if its actual rate of variation with heat at the instant in question were a constant coefficient, expressing the ratio of the whole pressure to the whole actual heat present.

The combination of these principles, expressed symbolically, gives the following result:

$$H_{A,B} \text{ (for } Q = \text{const.}) = Q\int_{V_A}^{V_B} \frac{dP}{dQ} dV = \int_{V_A}^{V_B} P\,dV + S_B - S_A;$$

whence we deduce the following general value for the potential of molecular action:—

$$S = \int\left(Q\frac{dP}{dQ} - P\right)dV + \phi \,.\, Q, \qquad . \qquad . \quad (5.)$$

in which $\phi . Q$ denotes some function of the total actual heat not depending on the density of the substance. This value being introduced into equation (2), produces the following:—

$$H_{A,B}\int_{V_A}^{V_B} P\,dV = Q_B - Q_A + S_B - S_A$$

$$= Q_B - Q_A + \phi \,.\, Q_B - \phi \,.\, Q_A + \int_{V_A}^{V_B}\left(Q\frac{dP}{dQ} - P\right)dV = \Psi_B - \Psi_A. \quad (6.)$$

The symbol $\Psi = Q + S$ is used to denote the sum of the actual energy of heat, and the potential energy of molecular action, present in the substance in any given condition.

The above is the General Equation of the Expansive Action of Heat in a Homogeneous Substance, and is the symbolical expression of the Geometrical Theorems I. and II. combined.

When the variations of actual heat and of volume become indefinitely small, this equation takes the following differential form:—

$$\left.\begin{aligned} d\,.\,\Psi &= d\,.\,H - P\,dV = dQ + d\,.\,S \\ &= \left(1 + \phi' \,.\, Q + Q\frac{d^2}{dQ^2}\int P\,dV\right)dQ + \left(Q\frac{dP}{dQ} - P\right)dV, \\ \text{otherwise} & \\ d\,.\,H &= \frac{K_v}{k} \,.\, dQ + Q\frac{dP}{dQ} \,.\, dV. \end{aligned}\right\} \quad (7.)$$

The coefficient of $d\,Q$ in the above expressions, viz.—

$$\frac{K_v}{k} = 1 + \phi' \cdot Q + Q \frac{d^2}{d\,Q^2} \int P\,d\,V, \qquad (8.)$$

is the ratio of the apparent specific heat of the substance at constant volume to its real specific heat; that is, the ratio of the whole heat consumed in producing an indefinitely small increase of actual heat, to the increase of actual heat produced.

These general equations are here deduced independently of any special molecular hypothesis, as they also have been, by a method somewhat different, in the sixth section of a paper previously referred to. Equations equivalent to the above have also been deduced from the hypothesis of molecular vortices, in the paper already mentioned, and in a paper on the Centrifugal Theory of Elasticity. (*See p. 49.*)

9. *First Corollary from Proposition* II.—THEOREM. *If a succession of isothermal curves corresponding to quantities of heat diminishing by equal small differences* $\delta\,Q$, *be drawn across any pair of curves of no transmission, they will cut off a series of equal small quadrilaterals.*

Second Corollary.—THEOREM. *In Fig.* 6, *let* A D M, B C N *be any two curves of no transmission, indefinitely prolonged in the direction of* X, *and let*

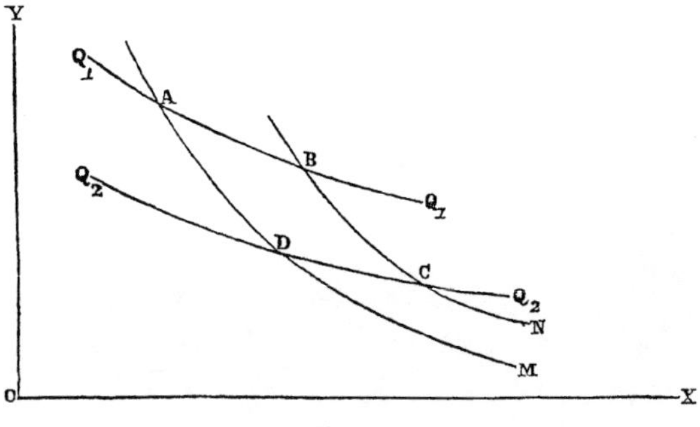

Fig. 6.

any two isothermal curves $Q_1\,Q_1$, $Q_2\,Q_2$, *corresponding respectively to any two quantities of actual heat* Q_1, Q_2, *be drawn across them. Then will the indefinitely-prolonged areas* M A B N, M D C N, *bear to each other the simple ratio of the quantities of actual heat,* Q_1, Q_2.

Or, denoting those areas respectively by H_1, H_2—

$$\frac{H_2}{H_1} = \frac{Q_2}{Q_1}. \qquad (9.)$$

This corollary is the geometrical expression of the law of the maximum efficiency of a perfect thermodynamic engine, already investigated by other methods. In fact, the area M A B N represents the whole heat expended, or the latent heat of expansion, the actual heat at which heat is received being Q_1; M D C N, the heat lost, or the latent heat of compression, which is carried off by conduction at the actual heat Q_2; and A B C D (being the indicator-diagram of such an engine), the motive power produced by the permanent disappearance of an equivalent quantity of heat; and the *efficiency* of the engine is expressed by the ratio of the heat converted into motive power to the whole heat expended, viz.:—

$$\frac{\text{A B C D}}{\text{M A B N}} = \frac{H_1 - H_2}{H_1} = \frac{Q_1 - Q_2}{Q_1}. \qquad (10.)$$

10. *Third Corollary (of Thermodynamic Functions).*

If the two curves of no transmission in Fig. 6, A D M, B C N, be indefinitely close together, the ratio of the heat consumed in passing from one of those curves to the other to the actual heat present, will be the same, whatever may be the form and position of the curve indicating the mode of variation of pressure and volume, provided it intersects the two curves of no transmission at a finite angle; because the area contained between this connecting curve and the two indefinitely-prolonged curves of no transmission will differ from an area whose upper boundary is an isothermal curve, by an indefinitely small area of the second order.

To express this symbolically, let

$$\frac{\delta H}{Q} = \delta F,$$

be the ratio in question, for a given indefinitely-close pair of curves of no transmission. Let the change from one of these curves to the other be made by means of any indefinitely-small changes of actual heat and of volume, δQ, δV. Then by the general equation (7), the following quantity—

$$\delta F = \frac{\delta H}{Q} = \left\{ \frac{1 + \phi' \cdot Q}{Q} + \frac{d^2}{d\,Q^2} \int P\,d\,V \right\} \delta Q$$

$$+ \frac{d\,P}{d\,Q} \cdot \delta V = \frac{K_v}{k\,Q} \cdot \delta Q + \frac{d\,P}{d\,Q} \delta V, \qquad (11.)$$

is constant for a given pair of indefinitely-close curves of no transmission, and is, therefore, the complete variation of a function, having a peculiar constant value for each curve of no transmission, represented by the following equation:—

$$F = \int \frac{d\,H}{Q} = \int \frac{1 + \phi' \cdot Q}{Q} d\,Q + \int \frac{d\,P}{d\,Q} d\,V. \quad . \quad (12.)$$

This function, which I shall call a *thermodynamic function*, has the following properties :—

$$H = \int Q\,d\,F, \quad . \quad . \quad . \quad . \quad (13.)$$

is equivalent to the general equation (6);

$$d\,F = 0, \quad . \quad . \quad . \quad . \quad (14.)$$

is the equation common to all curves of no transmission; and

$$F = \text{a given constant}, \quad . \quad . \quad . \quad (14\text{A}.)$$

is the equation of a particular curve of no transmission.

11. PROPOSITION III.—PROBLEM. *Let it be supposed that for a given substance, the forms of all possible isothermal curves are known, but of only one curve of no transmission; it is required to describe, by the determination of points, another curve of no transmission, passing through a given point, situated anywhere out of the known curve.*

(Solution: see Fig. 7.) Let L M be the known curve of no transmission; B the given point. Through B draw an isothermal curve Q_1 A B Q_1, cutting L M in A. Q_1 being the quantity of heat to which this curve corresponds, draw, indefinitely near to it, the isothermal curve $q_1\,q_1$, corresponding to the quantity of heat $Q_1 - \delta\,Q$, where $\delta\,Q$ is an indefinitely small quantity. Draw any other pair of indefinitely close isothermal

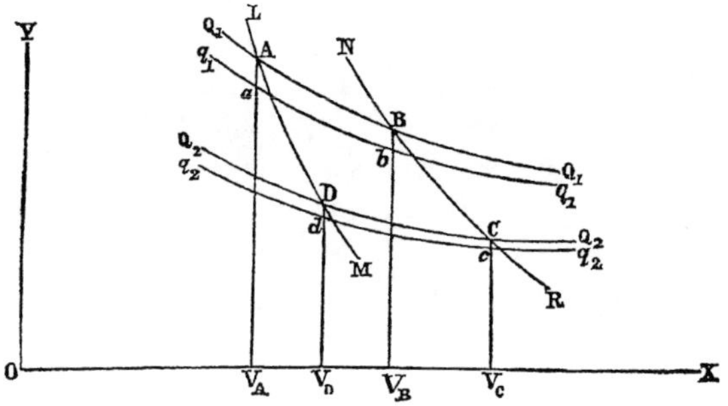

Fig. 7.

curves $Q_2\,Q_2$, $q_2\,q_2$, corresponding to the quantities of heat $Q_2\,Q_2 - \delta\,Q$; $\delta\,Q$ being the same as before. Let D be the point where the isothermal curve $Q_2\,Q_2$ cuts the known curve of no transmission. Draw the ordinates A V_A, B V_B parallel to O Y, enclosing, with the isothermal curves of Q_1 and $Q_1 - \delta\,Q$, the small quadrilateral A B *b a*. Draw the ordinate D V_D

parallel to O Y, intersecting the isothermal curve of $Q_2 - \delta Q$ in *d*. Lastly, draw the ordinate $C V_C$ in such a position as to cut off from the space between the isothermal curves of Q_2 and $Q_2 - \delta Q$ a quadrilateral D C *c d*, of area equal to the quadrilateral A B *b a*.

Then will C, where the last ordinate intersects the isothermal curve of Q_2, approximate indefinitely to the position of a point in the curve of no transmission passing through the given point B, when the variation of actual heat δQ is diminished without limit. And thus may be determined to as close an approximation as we please, any number of points in the curve of no transmission N B R which passes through any given point B, when any one curve of no transmission L M is known.

(Demonstration.) For when the variation δQ diminishes indefinitely, the curves $q_1 q_1$, $q_2 q_2$, approach indefinitely towards the curves $Q_1 Q_1$, $Q_2 Q_2$ respectively; and the small quadrilaterals bounded endways by the ordinates approximate indefinitely to the small quadrilaterals bounded endways by the curves of no transmission; which latter pair of quadrilaterals are equal, by the first corollary of Proposition II.

The *symbolical expression* of this proposition is as follows:—

Let V_A, V_B, V_C, V_D, be the volumes corresponding to the four points of intersection of a pair of isothermal curves with a pair of curves of no transmission; A and B being on the isothermal curve of Q_1, C and D on that of Q_2, A and D on one of the curves of no transmission, B and C on the other; then

$$\left.\begin{aligned} &\int_{V_A}^{V_B} \frac{dP}{dQ}\, dV \text{ (for } Q = Q_1) = \int_{V_D}^{V_C} \frac{dP}{dQ}\, dV \text{ (for } Q = Q_2), \\ &\text{or} \\ &F_B - F_A = F_C - F_D. \end{aligned}\right\} \quad (15.)$$

The second form of this equation is in the present case identical, because

$$F_D = F_A; \; F_C = F_B.$$

12. Proposition IV.—Problem (see Fig. 8). *The forms of all isothermal curves for a given substance being given, let* E F *be a curve of any form, representing an arbitrarily assumed succession of pressures and volumes. It is required to find, by the determination of points, a corresponding curve passing through a given point* B, *such that the quantity of heat absorbed or emitted by the substance, in passing from any given isothermal curve to any other, shall be the same, whether the pressures and volumes be regulated according to the original curve* E F, *or according to the curve passing through the point* B.

(Solution.) The process by which the latter curve is to be deduced from the former is precisely the same with that by which one curve of no transmission is deduced from another in the last problem.

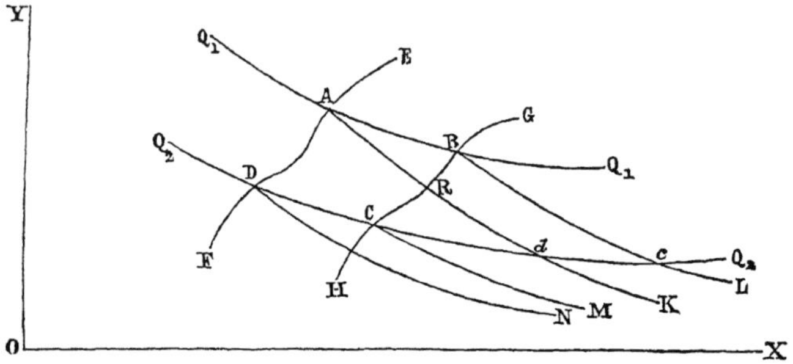

Fig. 8.

(Demonstration.) Let G B H be the required curve. This curve, and the curve E F, in their relation to each other, may be called *Curves of Equal Transmission.* Through B draw the isothermal curve $Q_1 Q_1$, intersecting the curve E F in A. Draw also any other isothermal curve $Q_2 Q_2$, intersecting E F in D and G H in C. Through A, B, C, D, respectively, draw the four indefinitely-prolonged curves of no transmission, A K, intersecting $Q_2 Q_2$ in *d*, B L, intersecting $Q_2 Q_2$ in *c*, C M, and D N. Conceive the whole space between the isothermal curves $Q_1 Q_1$, $Q_2 Q_2$, to be divided by other isothermal curves, into a series of indefinitely-narrow stripes, corresponding to equal indefinitely-small variations of actual heat. Then, by the construction of the solution, the quadrilaterals cut from those stripes by the pair of curves E F, G H are all equal; and so also are the quadrilaterals cut from the stripes by the pair of curves of no transmission, A K, B L. Therefore the area A B C D is equal to the area A B *c d*. The indefinitely-prolonged areas, M C D N, L *c d* K, are evidently equal; therefore, adding this pair of equal areas to the preceding, the pair of indefinitely-prolonged areas L B A K, M C B A D N are equal. Subtracting from each of these areas the part common to both, A B R, and adding to each the indefinitely-prolonged area K R C M, we find, finally, that the indefinitely-prolonged areas K A D N, L B C M are equal.

But the former of those areas (by Prop. I.) represents the mechanical equivalent of the heat absorbed by the substance in passing from the actual heat Q_2 to the actual heat Q_1 through a series of pressures and volumes represented by the co-ordinates of the curve E F; and the latter, the corresponding quantity for the curve G H; therefore those curves are, with respect to each other, *curves of equal transmission,* which was required.

The algebraical expression of this result is that the equation (15) holds for any pair of curves of equal transmission, as well as for a pair of curves of no transmission; or, in other terms, let F_A, F_B, F_C, F_D be the thermodynamic functions for the curves of no transmission passing through the four points where a pair of isothermal curves cut a pair of curves of equal transmission: A, B being on the upper isothermal curve; C, D on the lower; A, D on one curve of equal transmission, B, C on the other: then,

$$F_B - F_A = F_C - F_D. \quad . \quad . \quad . \quad (16.)$$

13. Proposition V.—Theorem. *The difference between the quantities of heat absorbed by a substance in passing from one given amount of actual heat to another, at two different constant volumes, is equal to the difference between the two latent heats of expansion in passing from one of those volumes to the other at the two different amounts of actual heat respectively, diminished by the corresponding difference between the quantities of expansive power given out.*

(Demonstration: see Fig. 9.) Let $Q_1 Q_1$ be the isothermal curve of the higher amount of actual heat; $Q_2 Q_2$ that of the lower. Let V_A, V_B be

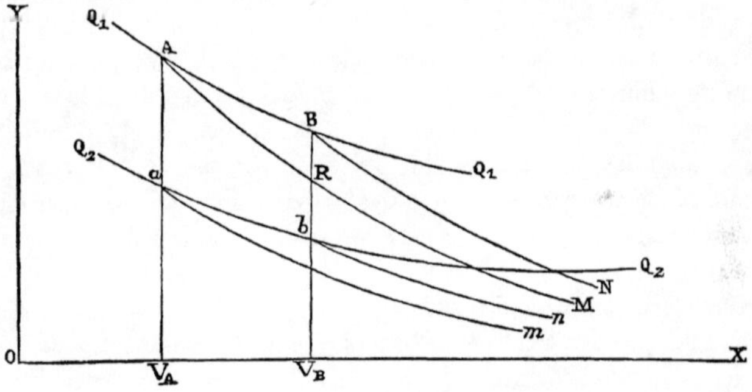

Fig. 9.

the two given volumes. Draw the two ordinates $V_A\,a\,A$, $V_B\,b\,B$, and the four indefinitely-prolonged curves of no transmission AM, *am*, BN, *bn*. the quantities of heat absorbed, in passing from the actual heat Q_2 to the actual heat Q_1, at the volumes V_A and V_B, are represented respectively by the indefinitely-prolonged areas $MAam$, $NBbn$. Then adding to each of those areas the indefinitely-prolonged area $nbBAM$ (observing that the space below the intersection R is to be treated as negative), we find for their difference

$$NBbn - MAam = NBAM - nbBAam = (NBAM - nbam) \\ - (V_B BAV_A - V_B baV_A);$$

but NBAM and *nbam* represent the latent heats of expansion from V_A to V_B, at the actual heats Q_1 and Q_2 respectively; and $V_B BAV_A$ and

$V_B\, b\, a\, V_A$ represent the power given out by expansion from V_A to V_B at the actual heats Q_1 and Q_2 respectively; therefore, the proposition is proved. Q. E. D.

This proposition, expressed symbolically, is as follows: ΔQ being the difference of actual heat, $Q_1 - Q_2$, let $\Delta\ (Q + S_A)$ be the heat absorbed in passing from Q_2 to Q_1 at the volume V_A, and $\Delta\ (Q + S_B)$ the corresponding quantity at the volume V_B; ΔS_A and ΔS_B representing quantities of potential energy stored up in altering molecular arrangement. Then

$$\Delta (S_B - S_A) = \Delta \left(Q \frac{d}{d Q} - 1 \right) \int_{V_A}^{V_B} P\, d V. \qquad (17.)$$

14. *Of Curves of Free Expansion.*

In all the preceding propositions, the whole motive power developed by an elastic substance in expanding is supposed to be communicated to external bodies; to a piston, for example, which the substance causes to move and to overcome the resistance of a machine.

Let us now suppose that as much as possible of the motive power developed by the expansion is expended in agitating the particles of the expanding substance itself, by whose mutual friction it is finally reconverted into heat (as when compressed air escapes freely from a small orifice); and let us examine the properties of the curves which, on a diagram of energy, represent the law of expansion of the substance under these circumstances, and which may be called *Curves of free Expansion.*

15. Proposition VI.—Theorem. *If from two points on a curve of free expansion there be drawn two straight lines perpendicular to and terminating at the axis of ordinates, and also two curves of no transmission, indefinitely prolonged away from the origin of co-ordinates; then the area contained between the curve of free expansion, the two straight lines, and the axis of ordinates, will be equal to the area contained between the curve of free expansion and the two indefinitely-prolonged curves of no transmission.*

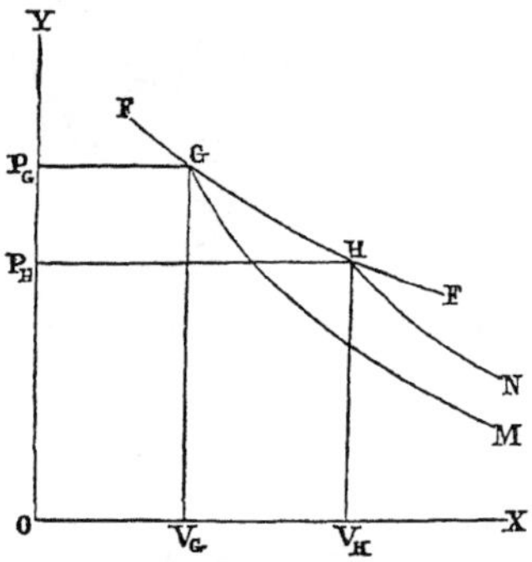

Fig. 10.

(Demonstration.) Let F F (Fig. 10) be a curve of free expansion; G H any two points in it; $G V_G$, $H V_H$ ordinates; $G P_G$, $H P_H$ lines perpendicular to O Y; G M, H N curves of no transmission, indefinitely prolonged in the direction of X. Then the indefinitely-prolonged area M G H N represents the heat which would have to be communicated to the substance, if the motive power developed were entirely transferred to external bodies, while the area $V_G\,G\,H\,V_H$ represents that motive power. The excess of the rectangular area $P_H\,H\,V_H\,O$ above the area $P_G\,G\,V_G\,O$, is the power necessarily given out by the elastic fluid in passing from a vessel in which the pressure is P_G and volume V_G, to a vessel in which the pressure is P_H and volume V_H. The remainder of the expansive power, represented by the area $P_G\,G\,H\,P_H$, by the mutual friction of the particles of the expanding substance, is entirely reconverted into heat, and is exactly sufficient (by the definition of the curve of free expansion) to render the communication of heat to the substance unnecessary; from which it follows, that this area is equal to the area M G H N. Q. E. D.

The equation of a curve of free expansion is

$$d\,(\Psi + P\,V) = 0. \quad . \quad . \quad . \quad (17\ \text{A.})$$

16. *Corollary.*—In Fig. 11, the same letters being retained as in the last figure, through G draw an isothermal curve $Q_1\,Q_1$, which the line $P_H H$ produced cuts in h;

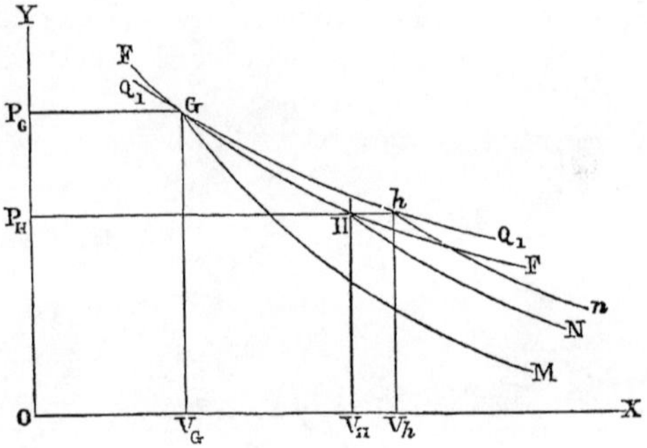

Fig. 11.

and from h draw the indefinitely-prolonged curve of no transmission, *hn*. Then because, by the proposition just proved, the areas $P_G\,G\,H\,P_H$ and M G H N are equal, it follows that the indefinitely-prolonged area, M G h n, which represents the latent heat of expansion at the constant actual heat Q_1, from the volume V_G to the volume V_h, exceeds $P_G\,G\,h\,P_H$, by the indefinitely-prolonged area N H h n, which represents the heat which the substance would give out, in falling, at the pressure P_H, from the actual heat Q_1 to the actual heat corresponding to the point H on the curve of free expansion passing through G. Subtracting from this area the excess of the rectangle $P_H\,V_h$ above the rectangle $P_G\,V_G$, we obtain the excess of the area M G h n above the area $V_G\,G\,h\,V_h$.

This conclusion may be thus expressed:—Let Q_2 be the actual heat for the point H; $\frac{K_P}{k}$ the ratio of specific heat at the constant pressure P_H to real specific heat; then

$$\left.\begin{aligned} &\int_{Q_2}^{Q_1} \frac{K_P}{k} dQ - P_H V_h + P_G V_G \\ &= \left(Q\frac{d}{dQ} - 1\right)\int_{V_G}^{V_h} P\,dV \text{ (for } Q = Q_1\text{);} \\ &\text{otherwise:—} \\ &\int_{Q_2}^{Q_1} \frac{K_P}{k} dQ + \int_{P_H}^{P_G} V\,dP = Q_1(F_h - F_G) \end{aligned}\right\} \quad (18.)$$

Equation (18) may be used, either to find points in the curve of free expansion which passes through G, when the isothermal curves and the curves of no transmission are known; or to deduce theoretical results from experiments on the form of curves of free expansion, such as those which have been for some time carried on by Mr. Joule and Professor William Thomson.

Considered geometrically, these experiments give values of the area N H *h n*. The area

$$P_G G h P_H = \int_{P_H}^{P_G} V\,dP$$

is known in each case from previous experiments on the properties of the gas employed; and this area, by Proposition VI., is equal to the area M G *h* H N; to which, adding the area N H *h n*, ascertained by experiment, we obtain the area M G *h n*, that is, the latent heat of expansion from the volume V_G to the volume V_h, at the constant actual heat Q_1, denoted symbolically by

$$H = Q_1 \frac{d}{dQ}\int_{V_G}^{V_h} P\,dV = Q_1(F_h - F_G).$$

Now the problem to be solved is of this kind. We know the *differences* of actual heat corresponding to a certain series of isothermal curves for the substance employed; and we have to ascertain the *absolute quantities* of actual heat corresponding to those curves. Of the above expression for the area M G *h n*, therefore, the factor Q_1 is to be determined, while the other factor, being the difference between two thermodynamic functions, is known; and the experiments of Messrs. Thomson and Joule, by giving the value of the product, enable us to calculate that of the unknown factor, and thence to determine the point on the thermometric scale corresponding to absolute privation of heat.

17. Proposition VII.—Problem. *To determine the ratio of the apparent specific heats of a substance at constant volume and at constant pressure, for a given pressure and volume; the isothermal curves and the curves of no transmission being known.*

(Solution.) In Fig. 12, let A be the point whose co-ordinates represent the given volume V_A and pressure P_A; Q A Q the isothermal curve passing

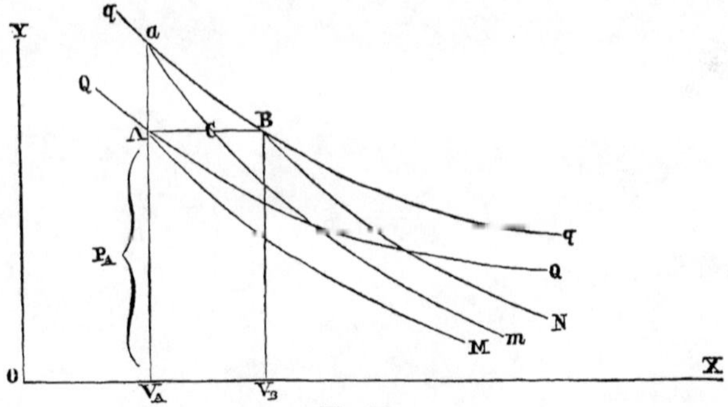

Fig. 12.

through A; $q\,q$ another isothermal curve, very near to Q Q. Through A draw the ordinate $V_A\,A\,a$ parallel to O Y, cutting $q\,q$ in a; draw also A B parallel to O X, cutting $q\,q$ in B. From A, a, B, draw the three indefinitely-prolonged curves of no transmission A M, $a\,m$, B N.

Then the heat absorbed in passing from the actual heat Q to the actual heat q, at the constant volume V_A, is represented by the indefinitely-prolonged area $\mathrm{MA}am$, while at the constant pressure P_A it is represented by the area M A B N. Let the curve $q\,q$ be supposed to approximate indefinitely to Q Q. Then will the three-sided area $\mathrm{A}a\mathrm{B}$ diminish indefinitely as compared with the areas between the curves of no transmission A M, $a\,m$, B N; and, consequently, the area M A B N will approximate indefinitely to the sum of the areas $\mathrm{MA}am$ and $ma\mathrm{BN}$; the ultimate ratio of which sum to the area $\mathrm{MA}am$ is therefore the required ratio of the specific heats. Now $ma\mathrm{BN}$, as $q\,q$ approaches Q Q, approximates indefinitely to the latent heat of the small expansion $V_B - V_A$ at the actual heat Q, and this small expansion bears ultimately to the increment of pressure $P_a - P_A$, the ratio of the subtangent of the isothermal curve Q Q to its ordinate at the point A.

The symbolical expression of this proposition is as follows:—Let δQ denote the indefinitely-small difference of actual heat between the isothermal curves Q Q, $q q$; δV the indefinitely-small variation of volume $V_B - V_A$; δP the indefinitely small variation of pressure $P_a - P_A$;

$$\frac{K_V}{k}\delta Q, \frac{K'_P}{k}\delta Q$$

the quantities of heat required to produce the variation δQ, at the constant volume V_A, and at the constant pressure P_A respectively.

Then

$$\delta V = \frac{\delta P}{-\frac{dP}{dV}} = \frac{\frac{dP}{dQ}}{-\frac{dP}{dV}} \cdot \delta Q;$$

and

$$\left.\begin{array}{l} \frac{K_P}{k} \cdot \delta Q = \frac{K_V}{k} \cdot \delta Q + Q\frac{dP}{dQ} \cdot \delta V = \left\{\frac{K_V}{k} + \frac{Q\left(\frac{dP}{dQ}\right)^2}{-\frac{dP}{dV}}\right\}\delta Q; \\ \text{consequently,} \\ \frac{K_P}{K_V} = 1 + \frac{k}{K_V} \cdot \frac{Q\left(\frac{dP}{dQ}\right)^2}{-\frac{dP}{dV}} \end{array}\right\} \quad (19.)$$

equations agreeing with equation (31) of a paper on the Centrifugal Theory of Elasticity before referred to.

18. *First Corollary.*—As the curves A M, $a m$, B N approximate indefinitely towards parallelism, and the point a towards C, where am intersects A B, the ratio of the areas M A B N : M A $a m$, approximates indefinitely to that of the lines AB:AC, which are ultimately proportional, respectively, to the subtangents of the isothermal curve and the curve of no transmission passing through A. Therefore,

$$\frac{K_P}{K_V} = \frac{\text{Subtangent of Isothermal Curve}}{\text{Subtangent of Curve of No Transmission}}. \quad (20.)$$

19. *Second Corollary.—Velocity of sound.* The subtangents of different curves at a given point on a diagram of energy being inversely proportional to the increase of pressure produced by a given diminution of volume according to the respective curves, are inversely proportional to the squares of the respective velocities with which waves of condensation and rarefaction will travel when the relations of pressure to volume are expressed by the different curves. Therefore, if there be no sensible transmission of heat between the particles of a fluid during the passage of sound, the

square of the velocity of sound must be greater than it would have been had the transmission of heat been instantaneous in the ratio of the subtangent of an isothermal curve to that of a curve of no transmission at the same point, or of the specific heat at constant pressure to the specific heat at constant volume.

This is a geometrical proof of Laplace's law for all possible fluids. The same law is deduced from the hypothesis of molecular vortices in the paper before referred to on the Centrifugal Theory of Elasticity.

20. PROPOSITION VIII.—PROBLEM. *The isothermal curves for a given substance being known, and the quantities of heat required to produce all variations of actual heat at a given constant volume; it is required to find any number of points in a curve of no transmission passing through a given point in the ordinate corresponding to that volume.*

(Solution.) In Fig. 13, let $V_A A_1$ be the given ordinate; $Q_1 Q_1$, $A_2 Q_2$ isothermal curves meeting it in A_1, A_2, respectively; and let it be required,

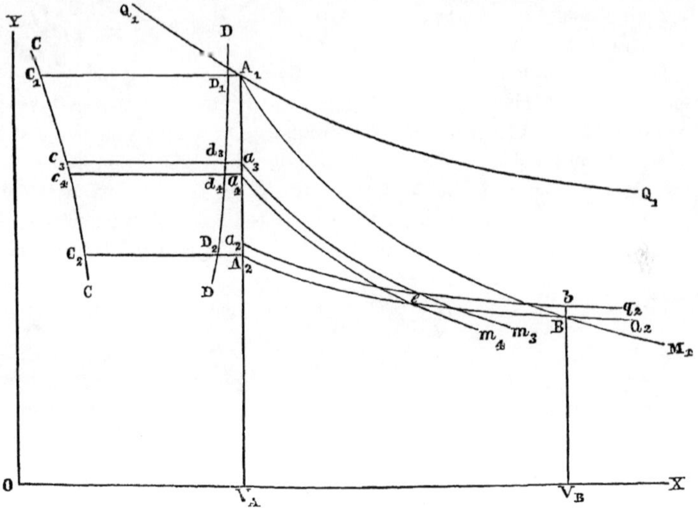

Fig. 13.

for example, to find the point where the curve of no transmission passing through A_1 intersects the isothermal curve $A_2 Q_2$. On the line $V_A A_2 A_1$, as an axis of abscissæ, describe a curve C C, whose ordinates (such as $A_2 C_2$, $a_4 c_4$, &c.) are proportional to the specific heat of the substance at the constant volume V_A, and at the degrees of actual heat corresponding to the points where they are erected, divided by the corresponding rate of increase of pressure with actual heat; so that the area of this curve between any two ordinates (*e.g.*, the area $a_4\, c_4\, c_3\, a_3$) may represent the mechanical equivalent of the heat absorbed in augmenting the actual heat from the amount corresponding to the lower ordinate to that corresponding to the higher (*e.g.*, from the amount corresponding to a_4 to that corresponding to a_3).

Very near to the isothermal curve $A_2 Q_2$, draw another isothermal curve $a_2 q_2$, and let the difference of actual heat corresponding to the interval between these curves be δQ. Draw a curve D D, such that the part cut off by it from each ordinate of the curve C C shall bear the same proportion to the whole ordinate which the difference δQ bears to the whole actual heat corresponding to the ordinate; for example, let

$$\overline{A_1 C_1} : \overline{A_1 D_1} :: Q_1 : \delta Q$$
$$\overline{A_2 C_2} : \overline{A_2 D_2} :: Q_2 : \delta Q, \text{ \&c.}$$

Then draw an ordinate $V_B B b$, parallel to O Y, cutting off from the space between the isothermal curves $A_2 Q_2$, $a_2 q_2$, a quadrilateral area $A_2 B b a_2$ equal to $A_1 D_1 D_2 A_2$, the area of the curve D D between the ordinates at A_1 and A_2.

Then, if the difference δQ be indefinitely diminished, the point B will approximate indefinitely to the intersection required of the isothermal curve $A_2 Q_2$ with the curve of no transmission passing through A_1; and thus may any number of points in this curve of no transmission be found.

(Demonstration.) Let $A_1 M_1$ be the curve of no transmission required. Let $a_3 c_3$, $a_4 c_4$ be any two indefinitely-close ordinates of the curve C C, corresponding to the mean actual heat $Q_{3,4}$. Let $a_3 m_3$, $a_4 m_4$ be curves of no transmission, cutting the curves $a_2 q_2$, $A_2 Q_2$, so as to enclose a small quadrilateral area *e*. Then, by the construction, and Proposition I.,

$$\text{The area } a_3 c_3 c_4 a_4 = \text{the indefinitely-prolonged area } m_3 a_3 a_4 m_4;$$

and by the first corollary of the second proposition and the construction,

$$\frac{\text{the area } e}{m_3 a_3 a_4 m_4} = \frac{\delta Q}{Q_{3,4}} = \frac{\text{area } a_3 d_3 d_4 a_4}{\text{area } a_3 c_3 c_4 a_4}.$$

Therefore, the area $e =$ the area $a_3 d_3 d_4 a_4$; but the area $A_1 D_1 D_2 A_2$ is entirely made up of such areas as $a_3 d_3 d_4 a_4$, to each of which there corresponds an equal area such as *e*; and when the difference δQ is indefinitely diminished, the area $A_2 B b a_2$ approximates indefinitely to the sum of all the areas such as *e*, that is, to equality with the area $A_1 D_1 D_2 A_2$. Q.E.D.

The symbolical expression for this proposition is found as follows:—

$$\text{The area } A_1 D_1 D_2 A_2 \text{ ultimately} = \delta Q \cdot \int_{Q_2}^{Q_1} \frac{K_v}{k Q} \cdot d Q \text{ (for } V = V_A);$$

$$\text{The area } A_2\, B\, b\, a_2 \text{ ultimately} = \delta Q \,.\, \int_{V_A}^{V_B} \frac{d\,P}{d\,Q}\, d\,V \;(\text{for } Q = Q_2)\,;$$

divide both sums by δQ and equate the results; then,

$$\int_{V_A}^{V_B} \frac{d\,P}{d\,Q}\, d\,V \;(\text{for } Q = Q_2) - \int_{Q_2}^{Q_1} \frac{K_v}{\mathfrak{k}\,Q}\, d\,Q \;(\text{for } V = V_A), \quad (21.)$$

which denotes the equality of two expressions for the difference, $F_1 - F_2$, between the thermodynamic functions for the curve of no transmission $A_1\,M$, and for that passing through the point A_2.

When the relations between pressure, volume, and heat, for a given substance, are known, the equation (21) may be transformed into one giving the volume V_B corresponding to the point at which the required curve of no transmission cuts the isothermal curve of Q_2.

Suppose, for instance, that for a perfect gas

$$P\,V = N\,Q \text{ sensibly; and } \frac{K_v}{\mathfrak{k}} = 1 \text{ sensibly;} \quad . \quad (22.)$$

N being a constant (whose value for simple gases and for atmospheric air and carbonic oxide is about 0·41); then the thermodynamic function for a perfect gas is sensibly

$$F = \text{hyp. log. } Q + N \text{ hyp. log. } V; \quad . \quad . \quad (22\,\text{A.})$$

and equation (21) gives, for the equation of a curve of no transmission,

$$\frac{V_B}{V_A} = \left(\frac{Q_1}{Q_2}\right)^{\frac{1}{N}}; \quad . \quad . \quad . \quad . \quad (23.)$$

whence

$$\frac{P_B}{P_A} = \left(\frac{V_B}{V_A}\right)^{-1-N}. \quad . \quad . \quad . \quad . \quad (24.)$$

Equations (23) and (24) are forms of the equation of a curve of no transmission for a perfect gas, according to the supposition of Mayer; and are approximately true for a perfect or nearly perfect gas on any supposition.

According to the hypothesis of molecular vortices, the relations between pressure, volume, and actual heat, for a perfect gas, are expressed by these equations:—

$$P\,V = N\,Q + h; \quad \frac{K_v}{\mathfrak{k}} = 1 + \frac{N^2\,h\,Q}{(N\,Q + h)^2}; \quad . \quad (25.)$$

where h is a very small constant, which is inversely proportional to the specific gravity of the gas, and whose value, in the notation of papers on the hypothesis in question, is

$$h = \mathrm{N}\, \mathfrak{k}\, \kappa, \quad . \quad . \quad . \quad . \quad (25\text{A.})$$

κ being the height, on the scale of a perfect gas-thermometer, of the point of absolute cold above the absolute zero of gaseous tension. Hence we find, for the thermodynamic function of a perfect gas,

$$\mathrm{F} = \text{hyp. log. } \mathrm{Q} - \frac{\mathrm{N}\, h}{\mathrm{N\, Q} + h} + \mathrm{N} \text{ hyp. log. } \mathrm{V}, \quad . \quad (26.)$$

and for the equation of a curve of no transmission,

$$\frac{\mathrm{V_B}}{\mathrm{V_A}} = \left(\frac{\mathrm{Q_1}}{\mathrm{Q_2}}\right)^{\frac{1}{\mathrm{N}}} \cdot e^{\left\{\frac{h}{\mathrm{N\, Q_2} + h} - \frac{h}{\mathrm{N\, Q_1} + h}\right\}} \quad . \quad . \quad (27.)$$

For all practical purposes yet known, these equations may be treated as sensibly agreeing with equation (23), owing to the smallness of h as compared with N Q.

Section III.—Of the Efficiency of Thermodynamic Engines, Worked by the Expansion and Condensation of Permanent Gases.

21. The *efficiency* of a thermodynamic engine is the proportion of the whole heat expended which is converted into motive power; that is to say, the ratio of the motive power developed to the mechanical equivalent of the whole heat consumed.

To determine geometrically the efficiency of a thermodynamic engine, it is necessary to know its true indicator-diagram; that is to say, the curve whose co-ordinates represent the successive volumes and pressures which the elastic substance working the engine assumes during a complete revolution. This true indicator-diagram is not necessarily identical in figure with the diagram described by the engine on the indicator-card; for the abscissæ representing volumes in the latter diagram include not only the volumes assumed by that portion of the elastic substance which really performs the work by alternately receiving heat while expanding, and emitting heat while contracting, in such a manner as permanently to transform heat into motive power, but also the volumes assumed by that portion of the elastic substance, if any, which acts merely as a *cushion* for transmitting pressure to the piston, undergoing, during each revolution, a series of changes of pressure and volume, and then the same series in an order exactly the reverse of the former order, so as to transform no heat permanently to power.

The thermodynamic engines to be considered in the present section, are those in which the elastic substance undergoes no change of condition. We shall, in the first place, investigate the efficiency of those which work *without* the aid of the contrivance called an "economiser" or "regenerator," and afterwards, those which work *with* the aid of that piece of apparatus.

22. LEMMA.—PROBLEM. *To determine the true from the apparent indicator-diagram of a thermodynamic engine; the portion of the elastic substance which acts as a* cushion *being known, and the law of its changes of pressure and volume.*

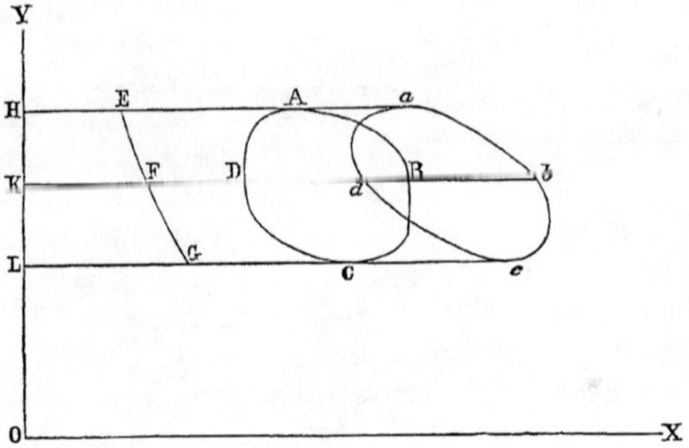

Fig. 14.

(Solution.) In Fig. 14, let $abcd$ be the apparent indicator-diagram. Parallel to O X draw $\overline{\mathrm{H}a}$ and $\overline{\mathrm{L}c}$, touching this diagram in a and c respectively; then those lines will be the lines of maximum and minimum pressure. Let $\overline{\mathrm{HE}}$ and $\overline{\mathrm{LG}}$ be the volumes occupied by the *cushion* at the maximum and minimum pressures respectively: draw the curve E G, such that its co-ordinates shall represent the changes of volume and pressure undergone by the cushion during a revolution of the engine. Let $\mathrm{KF}db$ be any line of equal pressure, intersecting this curve and the apparent indicator-diagram; so that $\overline{\mathrm{K}b}$, $\overline{\mathrm{K}d}$ shall represent the two volumes assumed by the whole elastic body at the pressure $\overline{\mathrm{OK}}$, and $\overline{\mathrm{KF}}$ the volume of the cushion at the same pressure. On this line take

$$\overline{b\mathrm{B}} = \overline{d\mathrm{D}} = \overline{\mathrm{KF}};$$

then it is evident that B and D will be two points in the true indicator-diagram; and in the same manner may any number of points be found.

The area of the true diagram A B C D is obviously equal to that of the apparent diagram $abcd$.

23. PROPOSITION IX.—PROBLEM. *The true indicator-diagram of a thermodynamic engine worked by the expansion and contraction of a substance which does not change its condition, and without a regenerator, being given, it is required to determine the efficiency of the engine.*

(Solution.) In Fig. 15, let A *a a′* B *b′ b* A be the given true indicator-diagram. Draw two curves of no transmission, A M, B N, touching this figure at A and B respectively, and indefinitely produced towards X. Then during the process denoted by the portion A *a a′* B of the diagram the elastic substance is receiving heat, and the mechanical equivalent of the total quantity received is represented by the indefinitely-prolonged area M A *a a′* B N; during the process denoted by the portion B *b′ b* A of the diagram, the substance is giving out heat, and the mechanical equivalent of the total heat given out is represented by the indefinitely-prolonged area M A *b b′* B N; while the difference between those areas, that is, the area of the indicator-diagram itself, represents at once the heat which permanently disappears and the motive power given out. The EFFICIENCY of the engine is the ratio of this last quantity to the total heat received by the elastic substance during a revolution; that is to say, it is denoted by the fraction,

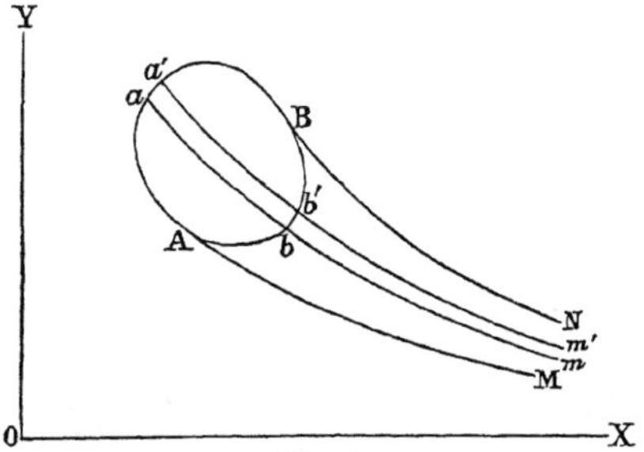

Fig. 15.

$$\frac{\text{area A } a\, a' \text{ B } b'\, b \text{ A}}{\text{area M A } a\, a' \text{ B N}}.$$

To express this result symbolically, find the limiting points A and B by combining the equation of the indicator-diagram with the general equation of curves of no transmission, viz. :—

$$d\,\mathrm{F} = 0.$$

Then draw two indefinitely-close and indefinitely-prolonged curves of no transmission, *a b m*, *a′ b′ m′*, through any part of the diagram, cutting out of it a quadrilateral stripe, *a b b′ a′*. Let Q_1 be the mean actual heat corresponding to the upper end *a a′* of this quadrilateral stripe; Q_2, that corresponding to the lower end, *b b′*.

The area of this indefinitely-narrow stripe, representing a portion of the heat converted into motive power, is found, according to the principles and notation of the third corollary to Proposition II. and of Proposition

III., by multiplying the difference between the actual heats by the difference between the thermodynamic functions for the curves of no transmission that bound the stripe, thus :—

$$\delta \mathrm{E} = (\mathrm{Q}_1 - \mathrm{Q}_2)\, \delta \mathrm{F} :$$

while the area of the indefinitely-prolonged stripe, $m\,a\,a'\,m'$, representing part of the total heat expended, is, according to the same principles,

$$\delta \mathrm{H}_1 = \mathrm{Q}_1\, \delta \mathrm{F} ;$$

and that of the indefinitely-prolonged stripe $m\,b\,b'\,m'$, representing part of the heat given out, is

$$\delta \mathrm{H}_2 = \mathrm{Q}_2\, \delta \mathrm{F}.$$

Integrating these expressions we find the following results :—

Whole heat expended,

$$\mathrm{H}_1 = \int_{\mathrm{F}_\mathrm{A}}^{\mathrm{F}_\mathrm{B}} \mathrm{Q}_1\, d\,\mathrm{F} ;$$

Heat given out,

$$\mathrm{H}_2 = \int_{\mathrm{F}_\mathrm{A}}^{\mathrm{F}_\mathrm{B}} \mathrm{Q}_2\, d\,\mathrm{F} ;$$

Motive power given out,

$$\mathrm{E} = \mathrm{H}_1 - \mathrm{H}_2 = \int_{\mathrm{F}_\mathrm{A}}^{\mathrm{F}_\mathrm{B}} (\mathrm{Q}_1 - \mathrm{Q}_2)\, d\,\mathrm{F} ;$$

Efficiency,

$$\frac{\mathrm{E}}{\mathrm{H}_1} = \frac{\int_{\mathrm{F}_\mathrm{A}}^{\mathrm{F}_\mathrm{B}} (\mathrm{Q}_1 - \mathrm{Q}_2)\, d\,\mathrm{F}}{\int_{\mathrm{F}_\mathrm{A}}^{\mathrm{F}_\mathrm{B}} \mathrm{Q}_1\, d\,\mathrm{F}} ; \qquad (28.)$$

formulæ agreeing with equation (28) of a paper on the Centrifugal Theory of Elasticity (*See p. 63*); it being observed that the symbol F in the last-mentioned paper denotes, not precisely the same quantity which is denoted by it in this paper, and called a thermodynamic function, but the product of the part of that function which depends on the volume by the real specific heat of the substance.

24. *First Corollary. Maximum efficiency between given limits of actual heat.* When the highest and lowest limits of actual heat at which the engine

can work are fixed, it is evident that the greatest possible efficiency of an engine without a regenerator will be attained when the whole reception of heat takes place at the highest limit, and the whole emission at the lowest; so that the true indicator-diagram is such a quadrilateral as is shown in Fig. 6, and referred to in the second corollary of Proposition II.; bounded above and below by the isothermal curves denoting the limits of actual heat, and, laterally, by any pair of curves of no transmission. The efficiency in this case, as has been already proved in various ways, is represented by

$$\frac{E}{H_1} = \frac{Q_1 - Q_2}{Q_1}, \quad . \quad . \quad . \quad . \quad (29.)$$

being the maximum efficiency possible between the limits of actual heat, Q_1 and Q_2.

25. *Second Corollary.*—PROBLEM. *To draw the diagram of greatest efficiency of a thermodynamic engine without a regenerator, when the extent of variation of volume is limited, as well as that of the variation of actual heat.*

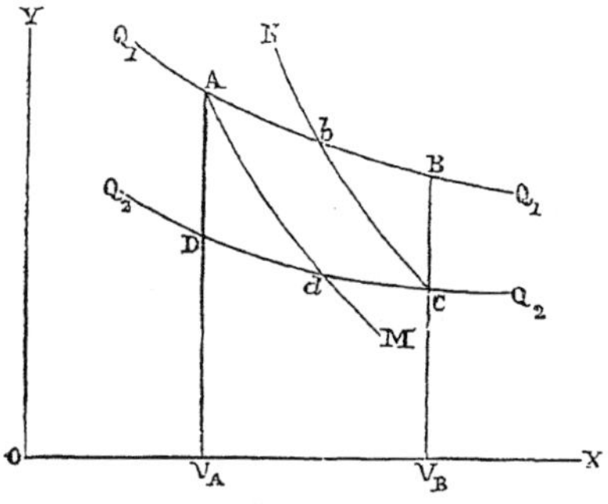

Fig. 16.

(Solution.) In Fig. 16, let $Q_1 Q_1$, $Q_2 Q_2$ be the isothermal curves denoting the limits of actual heat; V_A, V_B the limits of volume. Draw the ordinates V_A D A, V_B C B, intersecting the isothermal curves in the points A, B, C, D. Through A and C respectively draw the curves of no transmission, A M cutting $Q_2 Q_2$ in *d*, and C N cutting $Q_1 Q_1$ in *b*. Then will A *b* C *d* be the diagram required. An analogous construction would give the diagram of greatest efficiency when the variations of pressure and of actual heat are limited; as in the air-engine proposed by Mr. Joule.

26. *Of the use of the Economiser or Regenerator in Thermodynamic Engines.* As the actual heat of the elastic substance which works a thermodynamic engine requires to be alternately raised and lowered, it is obvious that unless these operations are performed entirely by compression and expansion, without reception or emission of heat (as in the case of maximum efficiency described in the first corollary of Proposition IX.), part, at least, of the heat emitted during the lowering of the actual heat may be stored up, by being communicated to some solid conducting substance, and used again by being communicated back to the elastic substance, when its actual heat is being raised. The apparatus used for

this purpose is called an economiser or regenerator, and was first invented about 1816, by the Rev. Robert Stirling. In the air-engine proposed by him, it consisted of a sheet-metal plunger surrounded by a wire grating or network; in that of Mr. James Stirling, it is composed of thin parallel plates of metal or glass through which the air passes longitudinally, and in the engine of Captain Ericsson of several sheets of wire gauze.

A regenerator may be regarded as consisting of an indefinite number of strata with which the elastic substance is successively brought into contact; each stratum serving to store up and give out the heat required to produce one particular indefinitely-small variation of the actual heat of the working substance.

A *perfect regenerator* is an ideal apparatus of this kind, in which the mass of material is so large, the surface exposed so extensive, and the conducting powers so great as to enable it to receive and emit heat instantaneously without there being any sensible difference of temperature between any part of the regenerator and the contiguous portion of the working substance; and from which no appreciable amount of heat is lost by conduction or radiation. In theoretical investigations it is convenient, in the first place, to determine the saving of heat effected by a perfect regenerator, and afterwards to make allowance for the losses arising from the non-fulfiment of the conditions of ideally perfect action; losses which, in the present imperfect state of our knowledge of the laws of the conduction of heat, can be ascertained by direct experiment only.*

27. PROPOSITION X.—PROBLEM. *The true indicator-diagram of any thermodynamic engine being given, to determine the amount of heat saved by a perfect regenerator.*

(Solution.) Let A B C D (in Fig. 17) be the given indicator-diagram. Across it draw any two indefinitely-close isothermal curves, $q_1 q_1$ intersecting it in a, b, and $q_2 q_2$ intersecting it in d, c. To the stripe between those two curves, speaking generally, a certain layer or stratum of the regenerator corresponds, which receives heat from the working substance during the change from b to c, and restores the same amount of heat during the change from d to a. The amount of heat economised by the layer in question is thus found. Through the four points a, b, c, d, draw the indefinitely-prolonged curves of no transmission, ak, bl, cm, dn; then the smaller of the two indefinitely-prolonged areas, $lbcm$, $kadn$, represents the heat saved by the layer of the regenerator corresponding to the indefinitely-narrow stripe between the isothermal curves $q_1 q_1$ and $q_2 q_2$.

* It is true that the problem of the waste of heat in the action of the regenerator is capable of a hypothetical solution by the methods of Fourier and Poisson; and I have by these methods obtained formulæ which are curious in a mathematical point of view; but owing to our ignorance of the absolute values and laws of variation of the coefficients of conductivity contained in these formulæ, they are incapable of being usefully applied; and I therefore for the present refrain from stating them.

Draw two curves of no transmission, B L, D N, touching the diagram; and through the points of contact, B and D, draw the isothermal curves,

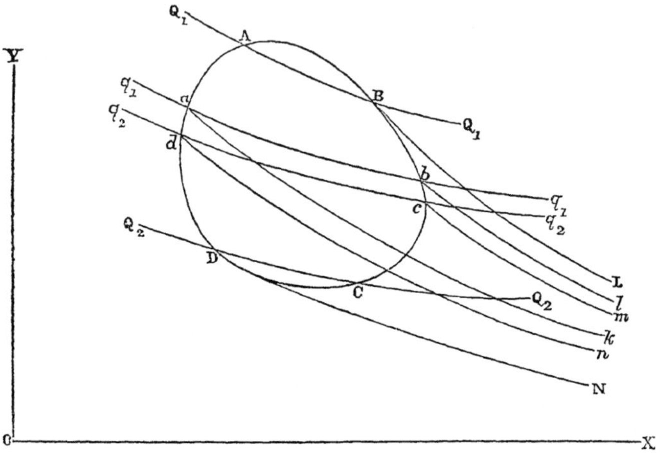

Fig. 17.

$Q_1\,Q_1$ cutting the diagram in A and B, and $Q_2\,Q_2$ cutting it in C and D. Then because, during the whole of the change from D through A to B, the working substance is receiving heat, and during the whole of the change from B through C to D, emitting heat, the regenerator can have no action above the isothermal curve $Q_1\,Q_1$, nor below the isothermal curve $Q_2\,Q_2$.

The whole of the diagram between these curves is to be divided by indefinitely-close isothermal curves into stripes like *abcd*; and the saving of heat effected by the layer of the regenerator corresponding to each stripe ascertained in the manner described, when the whole saving may be found by summation or integration.

The symbolical expression of this result is as follows: Let the points of contact, B, D, which limit the action of the regenerator, and the corresponding quantities of actual heat, Q_1, Q_2, be found, as in Proposition IX., by means of the equation $d\,\mathrm{F} = 0$.

Then, the saving of heat

$$= \int_{Q_2}^{Q_1} Q \frac{d\,.\,\mathrm{F}}{d\,Q} d\,Q = \int_{Q_2}^{Q_1} \left(\frac{\mathrm{K_v}}{k} + Q \frac{d\,\mathrm{P}}{d\,Q} \cdot \frac{d\,.\,\mathrm{V}}{d\,Q} \right) d\,Q, \quad (30.)$$

care being taken, when $\frac{d\,.\,\mathrm{F}}{d\,Q}$ has different values for the same value of Q, corresponding respectively to the two sides of the diagram, to choose the smaller in performing the integration.

28. *Corollary.*—It is evident that the regenerator acts most effectually when the outlines of the indicator-diagram from A to D, and from B to C, are portions of a pair of curves of *equal transmission* (determined as in Proposition IV.); for then, if the operation of the regenerator is perfect, the changes from B to C and from D to A will be effected without expenditure of heat; the heat transmitted from the working substance to a given stratum of the regenerator, during any part, such as bc, of the operation B C, being exactly sufficient for the corresponding part, da, of the operation D A. In this case $\frac{d \cdot F}{d Q}$ for each value of Q between Q_1 and Q_2, has the same value at either side of the diagram.

In fact, the effect of a perfect regenerator is, to confer upon any pair of curves of *equal transmission* the properties of a pair of curves of *no transmission.*

29. PROPOSITION XI.—THEOREM. *The greatest efficiency of a thermodynamic engine, working between given limits of actual heat, with a perfect regenerator, is equal to the greatest efficiency of a thermodynamic engine, working between the same limits of actual heat without a regenerator.*

(Demonstration.) In Fig. 18, let $Q_1 Q_1$, $Q_2 Q_2$ be the isothermal curves denoting the given limits of actual heat. Let A D, B C be a pair of curves of equal transmission of any form. Then by the aid of a perfect regenerator, the whole of the heat given out by the elastic substance during the operation B C may be stored up, and given out again to that substance in such a manner as to be exactly sufficient for the operation D A; so that the whole consumption of heat in one revolution by an engine whose

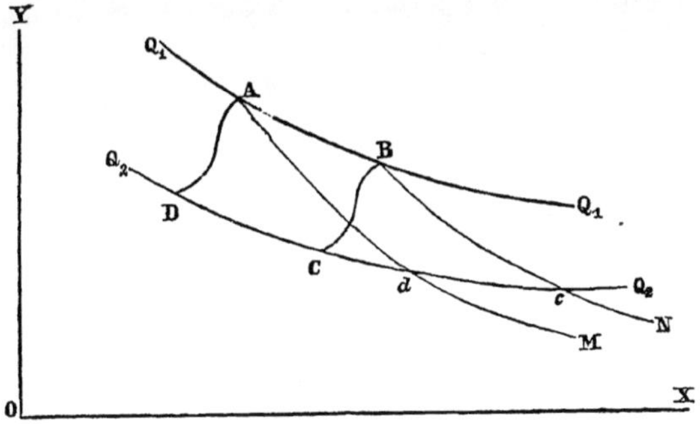

Fig. 18.

indicator-diagram is A B C D, may be reduced simply to the latent heat of expansion during the operation A B, which is represented by the indefinitely-prolonged area M A B N, A d M and B c N being curves of no transmission. The efficiency of such an engine is represented by

$$\frac{\text{the area A B C D}}{\text{the area M A B N}}.$$

Now the maximum efficiency of an engine without a regenerator, working between the same limits of actual heat, is represented by

$$\frac{\text{the area A B } c\, d}{\text{the area M A B N}} = \frac{Q_1 - Q_2}{Q_1};$$

and from the mode of construction of curves of equal transmission, described in Proposition IV., it is evident that

$$\text{the area A B C D} = \text{the area A B } c\, d;$$

hence the maximum efficiencies, working between the given limits of actual heat, Q_1 and Q_2, are equal, with or without a perfect regenerator. Q.E.D.

30. *Advantage of a Regenerator.*

It appears from this theorem that the advantage of a regenerator is, not to increase the maximum efficiency of a thermodynamic engine between given limits of actual heat, but to enable that amount of efficiency to be attained with a less amount of expansion, and, consequently, with a smaller engine.

Suppose, for instance, that to represent the isothermal curves, and the curves of no transmission, for a gaseous substance, we adopt the approximate equations already given in Article 20, viz. :—

$$\left.\begin{aligned} &\text{For the isothermal curve of Q, } P\,V = N\,Q; \\ &\text{For a curve of no transmission } \frac{V_2}{V_1} = \left(\frac{Q_2}{Q_1}\right)^{-\frac{1}{N}} = \left(\frac{P_2}{P_1}\right)^{\frac{-1}{1+N}}; \end{aligned}\right\} \quad (31.)$$

and let us compare the forms of the indicator-diagrams without and with a regenerator, for a perfect air-engine, working between given limits as to actual heat, defined by the isothermal curves $Q_1\,Q_1$, $Q_2\,Q_2$ in Fig. 19.

The amount of expansion at the higher limit of heat being arbitrary, let us suppose it to be from the volume V_A to the volume V_B, corresponding respectively to the points A and B, and to be the same in all cases, whether with or without a regenerator.

The engine being without a regenerator, the diagram corresponding to the maximum efficiency has but one form, viz., A B $c\,d$, where B c, A d are curves of no transmission. Hence, in this case, there must be an additional expansion, from the volume V_B to the volume

$$V_c = V_B \cdot \left(\frac{Q_1}{Q_2}\right)^{\frac{1}{N}}, \quad \cdot \quad \cdot \quad \cdot \quad (32.)$$

for the purpose merely of lowering the actual heat of the air without loss of heat; and the engine must be made large enough to admit of this expansion, otherwise heat will be wasted.

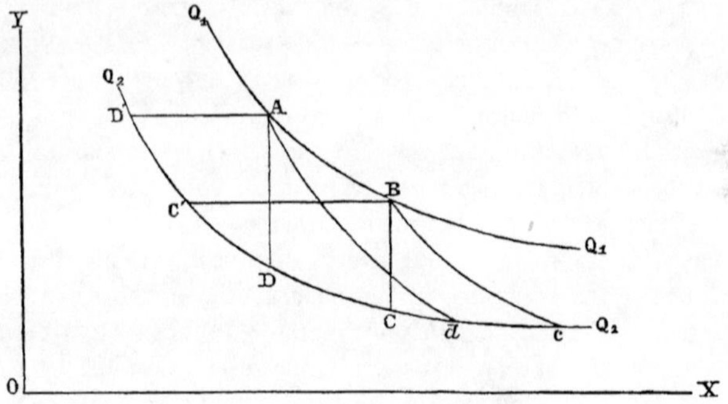

Fig. 19.

On the other hand, if the engine be provided with a perfect regenerator, any pair of curves of equal transmission passing through A and B will complete a diagram of maximum efficiency. The property of a pair of these curves being, as shown in Proposition IV., that the difference of their thermodynamic functions,

$$\Delta F \left(= \int \frac{dP}{dQ} dV, \text{ when Q is constant} \right),$$

is the same for every value of Q, it follows, that for a gas, according to the approximate equation (23), the property of a pair of curves of equal transmission is, that the volumes corresponding to the intersections of the two curves by the same isothermal curve, are in a ratio which is the same for every isothermal curve. Thus, let V_a, V_b be such a pair of volumes, then this equation

$$\frac{V_b}{V_a} = \frac{V_B}{V_A}, \quad \cdot \quad \cdot \quad \cdot \quad \cdot \quad \cdot \quad (33.)$$

defines a pair of curves of equal transmission. From this and from equation (31) it follows, that for such a pair of curves

$$\frac{P_b}{P_a} = \frac{P_B}{P_A}. \quad \cdot \quad \cdot \quad \cdot \quad \cdot \quad \cdot \quad (34.)$$

If one of the curves, or lines, of equal transmission is a straight line of equal volumes, that is, an ordinate A D parallel to O Y, then the other is an ordinate B C, parallel to O Y also. Then A B C D is the diagram of maximum efficiency for an air-engine with a perfect regenerator, when the air traverses the regenerator without alteration of volume; and by adopting this diagram, the additional expansion from V_B to V_C is dispensed with.

If one of the curves, or lines, of equal transmission is a straight line of equal pressures A D′ parallel to O X, then the other also is a straight line of equal pressures B C′. The diagram thus formed, A B C′ D′, is suitable, when the air, as in Ericsson's engine, has to traverse the regenerator without change of pressure.

It must be observed, that no finite mass, or extent of conducting surface, will enable a regenerator to act with the ideal perfection assumed in Propositions X. and XI., and their corollaries.

Owing to the want of a general investigation of the theory of the action of the regenerator based on true principles, those who have hitherto written respecting it have either exaggerated its advantages or unduly depreciated them. From this remark, however, must be excepted a calculation of the expenditure of heat in Captain Ericsson's engine, by Professor Barnard of the University of Alabama.*

31. *General Remarks on the preceding Propositions.*

The eleven preceding propositions, with their corollaries, are the geometrical representation of the theory of the mutual transformation of heat and motive power, by means of the changes of volume of a homogeneous elastic substance which does not change its condition. All these propositions are virtually comprehended in the first two, of which, perhaps, the most simple enunciations are the following :—

I. The mechanical equivalent of the heat absorbed or given out by a substance in passing from one given state as to pressure and volume to another given state, through a series of states represented by the co-ordinates of a given curve on a diagram of energy, is represented by the area included between the given curve and two curves of no transmission of heat drawn from its extremities, and indefinitely prolonged in the direction representing increase of volume.

II. If across any pair of curves of no transmission on a diagram of energy there be drawn any series of isothermal curves at intervals corresponding to equal differences of actual heat, the series of quadrilateral areas thus cut off from the space between the curves of no transmission will be all equal to each other.

These two propositions are the necessary consequences of the definitions of isothermal curves and curves of no transmission on a diagram of energy, and are the geometrical representation of the application to the particular

* *Silliman's Journal*, September, 1853.

case of heat and expansive power, of two axioms respecting Energy in the abstract, viz.:—

I. The sum of Energy in the Universe is unalterable.

II. The effect, in causing Transformation of Energy, of the whole quantity of Actual Energy present in a substance, is the sum of the effects of all its parts.

The application of these axioms to Heat and Expansive Power virtually involves the following definition of expansive heat:—

Expansive Heat is a species of Actual Energy, the presence of which in a substance affects, and in general increases, its tendency to expand.

And this definition, arrived at by induction from experiment and observation, is the foundation of the theory of the expansive action of heat.

Section IV.—Of Temperature, the Mechanical Hypothesis of Molecular Vortices, and the Numerical Computation of the Efficiency of Air-Engines.

32. In order to apply the propositions of the preceding articles to existing substances, besides experimental data sufficient for the determination, direct or indirect, of the isothermal curves and curves of no transmission, it is necessary also to know the relation, for the substance in question, between the quantity of heat actually present in it under any circumstances, and its *temperature;* a quantity measured by the product of the pressure, volume, and specific gravity of a mass of perfect gas, when in such a condition that it has no tendency to communicate heat to, or to abstract heat from, the substance whose temperature is ascertained.

The nature of the relation between heat and temperature has been discussed in investigations already published, as a consequence deducible from a hypothesis respecting the molecular constitution of matter, with the aid of data supplied by the experiments of Messrs. Thomson and Joule and of M. Regnault. Nevertheless, it seems to me desirable to add here a few words respecting the grounds, independent of direct experiment, for adopting the hypothesis of molecular vortices as a probable conjecture, the extent to which, by the aid of this hypothesis, the results of experiment were anticipated, and its use, in conjunction with the results of experiment, as a means of arriving at a knowledge of the true law of the relation between temperatures and total quantities of heat.

To introduce a hypothesis into the theory of a class of phenomena, is to suppose that class of phenomena to be, in some way not obvious to the senses, constituted of some other class of phenomena with whose laws we are more familiar. In thus framing a hypothesis, we are guided by some analogy between the laws of the two classes of phenomena: we conclude, from this analogy of laws, that the phenomena themselves are probably alike. This act of the mind is the converse of the process of ordinary physical reasoning; in which, perceiving that phenomena are alike, we conclude that their laws are analogous. The results, however, of the latter process of reasoning may be certainly true, while those of the former can never be more than probable; for how complete soever the analogy between the laws of two classes of phenomena may be, there will always remain a possibility of the phenomena themselves being unlike. A hypothesis, therefore, is incapable of absolute proof; but the agreement of its results with those of experiment may give it a high degree of probability.

The laws of the transmission of radiant heat are analogous to those of the propagation of a transverse oscillatory movement. The laws of thermometric heat are analogous to those of motion, inasmuch as both are convertible into mechanical effect; and motion, especially that of eddies in liquids and gases, is directly convertible into heat by friction. If, guided by these analogies, we assume as a probable hypothesis that heat consists in some kind of molecular motion, we must suppose that thermometric heat is such a molecular motion as will cause bodies to tend to expand; that is to say, a motion productive of centrifugal force. Thus we are led to the hypothesis of Molecular Vortices.

This hypothesis, besides the principles already enunciated, of the mutual transformation of heat and motive power in homogeneous substances, leads to the following special conclusion respecting the

RELATION BETWEEN TEMPERATURE AND ACTUAL HEAT:—

When the temperature of a substance, as measured by a perfect gas thermometer, rises by equal increments, the actual heat present in the substance rises also by equal increments—a principle expressed symbolically by the equation

$$Q = \mathfrak{k}\,(\tau - \kappa), \quad . \quad . \quad . \quad . \quad (35.)$$

where Q is the actual heat in unity of weight of a substance, τ its temperature, measured from the absolute zero of gaseous tension, κ the temperature of absolute cold, measured from the same point, and $\mathfrak{k}$ the real specific heat of the substance, expressed in terms of motive power.*

The enunciation of this law was originally an anticipation of the results of experiment; for when it appeared no experimental data existed by which its soundness could be tested.

Since then, however, one confirmation of this law has been afforded

* The hypothesis of Mayer amounts to supposing that $\kappa = 0$, or that the zero of gaseous tension coincides with the point of absolute cold.

by the experiments of M. Regnault, showing that the specific heat of atmospheric air is sensibly constant at all temperatures and at all densities throughout a very great range; and another, by the experiments of Messrs. Joule and Thomson, referred to in Proposition VI., on the thermic phenomena of gases rushing through small apertures, which not only verify the theoretical principle, but afford the means of computing approximately the position κ of the point of absolute cold on the thermometric scale.

According to this relation between temperature and heat, every isothermal curve on a diagram of energy is also a curve of equal temperature. The isothermal curve, for example, corresponding to a constant quantity of actual heat, Q, corresponds also to a constant absolute temperature,

$$\tau = \frac{Q}{\mathfrak{k}} + \kappa. \qquad (36.)$$

The curve of absolute cold is that of the absolute temperature κ.

Any series of isothermal curves at intervals corresponding to equal differences of heat, correspond to a series of equidistant temperatures.

Hence we deduce

PROPOSITION XII.—THEOREM. *Everything that has been predicated, in the propositions of the preceding articles, of the mutual proportions of quantities of actual heat and their differences, may be predicated also of the mutual proportions of temperatures as measured from the point of absolute cold, and their differences.*

The symbolical expression of this theorem is, that in all the equations of the preceding sections, we may make the following substitutions:—

$$\frac{Q_2}{Q_1} = \frac{\tau_2 - \kappa}{\tau_1 - \kappa}; \quad \frac{(\Delta, \delta, \text{ or } d)\, Q}{Q} = \frac{(\Delta, \delta, \text{ or } d)\, \tau}{\tau - \kappa} \quad . \qquad (36\text{A}.)$$

This theorem is not, like those which have preceded it, the consequence of a set of definitions. It is a law known by induction from experiment, aided by a hypothesis or conjecture, with the results of which those of experiment have been found to agree.

It is true that the theorem itself might have been stated in the form of a definition of degrees of temperature; but then induction from experiment would still have been required, to prove that temperature, as measured in the usual way, agrees with the definition.

By substituting symbols according to the above theorem, and making

$$\phi . Q = f . \tau,$$

the general equation of the expansive action of heat is made to take the following form:—

$$\Delta . \Psi = \Delta . \mathrm{H} - \int \mathrm{P}\, d\,\mathrm{V} = \Delta\,\mathrm{Q} + \Delta . \mathrm{S} = \mathrm{k} . \Delta \tau + \Delta f . \tau$$

$$+ \int \left\{ (\tau - \kappa) \frac{d\,\mathrm{P}}{d\,\tau} - \mathrm{P} \right\} . d\,\mathrm{V}, \quad . \quad . \quad (37.)$$

which agrees with the equation deduced directly from the hypothesis of molecular vortices, if we admit that

$$f . \tau = \mathrm{k}\,\mathrm{N}\,\kappa \left(\text{hyp. log. } \tau + \frac{\kappa}{\tau}\right)$$

and, consequently, $\quad . \quad (37\,\text{A.})$

$$f' . \tau = \mathrm{k}\,\mathrm{N} . \left(\frac{\kappa}{\tau} - \frac{\kappa^2}{\tau^2}\right)$$

The differential form of equation (37) is

$$d . \Psi = d . \mathrm{H} - \mathrm{P}\, d\,\mathrm{V} = d\,\mathrm{Q} + d . \mathrm{S} = \mathrm{K}_{\mathrm{v}}\, d\,\tau$$

$$+ \left\{ (\tau - \kappa) \frac{d\,\mathrm{P}}{d\,\tau} - \mathrm{P} \right\} d\,\mathrm{V}, \quad . \quad . \quad (38.)$$

where

$$\mathrm{K}_{\mathrm{v}} = \mathrm{k} + f' . \tau + (\tau - \kappa) \int \frac{d^2\,\mathrm{P}}{d\,\tau^2} . d\,\mathrm{V}.$$

The expression for the thermodynamic function denoted by F takes the form

$$\mathrm{F} = \int \frac{1 + \frac{1}{\mathrm{k}} . f' . \tau}{\tau - \kappa} . d\,\tau + \frac{1}{\mathrm{k}} \int \frac{d\,\mathrm{P}}{d\,\tau} . d\,\mathrm{V}; \quad . \quad (39.)$$

but a more convenient thermodynamic function, bearing the same relation to temperature as reckoned from the point of absolute cold, which the function F does to actual heat, is formed by multiplying the latter by the real specific heat k, thus:—

$$\Phi = \mathrm{k}\,\mathrm{F} = \int \frac{\mathrm{k} + f' . \tau}{\tau - \kappa} d\,\tau + \int \frac{d\,\mathrm{P}}{d\,\tau} d\,\mathrm{V}, \quad . \quad (40.)$$

which, being introduced into the general equation, transforms it to

$$\Delta . \Psi = \int (\tau - \kappa)\, d\,\Phi - \int \mathrm{P}\, d\,\mathrm{V}. \quad . \quad . \quad (40\,\text{A.})$$

33. *Of the Numerical Computation of the Efficiency of Air-Engines, with or without a perfect Regenerator.*

The relation between temperature and heat being known, the preceding propositions can be applied to determine the efficiency, and other circumstances relative to the working of thermodynamic engines. To exemplify this application of the theory, let the substance working the engine be atmospheric air, and let the real indicator-diagram be such as to develop the maximum efficiency between two given absolute temperatures τ_1 and τ_2, being a quadrilateral, as in Fig. 19, of which two sides are portions of the isothermal curves of those temperatures, and the other two portions of a pair of curves of equal transmission, of such a form as may be best suited to the easy working of the engine. Should these curves be curves of no transmission, a regenerator may be dispensed with. In every other case a regenerator is necessary, to prevent waste of heat; and for the present its action will be assumed to be perfect, as the loss which occurs from its imperfect action cannot be ascertained except by direct experiment.

In this investigation it is unnecessary to use formulæ of minute accuracy; and for practical purposes those will be found sufficient which treat air as a perfect gas, whose thermometric zero of pressure coincides with the point of absolute cold, viz.—

$$\left.\begin{array}{l} 272\tfrac{1}{2}^\circ \text{ Centigrade, or} \\ 490\tfrac{1}{2}^\circ \text{ Fahrenheit,} \end{array}\right\} \text{below melting ice;}^*$$

whose real specific heat is equal to its specific heat at constant volume, being

$$\mathfrak{k} = \mathrm{K_v} = \begin{cases} 234\cdot6 \text{ feet of fall per Centigrade degree, or} \\ 130\cdot3 \text{ feet of fall per degree of Fahrenheit;} \end{cases}$$

whose specific heat at constant pressure (as determined by M. Regnault) is $0\cdot238 \times$ the specific heat of liquid water; or

$$\mathrm{K_P} = \begin{cases} 330\cdot8 \text{ feet of fall per Centigrade degree, or} \\ 183\cdot8 \text{ feet of fall per degree of Fahrenheit;} \end{cases}$$

the ratio of these two quantities being

$$\frac{\mathrm{K_P}}{\mathrm{K_v}} = 1 + \mathrm{N} = 1\cdot41,$$

as calculated from the velocity of sound.

* This estimate of the position of the point of absolute cold is to be considered as merely approximate, recent experiments and calculations having shown that it may possibly be too high by about $1\tfrac{1}{2}^\circ$ Centigrade. It is, however, sufficiently correct for all practical purposes.

The volume occupied by an avoirdupois pound of air, at the temperature of melting ice, under the pressure of one pound on the square foot, as calculated from the experiments of M. Regnault, is

$$P_0 V_0 = 26214{\cdot}4 \text{ cubic feet.}$$

This represents also the length in feet of a column of air of uniform density and sectional area, whose weight is equal to its elastic pressure on the area of its section at the temperature of melting ice.

It will be found convenient, in expressing the temperature, as measured from the point of absolute cold, to make the following substitution :—

$$\tau - \kappa = T + T_0, \quad . \quad . \quad . \quad . \quad (41.)$$

where T represents the temperature as measured on the ordinary scale from the temperature of melting ice, and T_0 the height of the temperature of melting ice above the point of absolute cold, as already stated.

Then we have

$$N \mathfrak{k} = \frac{P_0 V_0}{T_0}. \quad . \quad . \quad . \quad . \quad (41\text{ A.})$$

According to these data, the equation of the isothermal curve of air for any temperature T is

$$P V = P_0 V_0 . \frac{T + T_0}{T_0} = N \mathfrak{k} (T + T_0). \quad . \quad (42.)$$

The thermodynamic functions are—

For quantities of actual heat,

$$F = \text{hyp. log. } Q + N \text{ hyp. log. } V;$$

For temperatures,

$$\Phi = \mathfrak{k} F + \text{constant} = K_v \{\text{hyp. log. } (T + T_0) + N \text{ hyp. log. } V\}$$

$$= K_v \text{ hyp. log. } (T + T_0) + \frac{P_0 V_0}{T_0} . \text{ hyp. log. } V; \quad (42\text{ A.})$$

consequently, the equation of any curve of no transmission is

$$\Phi = \text{constant};$$

otherwise

$$(T + T_0) . V^N = \text{const.}; \text{ or } P . V^{1+N} = \text{const.};$$

or

$$(T + T_0) . P^{\frac{N}{1+N}} = \text{constant}; \quad (43.)$$

in which

$$N = 0{\cdot}41,\ 1 + N = 1{\cdot}41,\ \frac{N}{1+N} = 0{\cdot}2908.$$

The maximum possible efficiency between any two temperatures T_1 and T_2 is given by the universal formula,

$$\frac{E}{H_1} = \frac{H_1 - H_2}{H_1} = \frac{T_1 - T_2}{T_1 + T_0}. \quad . \quad . \quad . \quad (44.)$$

The latent heat of expansion of unity of weight of air at a given constant temperature T_1, from the volume V_A to the volume V_B, is sensibly equivalent simply to the expansive power developed, being given by the following formula :—

$$H_1 = (T_1 + T_0) \cdot (\Phi_B - \Phi_A) = P_0 V_0 \cdot \frac{T_1 + T_0}{T_0} \cdot \text{hyp. log.} \frac{V_B}{V_A}$$

$$= \int_{V_A}^{V_B} P d V. \quad . \quad . \quad . \quad . \quad (45.)$$

Let V_a and V_b be the volumes corresponding to the points at which any isothermal curve intersects a given pair of curves of no transmission, or of equal transmission; then the ratio of these volumes,

$$\frac{V_b}{V_a}, \quad . \quad . \quad . \quad (46.)$$

is constant for every such pair of points on the given pair of curves; because the difference of the thermodynamic functions, which is proportional to the logarithm of this ratio, is constant.

Hence, if in Fig. 19 A, two isothermal curves, $T_1 T_1$, $T_2 T_2$, be the upper and lower boundaries of an indicator-diagram of maximum energy for an air-engine, A a D an arbitrary curve bounding the diagram at one side, and B the other limit of the expansion at the higher temperature; the fourth boundary, being a curve of equal transmission to A a D, may be described by this construction; draw any isothermal curve $t\,t$ cutting A a D in a, and make

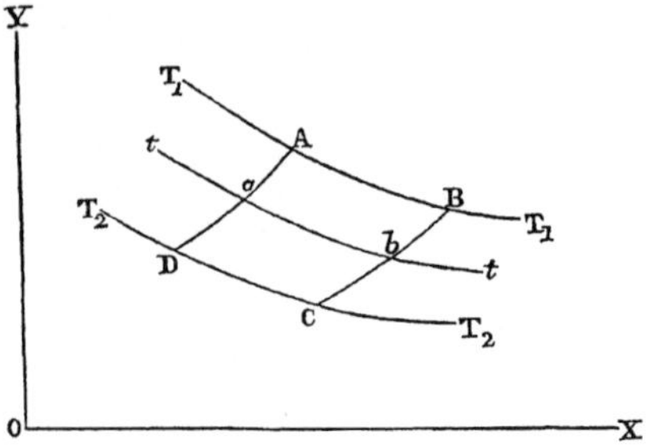

Fig. 19 A.

$$V_A : V_B :: V_a : V_b, \quad . \quad . \quad . \quad (47.)$$

then will b be a point in the curve sought, B b C.

Suppose, for example, that the form assumed for A a D is a hyperbola, concave towards O Y, and having the following equation:—

$$P_a = \frac{a}{\beta - V_a}, \quad . \quad . \quad . \quad (47\text{ A.})$$

in which a and β are two arbitrary constants; and let the ratio $\frac{V_B}{V_A} = r$.

Then must the curve B b C be another hyperbola concave towards O Y, having for its equation

$$P_b = \frac{a}{r\beta - V_b}. \quad . \quad . \quad . \quad (47\text{ B.})$$

The total expenditure of heat, per pound of air per stroke, in a perfect air-engine, is the latent heat of expansion from V_A to V_B, given by equation (45).

The heat to be abstracted by refrigeration is the latent heat of compression from V_C to V_D; and is found by substituting in the same equation the lower temperature T_2 for the higher temperature T_1.

The indicated work per pound of air per stroke, being the difference between those two quantities, is found by multiplying the range of temperature by the difference of the thermodynamic functions Φ for the curves A D, B C, or by multiplying the latent heat of expansion by the efficiency, and has the following value:—

$$E = H_1 - H_2 = (T_1 - T_2) \cdot (\Phi_B - \Phi_A)$$

$$= P_0 V_0 \cdot \frac{T_1 - T_2}{T_0} \cdot \text{hyp. log.} \frac{V_B}{V_A}. \quad . \quad . \quad (48.)$$

The heat alternately stored up and given out by the regenerator (supposing it to work perfectly), is to be computed as follows:—Let the arbitrary manner in which volume is made to vary with temperature, on either of the curves D a A, C b B, be expressed by an equation

$$V = \Psi \cdot T,$$

then the thermodynamic function Φ takes the form

$$\Phi = K_V \text{ hyp. log.} (T + T_0) + \frac{P_0 V_0}{T_0} \text{hyp. log.} \Psi \cdot T;$$

and the total heat stored up and given out per pound of air per stroke, is

$$\int_{T_2}^{T_1} (T + T_0) \frac{d\,.\,\Phi}{d\,T}\, d\,T = K_v (T_1 - T_2)$$

$$+ \frac{P_0 V_0}{T_0} \int_{T_2}^{T_1} \frac{(T + T_0)\, \Psi' . T}{\Psi . T}\, d\,T. \quad . \quad . \quad (49.)$$

For example, if, as before,

$$P_a = \frac{a}{\beta - V_a}$$

be the equation of the curve D A, then

$$V_a = \frac{\beta (T + T_0)}{T + T_0 \left(1 + \frac{a}{P_0 V_0}\right)},$$

and the heat stored up per pound of air per stroke, is

$$K_v (T_1 - T_2) + a\,.\,\text{hyp. log.} \left\{ \frac{T_1 + T_0 \left(1 + \frac{a}{P_0 V_0}\right)}{T_2 + T_0 \left(1 + \frac{a}{P_0 V_0}\right)} \right\} \quad (49\text{ A.})$$

33 A. *Numerical Examples.*

To illustrate the use of these formulæ, let us take the following example :—

Temperature of receiving heat,

$$T_1 = 343^\circ{\cdot}3 \text{ Centigrade.}$$
$$T_1 + T_0 = 615^\circ{\cdot}8 \text{ Centigrade.}$$

Temperature of emitting heat,

$$T_2 = 35^\circ{\cdot}4 \text{ Centigrade.}$$
$$T_2 + T_0 = 307^\circ{\cdot}9 \text{ Centigrade.}$$

Ratio of Effective Expansion,

$$\frac{V_B}{V_A} = \frac{V_C}{V_D} = \frac{P_A}{P_B} = \frac{P_D}{P_C} = \frac{3}{2}.$$

From these data are computed the following results :—

Maximum Efficiency,—

$$\frac{307^{\circ}\cdot 9}{615^{\circ}\cdot 8} = \frac{1}{2}.$$

Heat expended, or latent heat of expansion,—

$$H_1 = P_0 V_0 \times \frac{615\cdot 8}{272\cdot 5} \times \text{hyp. log.} \frac{3}{2}$$

= 24020 foot-pounds per pound of working air per stroke.

Heat abstracted by refrigeration,—

$$H_2 = P_0 V_0 \times \frac{307\cdot 9}{272\cdot 5} \times \text{hyp. log.} \frac{3}{2}$$

= 12010 foot-pounds per pound of working air per stroke.

Work performed,—

$$H_1 - H_2 = P_0 V_0 \times \frac{307\cdot 9}{272\cdot 5} \times \text{hyp. log} \frac{3}{2}$$

= 12010 foot-pounds per pound of working air per stroke.

To exemplify the computation of the heat stored by a perfect regenerator, let it be supposed, in the first place, that the indicator-diagram resembles A B C′ D′ in Fig. 19, where the curves of equal transmission are represented by a pair of lines of constant pressure. Then the heat to be stored is

$K_P (T_1 - T_2) = 101{,}800$ foot-pounds per pound of working air per stroke.

Secondly, let the diagram resemble A B C D in Fig. 19, where the curves of equal transmission are represented by a pair of lines of constant volume. Then the heat to be stored is

$K_V (T_1 - T_2) = 72{,}233$ foot-pounds per pound of working air per stroke.

Thirdly, let the curves of equal transmission, as in a recent example, be hyperbolas, concave towards O Y, and let the arbitrary constant *a* have the following value,—

$$a = P_0 V_0 = 26214\cdot 4 \text{ foot-pounds};$$

then the heat to be stored, according to equation (49 A), is

$$72{,}233 + 26214\cdot 4 \times \text{hyp. log.} \frac{888\cdot 3}{580\cdot 4} = 72{,}233 + 11{,}157$$

= 83,390 foot-pounds per pound of working air per stroke.

The large proportions borne by these quantities to the whole heat expended, show the importance of efficient action in the regenerator to economy of fuel. The quantity of heat to be stored, however, becomes smaller, as the curves of equal transmission approach those of no transmission, for which it is null. The additional expansion requisite in this last case is found by the following computation:—

$$\frac{V_C}{V_B} = \frac{V_D}{V_A} = \left(\frac{T_1 + T_0}{T_2 + T_0}\right)^{\frac{1}{N}} = 2^{\frac{1}{0 \cdot 41}} = 5 \cdot 423,$$

the result of which shows the great additional bulk of engine required, in order to obtain the maximum efficiency without a regenerator.

Supposing one pound of coal, by its combustion, to be capable of communicating heat to the air working in an engine corresponding with the above example, to an amount equivalent to

6,000,000 foot-pounds

(an amount which would evaporate about 7 lbs. of water), the maximum theoretical duty of one pound of such coal in such an engine, without waste of heat or power, would be

3,000,000 foot-pounds,

corresponding to

$\frac{3,000,000}{12,010} = 249$ strokes of a pound of working air, with the effective expansion $\frac{3}{2}$.

The deductions to be made from this result in practice must, of course, be determined by experience.

Section V.—Propositions Relative to a Heterogeneous Mass, or Aggregate, especially in Vapour-Engines.

34. The heterogeneous mass to which the present investigation refers, is to be understood to mean an aggregate of portions of different ingredients, in which each ingredient occupies a space, or a number of spaces, of sensible magnitude.

The results arrived at are not applicable to mixtures in which there is a complete mutual diffusion of the molecules of the ingredients, so that every space of appreciable magnitude contains every ingredient in a fixed proportion. A mixture of this kind, when the relations between its pressure, volume, heat, and temperature are known, may be treated, so far as regards the expansive action of heat, as a homogeneous substance.

The ingredients of an aggregate are heterogeneous with respect to the expansive action of heat, when either their specific heats, or their volumes for unity of weight at a given pressure and temperature, or both these classes of quantities, are different.

Hence a portion of a liquid, and a portion of its vapour, enclosed in the same vessel, though chemically identical and mutually transformable, are heterogeneous, and are to be treated as an aggregate, with respect to the expansive action of heat.

M. Clausius and Professor William Thomson have applied their formulæ to the aggregate composed of a liquid and its vapour, and have pointed out certain relations which must exist between the pressure and density of a liquid and its vapour, and the latent heat of evaporation.

I shall now apply the geometrical method of this paper to the theory of the expansive action of heat in an aggregate, especially that consisting of a liquid and its vapour. The total volumes are, for the present, supposed not to be large enough to exhibit any appreciable differences of pressure due to gravitation.

35. Proposition XIII.—Theorem. *In an aggregate in equilibrio, the pressure of each ingredient must be the same; and the quantity of heat in unity of weight of each ingredient must be inversely proportional to its real specific heat; that is to say, the temperature must be equal.*

The following is the symbolical expression of this theorem, with certain conclusions to which it leads:—

Let $\tau - \kappa$ be the common temperature of the ingredients, as measured from the point of absolute cold;

P, their common pressure;

n_1, n_2, n_3, &c., their proportions by weight, in unity of weight of the aggregate;

v_1, v_2, v_3, &c., the respective volumes of unity of weight of the several ingredients.

V, the volume of unity of weight of the aggregate;

q_1, q_2, q_3, &c., the respective quantities of actual heat in unity of weight of the several ingredients;

$\mathfrak{k}_1, \mathfrak{k}_2, \mathfrak{k}_3$, &c., their respective real specific heats;

Q, the quantity of heat in unity of weight of the aggregate;

Φ, a thermodynamic function for the aggregate.

Then these quantities are connected by the following equations:—

$$\Sigma \,.\, n = 1. \qquad (50.)$$

$$V = \Sigma \,.\, n v. \qquad (51.)$$

$$\tau - \kappa = \frac{q_1}{\mathfrak{k}_1} = \frac{q_2}{\mathfrak{k}_2} = \frac{q_3}{\mathfrak{k}_3} = \&c. \qquad (52.)$$

$$Q = \Sigma \,.\, n q = (\tau - \kappa) \,.\, \Sigma \,.\, n \mathfrak{k}. \qquad (53.)$$

$$\Phi = \int \frac{\Sigma \,.\, n (\mathfrak{k} + f' \,.\, \tau)}{\tau - \kappa} \,.\, d\tau + \int \frac{dP}{d\tau} \,.\, dV. \qquad (54.)$$

It is evident that all these equations hold, whether the proportions of the ingredients n_1, &c., are constant, as in an aggregate of chemically distinct substances, or variable, as in the aggregate of a liquid and its vapour.

Let δH be the heat which disappears in consequence of a small expansion of aggregate at constant temperature, represented by

$$\delta V = \Sigma \,.\, \delta u, \qquad (55.)$$

δu representing any one of the parts arising from the changes undergone by the different ingredients, of which the whole expansion of the aggregate δV is made up.

Then

$$\delta H = \Sigma \left\{ (\tau - \kappa) \frac{dP}{d\tau} \,.\, \delta u \right\}; \qquad (56.)$$

but the pressure P is the same for every ingredient, as well as the temperature; therefore, the factor $(\tau - \kappa) \frac{dP}{d\tau}$ is the same for every ingredient, and, consequently, for the whole aggregate; that is to say,

$$\delta H = (\tau - \kappa) \frac{dP}{d\tau} \,.\, \delta V = (\tau - \kappa)\, \delta \Phi. \qquad (57.)$$

This equation shows that the relation of *temperature* to the mutual transformation of heat and expansive power is the same in an aggregate as in a homogeneous substance.

Consequently, if we define *isothermal curves* for an aggregate to be *curves of constant temperature*, we arrive at the following conclusion:—

PROPOSITION XIV.—THEOREM. *Isothermal curves on the diagram of energy of an aggregate have the same properties, with reference to the mutual transformation of heat and expansive power, with those on the diagram of energy of a homogeneous substance.*

It is unnecessary to enunciate separately a similar proposition for curves of no transmission; for the demonstration of Proposition I., on which all their properties depend, is evidently applicable to an aggregate, constituted in any manner.

Hence it appears, that if the isothermal curves for an aggregate be drawn according to the above definition, all the propositions proved in this paper respecting homogeneous substances become true of the aggregate.

36. PROPOSITION XV.—THEOREM. *Every isothermal line for an aggregate of a liquid and its vapour, is a straight line of equal pressure, from the volume corresponding to complete liquefaction to the volume corresponding to complete evaporation.*

This is a fact known by experiment. The theorem is equivalent to a statement, that the pressure of a liquid and its vapour in contact with each other is a function of the temperature only.

Corollary.—THEOREM. *At any given temperature, the volume of an aggregate of liquid and vapour is arbitrary between and up to the limits of total liquefaction and total evaporation.*

To express this symbolically, let P be the pressure of an aggregate of liquid and vapour corresponding to the absolute temperature τ; and unity of weight being the quantity of the aggregate under consideration, let v be the volume corresponding to complete liquefaction, v' that corresponding to complete evaporation, and V the actual volume at any time; let n be the proportion of liquid, and $1 - n$ that of vapour, corresponding to the aggregate volume V; then

$$V = n v + (1 - n) v', \quad . \quad . \quad . \quad (58.)$$

and V may have any value not less than v nor greater than v', while P and τ remain constant; the proportion of liquid, n, being regulated according to the foregoing equation.

37. PROPOSITION XVI.—PROBLEM. *The density of a liquid and of its vapour, when in contact at a given temperature, being given, and the isothermal lines of the aggregate, it is required to determine the latent heat of evaporation of unity of weight of the fluid.*

(Solution.) The densities of the liquid and of its vapour are respectively the reciprocals of the volumes of total liquefaction and total evaporation of unity of weight, above-mentioned. In Fig. 20, let the abscissæ Ov, Ov' represent these volumes, and the equal ordinates, vA, v'B, the pressure corresponding to the given temperature; so that A B, parallel to O X, is the isothermal line of the aggregate for that temperature. Suppose two curves of no transmission AM, BN, to be drawn from A and B respectively, and indefinitely prolonged towards X; then the indefinitely-prolonged area M A B N represents the mechanical equivalent of the latent heat sought, and this area is to be computed in the following manner:—Draw a second isothermal line, $a\,b$, indefinitely near to A B, at an interval

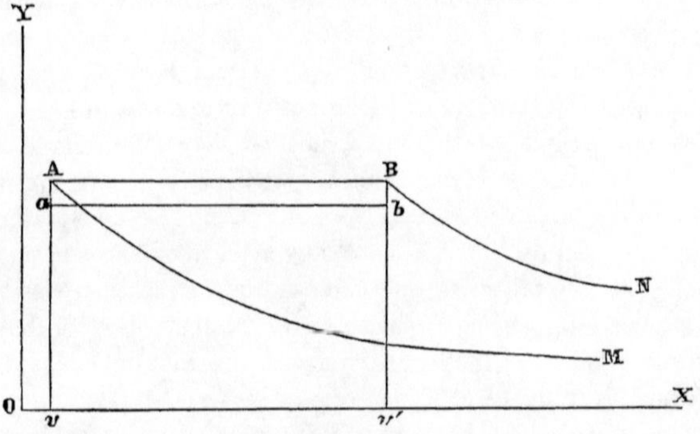

Fig. 20.

corresponding to the indefinitely-small difference of temperature $d\tau$; then, ultimately,

$$d\tau : \tau - \kappa : : \text{area ABba} : \text{area MABN};$$

or, symbolically,

$$\text{L} = \text{latent heat of evaporation} = (\tau - \kappa)\frac{d\,\text{P}}{d\,\tau}(v' - v). \quad (59.)$$

This is simply the application of Propositions I. and II. to the aggregate of a liquid and its vapour, *mutatis mutandis*.

(*Remarks.*)—The existence of a necessary relation between the density, pressure, and temperature of a vapour and its liquid in contact, and the latent heat of evaporation, was first shown by Carnôt. If for $\tau - \kappa$ in the preceding equation be substituted, according to Professor Thomson's notation, $\frac{\text{J}}{\mu}$, J being "Joule's equivalent" and μ "Carnôt's function," the equation is transformed into that deduced by Messrs. Clausius and Thomson from the combination of Carnôt's theory with the law of the mechanical convertibility of heat.

38. *Corollary.*—The volume occupied by unity of weight of vapour at saturation may be computed from its latent heat of evaporation by means of the inverse formula,—

$$v' - v = \frac{\text{L}}{(\tau - \kappa)\frac{d\,\text{P}}{d\,\tau}}; \quad . \quad . \quad . \quad (60.)$$

the latent heat, L, being, of course, always stated in units of motive power.

The want of satisfactory experiments on the density of vapours of any kind, has hitherto prevented the use of the direct formula (59).

It is otherwise, however, with the inverse formula (60), at all events in the case of steam; for, so far as we are yet able to judge, the experiments of M. Regnault have determined the latent heat of evaporation of water with accuracy throughout a long range of temperature.

M. Clausius, applying to those experimental data a formula founded on the supposition of Mayer (that is to say, similar to the above, with the exception that κ is supposed = 0), has calculated the densities of steam at certain temperatures, so as to show how much they exceed the densities calculated from the pressures and temperatures on the supposition that steam is a perfect gas. From these calculations he concludes, that either the supposition of Mayer is erroneous, or steam deviates very much at high densities from the condition of a perfect gas.

In the following table, the value of κ is supposed to be 2°·1 Centigrade, and use has been made of the formula for calculating the pressure of steam and other vapours at saturation, first published in the *Edinburgh New Philosophical Journal* for July, 1849 (*See p. 1*), viz.—

$$\log. \mathrm{P} = \alpha - \frac{\beta}{\tau} - \frac{\gamma}{\tau^2}. \qquad . \quad . \quad . \quad (61.)$$

This table exhibits, side by side, the volume in cubic feet occupied by one pound avoirdupois of steam at every twentieth Centigrade degree, from − 20° to + 260° (that is, from −4° to + 500° Fahrenheit)—first, as extracted from a table for computing the power of steam-engines, in the *Transactions of the Royal Society of Edinburgh*, Vol. XX. (*See p. 282*), which was calculated on the supposition that steam is a perfect gas; and, secondly, as computed by equation (60) from the latent heat of steam as determined by M. Regnault. The excess of the former quantity above the latter is also given in each case, with its ratio to the second value of the volume.

For convenience sake, a column is added containing the pressures of steam corresponding to the temperatures in the table in pounds per square foot.

The fourth column of this table could easily be extended and filled up, so as to replace the column of volumes of steam for every fifth Centigrade degree in the table previously published; but it would be unadvisable to do so at present, for the following reasons:—

First, the value of the constant κ is still uncertain.*

Secondly, the results of M. Regnault's direct experiments, on the density of steam and other vapours, may soon be expected to appear.

* It is probable that κ may be found to be inappreciably small; in which case the numbers in column (4) will have to be diminished to an extent varying from $\frac{1}{150}$ to $\frac{1}{350}$ of their amount.

TABLE OF COMPUTED VOLUMES OF 1 LB. AVOIRDUPOIS OF STEAM.

Temperature.		Volume supposed a Perfect Gas.	Volume computed from Latent Heat.	Difference.	Ratio of Difference to lesser Value of Volume.	Pressure.
Fahrenheit.	Centigrade.					
Deg.	Deg.	Cubic feet.	Cubic feet.	Cubic feet.		lb. per square foot.
− 4	−20	15757	15718	39	0·0025	2·4799
+32	0	3390·4	3377·2	13·2	0·0039	12·431
68	+20	936·81	931·50	2·31	0·0025	48·265
104	40	314·88	313·56	1·32	0·0042	153·34
140	60	123·65	122·63	1·02	0·0083	415·33
176	80	55·05	54·19	0·86	0·0158	988·67
212	100	27·166	26·478	0·688	0·0260	2116·4
248	120	14·596	14·076	0·520	0·0369	4149·3
284	140	8·420	8·004	0·416	0·0502	7557·0
320	160	5·158	4·838	0·320	0·0661	12931
356	180	3·326	3·071	0·255	0·0830	20979
392	200	2·241	2·033	0·208	0·1023	32512
428	220	1·568	1·396	0·172	0·1232	48425
464	240	1·134	0·990	0·144	0·1455	69680
500	260	0·843	0·722	0·121	0·1676	97275
Col. (1.)	(2.)	(3.)	(4.)	(5.)	(6.)	(7.)

Thirdly, it is possible that the values of the latent heat of evaporation of water, as deduced from M. Regnault's experiments, may still have to undergo some correction; because, according to the theoretical definition of the latent heat of evaporation, the liquid is supposed to be under the pressure of an atmosphere of its own vapour, which atmosphere, as it increases in bulk, performs work of some kind, such as lifting a piston; whereas, in M. Regnault's experiments, the water is pressed by an atmosphere of mingled steam and air, whose united pressure is that corresponding to the temperature of internal ebullition of the water; so that the pressure of the steam alone on the surface of the water, which regulates the superficial evaporation, may be less than the maximum pressure corresponding to the temperature of ebullition; and this steam, moreover, has no mechanical work to perform, except to propel itself along the passage leading to the calorimeter, and to agitate the water in the latter vessel. Under these circumstances, it is possible, though by no means certain, that the latent heat of evaporation of water, as deduced from M. Regnault's experiments, may be somewhat smaller than that which corresponds to the theoretical definition, especially at high pressures; and a doubt arises as to the precise applicability of the formulæ (59) and (60) to those experimental results, which cannot be solved except by direct experiments on the density of steam.

Notwithstanding this doubt, however, the preceding table must be regarded as adding a reason to those already known, for believing that saturated steam of high density deviates considerably from the laws of the perfectly gaseous condition.*

39. Proposition XVII.—Problem. *The isothermal lines for a liquid and its vapour, and the apparent specific heat of the liquid at all temperatures being given, and the expansion of the liquid by heat being treated as inappreciably small: to determine a curve of no transmission for the aggregate, passing through a given point on the ordinate whose distance from the origin approximately represents the volume of the liquid.*

(Solution.) In Fig. 21, let O v represent the volume of the liquid assumed to be approximately constant for all temperatures under consideration; let v A be an ordinate parallel to O Y, and let the heat

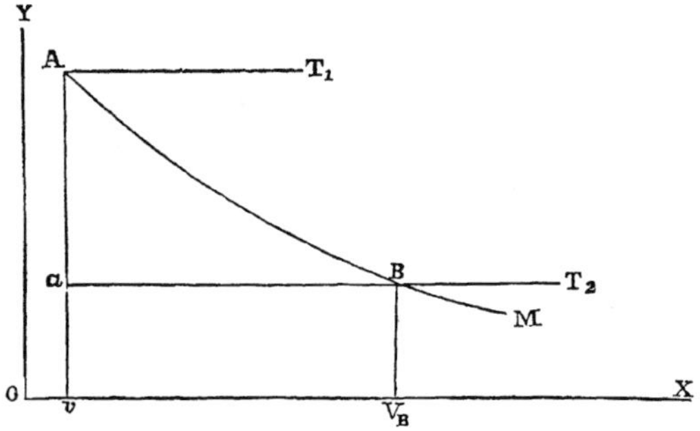

Fig. 21.

consumed by the liquid in passing from the temperature corresponding to any point on this ordinate to that corresponding to any other point, be known; let the isothermal lines for the aggregate of liquid and vapour, all of which are straight lines of equal pressure parallel to O X, such as A T_1, aBT_2, be known. Then to draw a curve of no transmission through any point A on the ordinate vA, the same process must be followed as in Proposition VIII.

To apply to this case the symbolical representation of Proposition VIII., viz., equation (21), let τ_1 be the absolute temperature corresponding to the point A (that is, to the isothermal line A T_1); τ_2 that corresponding to any lower isothermal line a B T_2; V_B the volume of the aggregate of liquid and vapour corresponding to the point B, where the curve sought, A M, intersects the latter isothermal line; K_L the apparent specific heat

* Evidence in favour of this opinion is afforded by the experiments recorded by Mr. C. W. Siemens (*Civil Engineer and Architect's Journal*). A remarkable cause, however, of uncertainty in all such experiments has lately been investigated by Professor Magnus (Poggendorff's *Annalen*, 1853, No. 8), viz., a power which solid bodies have of condensing, by attraction on their surfaces, appreciable quantities of gases.

of the liquid; then, making the proper substitutions of the symbols of temperature for those of heat, and observing that the operation

$$\int_{V_A}^{V_B} dV$$

is in this case equivalent to multiplication by $V_B - v$, we have

$$\Delta \Phi = \frac{dP}{d\tau}(V_B - v) \text{ (for } \tau = \tau_2) - \int_{\tau_2}^{\tau_1} \frac{K_L}{\tau - \kappa} d\tau, \quad . \quad (62.)$$

being an equation between two expressions for the difference between the thermodynamic functions Φ for the curve A B, and for that which passes through a.

If the specific heat of the liquid is approximately constant, this equation becomes

$$\Delta \Phi = \frac{dP}{d\tau}(V_B - v) \text{ (for } \tau = \tau_2) = K_L \text{ hyp. log. } \frac{\tau_1 - \kappa}{\tau_2 - \kappa}. \quad (63.)$$

40. *Corollary.*—PROBLEM. *The same data being given as in the preceding problem, and the expansion of the liquid by heat neglected, a mass of liquid, having been raised from the absolute temperature τ_2 to the absolute temperature τ_1, is supposed to be allowed to evaporate partially, under pressure, without receiving or emitting heat, until its temperature falls again to τ_2, at which temperature it is liquefied under constant pressure by refrigeration: it is required to find the power developed.*

(Solution.) The power developed is represented by the area of the three-sided diagram of energy in Fig. 21, A B a; that is to say, by

$$\int_{\tau_2}^{\tau_1} (V - v) \frac{dP}{d\tau} \cdot d\tau = \iint_{\tau_2}^{\tau_1} \frac{K_L}{\tau - \kappa} d\tau^2, \quad . \quad . \quad (64.)$$

which, if K_L is nearly constant, becomes

$$K_L \int_{\tau_2}^{\tau_1} \text{hyp. log. } \frac{\tau_1 - \kappa}{\tau - \kappa} \cdot d\tau = K_L \left\{ (\tau_1 - \kappa) - (\tau_2 - \kappa) \cdot \left(1 + \text{hyp. log. } \frac{\tau_1 - \kappa}{\tau_2 - \kappa}\right) \right\}. \quad . \quad . \quad (65.)$$

41. *Numerical Example.*

Let one pound avoirdupois of water be raised, in the liquid state, from $T_2 = 40°$ Centigrade to $T_1 = 140°$ Centigrade. Then

$$\tau_1 - \kappa = T_1 + T_0 = 140^\circ + 272\tfrac{1}{2}^\circ = 412\tfrac{1}{2}^\circ \text{ Centigrade.}$$
$$\tau_2 - \kappa = T_2 + T_0 = \ \ 40^\circ + 272\tfrac{1}{2}^\circ = 312\tfrac{1}{2}^\circ \text{ Centigrade.}$$

The mean apparent specific heat of liquid water between those temperatures is

$K_L = K_W$ (or Joule's equivalent) $\times$ 1·006 = 1398 feet per Cent. degree; consequently, the heat expended is equivalent to 139,800 foot-pounds.

The other numerical data are,—

$\dfrac{dP}{d\tau}$ at 40° Centigrade = 8·2075 lbs. per square foot per Cent. degree;

v = mean volume of 1 lb. of liquid water = 0·017 cubic foot nearly.

Let it be required to find, in the first place, V_B, the volume to which the water must be allowed to expand by partial evaporation under pressure, in order that its temperature may fall to 40° Centigrade; and, secondly, how much power will be developed in all, after the water has been totally reliquefied by refrigeration at constant pressure, at the temperature of 40°.

First, by the equation (63),

$$\Delta \Phi = \frac{dP}{d\tau}(V_B - v) = 1398 \times \text{hyp. log.} \frac{412\frac{1}{2}}{312\frac{1}{2}} = 402{\cdot}624;$$

$$\text{divide by } \frac{dP}{d\tau} = 8{\cdot}2075; \text{ then } V_B - v = 49{\cdot}055 \text{ cubic feet.}$$

$$\text{add } v = \ \ 0{\cdot}017 \quad ,,$$

Aggregate volume of water and steam at 40°, V_B = 49·072 ,,

As the volume of one pound of steam at 40° Centigrade, according to the fourth column of the table in Article 38, is 313·56 cubic feet, it appears from this calculation that somewhat less than one-sixth of the water will evaporate.

Secondly, it appears, from equation (65), that after the water has been restored to the liquid state by refrigeration at 40° Centigrade, the whole power developed—that is to say, the area A B a—will be

$$1398 \text{ foot-pounds} \times \left\{ 412^\circ{\cdot}5 - 312^\circ{\cdot}5\left(1 + \text{hyp. log.} \frac{412{\cdot}5}{312{\cdot}5}\right) \right\}$$

$$= 1398 \text{ ft. lbs.} \times 10^\circ \text{ Centigrade} = 13{,}980 \text{ ft. lbs.},$$

or one-tenth of the equivalent of the heat expended. The other nine-tenths constitute the heat abstracted during the reliquefaction at 40° Centigrade.

This calculation further shows, that in order that one pound of water and steam at 40° C. may be raised to 140° C. solely by compressing it into the liquid state, it must occupy at the commencement of the operation the volume $V_B = 49{\cdot}072$ cubic feet; and that the power expended in the compression will be as follows:—

	Foot-pounds.
Area of the curvilinear triangle ABa, Fig. 21, as already calculated,	13,980
Area of the rectangle $a\,B\,V_B\,v = P_2\,(V_B - v) =$	7,522
Total, ...	21,502

42. Proposition XVIII.—Problem. *Having the same data as in the last proposition, it is required to draw a curve of no transmission through any point on the diagram of energy for the aggregate of a liquid and its vapour.*

(Solution.) In Fig. 22, through the given point B draw the straight isothermal line A B corresponding to the absolute temperature τ_1, and cutting the ordinate corresponding to the volume of total liquefaction in A. Through A, according to the last proposition, draw the curve of no transmission, A D M. Let E D C be any other isothermal line, corresponding to the absolute temperature τ_2, and cutting the curve A M in D. Draw isothermal lines $a\,b$, $e\,d\,c$ at indefinitely-small distances from A B, E D C

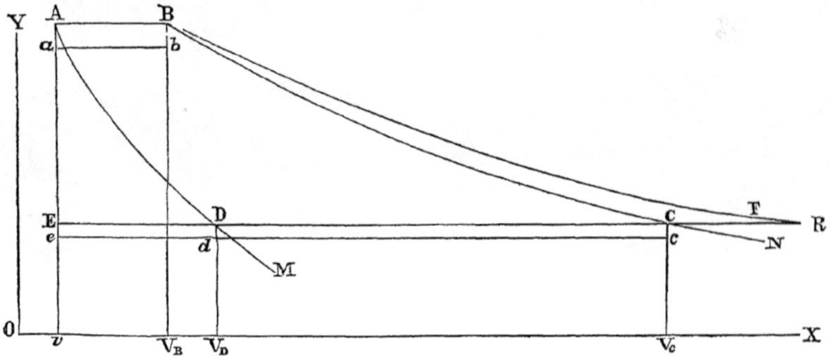

Fig. 22.

respectively, corresponding to the same indefinitely small difference of temperature $\delta\tau$. Draw the ordinates $V_D\,d\,D$, $V_B\,b\,B$; then draw the ordinate $V_C\,c\,C$ at such a distance from $V_D\,d\,D$, that the indefinitely-small rectangles D C $c\,d$, A B $b\,a$ shall be equal. Then, as the difference $\delta\tau$ is indefinitely diminished, C approximates indefinitely to a point on the required curve of no transmission, B N.

This is Proposition III. applied to aggregates, *mutatis mutandis*.

The symbolical representation of this proposition is as follows:—let P_1 and P_2 be the pressures of the aggregate of liquid and vapour corresponding respectively to the temperatures τ_1 and τ_2; then the following expressions for the difference between the thermodynamic functions Φ of the curves A M, B N are equal,

$$\Delta \Phi = \frac{d P_2}{d \tau} (V_C - V_D) = \frac{d P_1}{d \tau} (V_B - v). \quad . \quad (66.)$$

43. *Corollary.* (*Absolute Maximum Efficiency of Vapour-Engines.*)

If the volume V_B be that corresponding to complete evaporation at the temperature τ_1, that is to say, if

$$V_B = v',$$

then the curve B C N will represent the mode of expansion under pressure, of vapour of saturation in working an engine, and will be defined by the equation

$$V_C - V_D = \frac{\frac{d P_1}{d \tau} (v' - v)}{\frac{d P_2}{d \tau}}. \quad . \quad . \quad . \quad . \quad (67.)$$

If in this equation be substituted the value of $v' - v$, in terms of the latent heat of evaporation at the higher temperature, given by equation (60), it becomes

$$V_C - V_D = \frac{L_1}{(\tau_1 - \kappa) \frac{d P_2}{d \tau}}. \quad . \quad . \quad . \quad (68.)$$

In this case the diagram A B C D, Fig. 22, is evidently that of a vapour-engine working with the absolute maximum of efficiency between the absolute temperatures τ_1 and τ_2. The heat expended at each single stroke, per unit of weight of fluid, is the latent heat of evaporation at the higher temperature, or L_1; the area of the diagram is given by the following equation:—

$$E = (\tau_1 - \tau_2) \Delta \Phi = \frac{\tau_1 - \tau_2}{\tau_1 - \kappa} . L_1. \quad . \quad . \quad (69.)$$

This is the mechanical power developed at each single stroke by a unit of weight of the substance employed. The efficiency is represented by

$$\frac{E}{L_1} = \frac{\tau_1 - \tau_2}{\tau_1 - \kappa}, \quad . \quad . \quad . \quad . \quad . \quad (70.)$$

being the expression for the maximum efficiency of thermodynamic engines in general.

The conditions of obtaining this efficiency are the following:—

First, That the elevation of temperature from τ_2 to τ_1, during the operation represented by the curve D A on the diagram, shall be produced entirely by compression. The volume at which this heating by compression must commence is given, according to Proposition XVII., by the following equation:—

$$V_D = v + \frac{1}{\frac{d P_2}{d \tau}} \cdot K_L \text{ hyp. log. } \frac{\tau_1 - \kappa}{\tau_2 - \kappa} \quad . \quad . \quad . \quad (71.)$$

Secondly, That the expansive working of the vapour shall be carried on until the temperature falls, by expansion alone, to its lower limit; that is to say, until the volume reaches the following value, obtained by adding together equations (68) and (71):—

$$V_C = v + \frac{1}{\frac{d P_2}{d \tau}} \cdot \left\{ K_L \text{ hyp. log. } \frac{\tau_1 - \kappa}{\tau_2 - \kappa} + \frac{L_1}{\tau_1 - \kappa} \right\}. \quad (72.)$$

44. *Numerical Example.*

To exemplify this numerically, let the same data be employed as in Article 41, the substance working being one pound avoirdupois of water. These data, with some additional data deduced from them, are given in the following table:—

	At upper limit	At lower limit
	of Actual Heat.	
Temperature in Centigrade Degrees:—		
Above melting ice (T), . . .	140°	40°
Above zero of gaseous tension (τ), .	414·6	314·6
Above absolute cold ($\tau - \kappa$), . .	412·5	312·5
Pressure in pounds per square foot (P), . .	7557	153·34
„ „ per square inch, . .	52·5	1·065

Initial Volume of saturated steam, $V_B = v'_1 =$ 8·004 cubic ft. per lb.

Latent Heat of Evaporation:—

In degrees, applied to one pound of liquid water, 509°·1 Centigrade.

In foot-pounds (L_1), ... 707,445·36

From these data are deduced the following results:—

Absolute Maximum Efficiency; $\frac{100}{412{\cdot}5} = 0{\cdot}2424$

Duty of one pound of water; being the area of the diagram A B C D, . . 171,484·75 ft. lbs.

Volume at which the compression must commence; calculated as in Art. 41, . $V_D =$ 49·1 cubic ft. per lb.

Volume to which the expansion must be carried; calculated by equation (72), . $V_C =$ 258·1 cubic ft. per lb.

Ratio of Expansion, ... $= \frac{V_C}{V_B} = \frac{258 \cdot 1}{8 \cdot 004} = 32 \cdot 25.$

45. *Liquefaction of Vapour by Expansion under Pressure.*

In Fig. 22, let the abscissæ of the curve B F R indicate the volumes corresponding to complete evaporation at the pressures denoted by its ordinates. For most known fluids, a curve of no transmission, B C N, drawn from any point B of the curve of complete evaporation in the direction of X, falls within that curve; so that by expansion of saturated vapour under pressure, a portion in most cases will be liquefied.

To ascertain whether this will take place in any particular case, and to what extent, equation (60), which gives the volume of unity of weight of saturated vapour at the temperature τ_2, is to be compared with equation (72), which gives the volume at the same temperature of unity of weight of an aggregate of liquid and vapour, which has expanded under pressure from a state of complete evaporation at the temperature τ_1. The difference between the volumes given by these equations is as follows (neglecting, as usual, the expansibility in the liquid state) :—

$$v'_2 - V_C = \frac{1}{\frac{d P_2}{d \tau}} \cdot \left\{ \frac{L_2}{\tau_2 - \kappa} - \frac{L_1}{\tau_1 - \kappa} - K_L \cdot \text{hyp. log.} \frac{\tau_1 - \kappa}{\tau_2 - \kappa} \right\} \quad (73.)$$

That this quantity is almost always positive appears from the following considerations. The latent heat of evaporation L, is in general capable of being represented approximately by an expression of this form :

$$L = a - b(\tau - \kappa), \quad . \quad . \quad . \quad . \quad (74.)$$

(For water, $a = 796°$ Centigrade $\times K_W = 1{,}106{,}122$ ft. lbs. ; $b = 0 \cdot 695 \times K_W = 965 \cdot 772$ ft. lbs. per Centigrade degree).

Hence, the second factor in equation (73) is nearly equal to

$$\frac{a(\tau_1 - \tau_2)}{(\tau_1 - \kappa) \cdot (\tau_2 - \kappa)} - K_L \cdot \text{hyp. log.} \frac{\tau_1 - \kappa}{\tau_2 - \kappa} \quad . \quad (75.)$$

Now

$$\text{hyp. log.} \frac{\tau_1 - \kappa}{\tau_2 - \kappa} < \frac{\tau_1 - \tau_2}{\tau_2 - \kappa}.$$

Therefore, the expression (75) is positive so long as

$$\frac{a}{\tau_1 - \kappa} \text{ exceeds } K_L, \text{ the specific heat of the liquid.} \quad (75\text{ A.})$$

For water this condition is fulfilled for all temperatures lower than $523\frac{1}{2}°$ Centigrade (at which $\tau_1 - \kappa = 796°$ Centigrade); and there is reason to believe that it is fulfilled also for other fluids at those temperatures at which their vapours can be used for any practical purpose.

To determine the proportion of the fluid which is liquefied by a given expansion under pressure, we have the following formula, deduced from equation (58):—

$$n = \frac{v'_2 - V_c}{v'_2 - v}. \quad (76.)$$

As a numerical example, we may take the case of Art. 44, where saturated steam at 140° Centigrade is supposed to be expanded under pressure until its temperature falls to 40° Centigrade. The volume of one pound of water and steam at the end of the expansion has already been found to be

$$V_c = 258{\cdot}1 \text{ cubic feet.}$$

While, according to the table in Art. 38, the volume of a pound of steam at that temperature is

$$v'_2 = 313{\cdot}56 \text{ cubic feet.}$$

Consequently, the fraction liquefied by the expansion is

$$n = \frac{313{\cdot}56 - 258{\cdot}1}{313{\cdot}56 - 0{\cdot}016} = \frac{55{\cdot}46}{313{\cdot}544} = 0{\cdot}177.$$

This conclusion was arrived at contemporaneously and independently, by M. Clausius and myself, about four years since. Its accuracy was subsequently called in question, chiefly on the ground of experiments, which show that steam, after being expanded by being "wire-drawn," that is to say, by being allowed to escape through a narrow orifice, is superheated, or at a higher temperature than that of liquefaction at the reduced pressure. Soon afterwards, however, Professor William Thomson proved that those experiments are not relevant against the conclusion in question, by showing the difference between the *free expansion* of an elastic fluid, in which all the power due to the expansion is expended in agitating the particles of the fluid, and is reconverted into heat, and the expansion of the same fluid *under a pressure equal to its own elasticity*, when the power developed is all communicated to external bodies, such, for example, as the piston of an engine.

The free expansion of a vapour will be considered in the sequel.

46. *Efficiency of a Vapour-Engine without heating by compression.*

The numerical example of Art. 44 sufficiently illustrates the fact, that the strict fulfilment of the condition specified in Art. 43, as necessary to the attainment of the absolute maximum of efficiency of a vapour-engine, is impossible in practice.

Let us consider, in the first place, the effect of dispensing with the process D A, during which the fluid is supposed to have its high temperature restored solely by compression.

The effect of this modification is evidently to add to the heat expended that which is necessary to elevate the temperature of the liquid from τ_2 to τ_1, and to add to the power developed an amount represented by the area A D E (Fig. 22).

To express this symbolically, we have—

The latent heat of evaporation at τ_1, as before, ...	L_1
The additional heat expended (K_L being the *mean specific heat* of the liquid between τ_1 and τ_2), ...	$K_L(\tau_1 - \tau_2)$
Total heat expended, . .	$L_1 + K_L(\tau_1 - \tau_2)$ (77.)

Then, for the power developed, we have

The area A B C D, as in Art. 43, $= \frac{\tau_1 - \tau_2}{\tau_1 - \kappa} . L_1$,

the area A D E, as in Art. 40, equation (65),

$$= K_L \left\{ (\tau_1 - \kappa) - (\tau_2 - \kappa)\left(1 + \text{hyp. log.} \frac{\tau_1 - \kappa}{\tau_2 - \kappa}\right) \right\}$$

the sum of which quantities is the total power developed, . . (78.)

The efficiency may be expressed in the following form :—

$$\left. \begin{aligned} \frac{\text{Power developed}}{\text{Heat expended}} &= \frac{\tau_1 - \tau_2}{\tau_1 - \kappa} \\ &- \frac{K_L(\tau_2 - \kappa)\left\{\text{hyp. log.} \frac{\tau_1 - \kappa}{\tau_2 - \kappa} - \frac{\tau_1 - \tau_2}{\tau_1 - \kappa}\right\}}{L_1 + K_L(\tau_1 - \tau_2)}, \end{aligned} \right\} \quad (79.)$$

an equation which shows at once how far the efficiency falls short of the absolute maximum.

For a numerical example, the same data may be taken as in Arts. 41 and 44. Then the heat expended, per pound of steam, is thus made up:—

	Foot-pounds.
Latent heat of evaporation, as in Art. 44, ...	707,445·36
Heat required to raise the water 100° C., as in Art. 41, .	139,800·00
Total heat expended per lb. of water, .	847,245·36

The power developed consists of,—

	Foot-pounds.
The area A B C D, as in Art. 41, ...	171,484·75
The area A D E, as in Art. 41, ...	13,980·00
Total power developed per lb. of water,	185,464·75

Efficiency $\frac{185{,}484{\cdot}75}{847{,}245{\cdot}36}$	0·2189
Absolute maximum efficiency, as in Art. 44, . . .	0·2424
Loss of efficiency by omitting the heating by compression,	0·0235

or about one-tenth part of the absolute maximum.

47. *Efficiency of a Vapour-Engine with incomplete expansion.*

It is in general impossible in practice to continue the expansion of the vapour down to the temperature of final liquefaction; and from this cause a further loss of efficiency is incurred.

Let it be supposed, for example, that while the pressure of evaporation

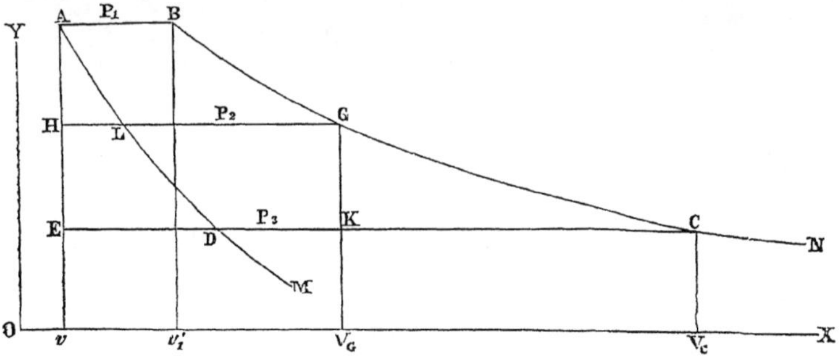

Fig. 23.

P_1 corresponds to the line A B, in Fig. 23, and the pressure of liquefaction, P_3, to the line E D C, the pressure at which the expansion terminates, P_2, corresponds to an intermediate line H L G. Let v A, v_1' B, as before, be the ordinates corresponding to complete liquefaction and to complete evaporation, at the pressure P_1.

Draw, as before, the curves of no transmission A M, B N, cutting H L G in L and G, and E D C in D and C; draw also the ordinate V_G K G, cutting E D C in K.

Then the expansion terminates at the volume V_G, and A B G K E is the indicator-diagram of the engine.

To find the power represented by this diagram, the area A L H is to be found as in Art. 40, the area A B G L as in Art. 43, and the rectangle H G K E by multiplying its breadth $V_G - v$ (found as in Art. 43) by its height H E, which is the excess of the pressure at the end of the expansion P_2, above the pressure of final liquefaction, P_3.

Hence, we have the following formula for the indicated power developed, per unit of weight of fluid evaporated:—

$$\left.\begin{aligned} E = \text{area ABGKE} = K_L \Big\{ (\tau_1 - \kappa) - (\tau_2 - \kappa) \\ \left(1 + \text{hyp. log.} \frac{\tau_1 - \kappa}{\tau_2 - \kappa}\right) \Big\} + L_1 \cdot \frac{\tau_1 - \tau_2}{\tau_1 - \kappa} \\ + (P_2 - P_3) \frac{1}{\frac{d P_2}{d \tau}} \left\{ \frac{L_1}{\tau_1 - \kappa} + K_L \text{ hyp. log.} \frac{\tau_1 - \kappa}{\tau_2 - \kappa} \right\} \end{aligned}\right\} \quad (80.)$$

The heat expended is of course $L_1 + K_L (\tau_1 - \tau_3)$.

To illustrate this numerically, let the fluid be water; let the temperature of evaporation be 140° Centigrade, and that of liquefaction 40°, as in the previous examples; and let the expansion terminate when the pressure has fallen to 100° Centigrade.

The numerical data in this case are the following:—

	1. During the evaporation.	2. At the end of the expansion.	3. During the final liquefaction.
Temperature in Cent. degrees:—			
Above melting ice,	140°	100°	40°
Above zero of gaseous tension, $\tau =$	414·6	374·6	314·6
Above absolute cold, $\tau - \kappa$,	412·5	372·5	312·5
Pressure, in lb. per square foot, P =	7557	2116·4	153·34
Pressure, in lb. per square inch,	52·5	14·7	1·065

	1. During the evaporation.	2. At the end of the expansion.	3. During the final liquefaction.
$\frac{dP}{d\tau}$ in lb. per square foot per Centigrade degree, . .	214·16	75·617	8·2075
Initial Volume of steam in cubic feet per lb., . . .	8·004		
Latent heat of evaporation, L_1, in foot-pounds per lb. of steam,	707,445·36		
Total heat expended, in foot-pounds per lb. of steam, . .	847,245·36		
Mean specific heat of liquid water—			
Between 40° and 140° Cent.,	1398 feet of fall.		
Between 100° and 140° Cent.,	1409 ,,		

Applying equation (80) to these data, we obtain the following results :—

	Foot-pounds.
Area A L H, ...	2,818
Area A B G L, ...	68,601
Area H G K E = $(P_2 - P_3) . (V_G - v)$ = 1963 lbs. per square foot × 24·58 cubic feet, . . . =	48,250
Total power developed by 1 lb. of water evaporated,	119,669
Efficiency = $\frac{119,669}{847,245}$	=0·1413
Efficiency computed in the last article, . . .	0·2189
Difference = loss of efficiency by incomplete expansion,	0·0776

Ratio of expansion $\frac{V_G}{v'} = \frac{24·60}{8·004} = 3·07$ nearly.

If the power of the same engine be now computed by the tables and formulæ published in the twentieth volume of the *Transactions of the Royal Society of Edinburgh* (*See p. 278*), which were calculated on the supposition that steam is sensibly a perfect gas, the following results are obtained :—

$$\text{Ratio of expansion, } \frac{24·60}{8·4204} = 2·921 = s \text{ in tables.}$$

	Foot-pounds.
"Action at full pressure" (P_1V_1 in tables), . . .	63,633
"Coefficient of gross action" (Z in tables) for the expansion 2·921, ...	1·98
Gross action (P_1V_1Z), ...	125,993
Deduct for back-pressure of liquefaction, $P_3V_G = 153{\cdot}34 \times 24{\cdot}6$, ...	3,772
Power developed per lb. of steam, .	122,221

This result is too large by about one forty-seventh part, a difference to be ascribed chiefly to the error of treating steam as a perfect gas. This difference, however, is not of material consequence in computing theoretically the power of a steam-engine, being less than the amount of error usually to be expected in such calculations.

48. My object in entering thus minutely into the theory of the efficiency of vapour-engines is, not so much to provide new formulæ for practical use, as to illustrate the details of the mechanical action of heat under varied and complicated circumstances, and to show with precision the nature and influence of the circumstances which prevent the production, by steam-engines, of the absolute maximum of efficiency corresponding to the temperatures between which they work.

To illustrate the results of these calculations with respect to the consumption of coal, let it be assumed, as in Art. 33, that each pound of coal consumed in the furnace communicates to the water, or air, or other elastic substance which performs the work, an amount of heat equivalent to 6,000,000 foot-pounds, which corresponds to a power of evaporating, in

	Efficiency.		Effect per pound of coal in foot-pounds.	
Absolute Theoretical Maximum, being the same for every perfect thermodynamic engine working between the same limits of temperature, $\frac{140^\circ - 40^\circ}{140^\circ + 272\frac{1}{2}^\circ}$	...	0·2424	...	1,454,400
Deductions:—				
For raising the temperature of the feed-water from 40° to 140° Cent.,	0·0235	...	141,000	...
For stopping the expansive working at 3·07 times the initial volume instead of 32 times, . . .	0·0776	...	465,600	...
		0·1011		606,600
Reduced Efficiency and Effect, .	...	0·1413	...	847,800

round numbers, about seven times its weight of water. Then the following calculation shows the theoretical indicated duty of one pound of such coal, when the limits of working temperature are 140° and 40° Centigrade, at the absolute maximum of theoretical efficiency, and at the reduced efficiency, computed in the preceding article, on the supposition that the expansive working ceases at the atmospheric pressure.

The last of these quantities corresponds to a consumption of about 2·34 lbs. of coal per indicated horse-power per hour.

The conditions of the preceding investigations are very nearly fulfilled in steam-engines with valves and steam-passages so large, and a velocity of piston so moderate, that the pressure in the cylinder during the admission of the steam is nearly the same with that in the boiler.

In many steam-engines, however, the steam is more or less "wire-drawn;" that is to say, it has to rush through the passages with a velocity, to produce which there is required a considerable excess of pressure in the boiler above that in the cylinder. The power developed during the expansion of the steam from the pressure in the boiler to that in the cylinder is not altogether lost; for, as already stated in Art. 45, it is expended in agitating the particles of the steam, and is ultimately converted into heat by friction, so that the steam begins its action on the piston in a superheated state; and both its initial pressure and its expansive action are greater than those of steam of saturation of the same density. The numerical relations of the temperature, pressure, and density of superheated steam are not yet known with sufficient precision to constitute the groundwork of a system of exact formulæ representing its action. Some general theorems, however, will be proved in the sequel, respecting superheated vapours, which may be found useful when the necessary experimental data have been obtained.

Calculation and experiment concur to prove that in Cornish single-acting engines the initial pressure of the steam in the cylinders is very much less than the maximum pressure in the boilers; generally, indeed, less than one-half.* It is doubtful, however, whether this arises altogether from wire-drawing in the steam-passages and valves; for when it is considered that in such engines, even at their greatest speed, the steam-valve remains shut nearly the whole of each stroke, being opened during a small portion of the stroke only, it may be regarded as probable that the sudden opening of this valve causes a temporary reduction of temperature and pressure in the boiler itself.

49. *Composite Vapour-Engines.*

The steam-and-ether engine of M. du Trembley is an example of what may be called a *composite vapour-engine*, in which two fluids are em-

* See Mr. Pole's work on the Cornish Engine, and Art. 36 of a paper on the Mechanical Action of Heat, *Trans. Roy. Soc. of Edin.*, Vol. XX. (*Vide p. 291.*)

ployed, a less and a more volatile; the heat given out during the liquefaction of the less volatile fluid serving to evaporate the more volatile fluid, which works an auxiliary engine, and is liquefied in its turn by refrigeration.

Let the efficiency of the engine worked by the less volatile fluid be expressed in the form

$$1 - \frac{1}{n};$$

so that $\frac{1}{n}$ is the fraction of the whole heat expended which is given out to the more volatile fluid. Let the efficiency of the engine worked by the more volatile fluid be

$$1 - \frac{1}{n'};$$

then the efficiency of the combined engines will be

$$1 - \frac{1}{nn'}. \quad . \quad . \quad . \quad . \quad (81.)$$

If both the engines are perfect thermodynamic engines, let τ_1 be the absolute temperature at which the first fluid is evaporated; τ_2 that at which it is condensed, and the second fluid evaporated; and τ_3 that at which the second fluid is condensed; then,

$$1 - \frac{1}{n} = 1 - \frac{\tau_2 - \kappa}{\tau_1 - \kappa}; \; 1 - \frac{1}{n'} = 1 - \frac{\tau_3 - \kappa}{\tau_2 - \kappa}; \; 1 - \frac{1}{nn'} = 1 - \frac{\tau_3 - \kappa}{\tau_1 - \kappa}; \quad (81 \text{ A.})$$

being equal to the theoretical maximum efficiency of a simple thermodynamic engine working between the limits of temperature τ_1 and τ_3.

Composite vapour-engines, therefore, have the same theoretical maximum efficiency with simple vapour-engines, and other engines moved by heat, working between the same temperatures; but they may, nevertheless, enable the same efficiency to be obtained with smaller engines.

50. *Curves of free expansion for nascent vapour.*

By *nascent vapour* is to be understood that which is in the act of rising from a mass of liquid. If this vapour be at once conducted to a condenser, without performing any work, and there liquefied at a temperature lower than that at which it was evaporated, its expansion, from the pressure of evaporation down to the pressure of liquefaction, will take place according to a law defined by a curve analogous, in some respects, but not in all, to the curve of free expansion for a homogeneous substance, referred to in Proposition VI. To determine theoretically the form of this curve, it is necessary to know the properties of the isothermal curves and curves of no transmission for the fluid in question in the gaseous state, when above the temperature of saturation for its pressure. Having these data, we can solve numerically the following problem :—

PROPOSITION XIX.—PROBLEM. *To draw the curve of free expansion for vapour nascent under a given pressure.*

(Solution.) In Fig. 24, let A B, parallel to O X, be the isothermal line of an aggregate of liquid and vapour at the pressure of evaporation P_1 corresponding to the temperature τ_1: let A v_1, B v_1' be ordinates parallel to O Y; so that v_1 is the volume of unity of weight of the liquid at this

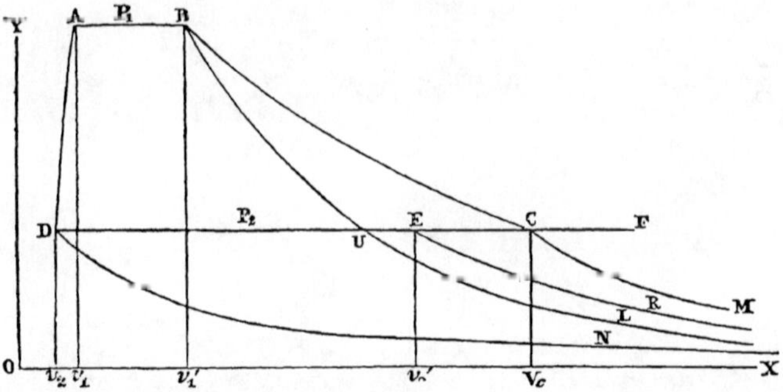

Fig. 24.

temperature, and v_1' that of unity of weight of the vapour at saturation. Let D F be a line drawn parallel to O X, at a distance representing any lower pressure P_2 corresponding to the temperature τ_2. It is required to find the point where the curve of free expansion drawn from B intersects D F.

Let v_2 be the volume of unity of weight of the liquid at the lower pressure and temperature, v_2 D an ordinate parallel to O Y, and D A a curve representing the law of expansion of the liquid as the pressure and temperature increase. Draw the curves of no transmission D N, B L indefinitely prolonged towards X; ascertain the indefinitely-prolonged area L B A D N; draw a curve of no transmission M C, cutting D F in C, such that the indefinitely-prolonged area M C D N shall be equal to the indefinitely-prolonged area L B A D N; then will C be the point required where the curve of free expansion B C intersects the line D F.

(Demonstration.) Unity of weight of the fluid being raised in the liquid state from the temperature τ_2 and corresponding pressure P_2, to the temperature τ_1 and corresponding pressure P_1; then evaporated completely at the latter pressure and temperature; then expanded without performing work, until it falls to the original pressure P_2; then cooled at this pressure till it returns to the original temperature τ_2, at which it is finally liquefied; the area A B C D represents the expansive power developed during this cycle of operations, which, as no work is performed, must be wholly expended in agitating the fluid, and reproducing by friction the heat consumed by the free expansion represented by the curve B C, which heat is measured by the indefinitely-prolonged area M C B L, which area is therefore equal to the area A B C D. Subtracting from each of these equal areas the common area B U C, and adding to each of the equal remainders the indefinitely-prolonged area L U D N, we form the areas M C D N, L B A D N, which are consequently equal. Q. E. D.

51. *Of the total heat of evaporation.*

The symbolical expression of the preceding proposition is formed in the following manner. The area L B A D N represents the *total heat of evaporation, at* the temperature τ_1, *from* the temperature τ_2, and is composed of two parts, as follows:—

$$\text{LBADN} = \int_{\tau_2}^{\tau_1} \text{K}_\text{L}\, d\tau + \text{L}_1, \quad . \quad . \quad (82.)$$

of which the first is the heat necessary to raise the liquid, whose specific heat is K_L, from τ_2 to τ_1, and the second is the latent heat of evaporation at τ_1.

Let v_2' be the volume of unity of weight of the vapour at the pressure P_2 and temperature of saturation τ_2; draw the ordinate v_2' E, meeting D F in E, through which point draw the indefinitely-prolonged curve of no transmission E R: then is the area M C D N divided into two parts, as follows:—

$$\text{MCDN} = \text{MCER} + \text{REDN} = \int_{\tau_2}^{\tau_\text{C}} \text{K}_\text{P}\, d\tau + \text{L}_2, \quad . \quad (83.)$$

in which equation τ_C denotes the temperature corresponding to the point C on the curve of free expansion, and K_P the specific heat of the vapour, at the constant pressure P_2 when above the temperature of saturation; so that the first term represents the heat abstracted in lowering the temperature of the vapour from τ_C to the temperature of saturation τ_2, at the constant pressure P_2; and the second term, the latent heat of evaporation at τ_2 abstracted during the liquefaction.

By equating the formulæ (82) and (83), the following equation is obtained:—

$$\int_{\tau_2}^{\tau_1} \text{K}_\text{L}\, d\tau + \text{L}_1 - \text{L}_2 = \int_{\tau_2}^{\tau_\text{C}} \text{K}_\text{P}\, d\tau, \quad . \quad . \quad (84.)$$

which is the symbolical solution of Proposition XIX., and shows a relation between the total heat of evaporation of a fluid, the free expansion of its vapour, and the specific heat of that vapour at constant pressure.

52. *Approximate law for a vapour which is a perfect gas.*

If the vapour of the fluid in question be a perfect gas, and of very great volume as compared with the fluid in the liquid state, the curve BC will be nearly a hyperbola, and will nearly coincide with the isothermal curve of the higher temperature τ_1, to which, consequently, τ_C will be nearly equal; and the following equation will be approximately true:

$$\int_{\tau_2}^{\tau_1} K_L \, d\tau + L_1 - L_2 = \int_{\tau_2}^{\tau_1} K_P \, d\tau, \qquad (85.)$$

which, when the difference between the higher and lower temperatures diminishes indefinitely, is reduced to the following:—

$$K_L + \frac{dL}{d\tau} = K_P, \qquad (86.)$$

that is to say:—

Corollary.—Theorem. *When a vapour is a perfect gas, and very bulky as compared with its liquid, the rate of increase of the total heat of evaporation with temperature is nearly equal to the specific heat of the vapour at constant pressure.*

This was demonstrated by a different process, in a paper read to the Royal Society of Edinburgh in 1850. It has not yet been ascertained, however, whether any vapour at saturation approaches sufficiently near to the condition of perfect gas to render the theorem applicable.

53. *Concluding Remarks.*

In conclusion, it may be observed, that the theory of the expansive action of heat embodied in the propositions of this paper contains but one principle of hypothetical origin—viz., Proposition XII., according to which the actual heat present in a substance is simply proportional to its temperature measured from a certain point of absolute cold, and multiplied by a specific constant; and that although existing experimental data may not be adequate to verify this principle precisely, they are still sufficient to prove that it is near enough to the truth for all purposes connected with thermodynamic engines, and to afford a strong probability that it is an exact physical law.

XXI.—ON FORMULÆ FOR THE MAXIMUM PRESSURE AND LATENT HEAT OF VAPOURS.*

1. It is natural to regard the pressure which a liquid or solid and its vapour maintain when in contact with each other and *in equilibrio*, as the result of an expansive elasticity in the vapour, balanced by an attractive force which tends to condense it on the surface of the liquid or solid, and which is very intense at that surface, but inappreciable at all sensible distances from it. According to this view, every solid or liquid substance is enveloped by an atmosphere of its own vapour, whose density close to the surface is very great, and diminishes at first very rapidly in receding from the surface; but at appreciable distances from the surface is sensibly uniform, being a function of the temperature and of the attractive force in question.

2. Many years since I investigated mathematically the consequences of this supposition, and arrived at the conclusion, that although it is impossible to deduce from it, in the existing condition of our knowledge of the laws of molecular forces, the exact nature of the relation between the temperature and the maximum pressure of a vapour, yet that if the hypothesis be true, it is probable that an approximate formula for the logarithm of that pressure for any substance will be found, by subtracting from a constant quantity, a converging series in terms of the powers of the reciprocal of the absolute temperature, the constant and the coefficients of the series being determined for each substance from experimental data. Such a formula is represented by

$$\text{Log. P} = \text{A} - \frac{\text{B}}{\tau} - \frac{\text{C}}{\tau^2} - \&\text{c.},$$

where P denotes the pressure, τ the absolute temperature, that is, the temperature as measured from the absolute zero of a perfect gas-thermometer, A the constant term, and B, C, &c., the coefficients of the converging series.

3. On applying this formula to M. Regnault's experiments on the pressure of steam, it was found that the first three terms were sufficient to

* Read before the British Association at Liverpool, September, 1854, and published in the *Philosophical Magazine*, December, 1854.

represent the results of these experiments with minute accuracy throughout their whole extent; that is to say, between the temperatures of

$$-30^\circ \text{ and } +230^\circ \text{ Centigrade}$$
$$= -22^\circ \text{ and } \quad 446^\circ \text{ Fahrenheit,}$$

and between the pressures of $\frac{1}{2200}$ of an atmosphere, and 82 atmospheres.

Formulæ of three terms were also found to represent the results of Dr. Ure's experiments on the vapours of alcohol and ether, and formulæ of two terms those of his experiments on the vapours of turpentine and petroleum, as closely as could be expected from the degree of precision of the experiments. A formula of two terms was found to represent accurately the results of M. Regnault's experiments on the vapour of mercury.

4. These formulæ, with a comparison between their results and those of the experiments referred to, were published in the *Edinburgh New Philosophical Journal* for July, 1849, in a paper the substance of which is summed up at its conclusion in the following proposition (*See p. 1*) ;—

If the maximum elasticity of any vapour in contact with its liquid be ascertained for three points on the scale of the air-thermometer, then the constants of an equation of the form

$$\text{Log. P} = \text{A} - \frac{\text{B}}{\tau} - \frac{\text{C}}{\tau^2}$$

may be determined, which equation will give, for that vapour, with an accuracy limited only by the errors of observation, the relation between the temperature (τ), *measured from the absolute zero, and the maximum elasticity* (P), *at all temperatures between those three points, and for a considerable range beyond them.*

5. In the case of water and mercury, the precision of the experimental data left nothing to be desired. I have, however, in the table of constants at the end of this paper, so far modified the coefficients for water and mercury as to adapt them to a position of the absolute zero (274° Centigrade, or 493°·2 Fahrenheit below the temperature of melting ice), which is probably nearer the truth than that employed in the original paper, which was six-tenths of a Centigrade degree lower. This modification, however, produces no practically appreciable alteration in the numerical results of the formulæ.

6. It was otherwise with respect to the other fluids mentioned, for which the experimental data were deficient in precision, so that the values of the constants could only be regarded as provisional.

7. A summary published in the *Comptes Rendus* for the 14th of August, 1854,* of the extensive and accurate experiments of M. Regnault on the

* See *Phil. Mag.*, Series 4, Vol. VIII., p. 269.

elasticities of the vapours of ether, sulphuret of carbon, alcohol, chloroform, and essence of turpentine, has now supplied the means of obtaining formulæ, founded on data as precise as it is at present practicable to obtain, for the maximum pressures of these vapours.

A synopsis of these formulæ, and of the constants contained in them, is annexed to this paper. The constants, as given in the table, are suited for millimètres of mercury as the measure of pressures, and for the scale of the Centigrade thermometer; but logarithms are given, by adding which to them they can be easily adapted to other scales.

The limited time which has elapsed since the publication of M. Regnault's experiments prevents my being yet able to bring the details of the investigation of the formulæ, and of the comparison of their results with those of experiment, into a shape suited for publication; but I shall here add some remarks on their degree of accuracy and the extent of their applicability.

8. M. Regnault explains, that his experiments were made by two methods; at low temperatures, by determining the pressure of the vapour *in vacuo;* at high temperatures, by determining the boiling-point under the pressure of an artificial atmosphere. For each fluid the pressures determined by both those methods were compared throughout a certain series of intermediate temperatures.

For all fluids in a state of absolute purity, the results of those two methods agreed exactly (as M. Regnault had previously shown to be the case for water).

The presence, however, of a very minute quantity of a foreign substance in the liquid under experiment was sufficient to make the pressure of the vapour *in vacuo* considerably greater than the pressure of ebullition at a given temperature; and it would appear, also, that a slight degree of impurity affects the accuracy even of the latter method of observation, although by far the more accurate of the two when they disagree.

9. The degree of precision with which it has been found possible to represent the results of the experiments by means of the formulæ, corresponds in a remarkable manner with the degree of purity in which, according to M. Regnault, the liquid can be obtained.

Sulphuret of Carbon, M. Regnault states, can easily be obtained perfectly pure. For this fluid, the agreement of the pressures computed by the formula with those determined by experiment throughout the whole range of temperature from $-16°$ Centigrade to $+136°$, is almost as close as in the case of steam.

Ether and *Alcohol* are less easy to be obtained perfectly pure. The discrepancies between calculation and experiment in these cases, though still small, are greater than for sulphuret of carbon.

For ether the formula may be considered as practically correct throughout the whole range of the experiments, from $- 20°$ Centigrade to $+ 116°$; but for alcohol below 0° Centigrade, the discrepancies, though absolutely small quantities, are large relatively to the entire pressures; and the formula can be considered applicable above this temperature only.

Essence of Turpentine has been discovered by M. Regnault to undergo a molecular change by continued boiling. For this fluid the agreement between the formula and the experiments is satisfactory above 40° Centigrade, and up to the limit of the experiments, 222°, but not below 40°.

It is impossible to obtain *Chloroform* free from an admixture of foreign substances. Accordingly, M. Regnault has found that the two methods of determining the pressure of the vapour of this fluid give widely different results, neither of which can be represented accurately by the formula now proposed below the temperature of 70° Cent. From this temperature, however, up to 130° Cent., the limit of the experiments, the agreement is close.

10. In the cases of alcohol and turpentine, the discrepancies between the formulæ and the experiments at very low temperatures are such as to indicate that they might be removed by introducing a fourth term into the formulæ, inversely proportional to the cube of the absolute temperature; but the trifling and uncertain advantage to be thus obtained would be outweighed by the inconvenience in calculation, and especially by the necessity for solving a cubic equation in computing the temperature from the pressure; whereas, with formulæ of three terms, it is only necessary to extract a square root, as the formula No. 2 shows.

11. Although, for the mere determination of the maximum pressure of a vapour at a given temperature, or its temperature at a given pressure, a table, or a curve drawn on a diagram may be sufficient, still there are many questions of thermodynamics respecting vapours for the solution of which a formula is essential.

Amongst these is the computation of the latent heat of evaporation, which is equivalent to the potential energy or work exerted by the vapour in overcoming external pressure, added to that exerted in overcoming molecular attraction. For *unity of weight* of a given substance, this is a function of the pressure, temperature, and density; but for a quantity of the substance such that its volume when evaporated exceeds its volume in the liquid or solid state by unity of cubic space, the latent heat of evaporation is simply the differential coefficient of the pressure with respect to the hyperbolic logarithm of the absolute temperature, as shown in the formula No. 3; so that, although the densities of the vapours of the seven fluids referred to in this paper are yet known by conjecture only, and not by direct experiment, we can, from the relation between the pressure and the temperature, determine accurately how much heat must be expended in the evaporation of so much of each of them as is necessary in order to propel a piston through a given space under a given *constant* pressure, and thus to solve many problems connected with engines driven by vapours of different kinds.

12. It is somewhat remarkable, that the coefficients of the reciprocal of the temperature (B) in the formulæ for ether, sulphuret of carbon, and alcohol, are nearly equal; as also those of the square of the reciprocal of the temperature (C) for ether and sulphuret of carbon.

In consequence of this, the pressure of the vapour of ether, and its latent heat for unity of space, as above defined, at a given temperature, exceed the corresponding quantities for sulphuret of carbon at the same temperature, in a ratio which is nearly, though not exactly, constant, and whose average value is somewhat less than 1·5.

SYNOPSIS OF THE FORMULÆ, &c.

Notation.

τ = absolute temperature = temp. Cent. + 274° C.
= temp. Fahr. + 461°·2 F.

P = maximum pressure of vapour at the absolute temperature τ.

v = volume of unity of weight of the liquid.

V = volume of unity of weight of saturated vapour.

L = latent heat of evaporation of unity of weight of the fluid expressed in units of work.

A, B, C, constants.

Formulæ.

1. To find the maximum pressure from the temperature,

$$\text{Com. log. P} = A - \frac{B}{\tau} - \frac{C}{\tau^2}.$$

2. To find the temperature from the maximum pressure,

$$\frac{1}{\tau} = \sqrt{\left\{\frac{A - \text{com. log. P}}{C} + \frac{B^2}{4\,C^2}\right\}} - \frac{B}{2\,C}.$$

3. To find the latent heat of evaporation (expressed in units of work) of so much of the fluid that its bulk when evaporated exceeds its bulk in the liquid state by an unit of space, that is to say, of the weight $\frac{1}{V - v}$ of fluid. In this formula the pressure must be expressed in units of weight per square unit.

$$\frac{L}{V - v} = \tau \frac{d P}{d \tau} = P \left(\frac{B}{\tau} + \frac{2 C}{\tau^2} \right) \times \text{hyp. log. } 10.$$

(Hyp. log. 10 = 2·30258509,

the common logarithm of which is 0·3622157.)

Units of work are reduced to units of heat (degrees in unity of weight of liquid water) by dividing by Joule's equivalent of the specific heat of liquid water, which has the following values, according to the units of temperature and length employed :—

		Logarithms.
Centigrade scale, and mètres, . .	423·54	2·6268969
Centigrade scale, and feet, . .	1389·6	3·1428898
Fahrenheit's scale, and feet, . .	772·0	2·8876173

CONSTANTS IN THE FORMULÆ FOR PRESSURES IN MILLIMÈTRES OF MERCURY, AND TEMPERATURES IN CENTIGRADE DEGREES.

FLUIDS.	A.	Log. B.	Log. C.	$\frac{B}{2C}$.	$\frac{B^2}{4C^2}$.
Ether, . . .	7·1284	3·0596504	4·7065130	0·011275	0·00012712
Sulphuret of carbon, .	6·8990	3·0520049	4·7078426	0·011044	0·00012197
Alcohol above 0° C., .	7·5259	3·0570610	5·2426805	0·0032610	0·000010634
Water, . . .	7·8143	3·1811430	5·0881857	0·0061934	0·000038358
Essence of turpentine above 40° C., . .	6·2522	2·9625209	5·3712157	0·0019511	0·0000038067
Chloroform above 70° C.,	5·8075	2·4007279	5·3919420	0·00051022	0·00000026032
Mercury up to 358° C.,	7·5243	3·4675637			

To adapt the formulæ to other scales of pressure, add the following logarithms to the constants A :—

For inches of mercury, ...	$\bar{2}\cdot59517$
For kilogrammes on the square mètre, . . .	1·13341
For pounds avoirdupois on the square foot, . .	0·44477

To adapt the formulæ to the scale of Fahrenheit's thermometer multiply B by 1·8, and C by $(1\cdot8)^2 = 3\cdot24$; that is to say,

Add to log. B, ...	0·2552725
Add to log. C, ...	0·5105450.

XXII.—ON THE DENSITY OF STEAM.*

1. THE object of the present paper is to draw a comparison between the results of the mechanical theory of heat, and those of the recent experiments of Messrs. Fairbairn and Tate on the density of steam, published in the *Philosophical Transactions* for 1860.

GENERAL EQUATION OF THERMODYNAMICS.

2. The equation which expresses the general law of the relations between heat and mechanical energy in elastic substances was arrived at independently and contemporaneously by Professor Clausius and myself, having been published by him in Poggendorff's *Annalen* for February, 1850, and communicated by me to the Royal Society of Edinburgh in a paper which was received in December, 1849, and read on the 4th of February, 1850 (*See p. 234*). The processes followed in the two investigations were very different in detail, though identical in principle and in results; Professor Clausius having deduced the law in question from the equivalence of heat and mechanical energy as proved experimentally by Mayer and Joule, combined with a principle which had been previously applied to the theory of substantial caloric by Sady Carnot, while by me the same law was deduced from the "hypothesis of molecular vortices," otherwise called the "centrifugal theory of elasticity."

3. Although, since the appearance of the paper to which I have referred, the notation of the general equation of thermodynamics has been improved and simplified in my own researches, as well as in those of others, I shall here present it, in the first place, precisely in the form in which I first communicated it to this Society, in order to show the connection between that equation in its original form, and the law of the density of steam, which has since been verified by the experiments of Messrs. Fairbairn and Tate. The equation, then, as it originally appeared in the twentieth volume of the *Transactions of the Royal Society of Edinburgh* (*See p. 249*) is as follows :—

* Read before the Royal Society of Edinburgh on April 28, 1862, and published in Vol. XXIII. of the *Transactions* of that Society.

$$\delta Q' = -\frac{\tau - \kappa}{C n M}\left\{\delta V\left(\frac{1}{V} - \frac{d U}{d V}\right) - \delta \tau \frac{d U}{d \tau}\right\}; \quad . \quad (1.)$$

in which the symbols have the following meanings :—

τ, the absolute temperature of an elastic substance as measured from the *zero of gaseous tension*, a point which was then estimated to be at 274°·6 Centigrade below that of melting ice, but which is now considered to be more nearly at 274° Centigrade, or 493°·2 Fahr. below that temperature;

κ, A constant, expressing the height on the thermometric scale of the temperature of total privation of heat above the zero of gaseous tension. This constant was then only known to be very small; according to later experiments, it is either null or insensible;

n M, The ideal or theoretical weight, in the perfectly gaseous state, of an unit of volume of the substance, under unity of pressure, at the temperature of melting ice;

C, The absolute temperature of melting ice, measured from zero of gaseous tension (that is to say, according to the best existing data C = 274° Centigrade, or 493°·2 Fahr.);

V, The actual volume of unity of weight of the substance;

δ V, An indefinitely small increment of that volume;

$\delta \tau$, An indefinitely small increment of temperature;

U, A certain function of the molecular forces acting in the substance;

$+ \delta Q'$, The quantity of heat which *appears*, or $- \delta Q'$, the quantity of heat which *disappears*, during the changes denoted by δ V and $\delta \tau$, through the actions of molecular forces, independently of heat employed in producing changes of temperature; such quantity of heat being expressed in equivalent *units of mechanical energy*.

The equation having been given in the above form, it is next shown (*See p. 252*), that the differential coefficients of the function U have the following values :—

$$\frac{d U}{d V} = \frac{1}{V} - C n M \frac{d P}{d \tau}; \quad . \quad . \quad . \quad . \quad . \quad (2.)$$

$$\frac{d U}{d \tau} = -\frac{\kappa}{\tau^2} - C n M \int d V . \frac{d^2 P}{d \tau^2}. \quad . \quad . \quad (3.)$$

4. The physical law of which the general equation just cited is the symbolical expression, may be thus stated in words :—*The mutual trans-*

formation of heat and mechanical energy during any indefinitely small change in the density and temperature of an elastic substance, is equal to the temperature, reckoned from the zero of absolute cold, multiplied by the complete differential of a certain function of the pressure, density, and temperature; which function is either nearly or exactly equal to the rate of variation with temperature of the work performed by indefinite expansion at a constant temperature.

5. It may be remarked that the quantity,—

$$\left.\begin{aligned}\phi &= \mathfrak{k} \text{ hyp. log. } \tau + \frac{1}{C\,n\,M}(\text{hyp. log. } V - U) \\ &= \mathfrak{k} \text{ hyp. log. } \tau + \int \frac{dP}{d\tau}\,dV\end{aligned}\right\} \quad . \quad (4.)$$

($\mathfrak{k}$ being the real specific heat of the substance in units of mechanical energy), is what, in later investigations, I have called the "*thermo-dynamic function;*" and that by its use, and by making $\kappa = 0$, equation (1) is reduced to the simplified form,

$$-\delta Q' = \tau\,\delta\phi - \mathfrak{k}\,\delta\tau; \quad . \quad . \quad . \quad (5.)$$

but the following notation is more convenient: Let δh denote the *whole heat absorbed by the substance,* not in units of mechanical energy, but in ordinary thermal units, and J the value of an ordinary thermal unit in units of mechanical energy, commonly called "Joule's equivalent," so that

$$J\,\delta h = \mathfrak{k}\,\delta\tau - \delta Q';$$

then the general equation of thermodynamics takes the form

$$J\,\delta h = \tau\,\delta\phi. \quad . \quad . \quad . \quad . \quad (6.)$$

6. For the purposes of the present paper, the most convenient form of the thermodynamic function is that given in the second line of equation (4); but it may nevertheless be stated, that in a paper read to this Society in 1855, and which now lies unpublished in their archives, it was shown that another form of that function, viz.:—

$$\phi = \left(\mathfrak{k} + \frac{P_0 V_0}{C}\right) \text{hyp. log. } \tau - \int \frac{dV}{d\tau}\,dP, \quad . \quad . \quad (7.)$$

was useful in solving certain questions; $\frac{P_0 V_0}{C}$ denoting the same thing with $\frac{1}{C\,n\,M}$ in equation (1).

Application of the General Equation of Thermodynamics to the Latent Heat and Density of Steam.

7. At the time when the general equation (1) was first published, sufficient experimental data did not exist to warrant its application to the computation of the density of a vapour from its latent heat. But very soon afterwards, various points, which had previously been doubtful, were settled by the experiments of Mr. Joule and Professor William Thomson; and in particular Mr. Joule, by his experiments, published in the *Philosophical Transactions* for 1850, finally determined the exact value of the mechanical equivalent of a British unit of heat, to which he had been gradually approximating since 1843, viz.:—

$$J = 772 \text{ foot-pounds};$$

and Messrs. Joule and Thomson in 1851, 1852, and 1853, made experiments on the free expansion of gases, especially dry air and carbonic acid, which established the very near, if not exact, coincidence of the true scale of absolute temperature with that of the perfect gas-thermometer; that is to say, those experiments proved that κ in the equation (1) is sensibly $= 0$. When, with a knowledge of these facts, equation (1) is applied to the phenomenon of the evaporation of a liquid under a constant pressure, and at a constant temperature, it takes the following form:—

$$J h = \tau \frac{d P}{d \tau} (V - v), \quad . \quad . \quad . \quad (8.)$$

where

J denotes Joule's equivalent, or 772 foot-pounds per British unit of heat (a degree of Fahrenheit in a pound of liquid water);

h, The heat which disappears during the evaporation of 1 lb. of the liquid; that is, its *latent* heat of evaporation in British units;

τ, The absolute temperature (= temperature on Fahrenheit's scale $+ 461°{\cdot}2$ Fahr.);

P, The pressure under which the evaporation takes place;

V, The volume of 1 lb. of the vapour;

v, The volume of 1 lb. of the liquid.

As the latent heat of evaporation of various fluids is much more accurately known by direct experiment than the volume or density of their vapours, the most useful form in which the equation (8) can be put, is that of a formula for computing the volume of a vapour from its latent heat, viz.:—

$$V = v + \frac{J h}{\tau \frac{d P}{d \tau}} \quad . \quad . \quad . \quad . \quad (9.)$$

8. Results of this formula were calculated by Messrs. Joule and Thomson, and by Professor Clausius for steam, and showed, as had been expected, a greater density and less volume than the law of the perfectly gaseous condition would give. Some results of the same kind, and leading to the same conclusion, were also computed by me, and published in the *Philosophical Transactions* for 1853-54. But for some years no attempt was made by any one to make a table for practical use of the volumes of steam from equation (9); because the scientific world were in daily expectation of the publication of direct experimental researches on that subject by M. Regnault.

9. At length, in the spring of 1855, having occasion to deliver to the class of my predecessor, Professor Gordon, a course of lectures on the mechanical action of heat, and finding it necessary to provide the students with a practical table of densities of steam founded on a more trustworthy basis than the assumption of the laws of the perfectly gaseous condition, I ventured upon the step of preparing a table of the densities of steam for every eighteenth degree of Fahrenheit's scale, from 86° to 410° inclusive, with the logarithms of those densities and their differences, arranged so as to enable the densities for intermediate temperatures to be calculated by interpolation. Those tables were published in a lithographed abstract of the course of lectures before mentioned, which is now out of print. The same tables, however, have since been revised, and extended to every ninth degree of Fahrenheit, from 32° to 428°, and have been printed at the end of a work *On Prime Movers*. An account of the original tables was read to the British Association in 1855.*

10. In the unpublished paper before mentioned as having been read to this Society in 1854, the densities of the vapours of ether and bisulphuret of carbon, under the pressure of one atmosphere, as computed by equation (9), are shown to agree exactly with those calculated from the chemical composition of those vapours.

11. The method of using equation (9) to calculate the volume of one pound of steam, is as follows:—

I. Calculate the *total heat of evaporation of steam* from 32°, at a given temperature T° on Fahrenheit's scale, by Regnault's well-known formula,

* The reason for using 9° Fahr. as the interval of temperature is, that it is equal to 5° Centigrade and to 4° Reaumur, so that the tables can be applied with ease to any one of those three scales.

$$1091{\cdot}7 + 0{\cdot}305\,(T^\circ - 32^\circ)\ . \qquad . \qquad . \quad (10.)$$

II. From that total heat subtract the heat required to raise 1 lb. of water from 32° to T° Fahr., viz.,

$$\int_{32^\circ}^{T} c\,d\,T\,;$$

c being the specific heat of water, the remainder will be the latent heat of evaporation of 1 lb. of steam at T°; that is to say,

$$h = 1091{\cdot}7 + 0{\cdot}305\,(T - 32^\circ) - \int_{32^\circ}^{T} c\,d\,T. \qquad . \quad (11.)$$

In computing the value of the integral in this formula, use has been made of an empirical formula founded on M. Regnault's experiments on the specific heat of water, as to which, see the *Transactions* of this Society for 1851, viz.:—

$$\int_{T'}^{T} c\,d\,T = T - T' + 0{\cdot}000000103\,\{(T - 39^\circ{\cdot}1)^3 - (T' - 39^\circ{\cdot}1)^3\} \quad (11\text{ A.})$$

III. The absolute temperature is given by the formula,

$$\tau = T + 461^\circ{\cdot}2 \text{ Fahr.} \qquad . \qquad . \qquad . \quad (12.)$$

IV. The value of $\tau \dfrac{d\,P}{d\,\tau}$ is deduced from the following formula, first published in the *Edinburgh Philosophical Journal* for July, 1849 (*See p. 1*) :—

$$\text{Com. log. } P = A - \frac{B}{\tau} - \frac{C}{\tau^2}\,; \ . \qquad . \qquad . \quad (13.)$$

from which it follows that

$$\tau \frac{d\,P}{d\,\tau} = 2{\cdot}3026\,P\left(\frac{B}{\tau} + \frac{2\,C}{\tau^2}\right); \qquad . \qquad . \quad (14.)$$

the values of the constants for steam being,—

A, for pounds of pressure on the square foot, .	8·2591
log. B for Fahrenheit's scale	=3·43642
log. C " "	=5·59873.

V. The volume of a pound of liquid water at the temperature T may be computed with sufficient accuracy for the present purpose by the following formula:—

$$v \text{ nearly} = 0{\cdot}00801\left(\frac{\tau}{500} + \frac{500}{\tau}\right). \qquad . \qquad . \quad (15.)$$

VI. These preliminary calculations having been made, the formula 9 can now be applied to the calculation of the volume of one pound of steam (making $J = 772$); and by this process the tables already mentioned were computed.

Comparison of the Results of Theory with those of Messrs. Fairbairn and Tate's Experiments.

12. The experiments of Messrs. Fairbairn and Tate on the density of steam are described in a paper which was read to the Royal Society of London, as the Bakerian Lecture, on the 10th of May, 1860, and published in the *Philosophical Transactions* for that year. The results of those experiments give what is called the "*relative volume*" of steam: that is, the ratio which its volume bears to that of an equal weight of water at the temperature of greatest density, 39°·1 Fahr.; but in the following table of comparison, each of those relative volumes is divided by 62·425, the weight of a cubic foot of water at 39°·1 in lbs., so as to give the volume of 1 lb. of steam in cubic feet. The numbers of the experiments are the same as in the original paper; those made at temperatures below 212° being numbered from 1 to 9, and those made at temperatures above 212° from 1′ to 14′.

Comparison of the Theory with the Experiments of Messrs. Fairbairn and Tate.

Number of Experiment.	Temperature Fahrenheit.	Volume of One Pound of Steam in Cubic Feet		Difference.	Difference ÷ Exper. Vol.
		By Theory.	By Exper.		
1	136°·77	132·20	132·60	−0·40	$-\frac{1}{330}$
2	155·33	85·10	85·44	−0·34	$-\frac{1}{250}$
3	159·36	77·64	78·86	−1·22	$-\frac{1}{65}$
4	170·92	60·16	59·62	+0·54	$+\frac{1}{110}$
5	171·48	59·43	59·51	−0·08	$-\frac{1}{744}$
6	174·92	55·18	55·07	+0·11	$+\frac{1}{501}$
7	182·30	47·28	48·87	−1·59	$-\frac{1}{30}$
8	188·30	41·81	42·03	−0·22	$-\frac{1}{191}$
9	198·78	33·94	34·43	−0·49	$-\frac{1}{70}$
1′	242·90	15·61	15·23	+0·12	$+\frac{1}{126}$
2′	244·82	14·77	14·55	+0·22	$+\frac{1}{66}$
3′	245·22	14·67	14·30	+0·37	$+\frac{1}{39}$
4′	255·50	12·39	12·17	+0·22	$+\frac{1}{55}$
5′	263·14	10·96	10·40	+0·56	$+\frac{1}{19}$
6′	267·21	10·29	10·18	+0·11	$+\frac{1}{93}$
7′	269·20	9·977	9·703	+0·274	$+\frac{1}{35}$
8′	274·76	9·158	9·361	−0·203	$-\frac{1}{47}$
9′	273·30	9·367	8·702	+0·665	$+\frac{1}{13}$
10′	279·42	8·539	8·249	+0·290	$+\frac{1}{28}$
11′	282·58	8·145	7·964	+0·181	$+\frac{1}{44}$
12′	287·25	7·603	7·340	+0·263	$+\frac{1}{28}$
13′	292·53	7·041	6·938	+0·103	$+\frac{1}{66}$
14′	288·25	7·494	7·201	+0·293	$+\frac{1}{24}$

REMARKS ON THE DIFFERENCES BETWEEN THE THEORETICAL AND EXPERIMENTAL RESULTS.

13. The differences between the theory and the experiments as to the volumes of steam at temperatures below 212° are, with few exceptions, of very small relative amount; and they are at the same time so irregular as to show that they must have mainly arisen from causes foreign to the data used in the theoretical computations.

14. Above 212° also, the differences show irregularity, especially in the case of experiments 8′ and 9′, where a *fall* of temperature is accompanied by a diminution instead of an increase in the volume of one pound of saturated steam, as determined by experiment. But still those differences, presenting as they do, in every case but one, an excess of the theoretical above the experimental volume, show that some permanent cause of discrepancy must have been at work; although they may not be regular enough to determine its nature and amount, nor large enough to constitute errors of importance in practical calculations relating to steam-engines.

15. So far as it is possible to represent those differences by anything like a formula, they agree, in a rough way, with a *constant excess* of about 0·24 of a cubic foot in the theoretical volume of one pound of steam above the experimental volume; and this also represents, in a rough way, the difference between the curves whose ordinates express respectively the results of the theoretical formula and those of an empirical formula deduced from the experiments, so far as those curves, as shown in a plate annexed to the paper referred to (*See Plate II.*), extend through the limits of actual experiment on steam, above 212°.

16. As the principles of the mechanical theory of heat may now be considered to be established beyond question, it is only in the data of the formula that we can look for causes of error in the theoretical calculation. I shall now consider those data, with reference to the probability of their containing numerical errors.

I. *Total Heat of Evaporation.*—It is very improbable that this quantity, as computed by M. Regnault's formula, involves any material error.

II. *Sensible Heat of the Liquid Water.*—The empirical formula from which this quantity is computed was determined from experiments by M. Regnault which agree extremely well amongst themselves. (For the investigation of the formula, see *Trans. Roy. Soc. Edin.*, Vol. XX., p. 441). The subtraction of the sensible heat from the total heat leaves the latent heat, upon which the increase of volume depends; hence, to account for an error in excess of the formula for the volume by means of an error in the computation of the sensible heat, it must be supposed that the specific heat of liquid water above 212° increases much more rapidly than M. Regnault's experiments show, so as to produce a correspondingly more rapid diminution in the latent heat of evaporation. It is easily computed, for example, that to account for an error in excess of 0·24 of a cubic foot in the volume of one pound of steam at 266°, by an error in defect in the sensible heat, we must suppose that error to amount to about 21 British thermal units per pound of water; but such an error is altogether improbable.

III. *Absolute Temperature.*—The position of the absolute zero may be considered as established with a degree of accuracy which leaves no room for any error sufficient to account for the differences now in question.

IV. *Function* $\tau \frac{dP}{d\tau}$.—The same may unquestionably be said of this function; which represents the mechanical equivalent of the latent heat of evaporation of so much water as fills one cubic foot more in the vaporous than in the liquid state.

V. *The volume of one pound of the liquid water* is itself too small to affect the question.

VI. The received value of the *mechanical equivalent of a unit of heat* cannot err by so much as $\frac{1}{100}$ part of its amount.

Conclusions.

17. It appears, then, that none of the data from which the theoretical calculations are made are liable to errors of a magnitude sufficient to account for the differences between the results of those calculations and the results of Messrs. Fairbairn and Tate's experiments, small as those differences are in a practical point of view. Neither does there appear to have been any cause of error in the mode of making the experiments. There remains only to account for those differences, the supposition that there was some small difference of molecular condition in the steam whose density was measured in the experiments of Messrs Fairbairn and Tate, above 212°, and the steam whose total heat of evaporation, as measured by M. Regnault, is the most important of the data of the theoretical formula,—a difference of such a nature as to make a given weight of steam in Messrs. Fairbairn and Tate's experiments occupy somewhat less space, and therefore require somewhat less heat for its production, than

the same weight of steam in M. Regnault's experiments at the same temperature. That difference in molecular condition, of what nature soever it may have been, was in all probability connected with the fact, that in the experiments of Messrs. Fairbairn and Tate the steam was at rest, whereas in those of M. Regnault it was in rapid motion from the boiler towards the condenser. It is obvious, however, that in order to arrive at a definite conclusion on this subject, further experimental researches are required.

XXIII.—ON THE SECOND LAW OF THERMODYNAMICS.*

1. It has long been established that all the known relations between heat and mechanical energy are summed up in two laws, called respectively the *first law* and the *second law* of thermodynamics: viz.—

First Law.—Quantities of heat and of mechanical energy are convertible at the rate very nearly of 772 foot-pounds to the British (or Fahrenheit-avoirdupois) unit, or 424 kilogrammetres to the French (or Centigrade-metrical) unit of heat.

Second Law.—The quantity of energy which is converted from one of those forms to the other during a given change of dimensions and condition in a given body, is the product of the absolute temperature into a function of that change, and of the kind and condition of the matter of the body.

By *absolute temperature* is here to be understood temperature measured according to a scale so graduated that the temperature of a homogeneous body shall vary in the simple proportion of the quantity of energy it possesses in the form of sensible or thermometric heat.

2. The laws of thermodynamics, as here stated, are simply the condensed expression of the facts of experiment. But they are also capable of being viewed as the consequence of the supposition, that the condition of bodies which accompanies the phenomena of sensible heat consists in some kind of motion amongst their particles.

3. The first law would obviously follow from the supposition of any kind of molecular motion whatsoever, and it therefore affords of itself no reason for preferring one supposition as to the kind of molecular motion which constitutes sensible heat to another.

4. But if there be molecular motions in bodies, it is certain that, although all such motions are capable of conversion into that which constitutes sensible heat, some of them are not accompanied by sensible heat. For example, the motion (supposed to be vibratory and wave-like) which constitutes radiance, whether visible or invisible, is not accompanied by sensible heat, and only produces sensible heat by its absorption; that

* Read before the British Association at Birmingham, September, 1865, and published in the *Philosophical Magazine*, October, 1865.

is, in the language of hypothesis, by its conversion into some other kind of motion; while, on the other hand, in the production of radiance sensible heat disappears.

5. The object of the present paper is to give an elementary proof of the proposition, *that the second law of thermodynamics follows from the supposition that sensible heat consists in any kind of steady molecular motion within limited spaces.*

6. The term "steady motion" is here used in the same sense as in hydrodynamics, to denote motion, whether of a continuous fluid or of a system of detached molecules, in which the velocity and direction of motion of a particle depend on its position only; so that each particle of the series of particles which successively pass through a given position assumes in its turn the velocity and direction proper to that position. In other words, steady motion may be defined as motion in a set of streams of invariable figure.

When steady motion takes place in matter that is confined within a limited space, the streams in which the particles move must necessarily return into themselves, and be *circulating streams*, being in that respect of the nature of whirls, eddies, or vortices.

7. Steady motion keeps unaltered the distribution of the density of the moving matter; and it therefore keeps unaltered the forces depending on such distribution, whether of the nature of pressure or of attraction. In that respect it differs from unsteady motion, such as vibratory and wave-like motion.

8. Conceive a limited space of any figure whatsoever to be filled with matter in a state of steady motion. The actual energy of any particle of that matter is the product of its mass into the half-square of its velocity; and the actual energy of the whole mass of matter is the sum of all those products; and because of the steadiness of the motion, the actual energy of the particle which at any instant whatsoever occupies a given position is some definite fraction of the whole actual energy, depending upon that position, and upon the distribution of matter within the space; but the scale of absolute temperature is defined as being so graduated that the whole actual energy of the matter within the space is the product of the absolute temperature, the mass of matter, and some function of the sort and distribution of the matter; therefore, *the half-square of the velocity of the particle which at any instant occupies a given position in the space considered, is equal to the absolute temperature multiplied by some function of that position, and of the sort and distribution of the matter.*

9. Suppose now that the dimensions of the limited space in which the moving matter is enclosed, and the distribution of that matter, undergo an indefinitely small change by the application of suitable forces, and that after that process the motion becomes steady as it was before.

Then the dimensions and position of each circulating stream will have been altered; and the work done in effecting that alteration will consist of energy converted between the forms of potential energy of the applied forces, and actual energy of the molecular motions—that is, between the forms of mechanical energy and of heat. Consider now a point in one of the circulating streams before the change, and let fall from it a perpendicular upon the same stream after the change. The work done in shifting the path of the particle which at any instant occupies that point, is the product of the perpendicular displacement of the stream into the force exerted along that perpendicular. But the perpendicular displacement of the stream is a function of the position of the point shifted, the distribution of matter in the space, and the change of dimensions and distribution; and the force is equal and opposite either to the centrifugal force of the particle or to one of its components, and is therefore proportional to the square of the velocity of the particle, and to some function of its position, and of the sort and distribution of matter in the body. Therefore, *the energy transformed in shifting the path of any particle is proportional to the square of its velocity, and to some function of its position, of the sort and distribution of matter in the space considered, and of the change in dimensions of that space and in the distribution of the matter.*

10. But the square of the velocity of the particle which at any instant occupies a given position has already been shown to be proportional to the absolute temperature, and to some function of that position and of the sort and distribution of the matter; therefore, *if sensible heat consists in any kind of steady molecular motion within limited spaces, the conversion of energy during any change in the dimensions of such spaces, and in the distribution of matter in them, is the product of the absolute temperature into some function of that change and of the sort and distribution of the matter.*

11. In a paper "On the Mechanical Action of Heat," published in the *Transactions of the Royal Society of Edinburgh* for 1850 (*See p. 234*), the author deduced the second law of thermodynamics, in the form above stated, from the hypothesis of a particular sort of steady molecular motion —viz., revolution in circular streams or vortices. In a paper "On the Centrifugal Theory of Elasticity," published in the same *Transactions* for 1851 (*See p. 49*), he deduced the same law from the hypothesis of steady molecular motion in circulating streams of any figure whatsoever, being a proposition substantially identical with that set forth in the present paper; but as the demonstration in the paper of 1851 involved tedious and intricate symbolical processes, he has written the present paper in order to show that the demonstration can be effected very simply.

12. It is obvious that the *steadiness* of the supposed molecular motions is the essential condition which makes the second law of thermodynamics deducible from a mechanical hypothesis; and that no kind of unsteady

motion, such as vibratory or wave-like motion, would lead to the same results. If, then, it be admitted as probable, that the phenomena of heat are due to unseen molecular motions, it must also be admitted, that while the motions which constitute radiance may be vibratory and wave-like, the motions which constitute sensible or thermometric heat must be steady and like those of circulating streams.

13. The function by which the absolute temperature is multiplied in calculating the conversion of energy between the mechanical and the thermic forms, is the variation of what the author has called the *metamorphic function*,* being one term of the *thermodynamic function*,† which corresponds to what Professor Clausius calls *entropie*.‡

APPENDIX.

The following is the symbolical expression of the demonstration given in the paper.

Let m stand for the specific properties of the sort of matter which is in a state of steady motion within a limited space;

f for the figures and dimensions of that space, and of the paths described by the particles contained in it; and δf for any indefinitely small change of such figures and dimensions;

p for the position, relatively to the centre of the matter contained in the space, of a point which is fixed so long as $\delta f = 0$. Because the motion is steady, each particle of matter which successively arrives at the point p assumes the velocity, direction, and curvature of motion proper to that point. Let v be that velocity, and r the radius of that curvature; then for a particle of mass unity, in the act of traversing p,

$$\text{Actual energy of mass } 1 = \frac{v^2}{2} = k\,\tau, \qquad . \qquad . \quad (1.)$$

where τ is a quantity upon whose uniformity throughout the space the steadiness of the motion depends, and k a function of (m, f, p); and

$$\text{Centrifugal force of mass } 1 = \frac{v^2}{r} = \frac{2\,k\,\tau}{r}; \qquad . \qquad . \quad (2.)$$

in which r, and consequently $\frac{2\,k}{r}$, are functions of (m, f, p).

* See "On the Science of Energetics," *page 209*.

† *Philosophical Transactions*, 1854.

‡ *Ueber verschiedene für die Anwendung bequeme Formen der Hauptgleichungen der mechanischen Wärmetheorie*, April, 1865.

Now, let the change denoted by δf take place, and let the steadiness of motion be restored: let δn be the length of a line drawn through the original position of the point p, so as to be perpendicular to the path of the stream of particles which formerly traversed p; and let $\widehat{rn}$ be the angle made by δn with r. Then δn and $\widehat{rn}$ are both functions of $(m, f, p, \delta f)$. Also, the work done, or energy converted, for a unit of mass at the point p, while the path of the particles that traverse p is shifted through δn, is as follows:—

$$\frac{v^2}{r} \cdot \delta n \cdot \cos \widehat{rn} = \frac{2\, k\, \tau \cdot \delta n \cdot \cos \widehat{rn}}{r} = \tau \times \text{function of } (m, f, p, \delta f). \quad (3.)$$

The energy converted during the change δf, throughout the whole space considered, is the sum of the quantities of energy converted for each unit of mass within the space. But τ by definition is uniform; and the sum of a set of functions of p is a function of f and m; therefore, the *whole energy converted* is

$$\tau \cdot \Sigma \cdot \frac{2\, k \cdot \delta n \cdot \cos \widehat{rn}}{r} = \times \text{ functions of } (m, f, \delta f); \quad (4.)$$

and because δf is indefinitely small, the preceding expression is equivalent to the following:

$$\text{Energy converted} = \tau \cdot \text{function } (m, f) \cdot \delta f = \tau \cdot \delta \mathrm{F}\, (m, f). \quad (5.)$$

Let τ be called *absolute temperature,* and this is the *second law of thermodynamics.* It is to be observed that f may be, and often is, a function of τ.

XXIV.—ON THE WANT OF POPULAR ILLUSTRATIONS OF THE SECOND LAW OF THERMODYNAMICS.*

1. THE science of thermodynamics is based on two laws, the first of which states the fact of the mutual convertibility of heat and mechanical energy, while the second shows to what extent the mutual conversion of those two forms of energy takes place under given circumstances. In the course of the last few years the first law has been completely "popularised;" it has been amply explained in books and lectures,† composed in a clear and captivating style, and illustrated by examples at once familiar and interesting, so as to make it easily understood by those who do not make science a professional pursuit.

2. The second law, on the other hand, although it is not less important than the first, and although it has been recognised as a scientific principle for nearly as long a time, has been much neglected by the authors of popular (as distinguished from elementary) works;‡ and the consequence is that most of those who depend altogether on such works for their scientific information remain in ignorance, not only of the second law, but of the fact that there is a second law; and knowing the first law only, imagine that they know the whole principles of thermodynamics. The latter is the worst evil of the two: "a little learning" is not "a dangerous thing" in itself, but becomes so when its possessor is ignorant of its littleness.

3. In the present paper I do not pretend to supply that want, but rather to point out its existence to authors who possess the faculty of popularising; in order that they, by means of lectures, writings, and lecture-room experiments, may convey a general knowledge of the nature and results of the second law of thermodynamics to those who feel an interest in science without making it a regular study.

4. Before considering how the second law can best be stated and explained, it may be well to point out how far it is possible to proceed

* From *The Engineer* of June 28, 1867.

† Such, for example, as Dr. Tyndall's *Heat Considered as a Mode of Motion.*

‡ The explanation of the second law of thermodynamics in Dr. Balfour Stewart's excellent treatise on heat is elementary; but it is not, nor does it profess to be, popular.

with the solution of questions in thermodynamics by means of the first law alone, without the aid of the second law. The first law informs us that when mechanical work is done by means of heat, a quantity of heat disappears, bearing a constant ratio to the quantity of work done—viz., that of one British unit of heat (or one degree of Fahrenheit in a pound of water) for every 722 foot-pounds of mechanical work done. In order, therefore, to calculate how much heat will disappear during a given change in the dimensions of a substance under the action of given forces, it is necessary to know the quantity of work done during such change; and the cases in which the expenditure of heat can be calculated by means of the first law alone, are those and those only, in which the work done can be directly measured; that is to say, in which the work is sensibly altogether *external*, or done against forces exerted between the body under consideration and other bodies, and in which no part or no sensible part of the work is *internal*, or done against forces exerted upon each other by the particles of the body, and therefore incapable of direct measurement.

5. The only phenomenon which fulfils that condition is the expansion of a perfect or sensibly perfect gas; that is, of a substance in a condition such that its pressure at a given temperature is proportional, or sensibly proportional, to its density simply. To illustrate this by an example: let us suppose one pound of atmospheric air (which, though not absolutely a perfect gas, may be treated as such for practical purposes) at the temperature of melting ice (32° Fahr.), to be contained in a cylinder whose sectional area is equal to one square foot, being confined in that cylinder by means of a piston loaded with a pressure amounting to 4233 pounds. Then, from the experiments of Regnault on the pressure and density of air, it is known that the length of the cylindrical space occupied by the air is 6·193 ft.

Next, let the load on the piston be gradually diminished until it is reduced to 2116½ lbs., being one-half of its original amount, and let the question to be solved be—*what quantity of heat must be communicated to the air, in order that its temperature may remain constant during the expansion which accompanies the diminution of pressure?* The solution is as follows:—As the temperature is constant, and the air is treated as perfectly gaseous, the product of the pressure on the piston into the volume of the air; that is to say, $4233 \times 6{\cdot}193 = 26215$, remains constant during the expansion; and the external work done in driving the piston against the gradually diminishing load is found by multiplying that product by the hyperbolic logarithm of 2, the rate of expansion, that is to say—

$$\text{Work done} = 26215 \times {\cdot}69315$$
$$= 18171 \text{ foot-pounds.}$$

That is the external work, and the internal work is practically inappreciable; therefore, that also is the mechanical equivalent of the quantity of heat required in order to keep the temperature of the air constant; and dividing 18171 by 772, that quantity of heat in ordinary British thermal units is found to be 23·54; that is, as much heat as would raise the temperature of 23·54 lbs. of water one degree of Fahrenheit; and such is the value of the latent heat of expansion of 1 lb. of air in doubling its volume at the constant temperature of 32° Fahr. Here, then, the solution has been obtained by means of the first law of thermodynamics only, without the aid of the second.

6. It is otherwise when the external work is accompanied by the performance of internal work to a practically important extent; in other words, when we deal with substances which cannot be treated as perfectly gaseous, such as fluids in the act of evaporating. For example, let there be, as before, a cylindrical vessel, whose sectional area is one square foot, and let it contain 1 lb. of water in the liquid state, at the temperature of 212° Fahr., which will occupy a length of the cylinder equal to ·017 of a foot. Let the water be confined by means of a piston, the load upon which, in order to be just sufficient to confine the water, must be equal to the mean atmospheric pressure on a square foot, that is to say, 2116·3 lbs. If additional heat be now communicated to the water, without altering the load on the piston, it is well known that its temperature does not rise, but that it passes by degrees into the state of steam, driving the piston before it. By the time the water has entirely assumed the state of steam, it occupies 26·36 ft. in length of the cylinder; so that the piston has been driven through 26·343 ft. against the constant load of 2116·3 lbs. Let the question proposed be, to calculate from those data the expenditure of heat. The *external* work has the following value:—

$$2116{\cdot}3 \times 26{\cdot}343 = 55750 \text{ foot-pounds,}$$

and this is equivalent to 72·2 units of heat nearly. But besides the external work done in driving the piston, there is internal work done in overcoming the cohesion of the particles of water; and that internal work is incapable of direct measurement.

7. Here it is that the second law of thermodynamics becomes useful; for it informs us how to deduce the whole amount of work done—internal and external—from the knowledge which we have of the external work. That law is capable of being stated in a variety of forms, expressed in different words, although virtually equivalent to each other. The most convenient form for the present purpose appears to be the following:—

To find the whole work, internal and external, multiply the absolute temperature at which the change of dimensions takes place, by the rate per degree at which the external work is varied by a small variation of the temperature.

By *absolute temperature* is meant temperature measured from the *absolute zero*, or point of total privation of heat, which is known by theoretical deduction from experimental data to be about 461°·2 below the ordinary zero of Fahrenheit's scale.

8. To apply the second law to the present problem, suppose the temperature at which the given increase of volume (viz., 26·343 cubic feet) takes place, to be lowered by one degree of Fahrenheit. Then, from tables and formulæ founded on Regnault's experiments, we know that the pressure of the steam is diminished by 42·05 lbs. on the square foot. Hence, the external work is diminished by reduction of temperature at the following rate:—

$$F = 42{\cdot}05 \times 26{\cdot}343 = 1107{\cdot}7 \text{ foot-pounds per degree of Fahrenheit.}$$

The absolute temperature is

$$t = 212^\circ + 461^\circ{\cdot}2 = 673^\circ{\cdot}2 \text{ Fahr.};$$

and, therefore, the whole work, internal and external, done during the evaporation of 1 lb. of water at the temperature of 212° Fahr. is,

$$tF = 673^\circ{\cdot}2 \times 1107{\cdot}7 = 745{,}800 \text{ foot-pounds very nearly.}$$

To reduce this quantity to British thermal units, divide by 772; the result is

$$745{,}800 \div 772 = 966,$$

being the *latent heat of evaporation of steam* at 212° Fahr.

The total work just calculated is made up as follows:—

	Foot-pounds.
External work, computed in Article 6, ...	55,750
Internal work, ...	690,050
Total work, ...	745,800

from which it appears that the external work done in evaporating water under the mean atmospheric pressure is less than $7\frac{1}{2}$ per cent. of the whole work; the remainder, or $92\frac{1}{2}$ per cent., being internal work.

9. The second law may also be applied to solve the inverse problem, of deducing from the expenditure of heat in a given process, and from the relations between pressure and temperature, the change of dimensions with which that process is accompanied; and such has been the use chiefly made of that law in the actual history of thermodynamics. Previous to the publication, in September, 1859, of the experiments of

Messrs. Fairbairn and Tate, there did not exist any accurate determinations of the density of saturated steam at different temperatures; and, therefore, some of the writers on thermodynamics found it necessary to calculate that density theoretically, by the help of the second law, from the latent heat of evaporation, which was very accurately known through the experiments of Regnault. The following is an example of such calculations:—

Latent heat of evaporation of 1 lb. of water at 212° Fahr., as known by experiment, in ordinary British thermal units,	966
Applying the first law, that is, multiplying by Joule's equivalent, ...	772
The value of that latent heat in foot-pounds of work is found to be	745752
Dividing that quantity of work by the absolute temperature, 673°·2 Fahr., the work per degree of absolute temperature is found to be, in foot-pounds, ...	1107·7
But the variation of pressure per degree, in pounds on the square foot, is	42·05

Therefore, the increase of volume of 1 lb. of water in the act of evaporating, at 212° Fahr., is

$$1107{\cdot}7 \div 42{\cdot}05 = 26{\cdot}343 \text{ cubic feet};$$

To which, adding the volume of the water in the liquid state	0·017 „
There is found the volume of 1 lb. of atmospheric steam, ...	26·360 cubic feet.

10. The example by which the second law of thermodynamics has here been illustrated, has been puposely chosen of a very simple kind; but that law enables the most complex questions respecting the expenditure of heat required to produce a given mechanical effect to be solved; and the solution is always effected in the same manner: that is to say, by deducing the total work, internal and external, from the manner in which a small variation of temperature affects the external work.

11. The *law of the efficiency of a perfect heat engine* may be stated thus: If the substance—for example, air or water—which does the work in a perfect heat engine receives all the heat expended at one fixed temperature, and gives out all the heat which remains unconverted into work at a

lower fixed temperature, the fraction of the whole heat expended which is converted into external work is expressed by dividing the difference between those temperatures by the higher of them, reckoned from the absolute zero. Now this is, in fact, the second law of thermodynamics expressed in other words; but whether the demonstration of that fact, that is, of the substantial identity of the second law, as stated in Article 7, with the law stated in the preceding sentence, is capable of being put in a popular form is doubtful, seeing that it involves the notion of limiting ratios. The applications, however, of the law of the efficiency of a perfect heat engine are very simple and easy. For example, it informs us that if the steam which drives an engine receives all the heat expended upon it at temperatures not exceeding 248° Fahr. (corresponding to the absolute temperature $248° + 461°{\cdot}2 = 709°{\cdot}2$), and if all the heat not converted into external work is given out by that steam at temperatures not below 104° Fahr. (being an ordinary temperature of condensation), the efficiency of that engine, being the fraction of the whole heat expended that is converted into external work, cannot possibly exceed the following value—

$$\frac{248 - 104}{248 + 461{\cdot}2} = \frac{144}{709{\cdot}2} = {\cdot}203.$$

The same law informs us that in order that the whole heat expended in a heat engine may be converted into external work, it is necessary that the temperature of the condenser or refrigerator should be the *absolute zero*—a temperature unattainable by human means. Thus, a knowledge of the second law of thermodynamics, as applied to the efficiency of heat engines, is a safeguard against the formation of projects for increasing the performance of such engines beyond the highest possible limits.

12. There seems, then, to be no difficulty in explaining and illustrating, in a popular way, the applications of the second law to various scientific and practical questions, and the agreement of its results with those of experiment, which agreement is the real proof of its being true. But it appears by no means so easy to demonstrate popularly the connection between the second law of thermodynamics with the idea of "heat as a mode of motion." That connection consists in the fact that the second law of thermodynamics necessarily follows from the established laws of dynamics when they are applied to the supposition that the sort of motion which constitutes heat is a *whirling or circulating motion of the particles of bodies*—a supposition otherwise called the "*hypothesis of molecular vortices.*" The original demonstrations of that fact, which appeared in February, 1850, and December, 1851 (*See p. 49*), involve algebraical processes that are quite beyond the range of a popular explanation; and to understand even the elementary proof without algebra, which was read to the British

Association in 1865 (*See p. 427*), requires the habit of scientific reasoning. It is much to be wished that some means could be devised for making that proposition as widely understood as the first law of thermodynamics now is; for as matters now stand, the popular knowledge of thermodynamics is, as regards the second law, eighteen years behind the actual state of that science.

XXV.—EXAMPLES OF THE APPLICATION OF THE SECOND LAW OF THERMODYNAMICS TO A PERFECT STEAM-ENGINE AND A PERFECT AIR-ENGINE.*

1. THE following examples are intended to illustrate the application of the second law of thermodynamics to a perfect steam-engine and a perfect air-engine, expending the same quantity of heat, and working between the same limits of temperature, viz. :—Quantity of heat expended per stroke, reduced to an equivalent quantity of mechanical work, in foot-pounds, 68420.

Temperatures.	Degrees of Fahrenheit. Ordinary scale.	Degrees of Fahrenheit. Absolute scale.
Upper limit,	266°	727°·2
Lower limit,	104°	565°·2
Difference, or range,	162°	162°·0

2. The phrase "perfect engine" is here used to denote an engine which realises the greatest quantity of mechanical work possible with the given expenditure of heat and between the given limits of temperature, being the limit towards which actual engines may be made to approach through the progress of practical improvements. Such an engine must fulfil the following conditions :—There must be no waste of heat through conduction or radiation, or of work through friction; and the whole heat expended must be communicated to the working substance at the higher limit of temperature. In other words, the elevation of the temperature of the working substance from the lower to the higher limit must be effected without any expenditure of heat; for whatever heat is expended in producing elevation of temperature, is either wholly or partially lost as regards the performance of mechanical work.

3. There are two ways of effecting the elevation of temperature without expenditure of heat. One is to raise the temperature by compression of the working substance in a non-conducting cylinder; the mechanical work necessary for that purpose being obtained by means of the expansion of the working substance in a non-conducting cylinder while its temperature is falling back from the higher to the lower limit. The other way is by conduction, viz. :—to make the working substance, while its temperature is falling, communicate its heat to a set of metal tubes, or to a network,

* From *The Engineer* of August 2, 1867.

called a "regenerator" or "economiser," from which the heat is given out again at the proper time to the working substance in order to raise its temperature. Supposing those two processes to be carried out in a theoretically perfect manner, the results as regards the economy of heat are exactly the same.* In the following examples the steam-engine will be supposed to act by compression, and the air-engine by the aid of a perfect regenerator.

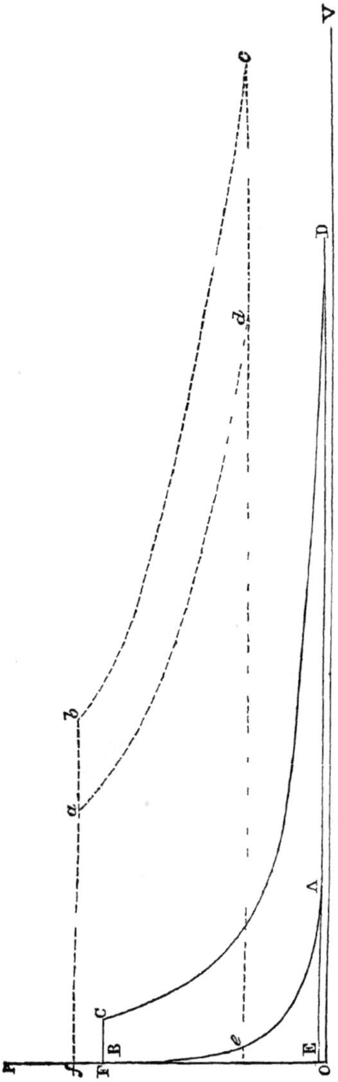

Fig. 1.

4. According to the second law of thermodynamics, the efficiency of each of those engines is the same, viz.:—

$$\frac{162}{727 \cdot 2} = 0 \cdot 2228;$$

and the work realised per stroke is also the same, viz.:—

$$68420 \times 0 \cdot 2228 = 15244 \text{ foot-pounds.}$$

The object of the following calculations is to show in detail what becomes of the difference between the whole energy expended in the form of heat and that obtained in the form of mechanical work, viz.;—

$$68420 - 15244 = 53176 \text{ foot-pounds,}$$

being the rejected or necessarily lost energy.

5. The diagrams of both engines are represented in the figure. Absolute pressures are supposed to be represented in pounds on the square foot by ordinates measured parallel to O P; volumes occupied by the working substance, in cubic feet, by distances measured parallel to O V.

* The fact that those results are the same is illustrated in the case of air-engines by numerical examples, which may be found in *A Manual of the Steam-Engine and other Prime Movers*, pages 347 to 369. In practice the regenerator answers best for an air-engine, because of the very large space required for the other process.

The diagram of the steam-engine is marked by full lines and capital letters; that of the air-engine by dotted lines and italic letters. In the case of the steam-engine the whole work done by the steam in driving the piston is represented by the area F C D E; the work expended in compressing part of the steam and feeding the boiler by the area F B A E; and the indicated work by the area B C D A.

In the case of the air-engine the whole work done by the air in driving the piston is represented by the area *f b c d;* the work expended in raising the pressure of the air and feeding the engine by the area *f a d e;* and the indicated work by the area *b c d a.*

6. To begin in detail with the steam-engine. From the limits of temperature given in Article 1, it is found, by means of the proper table or formula, that the limits of pressure are as follows:—

	Lb. on the Sq. In.	Lb. on the Sq. Ft.
Absolute pressure of admission, O F,	39·25	5652
Absolute back pressure, O E, . .	1·06	152·6

7. The volume of water in the liquid state which is used per stroke is represented by F B, a distance too small to be seen in the diagram; the volume of the same water when in the state of steam at the higher limit of pressure by F C; and, consequently, the increase of volume of the water in the act of evaporating by B C. That increase of volume is produced by communicating to the water the whole heat expended. Now 68420 foot-pounds is the mechanical value of the latent heat of evaporation, under the higher of the given pressures, of so much water as fills one cubic foot more in the state of steam than it does in the liquid state;* so that B C represents one cubic foot. The steam then expands in a non-conducting cylinder, without receiving or giving out heat, until its pressure and temperature fall to their lower limits. C D represents the expansion curve, and E D the volume occupied at the end of the expansion, partly by steam and partly by a small quantity of water which spontaneously liquefies during the expansion. Part of the steam represented in volume by D A, is then condensed by conduction of heat, at the lower limit of temperature; and a volume of steam represented by A E, less the volume of the water in the liquid state, is left uncondensed, in order that by its compression into the liquid state heat enough may be produced to raise

* Formula for that latent heat, in units of work:—

$$L - t\frac{dp}{dt};$$

where p is the absolute pressure and t the absolute temperature.

the temperature of the feed-water to the higher limit. The curve of compression, showing how the pressure increases as the volume is diminished, is represented by A B, and this completes the double stroke or revolution of the engine. The following are the successive volumes occupied by the steam at different periods, neglecting the volume of the liquid water, as being so small compared with that of the steam that it is unnecessary to take it into account for the present purpose.*

	Cubic Feet.
Volume on admission, F C, taken as sensibly equal to B C, ...	1·00
Volume at end of expansion, E D, ...	24·68
Volume at beginning of compression, E A, ...	4·08

The weight of water expended per stroke is 0·0956 lb.

8. The areas shown in the diagram, as computed by the proper formulæ, given in the footnote to the preceding article, are as follows :—

* The following are formulæ for these volumes, and for the work represented by the areas in the diagram, first demonstrated by the author in the *Philosophical Transactions* for December, 1853, and given also in *A Manual of the Steam-Engine and other Prime Movers.* Let t and t' denote the higher and lower limits of absolute temperature; p and p' the corresponding pressures; L and L′ the corresponding values of $p \frac{d}{d\,t}$; J, Joule's equivalent of the specific heat of liquid water (772 foot-pounds per degree of Fahrenheit in a pound of water); H, the value in foot-pounds of the latent heat of evaporation of 1 lb. of water at the higher limit ($= 745{,}800 - 0{\cdot}7\,\mathrm{J}\,(\mathrm{T} - \mathrm{A})$ nearly, where A is the boiling point under one atmosphere); then—

$$\frac{\mathrm{E\,A}}{\mathrm{B\,C}} = \frac{\mathrm{J}\,t'\,\mathrm{L}}{\mathrm{H\,L'}}\ \text{hyp. log.}\ \frac{t}{t'};$$

$$\frac{\mathrm{E\,D}}{\mathrm{B\,C}} = \frac{\mathrm{E\,A}}{\mathrm{B\,C}} + \frac{t'\,\mathrm{L}}{t\,\mathrm{L'}};$$

$$\mathrm{F\,C\,D\,E} = \frac{\mathrm{J\,L}}{\mathrm{H}}\left\{t - t'\left(1 + \text{hyp. log.}\ \frac{t}{t'}\right)\right\} + \frac{\mathrm{L}\,(t - t')}{t};$$

$$\mathrm{F\,B\,A\,E} = \frac{\mathrm{J\,L}}{\mathrm{H}}\left\{t - t'\left(1 + \text{hyp. log.}\ \frac{t}{t'}\right)\right\};$$

$$\mathrm{B\,C\,D\,A} = \frac{\mathrm{L}\,(t - t')}{t}.$$

All these formulæ were independently demonstrated by Clausius in 1855. (See his *Abhandlungen über die mechanische Wärmetheorie.*) If the water at the lower limit of pressure in the example were all in the state of steam, it would occupy 29·9 cubic feet. The difference between this and 24·68—viz., 5·22 cubic feet, shows what proportion of the steam is spontaneously liquefied during the given expansion in a non-conducting cylinder.

	Foot-pounds.
Total work obtained through the action of the steam in driving the piston, allowing for back pressure, F C D E = ...	16690
Work expended through compressing part of the steam into the liquid state, F B A E = ...	1446
Difference, being the indicated work per stroke, B C D A = ...	15244

as already calculated in Article 4, by the second law of thermodynamics. The lost or rejected heat,

$$68420 - 15244 = 53176 \text{ foot-pounds,}$$

is the heat abstracted during the condensation of the volume of steam represented by A D; and it is impossible to get back any part of that heat, because it is all abstracted at the lower limit of temperature, and heat will not pass from a colder to a hotter body.

9. To proceed now to the case of the air-engine. The limits of pressure do not, as in the case of steam, depend on the limits of temperature, but may be fixed according to convenience. In general, the lower limit of absolute pressure is the atmospheric pressure, which we will estimate in the present example at 2116·3 lbs. on the square foot, or 14·7 lbs. on the square inch; and this is represented by O *e* in the diagram. The upper limit of absolute pressure being arbitrary, we will assume it in the present case to be three atmospheres, or 6348·9 lbs. on the square foot, or 44·1 lbs. on the square inch. It will presently be seen that the weight of air to be used per stroke depends on the proportion borne by the upper limit of pressure to the lower. That weight of air—to be afterwards determined—is drawn into the air-pump at the atmospheric pressure and at the lower limit of temperature; and it occupies a volume represented in the diagram by *e d*. It is then compressed into a smaller volume, represented by *f a*, so that it rises to the higher limit of pressure, and transferred from the pump to the working cylinder; and, in order that this compression and transfer may cause the smallest possible expenditure of work, the air must be kept during the compression at the lower limit of temperature by means of a proper refrigerating apparatus for abstracting all the heat that the compression generates. None of that heat can be got back, for it is all abstracted at the lower limit of temperature. The temperature, then, being uniform, the volume varies inversely as the absolute pressure; *f a*, in the present example, is one-third of *e d*, and the curve *d a* is a common hyperbola. The work expended in the air-pump, which is the exact

equivalent of the heat generated there and abstracted by the refrigerating apparatus, is represented by the area *f a d e*, and is computed in foot-pounds by multiplying together the following factors:—The constant 53·15; the hyperbolic logarithm of the ratio compression (hyp. log. 3 = 1·0986); the absolute temperature at which the compression takes place in degrees of Fahrenheit (in this case 565°·2 Fahr.); and the weight of air used, in pounds.

10. The air on its way from the pump to the working cylinder passes, without change of pressure, through a perfect regenerator, in which all the heat given out by the previous supply of air is stored up; and it thus rises to the higher limit of absolute temperature, and at the same time undergoes dilatation from the volume *f a* to the volume *f b*, which volumes are to each other in the ratio of the limits of absolute temperature; viz.:—

$$fa : fb :: 565{\cdot}2 : 727{\cdot}2 :: 1 : 1{\cdot}2867.$$

11. The next process is that the air continues to expand and drive the piston until its pressure falls to the lower limit, its volume at the same time increasing to that represented by *e c*; and it is then finally expelled, giving out to the regenerator to be used over again the heat corresponding to the difference between the upper and lower limits of temperature. In order that the greatest quantity of work possible may be obtained from the expansion represented by the curve *b c*, and in order also that the air during its expulsion may give out to the regenerator a quantity of heat sufficient to raise the temperature of the next supply of air to the higher limit, the temperature must, by the supply of a sufficient quantity of heat, be maintained uniform during the expansion represented by *b c*; and that quantity of heat constitutes the whole expenditure in a perfect air-engine. Such being the case, the volume varies inversely as the pressure; *e c*, in the present example, is three times *f b*, and the curve *b c* is a common hyperbola. It is evident, moreover, that the volumes represented by *e d* and *e c* are to each other in the ratio of the limits of absolute temperature, already given. The whole work obtained by the action of the air in the cylinder, which is the exact equivalent of the whole heat expended, is represented by the area *f b c e*, and is computed in foot-pounds by multiplying together the following factors. The constant 53·15; the hyperbolic logarithm of the ratio of expansion (hyp. log. 3 = 1·0986); the absolute temperature at which the expansion takes place (in this case 727°·2 Fahr.): and the weight of air used, in pounds.

12. The area *b c d a* represents the indicated work per stroke, being the excess of the work obtained in the cylinder above the work expended in the air-pump; and the proportion which it bears to *f b c e* (representing the whole expenditure of heat) is obviously that of the range of temperature

to the higher absolute temperature, as already stated in Article 3, viz.: $\frac{162}{727 \cdot 2} = 0 \cdot 2228$. In other words, the areas shown in the diagram represent the following quantities:—

	Foot-pounds.
Total expenditure of heat per stroke, *f b c e* = . . .	68420
Heat produced by compression and abstracted by the refrigerator, *f a d e* = ...	53176
Indicated work per stroke, *b c d a* = ...	15244

The indicated work per stroke may be calculated independently by multiplying together the following factors:—The constant 53·15; the hyperbolic logarithm of the ratio of expansion (hyp. 3 = 1·0986); the range of absolute temperature (162° Fahr.); and the weight of air used, in pounds.

13. The weight of air used per stroke is determined by the consideration, that it is to be sufficient to absorb when expanding to three times its initial volume at the absolute temperature 727°·2 Fahr., a quantity of heat equivalent to 68420 foot-pounds. The calculation is as follows:—

Constant factor depending on the nature of the gas in foot-pounds per degree of Fahrenheit, . . .	53·15
Multiply by the absolute temperature, . . .	727°·2 F.
Product in foot-pounds, being equal to the product of the absolute pressure in pounds on the square foot, and the volume of 1 lb. of air in cubic feet,	38651
Multiply by the hyperbolic logarithm of the rate of expansion, hyp. log. 3 = ...	1·0986
Product being the heat absorbed by each pound of air in foot-pounds, ...	42516
$\frac{\text{Heat expended } 68420}{\text{divided by } \quad 42516}$ = ...	1·6113 lb.

of air used per stroke. With a different ratio of expansion the weight of air required per stroke would be different, varying inversely as the logarithm of that ratio. Had any other gas been employed instead of air the only difference in this quantity would have been that arising from a different value of the constant factor. The following are the volumes occupied by the air at different periods :—

	Cubic Feet.
c d	22·87
f a	7·62
f b	9·81
c e	29·43

14. The following is a summary of the comparison between the steam-engine and the air-engine in the example given:—

Per Stroke.	Steam-Engine. Foot-pounds.	Air-Engine. Foot-pounds.	Difference.
Heat expended, ...	68420	68420	0
Heat rejected, ...	53176	53176	0
Heat transformed into mechanical work,	15244	15244	0
Efficiency, ...	0·2228	0·2228	0
Work obtained by expansion, . .	16690	68420	51730
Work expended in compression, . .	1446	53176	51730
Indicated work, ...	15244	15244	0

The quantities in the column headed "difference," represent a diminution of the work obtained by expansion, and an exactly equal saving in the work expended in compression, produced by the mutual attraction of the particles of water.

15. It may be useful to point out what effects would be produced by departing from that condition of maximum efficiency which requires that the elevation of temperature shall be produced without the expenditure of heat. In the case of the steam-engine there would be a gain of indicated work, represented by the area F B A E = 1446 foot-pounds, so that the indicated work per stroke would be 16690 foot-pounds. At the same time there would be an additional expenditure of heat, calculated as follows :—

Joule's equivalent of the specific heat of liquid water,	772
Number of degrees of Fahr. by which the temperature of the feed-water has to be raised, . . .	162°
Weight of water used per stroke (lb.), . . .	0·0956
Product of those three factors, being the additional expenditure of heat per stroke in foot-pounds, .	11956
Expenditure of heat as formerly calculated, . .	68420
Total, ...	80376
Efficiency, as diminished by omitting the compression, $\frac{16690}{80376} =$	0·208

being about fifteen-sixteenths of the greatest possible efficiency with the given limits of temperature.

16. In the case of the air-engine, suppose the regenerator omitted, so that the whole elevation of temperature of the air has to be produced by heat supplied from the furnace; then, in the present example, there would be an additional expenditure of heat, calculated as follows:—

Dynamical equivalent of the specific heat of air under constant pressure, in foot-pounds per degree of Fahr., ...	183·45
Elevation of temperature, ...	162° F.
Weight of air used per stroke, the rate of expansion being 3, ...	1·6113 lb.
Product of those three factors, being the additional expenditure of heat in foot-pounds, . . .	47886
Expenditure of heat as formerly calculated, . .	68420
Total, ...	116306
Efficiency, as diminished by altogether omitting the regenerator $\frac{15244}{116306} =$	0·131 nearly,

or about six-tenths of the greatest possible efficiency with the given limits of temperature.

17. It is evident that the regenerator, or some equivalent apparatus, cannot be omitted consistently with economy of heat in an air-engine. An actual regenerator, however, will never succeed in storing and giving out the whole supply of heat required for the elevation of temperature. Suppose, for the sake of illustration, that in the present example the regenerator stores and gives out *nine-tenths* of the heat required, leaving *one-tenth* to be supplied from the furnace, then we have the following expenditure of heat per stroke:—

	Foot-pounds.
Expenditure as originally calculated, ...	68420
Additional expenditure through imperfect action of the regenerator, ...	4789
Total, ...	73209

Efficiency, as diminished by imperfect action of the regenerator $\frac{15244}{73209} =$ . . . . . . 0·208 nearly,

or about fifteen-sixteenths of the greatest possible efficiency with the given limits of temperature.

18. The law which the preceding examples illustrate leads to the conclusion that if by means of air-engines greater economy of fuel than in steam-engines is to be attained, it must be by the following means:— Working with a greater range of absolute temperature than is practicable or safe in steam-engines, and using the products of combustion directly to drive the piston, so as to save nearly the whole of the heat that is wasted in a steam boiler, or in an air-engine in which the products of combustion are not so used; and it is probable that the latter is the more easily practicable of the two means of economising heat.

19. The following examples illustrate the relation between the efficiency of a heat engine and the consumption of fuel per indicated horse-power per hour. The fuel is supposed to be nearly pure carbon, the mechanical equivalent of the whole heat produced by the combustion of one pound of it being 11,000,000 foot-pounds.

Efficiency.	Indicated Work per Pound of Carbon.	Pound Carbon per Indicated Horse-power per Hour.
1	11,000,000	0·18
0·5	5,500,000	0·36
0·25	2,750,000	0·72
0·20	2,200,000	0·90
0·15	1,650,000	1·20
0·125	1,375,000	1·44
0·100	1,100,000	1·80
0·075	825,000	2·40
0·050	550,000	3·60
0·025	275,000	7·20

SUPPLEMENT.

20. I think it desirable to add the following explanations and illustrations of the principle, that in an air-engine, under all circumstances whatsoever, *the heat produced by the compression of the air is wholly and unavoidably lost*—a principle which is a necessary consequence of the fact that heat never passes directly from a colder to a hotter body.

21. The heat produced by the compression must either be abstracted from the air as fast as it is produced, so as to keep the temperature of the air constant during the compression, or allowed to accumulate and raise the temperature of the air. In the former case (which is that described in Article 9) it is at once evident that the heat, being abstracted by means of an external substance, such as water or air, that is colder than the lowest temperature of the working substance, can never be transmitted to the working substance again.

22. In the latter case, when the heat produced by the compression is not abstracted, but allowed to accumulate and raise the temperature of the air, the air is passed through the regenerator or economiser on its way

towards the working cylinder at an increased temperature higher than the lower limit. Therefore, the air which has done its work and is escaping through the regenerator in the opposite direction is not reduced below the same increased temperature, for it is only through the entering air being at a lower temperature that any heat can be abstracted from the escaping air. Therefore, the escaping air carries off to waste a quantity of heat equal to the quantity of heat produced by compression and allowed to accumulate in the compressed air, so that the effect of such accumulation is neutralised, and the total expenditure of heat remains unaltered.

23. Moreover, the increase of temperature of the air undergoing compression causes an increase in the quantity of work expended in compressing it above the quantity of work which would be required if the temperature were kept constant; and thus the indicated work is diminished, and the engine ceases to be one of maximum efficiency between the given limits of temperature.

24. To illustrate this by an example, I will suppose that the compression of the air takes place in a non-conducting cylinder, and is carried to an extent such that the temperature of the air is raised from the lower limit to the higher limit by means of the compression alone. Then it is known that the pressure of the air, instead of varying inversely as the volume simply, varies inversely as that power of the volume whose index is 1·408, while the absolute temperature varies inversely as that power of the volume whose index is 0·408. In Fig. 2 (as in the former figure) let volumes in cubic feet be represented by distances parallel to O V, and pressures in pounds on the square foot by ordinates parallel to O P, so that the areas of diagrams represent quantities of work in foot-pounds. Let O E represent the original

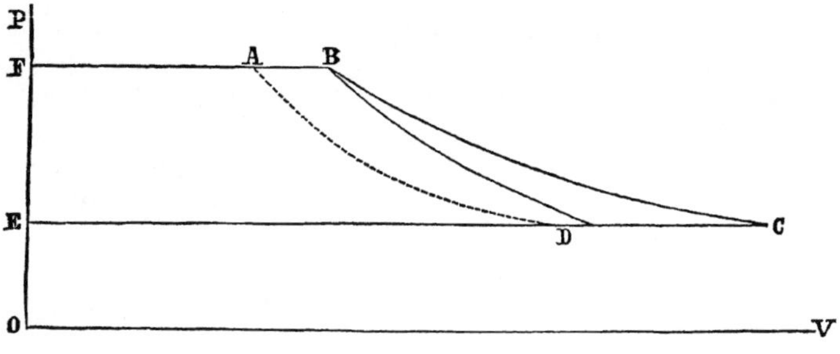

Fig. 2.

absolute pressure, E D the original volume of the air, and O F the increased pressure at the end of the compression. Then, if the temperature were kept uniform, the compression curve of the diagram would be a common hyperbola D A, such that $OF \times FA = OE \times ED$. But the rise of temperature causes the compression curve to assume a steeper figure, D B, of which the law has already been stated. The absolute temperature at the end of the compression is increased in the ratio $\frac{FB}{FA}$; and the work done in driving the compressing pump is represented by the area of the diagram E F B D, and is equivalent to the quantity of heat required in order to produce the rise of temperature from the lower to the higher limit under constant pressure. Now, taking the same data as before, viz. :—

Temperatures.	Degrees of Fahrenheit. Ordinary scale.	Absolute scale.
Upper limit, ...	266°	727°·2
Lower limit, ...	104°	565°·2
Difference, ...	162°	162°·0

Ratio in which the absolute temperature is to be increased,

$$\frac{FB}{FA} = \frac{727^\circ{\cdot}2}{565^\circ{\cdot}2} = 1{\cdot}2867;$$

Dynamical value of the specific heat of air under constant pressure, 183·45 foot-pounds per degree of Fahrenheit.

Original pressure, O E = 2116·3 lbs. on the square foot;

Original volume of one pound of air, 14·19 cubic feet;

We obtain the following results :—

Ratio of volumes, $\frac{ED}{FB} = 1{\cdot}8546$;

Ratio of pressures, $\frac{OF}{OE} = \frac{ED}{FA} = \frac{ED}{FB} \cdot \frac{FB}{FA} = 1{\cdot}8546$

$$\times\ 1{\cdot}2867 = 2{\cdot}3862;$$

Pressure at end of compression O F = 5050 pounds on the square foot;

Volume of one pound of air at end of compression, 7·65 cubic feet;

Work done in compressing pump per pound of air,

$$162^\circ \times 183{\cdot}45 = 29719 \text{ foot-pounds.}$$

25. The air being now at the upper limit of temperature, let it be transferred to the working cylinder, and there expanded without allowing

the temperature to fall below that limit, until the pressure falls to the atmospheric pressure O E, and then let it be discharged. The expansion-curve B C will, as in Article 11, be a common hyperbola, so that the ratio of expansion will be

$$\frac{\mathrm{E\,C}}{\mathrm{B\,F}} = \frac{\mathrm{O\,F}}{\mathrm{O\,E}} = 2{\cdot}3862,$$

and the volume of one pound of air at the end of the expansion will be 18·26 cubic feet. The area of the diagram F B C E will represent at once the work done in driving the piston, and the whole expenditure of heat. The method of computing that area has already been stated in Article 11, and the calculation in the present case is as follows, *for each pound of air used*:—

Constant factor, ...	53·15
Absolute temperature at which the expansion takes place, ...	727°·2 F.
Hyperbolic logarithm of the ratio of expansion; hyp. log. 2·3862 = ...	0·8697
Product of those three factors, being the work done and heat expended in the working cylinder, per pound of air used, in foot-pounds, ...	33615

26. The air finally escapes at the higher limit of temperature, and therefore carries to waste the whole of the heat which was employed in raising its temperature, having been produced by compression. The value of that heat in foot-pounds per pound of air has been already found, in Article 24, to be 29719.

27. The indicated work represented by the area B C D is consequently, in foot-pounds per pound of air,

$$33615 - 29719 = 3896;$$

and the efficiency of the engine,

$$\frac{3896}{33615} = 0{\cdot}116 \text{ nearly},$$

or about one-half of the greatest possible efficiency with the given limits of temperature.

28. In order that the present example may be the more easily compared with the former examples, we will suppose that the heat to be expended per stroke is, as before, equivalent to 68420 foot-pounds.

Then the expenditure of air per stroke required in order to take up that heat during its expansion is found to be

$$\frac{68420}{33615} = 2{\cdot}0354 \text{ lbs.};$$

whence the following results are obtained :—

	Foot-lbs.
Waste heat per stroke equivalent to the work of driving the compressing pump, $162^\circ \times 183{\cdot}45 \times 2{\cdot}0354 =$	60490
Indicated work per stroke, $3896 \times 2{\cdot}0354 =$ . .	7930

Efficiency, $\frac{7930}{68420} = 0{\cdot}116$, as already found.

Volumes successively occupied by the air in cubic feet per stroke:

$$ED = 28{\cdot}89;\ FB = 15{\cdot}58;\ EC = 37{\cdot}17.$$

29. The case of Article 9—where the whole of the heat generated in the compressing pump is abstracted as fast as it is produced—and that of the example just described—in which the whole of that heat is at first employed in raising the temperature of the air while an equal quantity of heat goes to waste with the escaping air that has done its work—form two extremes, and between those extreme cases there may lie an indefinite number of intermediate cases, in which part of the heat generated by compression is abstracted at once, and part employed in raising the temperature of the air. It will be readily understood that in all those intermediate cases the result is the same as in the two extreme cases—the whole of the heat generated by the compression of the air goes to waste either at once or with the escaping air, and none of it is available for conversion into indicated work; nor would it be possible that it should become so available, unless it were possible for heat to be directly transferred from a colder body to a hotter body.

XXVI.—ON THE WORKING OF STEAM IN COMPOUND ENGINES.*

1. *Principal Kinds of Compound Engines.*—By a compound steam-engine is meant one in which the mechanical action of the steam commences in a smaller cylinder and is completed in a larger cylinder. Those cylinders are respectively called, for convenience, the high-pressure cylinder and the low-pressure cylinder. Two classes of compound engines will be considered—first, those in which the steam passes directly or almost directly from the high-pressure to the low-pressure cylinder, the forward stroke of the latter cylinder taking place either exactly or nearly at the same time with the return stroke of the former cylinder; and, secondly, those in which the steam, on its way from the high-pressure to the low-pressure cylinder, is stored in a reservoir, so that any convenient fraction of a revolution (such, for example, as a quarter revolution) may intervene between the ends of the strokes of the cylinders. As to the latter class of engines, reference may be made to a paper by Mr. E. A. Cowper in the *Transactions of the Institution of Naval Architects* for 1864, page 248. Sometimes, especially in the first class of compound engines (those without reservoirs), there are a pair of low-pressure cylinders whose pistons move together, and which act like one cylinder divided into two parts.

2. *Advantages of Compound Engines.*—As regards the theoretical efficiency of the steam, the compound engine possesses no advantage over an engine with a single cylinder of the dimensions of the low-pressure cylinder, working with the same pressure of steam and the same rate of expansion. The advantages which it does possess are the following:—First, in point of strength, the action of the steam when at its highest pressure takes place, in the compound engine, upon a comparatively small piston, thus diminishing the amount of the greatest straining force exerted on the mechanism and framing; secondly, in point of economy of heat and steam, in a single-cylindered engine it is necessary, in order to prevent liquefaction and re-evaporation of the steam, and consequent waste of heat, that the whole metal of the cylinder should be kept, by means of a steam jacket, at a temperature equal to that of the steam when first

* From *The Engineer* of March 11, 1870.

admitted; whereas, in a compound engine, it is the smaller or high-pressure cylinder only which has to be kept at so high a temperature, it being sufficient to keep the larger or low-pressure cylinder at the temperature corresponding to the pressure at which the steam passes from the high-pressure to the low-pressure cylinder. Thirdly, in point of economy of work: the whole of the force exerted by the piston rod upon the crank in a single-cylindered engine takes effect in producing friction at the bearings; whereas, in compound engines, the mechanism can be so arranged that the forces exerted by the piston rods on the bearings shall, to a certain extent, balance each other, thus diminishing the friction. When there are a pair of low-pressure cylinders with a high-pressure cylinder between them (as in the engines of H.M.S. "Constance," by Messrs. Randolph, Elder, and Co.) the balance can be made almost perfect. These remarks apply not only to the forces due to the pressure of the steam, but to those produced by the reaction or inertia of the pistons and of the masses which move along with them. The advantages which have been stated are obviously greatest with high rates of expansion.

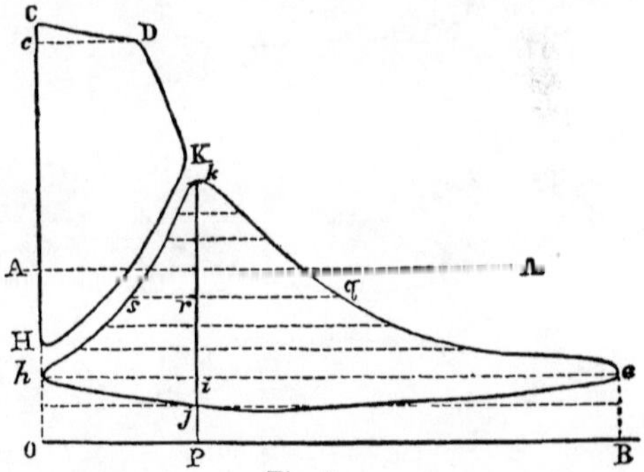

Fig. 1.

3. *Combination of Diagrams.*—When the diagrams of the high and low-pressure cylinders of a compound engine are taken by means of one indicator they have the same length of base; and when arranged in the customary way for inspection they present appearances which are represented in Fig. 1 for engines without reservoirs, and in Fig. 2 for engines with reservoirs. In each Fig. A A is the atmospheric line, O B the zero line of absolute pressure, and the length O P on that line is the common length of the diagrams of both cylinders, as originally drawn. The diagram of the high-pressure cylinder is represented in Fig. 1 by C D K H, and in Fig. 2 by C D K L; that of the low-pressure cylinder, as drawn by the indicator, is represented in Fig. 1 by *k i h*, and in Fig. 2 by *l i g h*. In combining the diagrams of the two cylinders into one diagram, it is to be borne in mind that when the area of a diagram is considered as representing the work done by the steam on the piston at one stroke, the length of the base of the diagram is to be considered as representing the *effective* capacity of the cylinder: that is, the space swept through by the piston at one stroke. Hence, in order to prepare the

diagram of the low-pressure cylinder for combination with that of the high-pressure cylinder, the lengths of its base, and of every line in it

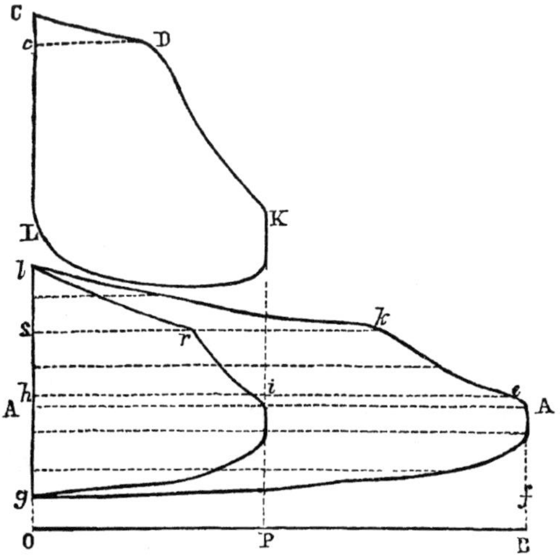

Fig. 2.

parallel to its base, are to be increased in the ratio in which the effective capacity of the low-pressure cylinder is greater than that of the high-pressure cylinder.* (When there are a pair of low-pressure cylinders combined with one high-pressure cylinder, they are equivalent to one low-pressure cylinder of double the capacity.) In each of the Figs. 1 and 2, then, the base O P is, in the first place, to be produced to B, making O B greater than O P in the proportion above-mentioned. To complete the preparation of the low-pressure diagram draw, in each case, a series of lines across it parallel to the base O B, such as the dotted lines in each Fig., of which one is marked *h i e*. Let *c* denote the ratio O B ÷ O P. Then, in the case of an engine without a reservoir (Fig. 1) draw P *k* perpendicular to O B, cutting all the parallel dotted lines, and on each of those lines (such as *s r q*) lay off $sq = c \cdot rs$. A curve *k q e j*, drawn through

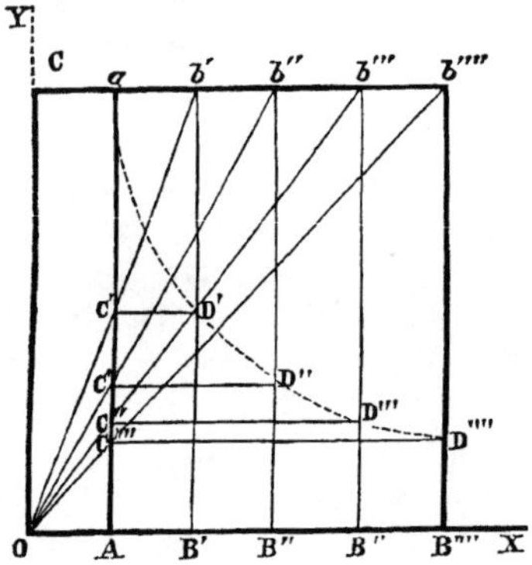

Fig. 3.

* *Manual of the Steam-Engine and other Prime Movers*, page 334.

the points, such as q, thus found, will be the required boundary of the enlarged low-pressure diagram, $k\,q\,e\,j\,h\,s\,k$, which, being joined on to the

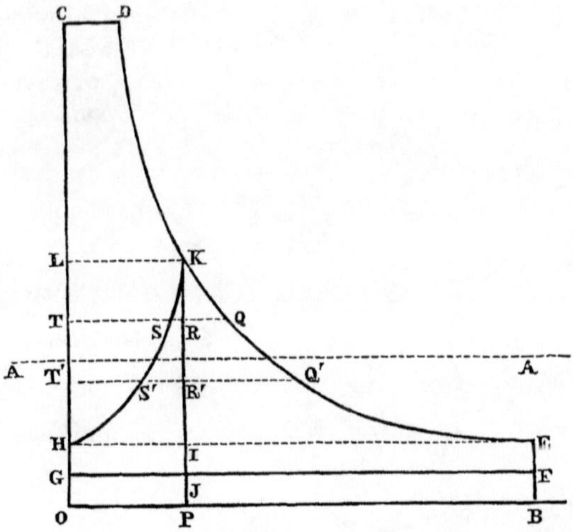

Fig. 4.

high-pressure diagram C D K H, makes the combined diagram. When the engine has a reservoir, draw O l (Fig. 2) perpendicular to O B, and crossing all the parallel dotted lines, and on each of those lines (such as

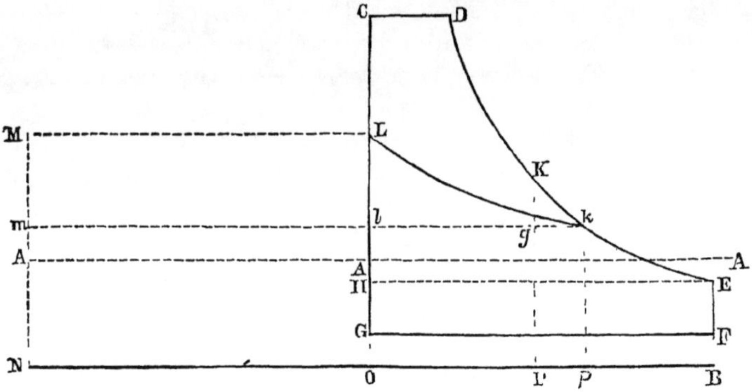

Fig. 5.

$s\,r\,k$) lay off $s\,k = c \cdot s\,r$. A curve $l\,k\,e\,f\,g$, drawn through the points, such as k, thus found, will be the required boundary of the enlarged low-pressure diagram, which, being joined on to the high-pressure diagram C D K L, makes the combined diagram. In a theoretically perfect engine, in which the steam passed from the high-pressure to the low-pressure cylinder without change of pressure or temperature, the two diagrams would join exactly at the boundaries K L and $k\,l$ in Fig. 2, or K H and $k\,h$ in Fig. 1, so as to form one diagram identical with that produced by the same quantity of steam working between the same limits of pressure in the larger cylinder only. But in actual engines there is sometimes a gap between the high and low-pressure diagrams, as in the Figs. 1 and 2; and sometimes, when the steam reservoir is heated, they overlap each other.

4. *Rates of Expansion.*—In Figs. 1 and 2 let D be the point where the cut-off takes place in the high-pressure cylinder; draw D c parallel to B O; then $\frac{\mathrm{OP}}{c\mathrm{D}}$ is the rate of expansion in the high-pressure cylinder; $\frac{\mathrm{OB}}{\mathrm{OP}}$ is the total increase of volume in passing from the high-pressure to the low-pressure cylinder; and the product of those quantities,

$$\frac{\mathrm{OP}}{c\mathrm{D}} \cdot \frac{\mathrm{OB}}{\mathrm{OP}} = \frac{\mathrm{OB}}{c\mathrm{D}},$$

is the total rate of expansion.

5. *Construction of Theoretical Expansion-Diagrams for Proposed Engines.*—In constructing the theoretical diagram of a proposed steam-engine, certain well-known assumptions are made in order to simplify the figure and the calculations founded upon it. In the first place, the pressure of the steam during its admission is assumed to be constant, so that the uppermost boundary of the diagram, as in Figs. 4 and 5, is a straight line, C D, parallel to the zero line, O B, the height O C representing the absolute pressure of admission. Secondly, the back pressure is assumed to be constant; so that the lower boundary of the diagram also is a straight line F G parallel to O B, the height O G representing the mean absolute back pressure, as estimated from the results of experience. Thirdly, it is commonly assumed that at the beginning of the forward stroke the pressure rises suddenly from the back pressure to the pressure of admission, so that the first end-boundary of the diagram is a straight line G C perpendicular to O B. Fourthly, it is assumed that at the end of the forward stroke the pressure falls suddenly from the pressure at the end of the expansion (or *final pressure*) to the back pressure, so that the second end-boundary of the diagram is a straight line, E F, perpendicular to O B. Fifthly, for the expansion curve (D E in Figs. 4 and 5), which completes the boundaries of the diagram, there is assumed a line of the hyperbolic class. Thus the area of an assumed theoretical diagram of the work of the steam in a proposed engine is made up of a hyperbolic area C D E H, and a rectangular area E F G H. The form of the expansion curve depends on a number of circumstances, such as the initial pressure and temperature of the steam,

the proportion of water (if any) admitted along with it in the liquid state, the communication of heat between the steam and the metal of the cylinder, the communication of additional heat to the steam during its expansion by the help of a steam jacket. Writers on thermodynamics have determined the exact form of that curve in various cases, such as that of steam originally dry, expanding in a non-conducting cylinder; that of steam originally containing a given proportion of moisture expanding in a non-conducting cylinder; that of steam originally dry supplied during the expansion with heat just enough to keep any part of it from condensing; that of steam supplied during the expansion with heat sufficient to keep it at a constant temperature. For elementary methods of approximating to the results of the exact methods in such cases, see *The Engineer* of the 5th January, 1866. For most practical purposes the common hyperbola forms a good approximation to the true expansion curve, and it is convenient because of the simplicity of the processes for finding its figure, whether by calculation or by construction. To find by calculation the series of absolute pressures corresponding to a given series of volumes assumed by the steam, on the supposition that the expansion curve is a common hyperbola, multiply the initial absolute pressure by the initial volume; divide the product by any one of the given series of volumes; the quotient will be the corresponding absolute pressure. To find a series of points in the common hyperbola, in Fig. 3, draw the two axes O X and O Y perpendicular to each other; O X to form a scale of volumes, and to represent the zero line of absolute pressure; O Y to form a scale of absolute pressures. On O X lay of O A to represent the initial volume of the steam; also O B′, O B″, &c., to represent a given series of volumes occupied by the same steam during its expansion. On O Y lay off O C to represent the initial absolute pressure of the steam. Through C draw the straight line C *a b′ b″*, &c., parallel to O X, and through the points A, B′, B″, &c., draw the series of straight lines A *a*, B′ *b′*, B″ *b″*, &c., parallel and equal to O C. From O draw the series of diverging straight lines O *b′*, O *b″*, &c., and mark the series of points C′, C″, &c., where they cut A *a*. From these points, and parallel to O X, draw the series of straight lines C′ D′, C″ D″, &c., and mark the series of points D′, D″, &c., where they cut the series of straight lines B′ *b′*, B″ *b″*, &c. These points, together with the point *a*, will be points in the required hyperbola, *a* D′ D″, &c., which is taken as an approximation to the expansion curve.

6. *Calculation of Mean Absolute Pressure and of Indicated Work in a Theoretical Diagram.*—Suppose that in Fig. 3 O B″″ represents the final volume of the steam, so that D″″ is the end of the expansion curve, and that B″″ D″″ represents the final absolute pressure. The intermediate volumes O B′, O B″, &c., are to be so chosen that the points B′, B″, &c., shall divide A B″″ into an even number of equal intervals. Multiply the

series of absolute pressures represented by A a, B′ b', &c., by "Simpson's multipliers," which are, for the initial and final pressures, 1; and for the intermediate pressures, 4 and 2 alternately; so that, for example, for four intervals and five absolute pressures, the multipliers are 1, 4, 2, 4, 1; for six intervals and seven absolute pressures, 1, 4, 2, 4, 2, 4, 1, and so on. Add the products together; divide the sum by three times the number of intervals. Multiply the quotient by the rate of expansion less one; to the product add the initial absolute pressure; divide the sum by the rate of expansion; the quotient will be the required mean absolute pressure nearly.—*Example:* Rate of expansion, 5; expansion divided into 8 equal intervals; initial absolute pressure, 37·8 lbs. on the square inch.

	Absolute Pressures.	Multipliers.	Products.
Initial	37·80	1	37·80
Intermediate	$37{\cdot}8 \times \frac{2}{3} = 25{\cdot}20$	4	100·80
	$37{\cdot}8 \times \frac{1}{2} = 18{\cdot}90$	2	37·80
	$37{\cdot}8 \times \frac{2}{5} = 15{\cdot}12$	4	60·48
	$37{\cdot}8 \times \frac{1}{3} = 12{\cdot}60$	2	25·20
	$37{\cdot}8 \times \frac{2}{7} = 10{\cdot}80$	4	43·20
	$37{\cdot}8 \times \frac{1}{4} = 9{\cdot}45$	2	18·90
	$37{\cdot}8 \times \frac{2}{9} = 8{\cdot}40$	4	33·60
Final	$37{\cdot}8 \times \frac{1}{5} = 7{\cdot}56$	1	7·56

Divide by 8 intervals × 3 = 24	) 365·34 sum
Quotient	15·2225
Multiply by rate of expansion 5 − 1	4
Product	60·89
Add initial absolute pressure . .	37·80
Divide by rate of expansion . . 5)	98·69 sum
Mean absolute pressure nearly	19·738

The remainder left after subtracting the back pressure from the mean absolute pressure is the *mean effective pressure*, which, being multiplied by the area of the piston, and by the distance moved through by the piston in a given time, gives the indicated work of the steam in that time.*

* A well-known formula for the ratio of the mean to the initial absolute pressure is $\frac{1 + \text{hyp. log } r}{r}$, r being the rate of expansion. For a graphic approximate solution of the same question, see *The Engineer* for April 13, 1866.

7. *Theoretical Combined Diagrams.*—By the process described in the preceding section there may be constructed the approximate theoretical diagram of steam working with a given initial pressure and a given rate of expansion, and against a given mean back pressure. In the case of a proposed compound engine, that theoretical diagram is to be regarded as the combined diagram of the two cylinders, and it is to be divided into two parts, representing the parts of the indicated work done in the two cylinders respectively. In each of the two Figs. 4 and 5, the theoretical combined diagram is represented by C D E F G, O C being the initial, and B E the final absolute pressure, O G = B F the mean back pressure, A A the atmospheric line, O B the zero line of absolute pressure, H E = F G = O B the effective capacity of the large cylinder, C D the initial volume of steam admitted per stroke, and $\frac{EH}{CD}$ the total rate of expansion. The dividing line which marks the boundary between the high-pressure and low-pressure theoretical diagrams is represented in Fig. 4 by K H, and in Fig. 5 either by *l* L, or by *k l*, or by some line near those lines, as will afterwards be more fully explained.

8. *Theoretical Diagrams of a Compound Engine without a Reservoir.*—When the steam passes directly, without loss of pressure or of heat, from the high-pressure to the low-pressure cylinder, the dividing line of the theoretical compound diagram is found by the following process. In O B (Fig. 4) lay off O P to represent the effective capacity of the high-pressure cylinder. Through P, parallel to O C, draw the straight line P J K, cutting the back-pressure line in J, and the expansion curve in K; then K will be one end of the dividing line. Through the lower end E of the expansion curve, and parallel to B O, draw E H, cutting O C in H; then H will be the other end of the dividing line. To find intermediate points, draw, parallel to O B, a series of straight lines, such as T R Q, T′ R′ Q′, across the part of the diagram which lies below the point K, and in each of those lines, for example, in Q R T, lay off R S, bearing the same proportion to R Q that P O bears to P B; the points thus marked, such as S and S′, will be in the required dividing line K H. The areas of the two parts of the theoretical diagram, C D K H, and K E F G H, being measured by ordinary methods, will show the comparative quantities of work done in the high-pressure and low-pressure cylinders respectively. The advantages of the compound engine in point of diminution of stress and friction are most fully realised when those quantities of work are equal; that is, when the line K H divides the area C D E F G into two equal parts; for then the mean values of the forces exerted through the two piston-rods are equal; hence the proportion borne by the effective capacity of the high-pressure cylinder to that of the low-pressure cylinder ought to be chosen so as to realise that condition as nearly as possible. An exact rule for that pur-

pose would be too complex to be useful for practical purposes. The following empirical rule has been found by trial to give a good rough approximation to the required result in ordinary cases of compound engines without reservoirs: Make the ratio in which the low-pressure cylinder is larger than the high-pressure cylinder, equal to the square of the cube root of the total rate of expansion; for example, if the total rate of expansion is to be 8, let the low-pressure cylinder be four times the capacity of the high-pressure cylinder (this rule was first given in *Ship-building, Theoretical and Practical,* page 275). When a table of squares and cubes is at hand, look for the total rate of expansion in the column of cubes, the required ratio will be found in the column of squares.

9. *Theoretical Diagrams of a Compound Engine with a Reservoir.*—To realise theoretical perfection in the working of an engine with an intermediate steam reservoir, that reservoir should be absolutely non-conducting, so that the steam may pass from it into the low-pressure cylinder at exactly the same pressure and volume at which it is received from the high-pressure cylinder. Supposing this condition to be realised, let O p in Fig. 5, represent the volume of steam admitted into the low-pressure cylinder at each stroke, so that $\frac{\mathrm{OB}}{\mathrm{O}p}$ is the rate of expansion in that cylinder; then O p will also represent the effective capacity of the high-pressure cylinder, and $\frac{\mathrm{O}p}{\mathrm{CD}}$ will be the rate of expansion in it; and if $p\,k$ be drawn parallel to O C, so as to cut the expansion curve in k, this point will be one end of the required dividing line. To find other points on that line under the same theoretical conditions, combined with the supposition that the forcing of the steam into and its delivery out of the reservoir take place at certain times, produce B O, making O N of a length representing the capacity of the reservoir: then in O C lay off O L greater than $p\,k$, in the proportion in which N p is greater than N O; L will be the other end of the dividing line k L, which line will be an expansion curve for steam of the initial volume represented by N O = M L, and initial absolute pressure represented by O L = N M, and may be constructed by the method of Sec. 5. The high-pressure diagram will be C D k L, and its lower boundary, k L, will represent the increase of pressure during the process of forcing the steam from the high-pressure cylinder into the reservoir; the low-pressure diagram will be L k E F G, and its upper boundary L k will represent the diminution of pressure during the process of delivering the steam from the reservoir into the low-pressure cylinder. But, in reality, the entrance of the steam into and its delivery from the reservoir take place partly at the same time, and the metal of the reservoir abstracts heat from the entering steam, and gives heat back to the escaping steam; the practical result, as shown by the diagrams

published in Mr. Cowper's paper already referred to, being that the pressure of the steam in the reservoir is nearly constant, so that the upper boundary of the low-pressure diagram nearly coincides with a straight line, $l\,k$, parallel to O B. The same straight line also coincides nearly with the lower boundary of the high-pressure diagram.

In the engines experimented on by Mr. Cowper, the effective capacity of the high-pressure cylinder was somewhat smaller than the volume of steam admitted at each stroke into the low-pressure cylinder, being represented, for example, by O P instead of by $O\,p$; and the final pressure P K in the high-pressure cylinder was greater than the pressure in the reservoir.

The high-pressure diagram was thus made to resemble $C\,D\,K\,q\,l$ in Fig. 5, leaving a sort of *notch*, $K\,q\,k$, between it and the low-pressure diagram $l\,K\,E\,F\,G$; but it appears that this loss of area was compensated by the effect of the steam-jacket enveloping the reservoir, which, by imparting additional heat to the steam, caused the low-pressure diagram to be of a fuller form in the part $k\,E$ than that bounded by the theoretical expansion curve.

In this case a rough approximation to an equal division of work between the high and low-pressure cylinders may be obtained by making the rate of expansion in the low-pressure cylinder equal to the square root of the total rate of expansion.

XXVII.—ON THE THEORY OF EXPLOSIVE GAS-ENGINES.*

1. *Thermodynamical Propositions.*—In calculations respecting the practical use of heat engines, it is convenient to employ rules in which the pressures and volumes alone of the working substance are taken explicitly into account, so as to avoid the necessity for computing temperatures. Such rules exist in the case of the steam-engine. The object of the present communication is to explain a similar set of rules applicable to explosive gas-engines. They are based mainly on the following established propositions in thermodynamics: Let k denote the ratio in which the specific heat of a substance in the perfectly gaseous state under constant pressure, exceeds the specific heat of the same substance at constant volume. Then—

First proposition.—When a mass of that substance passes from the absolute pressure p and volume v, to the absolute pressure p' and volume v', the dynamical equivalent of the *sensible heat* absorbed by it (that is, heat employed in producing elevation of temperature, as distinguished from heat which disappears in doing work) has the following value:

$$\frac{p' v' - p v}{k - 1}.$$

The pressures and volumes are supposed to be given in such measures that the product of a pressure and volume may be expressed in units of work. For example, if volumes are given in cubic feet, pressures should be given in pounds on the square foot, in order that the product of a pressure and a volume may be expressed in foot-pounds.

Second proposition.—When a mass of the same substance performs work by expanding without transfer of heat, the pressure falls in such a manner that $p v^k$ is a constant quantity.

The value of k for atmospheric air is 1·408; it is very nearly the same for oxygen and nitrogen; and it does not differ much from 1·4 in the gaseous mixture resulting from the explosion of coal gas and air in the ordinary proportions: a mixture of which about three-fourths consists of nitrogen. Consequently, throughout this communication, $1·4 = \frac{7}{5}$ will

* From *The Engineer* of July 27, 1866.

be taken as a value of k, sufficiently near to the truth for practical purposes where minute accuracy is neither necessary nor possible; so that

$$\frac{1}{k-1} = \frac{5}{2};\ \frac{k}{k-1} = \frac{7}{2}.$$

2. *Rules as to Heat and Expansion.*—The following rules are the immediate consequences of the two propositions just stated:

I. A mass of a gaseous mixture, occupying the constant volume v, has its pressure increased from p to p'; the quantity of heat in units of work required to effect that change is

$$\frac{5}{2} v\,(p' - p).$$

II. A mass of a gaseous mixture, under the constant pressure p, has its volume increased from v to v'. The quantity of heat employed in this case to produce rise of temperature is, as before, in units of work, $\frac{5}{2} p\,(v' - v)$; and at the same time the work done through the expansion is $p\,(v' - v)$, and an equivalent quantity of heat disappears; so that the whole quantity of heat required, in units of work, is

$$\frac{7}{2} p\,(v' - v).$$

III. A mass of a gaseous substance performs work by expanding from the volume v to the greater volume $r\,v$ without transfer of heat, r being the rate of expansion. Then, if the original absolute pressure is p_1, the final absolute pressure will be

$$p_2 = p_1 r^{-\frac{7}{5}}$$

The following table gives some results of this rule:—

Rate of Expansion. r	Cut-off. $\frac{1}{r}$	$\frac{\text{Final Pressure.}}{\text{Initial Pressure.}}$ $\frac{p_2}{p_1} = r^{-\frac{7}{5}}$
5	0·2	0·105
4	0·25	0·144
$3\frac{1}{3}$	0·3	0·185
$2\frac{6}{7}$	0·35	0·230
$2\frac{1}{2}$	0·4	0·277
$2\frac{2}{9}$	0·45	0·327
2	0·5	0·379

The exact calculation of $r^{-\frac{7}{5}}$ requires the aid of logarithms. In the absence of logarithms an approximate value may be computed by the following empirical formula :—

$$\frac{p_2}{p_1} = 0{\cdot}54 \left(\frac{1}{r} + \frac{1}{r^2}\right) - 0{\cdot}025 \text{ nearly;}$$

which is correct to about one per cent. when r *is not less than* 2, *nor greater than* 7; but should not be used beyond those limits.

3. *Diagram.*—The general character of the indicator-diagram of an explosive engine is shown by the lines marked A C E G H A in the figure.

The base of the figure, O V, represents a scale of volumes, on which O B may be taken to denote one cubic foot of a suitable explosive mixture introduced into a cylinder at the atmospheric pressure represented by the ordinates O A = B C. In symbols, let p_0 stand for the atmospheric pressure: then A C will represent the line drawn by the indicator-pencil during the introduction of one cubic foot of the explosive mixture. Suppose that the admission is now cut off, and the mixture fired by a spark; and suppose also that the time occupied by the explosion is very small, compared with the time occupied by a stroke of the piston: then the sudden increase of pressure produced by the explosion may be approximately represented by C E, the absolute pressure immediately afterwards being represented by B E. The gaseous mixture of products of the explosion then expands, driving the piston before it; let E G be the expansion curve, so that O F is the final volume, and F G the final absolute pressure of the gas. G H represents the fall to the atmospheric pressure upon opening the eduction valve, and H A the expulsion of the gaseous mixture against the atmospheric pressure, so that the work done by each cubic foot of explosive mixture is represented by the area C E G H C.

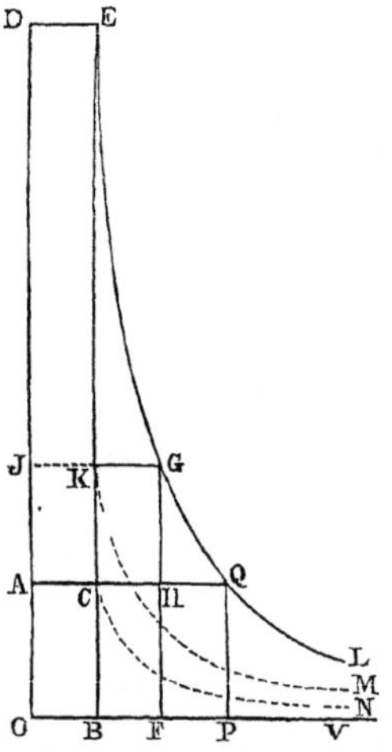

4. *Total and Available Heat of Explosion.*—The total heat of explosion may be calculated theoretically from the composition of the explosive mixture employed, by the aid of data obtained from such experiments as those of Favre and Silbermann. For example, according to information given by Dr. Letheby (*Engineer*, 28th June, 1866, p. 448), the mixture

found to answer best in Lenoir's gas-engine is composed of eight parts by volume of air to one of common coal gas. From Dr. Letheby's analysis of the gas, and the known values of the total heat of combustion of its constituents, it appears that the total heat of explosion of *one cubic foot of the mixture* is equivalent nearly to 56,900 foot-pounds.

To find the available heat of explosion, it is necessary to have recourse to experiments on actual gas-engines. Let p_1 be the absolute pressure immediately after explosion; then, according to Rule I. of Article 2, the available heat of the explosion, in units of work, per cubic foot of explosive mixture, is expressed by

$$\frac{5}{2}(p_1 - p_0).$$

Now, from experiments quoted by Dr. Letheby, it appears that p_1 = about 5 atmospheres on an average, so that

$$p_1 - p_0 = 4 \text{ atmospheres} = 8464 \text{ lbs. on the square foot};$$

and, consequently, the available heat of explosion per cubic foot is

$$\frac{5}{2} \times 8464 = 21160 \text{ foot-pounds.}$$

The difference between this and the total heat of explosion represents the loss which occurs through conduction and imperfect combustion. The ratio of the available to the total heat, viz.:—

$$\frac{21160}{56900} = 0{\cdot}372,$$

may be called the *efficiency of the explosion.*

In the diagram the available heat of explosion is represented by $\frac{5}{2} \times$ the area of the rectangle A D E C; and according to an established proposition in thermodynamics, it may also be represented as follows (*Philosophical Transactions* for 1854) (*See p. 339*):—Through E and C draw a pair of *adiabatic curves*, E L and C N; that is, curves of expansion without transfer of heat; then the heat required to produce the rise of pressure C E at the constant volume O B is represented by the limit to which the area N C E L between those curves approaches as the curves are prolonged indefinitely towards N and L.

5. *Final Pressure.*—The pressure at the end of the expansion represented by F G may be approximately computed by Rule III. of Article 2. For example, if the rate of expansion is 2, the table shows that $p_2 = 0{\cdot}379\, p_1$; so that if p_1 = 5 atmospheres = 10580 lbs. on the square foot, we

shall probably have, with the expansion 2, the final absolute pressure $p_2 \doteq 4010$ lbs. on the square foot nearly, or 1894 lbs. on the square foot above the mean atmospheric pressure.

The following rule serves to determine the rate of expansion r_1 required in order to make the final pressure be equal to the atmospheric pressure, or nearly so :—

$$r_1 = \left(\frac{p_1}{p_0}\right)^{\frac{5}{7}}; \quad . \quad . \quad . \quad . \quad . \quad \text{(IV.)}$$

and this is the rate of expansion which realises the greatest indicated work. For example, let $\frac{p_1}{p_0} = 5$; then $r_1 = 5^{\frac{5}{7}} = 3{\cdot}16$ nearly. In the diagram this ratio is represented by $\frac{\text{Q A}}{\text{A C}} = \frac{\text{O P}}{\text{O B}}$; Q being the point where the line of atmospheric pressure cuts the expansion curve.

The expansion curve is here assumed to coincide sensibly with an adiabatic curve.

6. *Indicated Work.*—Draw G K J parallel to V O. Then the area C E G H C, representing the work done by each cubic foot of explosive mixture, consists of a rectangular part CKGH, and a triangular part KEG.

The work represented by the rectangular part is simply $(r - 1)(p_2 - p_0)$.

The work represented by the triangular part is determined by the aid of Rules I. and II. of Article 2 as follows:—Conceive an adiabatic curve K M to pass through the point K. Then the area K E G is the difference between the limits of the indefinitely-prolonged areas M K E L and M K G L. But, according to a principle stated in Article 4, the limit of the area M K E L represents the quantity of heat required to produce the increase of pressure K E at the constant volume O B; and, according to Rule I., that quantity of heat, in units of work, is expressed by

$$\frac{5}{2} \times \text{rectangle J D E K} = \frac{5}{2}(p_1 - p_2);$$

also, according to the same principle, the limit of the area M K G L represents the quantity of heat required to produce the increase of volume K G at the constant pressure F G; and, according to Rule II., that quantity of heat, in units of work, is expressed by

$$\frac{7}{2} \times \text{rectangle B K G F} = \frac{7}{2}(r - 1)p_2;$$

whence the work represented by the area K E G is found to be

$$\frac{5}{2}(p_1 - p_2) - \frac{7}{2}(r - 1)p_2;$$

and, combining with this the work represented by the rectangle C K G H, the whole work per cubic foot of explosive mixture is found to be expressed as follows:—

$$W = \frac{5}{2}(p_1 - p_2) - \frac{7}{2}(r - 1)p_2 + (r - 1)(p_2 - p_0). \quad . \quad (V.)$$

For example, in the previous calculations we have $r = 2$; $p_1 = 10580$; $p_2 = 4010$; $p_0 = 2116$; and, consequently, $W = 16425 - 14035 + 1894 = 4284$ foot-pounds per cubic foot of explosive mixture.

The mean effective pressure (p_e) is given by dividing the work done by the space swept by the piston; that is to say,

$$p_e = \frac{W}{r}. \quad . \quad . \quad . \quad . \quad (VI.)$$

Thus, in the example already given

$$p_e = \frac{4284}{2} = 2142 \text{ lbs. on the square foot.}$$

7. *Efficiency.*—The term "*efficiency of the expansion*" may be used to denote the ratio of the work done to the available heat of explosion; that is to say,

$$\frac{2\,W}{5\,(p_1 - p_0)}. \quad . \quad . \quad . \quad . \quad (VII.)$$

Its value in the example is

$$\frac{4284}{21160} = 0{\cdot}203 \text{ nearly.}$$

If the efficiency of the expansion be multiplied by the efficiency of the explosion, already mentioned in Article 4, the product is the resultant efficiency of the heat, whose value, in the example, is

$$0{\cdot}203 \times 0{\cdot}372 = 0{\cdot}075 \text{ nearly;}$$

so that $7\frac{1}{2}$ per cent. of the whole heat of explosion is converted into mechanical work.

8. *Greatest Efficiency.*—As already stated in Article 5, the greatest efficiency of the expansion occurs when the final pressure is equal to the atmospheric pressure. The diagram of work is then represented by C E Q, and the quantity of work per cubic foot of explosive mixture is found by making $p_2 = p_0$ in formula (V.) Let W_1 be that quantity of work; then,

$$W_1 = \frac{5}{2}(p_1 - p_0) - \frac{7}{2}(r_1 - 1)\, p_0 . \qquad \text{(VIII.)}$$

The corresponding value of the efficiency of the explosion is—

$$\frac{W_1}{5\,(p_1 - p_0)} = 1 - \frac{7\,(r_1 - 1)\, p_0}{5\,(p_1 - p_0)} . \qquad \text{(IX.)}$$

In the example chosen we have $r = 3{\cdot}16$, and, consequently, the work per cubic foot of the mixture is

$$W_1 = 21160 - 16000 = 5160 \text{ foot-pounds per cubic foot;}$$

the mean effective pressure,

$$\frac{5160}{3{\cdot}16} = 1636 \text{ lbs. on the square foot;}$$

and the efficiency of the expansion,

$$\frac{2\, W_1}{5\,(p_1 - p_0)} = \frac{5160}{21160} = 0{\cdot}244.$$

The resultant efficiency of the heat is

$$0{\cdot}244 \times 0{\cdot}372 = 0{\cdot}09,$$

so that nine per cent. of the whole heat of explosion is converted into mechanical work.

9. *Remarks.*—In the preceding calculations of work and efficiency no deduction is made for friction, nor for any increase in the back pressure which may arise from resistance to the escape of the waste gases. The allowances to be made for such losses can be deduced from practical trials alone. Hence, the results of the formulæ are theoretical limits, which may be aimed at in practice, but probably cannot be absolutely attained.

XXVIII.—ON THE EXPLOSIVE ENERGY OF HEATED LIQUIDS.*

1. *Reference to Theoretical Investigations.*—In contemplation of the revival of the application (first invented by Perkins) of the sudden evaporation of highly heated liquid water, in order to propel projectiles, it may be useful to give a summary of the rules for calculating the utmost theoretical effect of a given fluid when so employed, under given circumstances. For the theoretical deduction of those rules from the laws of thermodynamics I have to refer to two independent investigations, made respectively by myself and by Clausius; the former published in the *Philosophical Transactions* for 1854 (*See p. 339*); the latter in *Poggendorff's Annalen* for 1856. The rules themselves, with some tables of their results, having reference to the bursting of steam-boilers, have also been published in the *Transactions of the Institution of Engineers in Scotland* for 1863–4, Vol. VII., page 8; and in the *Philosophical Magazine* for 1863, Vol. XXVI., pages 338 and 436. In a subsequent communication I propose to consider the case when the fluid passes into the state of vapour before its admission into the gun.

2. *General Formulæ for all Fluids.*—Suppose a closed boiler to be entirely filled with a fluid in the liquid state, at a certain absolute temperature t_1. Let the absolute temperature t_2, being lower than t_1, be the boiling point of that fluid in a boiler open to the atmosphere. Let a given mass of the liquid be made to escape from the boiler, and to perform work by expanding partly or wholly, as the case may be, into the state of vapour, and driving a solid body (such as a bullet) before it, until its pressure falls to that of the surrounding atmosphere, and its absolute temperature (consequently) to t_2. Then the energy exerted by that mass of fluid is equivalent to the raising of its own weight to the height given by the following equation:—

$$U = K t_2 (n - 1 - \text{hyp. log. } n), \quad . \quad . \quad (1.)$$

in which K denotes the dynamical value of the specific heat of the fluid in

* From *The Engineer* of November 11, 1870.

the liquid state; and $n = \frac{t_1}{t_2}$, the ratio in which the initial absolute temperature is greater than the final. Moreover, the following formula gives the excess of the space filled by each unit of weight of the fluid at the end of the expansion, above the space filled by an unit of weight of the liquid:

$$s = \frac{\text{K hyp. log. } n}{\frac{d p_2}{d t_2}} \quad . \quad . \quad . \quad . \quad (2.)$$

in which $\frac{d p_2}{d t_2}$ denotes the rate at which the pressure of saturation varies with the boiling point, at the final temperature.

Absolute temperatures are given, as is well-known, by adding 461°·2 to temperatures on the ordinary Fahrenheit's scale, or 274° to temperatures on the ordinary Centigrade scale.

3. *Formulæ for Water.*—For water the values of the co-efficients in the formulæ are as follows, very nearly:—

$$\text{K } t_2 = 520{,}000 \text{ feet};\text{*}$$

$$\frac{\text{K}}{\frac{d p_2}{d t_2}} = 18{\cdot}38 \text{ cubic feet per lb.};\dagger$$

or to a rough approximation, about 1100 times the volume of the hot liquid water. Hence we have the following formulæ for water; energy of the explosion in foot-pounds per pound of water:—

$$\text{U} = 520{,}000\,(n - 1 - \text{hyp. log. } n). \quad . \quad . \quad (1 \text{ A.})$$

Space swept by the explosion, or final volume of the water and steam in cubic feet to the pound.

$$s = 18{\cdot}38 \text{ hyp. log. } n; \quad . \quad . \quad . \quad (2 \text{ A.})$$

or, in terms of the volume of the liquid water,

$$1100 \text{ hyp. log. } n, \text{ nearly.} \quad . \quad . \quad . \quad (2 \text{ B.})$$

4. *Examples.*—To illustrate the results of the preceding formulæ, the two following examples are given, in which the values assumed for n are respectively 2 and 2¼. The pressures corresponding to the temperatures given by those ratios are not known by experiment. The pressures given

* 158,500 metres, nearly.

† 1·147 cubic metres per kilogramme, nearly.

in the following table of results are calculated on the assumption that the formulæ which are found to be accurate up to the limits of experiment, are applicable also to temperatures far beyond those limits; hence, those pressures are to be viewed as in a great measure conjectural. This affects the safety of the boiler and of the gun; but not the energy of the explosion, nor the final volume of the fluid; for these two quantities vary with the temperature only.

	Example I.	Example II.
Ratio of initial to final absolute temperature,	2	$2\frac{1}{4}$
Final absolute temperature, Fahr.,	673°·2	673°·2
Final absolute temperature, Cent.,	374°	374°
Initial absolute temperature, Fahr.,	1346°·4	1514°·7
Initial absolute temperature, Cent.,	748°	841°·5
Initial temperature, ordinary scale, Fahr.,	885°·2	1053°·5
Initial temperature, ordinary scale, Cent.,	474°	567°·5
Energy of the explosion, foot-pounds per pound of water,	159,562	228,189
Final volume—cubic feet per pound of water and steam,	12·74	14·9
Final volume—ratio to initial volume of water, nearly	760	890
Conjectural absolute pressure in boiler, pounds on the square inch,	7,180	13,345
Ditto, ditto, in atmospheres,	490	908

The values of the energy of the explosion in the two examples agree very nearly with the least and greatest values found by experiment for the energy of the explosion of 1 lb. of gunpowder; hence the examples may be taken as showing the conditions which must be fulfilled in order that 1 lb. of heated water may produce the same effect as 1 lb. of gunpowder. In both examples the initial pressures are so high that the only safe form of boiler is a coil of tube of small bore compared with its thickness. This was the form employed by Perkins.

5. *Expenditure of Heat.*—The expenditure of heat required in order to produce the elevation of temperature of each unit of mass of liquid from

the temperature of the feed to that at which it escapes from the boiler is expressed in dynamical units as follows :—

$$H = K(t_1 - t_3), \quad . \quad . \quad . \quad . \quad (3.)$$

in which t_3 denotes the absolute temperature of the feed. Let this latter temperature bear the ratio n^1 to the absolute temperature of the atmospheric boiling point; then we may express the same expenditure of heat in the following manner :—

$$H = K\, t_2 (n - n^1); \quad . \quad . \quad . \quad (3\text{ A.})$$

and for water, the value of this in foot-pounds per pound is very nearly

$$H = 520{,}000\, (n - n^1). \quad . \quad . \quad . \quad (3\text{ B.})$$

The value of n^1 for water ranges, in ordinary cases, between 0·7 and 0·8. Assuming it to be 0·75, the expenditure of heat in the two preceding examples is found to have the values given in the following table :—

Heat expended	Example I.	Example II.
Foot-pounds per pound, . .	650,000	780,000
Units of evaporation, . . .	0·873	1·047

The difference between the quantities in the first line and the values of the energy of explosion, are the quantities of heat which go to waste with the escaping steam and water after the explosion, viz :—

Waste heat.	Example I.	Example II.
Foot-pounds per pound, . .	490,438	551,811

6. *Efficiency of the Explosion.*—This term may be used to express the ratio borne by the energy of the explosion to the whole expenditure of heat. Its value is as follows :—

$$\frac{U}{H} = \frac{n - 1 - \text{hyp. log. } n}{n - n^1}. \quad . \quad . \quad . \quad . \quad (4.)$$

And it is to be observed that this value depends solely on the ratios borne to the absolute temperature of the atmospheric boiling point, by two other absolute temperatures—viz., that of the feed water, and that of the liquid just before it escapes from the boiler. In the two examples the values of the efficiency of the explosion are respectively—

Example I.	Example II.
0·245	0·293

7. *Remarks.*—The preceding formulæ all proceed on the assumption that the specific heat of the liquid is sensibly constant. This is not perfectly

accurate, for the specific heat of every liquid increases slowly with the temperature. The effects of that increase are shown in the original theoretical investigations referred to at the commencement; but for practical purposes it is unnecessary to take them into account.

The formulæ also take no account of the retarding effect of friction on the bullet, nor of the inertia of the air which it drives before it in the barrel of the gun, nor of the loss of energy which may take place through the abstraction of heat from the water by the metal of the barrel: those being quantities which can be determined by direct experiment alone.

The initial temperatures assumed in the examples have been chosen so as to make the explosive energy of the water nearly equivalent to that of an equal weight of gunpowder. By choosing a lower initial temperature the initial pressure may be moderated; but the explosive energy of a given weight is at the same time diminished; and a greater mass of water must be used in order to obtain a given amount of energy, thus increasing the proportionate quantity of energy which is lost in propelling the explosive material itself, as the following section will show.

8. *Efficiency of Projection.*—This term may be used to denote the proportion which the energy of the bullet at the instant of its leaving the gun bears to the whole energy of the explosion.

Let m denote the ratio which the mass of the bullet bears to the mass of the explosive material; M the ratio which the whole mass that recoils bears to the mass of the explosive material; v the velocity of the bullet at the instant when the action of the explosion ceases, so that the energy of the bullet at that instant, *per unit of mass of explosive material*, is $\frac{m v^2}{2 g}$; then, neglecting friction and the inertia of the air, &c., it can be shown that the energy of the explosion of an unit of mass of explosive material is disposed of in the following manner:—

$$U = \frac{v^2}{2 g} \left\{ m + \mathrm{M} \left(\frac{2 m + 1}{2 \mathrm{M} + 1} \right)^2 + \frac{1}{3} \left(1 - \frac{2 m + 1}{2 \mathrm{M} + 1} + \left(\frac{2 m + 1}{2 \mathrm{M} + 1} \right)^2 \right) \right\}. \qquad (5.)$$

On the right hand side of this equation the first term $\frac{m v^2}{2 g}$ is the energy of the bullet.

The second,

$$\frac{\mathrm{M} v^2}{2 g} \left(\frac{2 m + 1}{2 \mathrm{M} + 1} \right)^2$$

is the energy of the mass which recoils.

And the third,

$$\frac{v^2}{6g}\left(1 - \frac{2m+1}{2M+1} + \left(\frac{2m+1}{2M+1}\right)^2\right)$$

is the energy of the projectile motion of the products of explosion, at the instant when they cease to act on the bullet. Hence, the *counter-efficiency* of projection, being the reciprocal of the efficiency, or in other words, the ratio in which the whole energy of the explosion is greater than that of the bullet, is expressed as follows:

$$c = 1 + \frac{M}{m}\left(\frac{2m+1}{2M+1}\right)^2$$
$$+ \frac{1}{3m}\left\{1 - \frac{2m+1}{2M+1} + \left(\frac{2m+1}{2M+1}\right)^2\right\} \quad . \quad . \quad (6.)$$

from which it appears that the energy lost through the projection of the products of explosion is greater, the greater the proportion $\frac{1}{m}$ borne by the mass of the explosive material to that of the bullet, and that when the proportionate weight M of the recoiling mass is very great, that lost energy is approximately equal to the fraction $\frac{1}{3m}$ of the energy of the bullet.

For example, let $m = 8$, and $M = 1000$; then the three terms of the counter-efficiency of projection have the values shown in the following equation to three places of decimals;

$$c = 1 + 0{\cdot}009 + 0{\cdot}041 = 1{\cdot}050;$$

that is to say, the energy lost in the recoil is 0·009, and the energy lost in projecting the products of explosion 0·041 of the energy of the bullet; the latter being by far the more important loss; and hence it is desirable not to increase unnecessarily the comparative weight of the explosive material.

9. *Ratio of Final Volume of Steam to Volume of Bullet.*—In section 3 of this communication, equations (2A) and (2B), expressions have been given for the space (s) filled by an unit of weight of the mixture of water and steam, when it has expanded until its pressure is equal to that of the atmosphere. Let w denote the heaviness of the material of which the bullet is made; then, m being, as before, the ratio of the mass of the bullet to the mass of the fluid which drives it, the ratio in which the final volume of the fluid exceeds the volume of the bullet is given by the following expression:—

$$\frac{ws}{m} = \frac{w}{m} \times 18{\cdot}38 \text{ hyp. log. } n; \quad . \quad . \quad . \quad (7.)$$

when w is stated in pounds to the cubic foot.

If lead be the material of the bullet, we have $w = 712$ nearly; and if iron or steel, $w =$ about 480. Hence are deduced the following formulæ:—

$$\left.\begin{array}{ll} \text{For lead, } \quad . \quad . & \dfrac{v s}{m} = \dfrac{13087 \text{ hyp. log. } n}{m}; \\ \text{For iron and steel, } & \dfrac{v s}{m} = \dfrac{8822 \text{ hyp. log. } n}{m}. \end{array}\right\} \quad (7\text{ A.})$$

When these formulæ are applied to the two examples given in section 4, the bullet being supposed, as in section 8, to have eight times the mass of the explosive material (so that $m = 8$), the following results are obtained:—

Ratio of final volume of water and steam to volume of bullet.

	Example I.	Example II.
Lead bullet, ...	1134	1327
Iron or steel bullet, . . .	764	894

Such would be the ratio which the volume of the gun-barrel would have to bear to that of the bullet, in order to render available the whole of the energy developed by the expansion of the steam. It is obvious that barrels of such dimensions are purely ideal, being many times longer than the greatest length that it is possible to use in practice. It therefore becomes necessary to limit the barrel to a practicable length, and to sacrifice part of the energy due to the expansion of the steam.

10. *Full-pressure Steam Gun.*—In the following investigation the supposition is made that the communication between the boiler and the gun-barrel remains full open during the whole time of the motion of the bullet along the barrel; and it is further assumed that at the instant when the bullet quits the muzzle the barrel is filled with fluid of uniform pressure and density, which, consequently, is at that instant moving with a velocity equal to that of the bullet (v). The pressure in the boiler must be higher than that in the barrel to the extent required in order that the expansion of the fluid in passing from the higher to the lower pressure may be sufficient to produce a velocity of outflow equal to that of the bullet; and so far the action of the fluid is expansive; but its action in driving the bullet is equal simply to that due to the difference between the pressure in the barrel and the atmospheric pressure, acting as in a non-expansive steam-engine. This apparatus may be called a "full-pressure steam gun."

The ratio in which the volume of the barrel exceeds that of the bullet is supposed to be fixed according to practical convenience.

11. *Calculation of Driving Pressure.*—Let b denote the volume of the bullet; B, that of the space through which the bullet sweeps in the

barrel. Let p_3 be the absolute intensity of the pressure which resists the motion of the bullet: being that of the atmosphere, with an addition, to be determined by experiment, for friction and for the inertia of the air expelled in front of the bullet. Let p_2 be the absolute intensity of the pressure of steam in the barrel. Then $p_2 - p_3$, is the effective intensity of the driving pressure. The weight of the bullet is $w\,b$, and the energy impressed on it is $\frac{w\,b\,v^2}{2\,g}$, being equal to the work of raising it to the height $\frac{v^2}{2\,g}$ due to its velocity of discharge. The energy exerted in driving the bullet is that due to the pressure whose effective intensity is $p_2 - p_3$, acting through a space of the volume B; therefore, by equating these quantities of energy as follows,—

$$(p_2 - p_3)\,\mathrm{B} = \frac{w\,b\,v^2}{2\,g};$$

we obtain the following formula for the required effective intensity of the driving pressure:—

$$p_2 - p_3 = w \cdot \frac{b}{\mathrm{B}} \cdot \frac{v^2}{2\,g}; \qquad \text{(8.)}$$

that is to say, *the excess of the absolute driving pressure* (p_2) *above the resisting pressure* (p_3) *is equivalent to the weight of a column of the metal of which the bullet is made, whose height is less than the height due to the velocity, in the same proportion in which the volume of the bullet is less than the volume of the space through which the bullet sweeps in the barrel.*

For example, let $v = 1605$ ft. per second; let us assume $\mathrm{B} = 100\,b$; and let the material of the bullet be lead, so that $w = 712$ lbs. per cubic foot; then we have the following results:—

Height due to velocity, $\frac{v^2}{2\,g} = 40{,}000$ ft.

Effective driving pressure, $p_2 - p_3$;	lb. on the square foot,	284,800
,, ,, ,,	lb. on the square inch,	1,978
Absolute driving pressure, if friction and the inertia of the air be neglected;	lb. on the square inch,	1,993

so that in this example we may conclude that the absolute intensity of the driving pressure required would be 2,000 lbs. on the square inch, or thereabouts; or between 135 and 136 atmospheres. For other proportions of the volume of the barrel to that of the bullet, the effective pressure required, of course, varies inversely as the volume of the barrel.

12. *Calculation of Pressure in Boiler.*—The pressure in the boiler must be such that a mass of water escaping from the boiler, and expanding from

that pressure until the pressure of the mixed water and steam falls to that in the barrel, shall acquire a velocity equal to that of the bullet. Hence, let t_2 denote the absolute temperature corresponding to the absolute driving pressure p_2, as found by means of suitable formulæ or tables; $t_1 = n\, t_2$, the absolute temperature of the water just before it escapes from the boiler, and K (as before) the dynamical value of the specific heat of liquid water; then (as in equation (1) of section 2), we have

$$\mathrm{K}\, t_2\,(n - 1 - \text{hyp. log. } n) = \frac{v^2}{2\,g}; \quad . \quad . \quad (9.)$$

and this transcendental equation is to be solved by approximation, so as to find n, and thence $t_1 = n\, t_2$. When the absolute temperature in the boiler has thus been found, the corresponding pressure p_1 may be calculated by the help of formulæ, or of tables.

In applying the foregoing rules to such examples as that already given in section 11, great uncertainty arises from a cause formerly referred to—viz., that the pressures and temperatures lie far beyond the range of the experiments from which the formulæ for the pressure and temperature of steam were deduced.

By the use of an already known formula,* the absolute temperature corresponding to the absolute pressure of 2,000 lbs. on the square inch is found to be 1104° Fahr. = 613° Cent.; corresponding to 643° Fahr., or 339° Cent. on the ordinary scales. The corresponding value of $\mathrm{K}\, t_2$ is 852,000 ft.; and by solving equation (9) by approximation we obtain the following results, which, however, are to a great extent conjectural: hyp. log. n = 0·292 nearly; n = 1·339 nearly; absolute temperature in boiler, t_1 = 1478° Fahr. nearly = 821° Cent. nearly (or, on the ordinary scales, 1017° Fahr., or 547° Cent.) Absolute pressure in boiler, about 11,770 lbs. on the square inch, or about 800 atmospheres.

13. *Expenditure of Water, and Heat-Efficiency.*—The weight of water expended per shot, supposing that there is no waste, is expressed by $\frac{\mathrm{B}}{s_2}$, in which s_2 denotes the volume filled by each unit of weight of the mixture of water and steam in the barrel; and the ratio which that weight bears to the weight $w\, b$ of the bullet is given by the following formula:—

$$\frac{1}{m} = \frac{\mathrm{B}}{w\, b\, s_2} = \frac{\mathrm{B}}{b\, w\, \mathrm{K}\,.\,\frac{d\, t_2}{d\, p_2}\,\text{hyp. log.}\cdot n}. \quad . \quad . \quad (10.)$$

For the reason already given the value of $\frac{d\, t_2}{d\, p_2}$ is very uncertain; but, as before, a conjectural value may be computed by means of the ordinary

* See *Manual of the Steam-Engine and other Prime Movers*, p. 237.

formula.* In the example already given we find the following results :—

$$s_2 = 0{\cdot}12 \text{ cubic foot per lb. of water;}$$

$$\frac{1}{m} = 1{\cdot}17;$$

$$m = 0{\cdot}854.$$

The dynamical equivalent of the expenditure of heat for each unit of weight of bullet is expressed as follows :—

$$H = \frac{K(t_1 - t_4)}{m}; \quad . \quad . \quad . \quad (11.)$$

in which t_4 is the temperature of the feed; and in the example given the value of this quantity of heat, subject to the causes of uncertainty already mentioned, is found to be about 810,000 foot pounds per pound weight of bullet. The energy of each pound weight of bullet is 4,000 foot-pounds, so that the efficiency and counter-efficiency of the gun are respectively as follows :—

Efficiency, ·049; Counter-efficiency, 20·25.

14. *Full-pressure Dry Steam Gun.*—If, instead of a coil of tube entirely filled with highly-heated liquid water, we assume it to be practicable to use a boiler having sufficient steam space to enable the gun to be supplied with dry steam, the calculation of the driving pressure required in the barrel is exactly the same with that already given in section 11, giving about 2,000 lbs. on the square inch in the example chosen. The density of the steam in the barrel, the weight expended per shot, the boiler pressure, the total expenditure of heat per shot, and the efficiency, may be calculated by means of known formulæ, subject to the uncertainty arising from the pressures and temperatures being beyond the range of previous experiments. The following are the results in the example :— $s_2 = 0{\cdot}24$ cubic foot per lb., showing that with the boiler quite full of liquid water half the fluid in the barrel is liquid,

$$m = 1{\cdot}72; \ \frac{1}{m} = 0{\cdot}582.$$

Boiler pressure, p_1, about 3,600 lbs. on the square inch; expenditure of heat per shot, 765,000 foot-pounds; counter-efficiency, 19; efficiency, 0·052.

15. *Remarks.*—The boiler-pressure, as well as the driving pressure required in order to produce a given velocity in a given bullet, varies inversely as the capacity of the barrel; and hence it is obvious that the safe and effective working of steam guns depends mainly on the practicability of making and using very long gun-barrels.

* See *Manual of the Steam-Engine and other Prime Movers*, p. 325.

PART III.

PAPERS RELATING TO WAVE-FORMS, PROPULSION OF VESSELS, STABILITY OF STRUCTURES, &c.

PART III.

PAPERS RELATING TO WAVE FORMS, PROPULSION OF VESSELS, STABILITY OF STRUCTURES, &c.

XXIX.—ON THE EXACT FORM OF WAVES NEAR THE SURFACE OF DEEP WATER.*

1. THE investigations of the Astronomer-Royal, and of some other mathematicians, on straight-crested parallel waves in a liquid, are based on the supposition that the displacements of the particles of the liquid are small compared with the length of a wave. Hence it has been very generally inferred that the results of those investigations are approximate only, when applied to waves in which the displacements, as compared with the length of a wave, are considerable.

2. In the present paper, I propose to prove that one of those results (viz., that in very deep water the particles move with a uniform angular velocity in vertical circles whose radii diminish in geometrical progression with increased depth, and, consequently, that surfaces of equal pressure, including the upper surface, are trochoidal) is exact for all displacements, how great soever.

3. I believe the trochoidal form of waves to have been first explicitly stated by Mr. Scott Russell; but no demonstration of its exactly fulfilling the conditions of the question has yet been published, so far as I know.

4. In *A Manual of Applied Mechanics* (first published in 1858), page 579, I stated that the theory of rolling waves might be deduced from that of the positions assumed by the surface of a mass of water revolving in a vertical plane about a horizontal axis; as the theory of such waves, however, was foreign to the subject of the book, I did not then publish the investigation on which that statement was founded.

5. Having communicated some of the leading principles of that investigation to Mr. William Froude in April, 1862, I learned from him that he had already arrived independently at similar results by a similar process, although he had not published them.

* Read before the Royal Society of London on November 27, 1862, and published in the *Philosophical Transactions* for 1863.

6. PROPOSITION I.—*In a mass of gravitating liquid, whose particles revolve uniformly in vertical circles, a wavy surface of trochoidal profile fulfils the conditions of uniformity of pressure—such trochoidal profile being generated by rolling, on the underside of a straight line, a circle whose radius is equal to the height of a conical pendulum that revolves in the same period with the particles of liquid.*

In Fig. 1 let B be a particle of liquid revolving uniformly in a vertical circle of the radius C B, in the direction indicated by the arrow

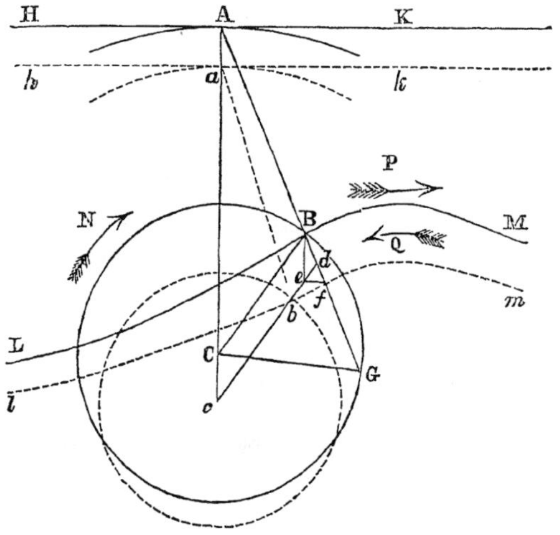

Fig. 1.

N; and let it make n revolutions in a second. Then the centrifugal force of B (taking its mass as unity) will be $4\,\pi^2\,n^2\,.\,\mathrm{C\,B}$.

Draw $\overline{\mathrm{C\,A}}$ vertically upwards, and of such a length that centrifugal force : gravity :: $\overline{\mathrm{C\,B}} : \overline{\mathrm{A\,C}}$; that is to say, make

$$\overline{\mathrm{A\,C}} = \frac{g}{4\,\pi^2\,n^2};$$

which is the well-known expression for the height of a revolving pendulum making n revolutions in a second.

Then $\overline{\mathrm{A\,C}}$ being in the direction of and proportional to gravity, and $\overline{\mathrm{C\,B}}$ in the direction of and proportional to centrifugal force, $\overline{\mathrm{A\,B}}$ will be in the direction of and proportional to the resultant of gravity and centrifugal force; and the surface of equal pressure traversing B will be normal to $\overline{\mathrm{A\,B}}$.

The profile of such a surface is obviously a trochoid L B M, traced by the point B, which is carried by a circle of the radius $\overline{CA}$ rolling along the underside of the horizontal straight line H A K. Q.E.D.

7. *Corollaries.*—The length of the wave whose period is one-nth of a second, is equal to the circumference of the rolling circle; that is to say (denoting that length by λ),

$$\lambda = 2\,\pi\,.\,\overline{CA} = \frac{g}{2\,\pi\,n^2}:$$

the period of a wave of a given length λ is given in seconds, or fractions of a second, by the equation

$$\frac{1}{n} = \sqrt{\frac{2\,\pi\,\lambda}{g}};$$

and the velocity of propagation of such wave is

$$n\,\lambda = \frac{g}{2\,\pi\,n} = \sqrt{\frac{g\,\lambda}{2\,\pi}} = \sqrt{g\,.\,\overline{CA}};$$

results agreeing with those of the known theory.

8. Proposition II.—*Let another surface of uniform pressure be conceived to exist indefinitely near to the first surface; then, if the first surface is a surface of continuity, so also is the second.*

By a surface of continuity is here meant one which always passes through the same set of particles of liquid, so that a pair of such surfaces contain between them a layer of particles which are always the same.

The perpendicular distance between a pair of surfaces of uniform pressure is in this case inversely proportional to the resultant of gravity and centrifugal force; that is to say, to the normal $\overline{AB}$. Hence, if a curve $l\,b\,f\,m$ be drawn indefinitely near to the curve L B M, so that the perpendicular distance between them, $\overline{Bf}$, shall everywhere be inversely proportional to the normal $\overline{AB}$, the second curve will also be the profile of a surface of uniform pressure.

Conceive now that the whole mass of liquid has, combined with its wave-motion, a uniform motion of translation, with a velocity equal and

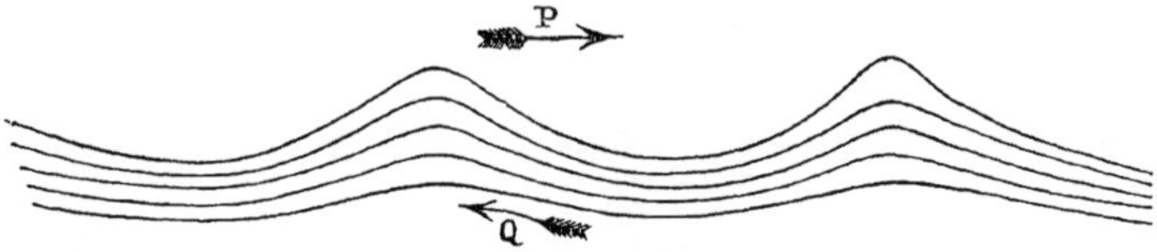

Fig. 2.

opposite to that of the propagation of the waves. The dynamical conditions of the mass are not in the least altered by this; but the forms of the waves are rendered stationary (as we sometimes see in a rapid stream), and, instead of a series of waves propagated in the direction shown by the arrow P, we have an *undulating current* running the reverse way, in the direction shown by the arrow Q. (This is further illustrated by Fig. 2.) According to a well-known property of curves described by rolling, the velocity of the particle B in that current is proportional to the normal A B, and is given by the expression $2\pi n \cdot \overline{AB}$.

Consider the layer of the current contained between the surfaces L B M and *l b m*. In order that the latter of those surfaces, as well as the former, may be a surface of continuity, it is necessary and sufficient that the thickness of the layer $\overline{Bf}$ at each point should be inversely as the velocity; and that condition is already fulfilled; for $\overline{Bf}$ varies inversely as $\overline{AB}$, and $\overline{AB}$ varies as the velocity of the current at B; therefore, L B M and *l b m* are not only a pair of surfaces of uniform pressure, but a pair of surfaces of continuity also. Q.E.D.

9. *Corollary.*—The surfaces of uniform pressure are identical with surfaces of continuity throughout the whole mass of liquid.

10. *Corollary.*—Inasmuch as the resultant of gravity and centrifugal force at B is represented by

$$g \cdot \frac{\overline{AB}}{\overline{AC}};$$

the excess of the uniform pressure at the surface *l b m* above that at the surface L B M is given by the expression

$$dp = w \cdot \frac{\overline{AB}}{\overline{AC}} \cdot \overline{Bf},$$

in which w is the heaviness of the liquid, in units of weight per unit of volume. By omitting the factor w, the pressure is expressed in units of height of a column of the liquid.

11. Proposition III.—*The profile of the lower surface of the layer referred to in the preceding proposition is a trochoid generated by a rolling circle of the same radius with that which generates the first trochoid; and the tracing-arm of the second trochoid is shorter than that of the first trochoid by a quantity bearing the same proportion to the depth of the centre of the second rolling circle below the centre of the first rolling circle, which the tracing-arm of the first rolling circle bears to the radius of that circle.*

At an indefinitely small depth $\overline{Aa}$ below the horizontal line H A K, draw a second horizontal line *h a k*, on the underside of which let a circle roll with a radius $\overline{ca} = \overline{CA}$, the radius of the first rolling circle; so that the indefinitely small depths $\overline{Cc} = \overline{Aa}$. To find the tracing-arm of the second rolling circle, draw $c\,d$ parallel to $\overline{\mathrm{C\,B}}$, the tracing-arm of the first circle; in $c\,d$ take $\overline{c\,e} = \overline{\mathrm{C\,B}}$, and cut off $\overline{e\,b} = \overline{e\,d}$; b will be the tracing-point, and $\overline{c\,b}$ the tracing-arm required; for, according to the principle laid down in the enunciation, we are to have

$$\overline{\mathrm{C\,B}} - \overline{c\,b} = \overline{e\,b} = \mathrm{C}\,c\,.\,\frac{\mathrm{C\,B}}{\mathrm{C\,A}}.$$

Let the second circle roll that b will trace a trochoid $l\,b\,m$. From b let fall $b\,f$ perpendicular to $\overline{\mathrm{A\,B}}$ produced; $\mathrm{B}\,f$ will be the indefinitely small thickness at B of the layer between the two trochoidal surfaces.

The proposition enunciated amounts to stating that $\mathrm{B}\,f$ is everywhere inversely proportional to the normal $\overline{\mathrm{A\,B}}$; so that $l\,b\,m$ is the profile of a surface of uniform pressure and of continuity.

To prove this, join $\overline{\mathrm{B}\,e}$ and $\overline{e\,f}$. Then $\mathrm{B}\,e$ is parallel to $\mathrm{A\,C}\,c$, and equal to $\mathrm{C}\,c$, and $d\,e\,f$ is evidently an isosceles triangle, $\overline{e\,f}$ being $= e\,d$. Let A B (produced if necessary) cut the circle of the radius C B in G; then C G is parallel to $e\,f$, and the indefinitely small triangle $\mathrm{B}\,e\,f$ is similar to the triangle A C G; consequently, $\overline{\mathrm{A\,C}} : \overline{\mathrm{A\,G}} :: \overline{\mathrm{B}\,e} = \overline{\mathrm{C}\,c} : \overline{\mathrm{B}\,f}$; or

$$\overline{\mathrm{B}\,f} = \overline{\mathrm{C}\,c} \cdot \frac{\overline{\mathrm{A\,G}}}{\overline{\mathrm{A\,C}}};$$

but, by a well-known property of the circle,

$$\overline{\mathrm{A\,G}} = \frac{\overline{\mathrm{A\,C}^2} - \overline{\mathrm{C\,B}^2}}{\overline{\mathrm{A\,B}}};$$

and, therefore,

$$\overline{\mathrm{B}\,f} = \overline{\mathrm{C}\,c} \cdot \frac{\overline{\mathrm{A\,C}^2} - \overline{\mathrm{C\,B}^2}}{\overline{\mathrm{A\,C}} \cdot \overline{\mathrm{A\,B}}};$$

that is to say, *the thickness of the layer varies inversely as the normal* $\overline{\mathrm{A\,B}}$; *and the second trochoid, $l\,b\,m$, is therefore the profile of a surface of uniform pressure and of continuity.* Q.E.D.

12. *Corollaries.*—The profiles of the surfaces of uniform pressure and of continuity form an indefinite series of trochoids, described by equal rolling circles, rolling with the same speed below an indefinite series of horizontal straight lines.

The tracing-arms of those circles (each of which arms is the radius of the circular orbit of the particles contained in the trochoidal surface which it traces) diminish in geometrical progression with increase of depth, according to the following laws:—

For convenience, let $\overline{Cc}$ be denoted by dk, $\overline{CB}$ by r, and $\overline{cb}$ by $r - dr$; then,

$$dr = dk \cdot \frac{r}{\overline{AC}} = dk \cdot \frac{r}{2\pi\lambda},$$

and the integration of this equation gives the following result:—

Let k denote the vertical depth of the centre of the generating circle of a given surface below the centre of the generating circle of the free upper surface of the liquid;

r_0 the tracing-arm of the free upper surface (= half the amplitude of disturbance);

r_1 the tracing-arm of the surface whose middle depth is k; then,

$$r_1 = r_0 e^{-\frac{k}{\overline{AC}}} = r_0 e^{-\frac{2\pi k}{\lambda}},$$

a formula exactly agreeing with that found for indefinitely small disturbances by previous investigators.

13. Proposition IV.—*The centres of the orbits of the particles in a given surface of equal pressure stand at a higher level than the same particles do when the liquid is still, by a height which is a third proportional to the diameter of the rolling circle and the tracing-arm or radius of the orbits of the particles, and which is equal to the height due to the velocity of revolution of the particles.*

If the liquid were still, the given surface of equal pressure would become horizontal. To find the level at which it would stand, we must first find what relation the mean vertical depth of a given layer of particles bears to the depth $\overline{Cc} = dk$ between the centres of the rolling circles that generate its boundaries.

The length of the arc of the curve L B M described in an indefinitely short interval of time dt is

$$2\pi n \cdot \overline{AB} \cdot dt,$$

and the thickness of the layer being

$$\overline{Bf} = dk \cdot \frac{\overline{AC}^2 - \overline{CB}^2}{\overline{AC} \cdot \overline{AB}},$$

let the product of those quantities be divided by the distance through which the centre of the rolling circle moves in the same time, viz.—

$$2\pi n \cdot \overline{AC} \cdot dt,$$

and the result will be the mean vertical depth of the layer, which being denoted by dk_0, we have

$$d k_0 = d k \,.\, \left(1 - \frac{\overline{\mathrm{C\,B}^2}}{\overline{\mathrm{A\,C}^2}}\right) = d k \,.\, \left(1 - \frac{r^2}{\overline{\mathrm{A\,C}^2}}\right) = d k \,.\, \left(1 - \frac{r_0^2}{\overline{\mathrm{A\,C}^2}} e^{-\frac{2k}{\mathrm{A\,C}}}\right).$$

The difference by which the mean vertical thickness of the layer falls short of the difference of level of the rolling circles of its upper and lower surfaces, is given by the following expression,

$$d k - d k_0 = \frac{r_0^2}{\overline{\mathrm{A\,C}^2}} e^{-\frac{2k}{\mathrm{A\,C}}} d k;$$

and this being integrated from ∞ to k, gives the depth of the position of a given particle, when the liquid is still, below the level of the centre of the orbit of the same particle when disturbed, viz.—

$$k_0 - k = \frac{r_0^2}{2\,\overline{\mathrm{A\,C}}} \,.\, e^{-\frac{2k}{\mathrm{A\,C}}} = \frac{r^2}{2\,\overline{\mathrm{A\,C}}} = \frac{\pi r^2}{\lambda},$$

or, *a third proportional to the diameter of the rolling circle and the radius of the orbit of the particle ; also*

$$\frac{r^2}{2\,\overline{\mathrm{A\,C}}} = \frac{4\,\pi^2 n^2 r^2}{2\,g}$$

is the height due to the velocity of revolution of the particles. Q.E.D.

13A. *Corollary.*—The mechanical energy of a wave is half actual and half potential,—half being due to motion, and half to elevation. In other words, the mechanical energy of a wave is double of that due to the motion of its particles only, there being an equal amount due to the mean elevation of the particles above their position when the water is still.

14. *Corollary.*—The crests of the waves rise higher above the level of still water than their hollows fall below it; and the difference between the elevation of the crest and the depression of the hollow is double of the quantity mentioned in Proposition IV., that is to say, it is

$$\frac{r^2}{\overline{\mathrm{A\,C}}} = \frac{2\,\pi\,r^2}{\lambda}.$$

15. *Corollary as to Pressures.*—An expression has already been given in Art. 10 for the difference of pressure at the upper and under surfaces of a given layer. Substituting in that expression the value of the thickness of the layer, we find

$$d\bar{p} = w \,.\, \frac{\overline{\mathrm{A\,B}}}{\overline{\mathrm{A\,C}}} \,.\, d k \,.\, \frac{\overline{\mathrm{A\,C}^2} - \overline{\mathrm{C\,B}^2}}{\overline{\mathrm{A\,C}} \,.\, \overline{\mathrm{A\,B}}} = w \,.\, d k \left(1 - \frac{\overline{\mathrm{C\,B}^2}}{\overline{\mathrm{A\,C}^2}}\right) = w \,.\, d k_0$$

(as the preceding corollary shows), being precisely the same as if the liquid were still; and hence it follows that *the hydrostatic pressure at each individual particle during wave-motion is the same as if the liquid were still.*

16. In Proposition III. it has been shown, by geometrical reasoning from the mechanical construction of the trochoid, that a wave consisting of trochoidal layers satisfies the condition of continuity. It may be satisfactory also to show the same thing by the use of algebraic symbols. For that purpose the following notation will be used:—

Let the origin of co-ordinates be assumed to be in the horizontal line containing the centre of the circle which is rolled to trace the profile of *cycloidal* waves, having cusps, and being (as Mr. Scott Russell long ago pointed out) the highest waves that can exist without breaking. In such waves, the tracing-arm, or radius vector, of the uppermost particles is equal to the radius of the rolling circle; and that arm diminishes for each successive layer proceeding downwards.

Let x and y be the co-ordinates of any particle, x being measured horizontally *against* the direction of propagation, and y vertically downwards.

Let k (as before) be the vertical co-ordinate of the centre of the given particle's orbit; h the horizontal co-ordinate of the same centre.

Let R be the radius of the rolling circle, a the angular velocity of the tracing-arm ($= 2\pi n$), so that

$$2\pi \mathrm{R} = \lambda$$

is the length of a wave, and

$$a \mathrm{R} = n \lambda \sqrt{\frac{g \lambda}{2\pi}} = \sqrt{g \mathrm{R}}$$

is the velocity of propagation.

Let θ denote the *phase* of the wave at a given particle, being the angle which its radius vector, or tracing-arm, makes with the direction of $+ y$, that is, with a line pointing vertically downwards.

Let t denote time, reckoned from the instant at which all the particles for which $h = 0$ are in the axis of y; then

$$\theta = a t + \frac{h}{\mathrm{R}}. \qquad (1.)$$

Then the following equations give the co-ordinates of a given particle at a given instant:

$$x = h + \mathrm{R}\, e^{-\frac{k}{\mathrm{R}}} \sin \theta; \qquad (2.)$$

$$y = k + \mathrm{R}\, e^{-\frac{k}{\mathrm{R}}} \cos \theta. \qquad (3.)$$

Let u and v denote the vertical and horizontal components of the velocity of the particle at the given instant; then

$$u = \frac{d\,x}{d\,t} = \quad a\,\mathrm{R}\,.\,e^{-\frac{k}{\mathrm{R}}}\cos\theta = \quad a\,(y-k); \quad . \quad (4.)$$

$$v = \frac{d\,y}{d\,t} = -\,a\,\mathrm{R}\,.\,e^{-\frac{k}{\mathrm{R}}}\sin\theta = -\,a\,(x-h). \quad . \quad (5.)$$

The well-known equation of continuity in a liquid in two dimensions is

$$\frac{d\,u}{d\,x} + \frac{d\,v}{d\,y} = 0; \quad . \quad . \quad . \quad . \quad (6.)$$

and from equations (4) and (5) it appears that we have in the present case,

$$\frac{d\,u}{d\,x} + \frac{d\,v}{d\,y} = a\left(-\frac{d\,k}{d\,x} + \frac{d\,h}{d\,y}\right) = a\left(-\frac{d\,k}{d\,x} + \frac{\mathrm{R}\,d\,\theta}{d\,y}\right). \quad (7.)$$

In the original formulæ, k and θ are the independent variables. When x and y are made the independent variables instead, we have, by well-known formulæ,

$$\left.\begin{aligned} \frac{d\,k}{d\,x} &= 1 \div \left\{\frac{d\,x}{d\,k} - \frac{d\,x}{d\,\theta}\cdot\frac{\frac{d\,y}{d\,k}}{\frac{d\,y}{d\,\theta}}\right\} = \frac{e^{-\frac{k}{\mathrm{R}}}\sin\theta}{1-e^{-\frac{2\,k}{\mathrm{R}}}} \\ \text{and} \\ \frac{d\,\theta}{d\,y} &= 1 \div \left\{\frac{d\,y}{d\,\theta} - \frac{d\,y}{d\,k}\cdot\frac{\frac{d\,x}{d\,\theta}}{\frac{d\,x}{d\,k}}\right\} = \frac{e^{-\frac{k}{\mathrm{R}}}\sin\theta}{\mathrm{R}\left(1-e^{-\frac{2k}{\mathrm{R}}}\right)} \end{aligned}\right\} \quad . \quad (8)$$

so that the equation of continuity (6) is exactly verified.

17. Another mode of testing algebraically the fulfilment of the condition of continuity is the following. It is analogous to that employed by Mr. Airy; but inasmuch as the disturbances in the present paper are regarded as considerable compared with the length of a wave, it takes into account quantities which, in Mr. Airy's investigation, are treated as inappreciable.

Consider an indefinitely small rhomboidal particle, bounded by surfaces for which the values of h and k are respectively h, $h + dh$, k, $k + dk$. Then the area of that rhomboid is

$$\left(\frac{dx}{dh} \cdot \frac{dy}{dk} - \frac{dx}{dk} \cdot \frac{dy}{dh}\right) dh \, . \, dk;$$

and the condition of continuity is that this area shall be at all times the same; that is to say, that

$$\frac{d}{dt}\left(\frac{dx}{dh} \cdot \frac{dy}{dk} - \frac{dx}{dk} \cdot \frac{dy}{dh}\right) = 0. \quad . \quad . \quad . \quad (9.)$$

Upon performing the operations here indicated upon the values of the co-ordinates in equations (2) and (3), the value of the quantity in brackets is found to be

$$1 - e^{-\frac{2k}{R}}; \quad . \quad . \quad . \quad . \quad (10.)$$

which is obviously independent of the time, and therefore fulfils the condition of continuity.

APPENDIX.

ON THE FRICTION BETWEEN A WAVE AND A WAVE-SHAPED SOLID.

Conceive that the trough between two consecutive crests of the trochoidal surface of a series of waves is occupied, for a breadth which may be denoted by z, by a solid body with a trochoidal surface, exactly fitting the wave-surface; that the solid body moves forward with a uniform velocity equal to that of the propagation of the waves, so as to continue always to fit the wave-surface; and that there is friction between the solid surface and the contiguous liquid particles, according to the law which experiment has shown to be at least approximately true—viz., varying as the surface of contact, and as the square of the velocity of sliding.

Conceive, further, that each particle of the liquid has that pressure applied to it which is required in order to keep its motion sensibly the same as if there were no friction; the solid body must of course be urged forwards by a pressure equal and opposite to the resultant of all the before-mentioned pressures.

The action, amongst the liquid particles, of pressures sufficient to overcome the friction, will disturb to a certain extent the motions of the liquid particles, and the figures of the surfaces of uniform pressure; but it will be assumed that those disturbances are small enough to be neglected, for the purposes of the present inquiry. The smallness of the pressures producing such disturbances, and consequently the smallness of those disturbances themselves, may be inferred from the fact, that the friction of a current of water over a surface of painted iron of a given area is equal to the weight of a layer of water covering the same area, and of a thickness which is only about ·0036 of the height due to the velocity of the current.

Those conditions having been assumed, let it now be proposed, *to find approximately the amount of resultant pressure required to overcome the friction between the wave and the wave-shaped solid.*

This problem is to be solved by finding the mechanical work expended in overcoming friction in an indefinitely small time $d\,t$, and dividing that work by the distance through which the solid moves in that time.

Taking, as before, as an independent variable the *phase* θ, being the angle which the tracing-arm $\overline{\mathrm{C\,B}} = r$ (Fig. 1) makes with a line pointing vertically downwards, the length of the elementary arc corresponding to an indefinitely small increment of phase $d\,\theta$ is

$$q\,d\,\theta,$$

where q is taken, for brevity's sake, to denote the normal A B.

The area of the corresponding element of the solid surface is

$$z\,q\,d\,\theta.$$

The velocity of sliding of the liquid particles over that elementary surface is

$$a\,q,$$

in which a, as before, denotes $\frac{d\,\theta}{d\,t}$, the angular velocity of the tracing-arm. Hence, let ρ denote the heaviness (or weight of unity of volume) of the liquid, and f its coefficient of friction when sliding over the given solid surface; the intensity of the friction per unit of area is

$$\frac{f\,\rho\,a^2\,q^2}{2\,g}.$$

That friction has to be overcome, during the time $d\,t$, through the distance

$$a\,q\,d\,t = q\,d\,\theta.$$

Multiplying now together the elementary area, the intensity of the friction, and the distance through which it is overcome in the time $d\,t$, we find the following value for the work performed in that time in overcoming the friction at the given elementary surface,

$$z\,q\,d\,\theta \times \frac{f\,\rho\,a^2\,q^2}{2\,g} \times q\,d\,\theta = \frac{f\,\rho\,a^2}{2\,g}\,.\,q^4\,z\,d\,\theta^2.$$

Now, during the time $d\,t$, the solid advances through the distance

$$a\,\mathrm{R}\,d\,t = \mathrm{R}\,d\,\theta$$

(R, as before, being the radius of the rolling circle); and dividing the elementary portion of work expressed above by that distance, we find the following value for an elementary portion of the pressure required to overcome the friction,

$$d\,\mathrm{P} = \frac{f\,\rho\,a^2}{2\,g}\,.\,\frac{q^4\,z}{\mathrm{R}}\,.\,d\,\theta. \quad . \quad . \quad . \quad (1.)$$

The total pressure required to overcome the friction is found by integrating the preceding expression throughout an entire revolution; that is to say,

$$\mathrm{P} = \frac{f\,\rho\,a^2\,z}{2\,g\,\mathrm{R}}\int_0^{2\pi} q^4\,d\,\theta. \quad . \quad . \quad . \quad (2.)$$

To obtain this integral the following value of the square of the normal q or AB is to be substituted,

$$q^2 = \mathrm{R}^2 + r^2 + 2\,\mathrm{R}\,r\,.\cos\theta,$$

whence,

$$\int_0^{2\pi} q^4\,d\,\theta = \mathrm{R}^4\int_0^{2\pi}\left(1 + \frac{2\,r^2}{\mathrm{R}^2} + \frac{r^4}{\mathrm{R}^4} + 4\left(1 + \frac{r^2}{\mathrm{R}^2}\right)\frac{r}{\mathrm{R}}\,.\,\cos.\,\theta\right.$$

$$\left. + \,4\,\frac{r^2}{\mathrm{R}^2}\cos^2\!.\,\theta\right)d\,\theta = 2\,\pi\,\mathrm{R}^4\left(1 + 4\,\frac{r^2}{\mathrm{R}^2} + \frac{r^4}{\mathrm{R}^4}\right),$$

and

$$\mathrm{P} = \frac{2\,\pi\,f\,\rho\,a^2\,\mathrm{R}^3\,z}{2\,g}\,.\left(1 + 4\,\frac{r^2}{\mathrm{R}^2} + \frac{r^4}{\mathrm{R}^4}\right). \quad . \quad . \quad (3.)$$

The following modification of this expression is sometimes convenient:—

Let $\mathrm{V} = a\,\mathrm{R}$ denote the velocity of advance of the solid;

$\lambda = 2\,\pi\,\mathrm{R}$, as before, its length, being the length of a wave;

$\mathrm{Sin}\,\beta = \frac{r}{\mathrm{R}}$ the sine of the greatest angle made by a tangent to the trochoidal surface with the direction of advance; then

$$P = \frac{f \rho V^2}{2g} . \lambda z (1 + 4 \sin^2 \beta + \sin^4 \beta).* \quad . \quad . \quad (4.)$$

It is to be observed that the resistance P, as determined by the preceding investigation, being deduced from the amount of work performed against friction, includes not only the longitudinal components of the direct action of friction on each element of the surface of the solid, but the longitudinal components of the excess of the hydrostatic pressure against the front of the solid above that against its rear, which is the indirect effect of friction. The only quantities neglected are those arising from the disturbances of the figures of the surfaces of equal pressure, which quantities are assumed to be unimportant, for reasons already stated. The consideration of such quantities would introduce terms into the resistance varying as the fourth and higher powers of the velocity.

NOTE.—Added October, 1862.

The investigation of Mr. Stokes (*Camb. Trans.*, Vol. VIII.) proceeds to the second degree of approximation in shallow water, and to the third degree in water indefinitely deep. In the latter case he arrives at the result, that the crests of the waves rise higher above the level of still water than the troughs sink below that level, by a height agreeing with that stated in Art. 14 of this paper, and that the profile of the waves is *approximately* trochoidal.

Mr. Stokes also arrives at the conclusion, that, when the disturbance is considerable compared with the length of a wave, there is combined with the orbital motion of each particle a *translation* which diminishes rapidly as the depth increases. No such translation has been found amongst the results of the investigation in the present paper; and hence

* This formula (neglecting $\sin^4 \beta$ as unimportant in practice) has been used to calculate approximately the resistance of steam-vessels, and its results have been found to agree very closely with those of experiment, and have also been used since 1858 by Mr. James R. Napier and the author with complete success in practice, to calculate beforehand the engine-power required to propel proposed vessels at given speeds. The formula has been found to answer approximately, even when the lines of the vessel are not trochoidal, by putting for β the mean of the values of the greatest angle of obliquity for a series of water-lines. The method of using the formula in practice, and a table showing comparisons of its results with those of experiment, were communicated to the British Association in 1861, and printed in the *Civil Engineer and Architect's Journal* for October of that year, and in part also in the *Mechanics' Magazine*, *The Artisan*, and *The Engineer*. The ordinary value of the coefficient of friction f appears to be about ·0036 for water gliding over painted iron. The quantity $\lambda z (1 + 4 \sin^2 \beta + \sin^4 \beta)$ corresponds to what is called, in the paper referred to, the *augmented surface.*

it would appear that Mr. Stokes's results and mine represent two different possible modes of wave-motion.*

The simplicity with which an exact result is obtained in the present paper, is entirely due to the following peculiarity:—Instead of taking for independent variables (besides the time) the *undisturbed* co-ordinates of a particle of liquid, there are taken two quantities, h and k, which are *functions* of those co-ordinates, of forms which are left indeterminate until the end of the investigation. h then proves to be identical with the undisturbed horizontal co-ordinate; but k proves to be a function of the undisturbed vertical co-ordinate, for which there is no symbol in our present notation, being the root of the transcendental equation

$$k_0 - k - \frac{r_0^2}{2\,\mathrm{R}} \cdot e^{-\frac{2k}{\mathrm{R}}} = 0,$$

in which k_0 is the undisturbed vertical co-ordinate (see Art. 13). Hence it is evident that, had k_0 instead of k been taken as the independent variable, the question of wave-motion considered in this paper could not have been solved except by a complex and tedious process of approximation.

* NOTE, added June, 1863.—The difference between the cases considered by Mr. Stokes and by me is the following:—In Mr. Stokes's investigation, the *molecular rotation* is null; that is to say

$$\tfrac{1}{2}\left(\frac{dv}{dx} - \frac{du}{dy}\right) = 0;$$

while in my investigation it is constant in each layer, being the following function of k,

$$\tfrac{1}{2}\left(\frac{dv}{dx} - \frac{du}{dy}\right) = \frac{a\, r_0^2\, e^{-\frac{2k}{\mathrm{R}}}}{\mathrm{R}^2 - r_0^2\, e^{-\frac{2k}{\mathrm{R}}}} \quad . \quad . \quad . \quad (11.)$$

From this last equation it follows that

$$\frac{d}{dt}\left(\frac{dv}{dx} - \frac{du}{dy}\right) = 0;$$

and therefore that the condition of continuity of pressure is verified.

XXX.—ON PLANE WATER-LINES IN TWO DIMENSIONS.
(See PLATES III. & IV.)*

SECTION I.—INTRODUCTION, AND SUMMARY OF KNOWN PRINCIPLES.

1. *Plane Water-Lines in two Dimensions defined.*—By the term "Plane Water-Line in two Dimensions" is meant a curve which a particle of liquid describes in flowing past a solid body, when such flow takes place in plane layers of uniform thickness. Such curves are suitable in practice for the water-lines of a ship, in those cases in which the vertical displacements of the particles of water are small compared with the dimensions of the ship; for in such cases the assumption that the flow takes place in plane layers of uniform thickness, though not absolutely true, is sufficiently near the truth for practical purposes, so far as the determination of good forms of water-line is concerned. As water-line curves have at present no single word to designate them in mathematical language, it is proposed, as a convenient and significant term, to call them *Neoïds* (from νηὸς, the Ionic genitive of ναῦς).

2. *General Principles of the Flow of a Liquid past a Solid.*—The most complete exposition yet published, so far as I know, of the principles of the flow of a liquid past a solid, is contained in Professor Stokes's paper "On the Steady Motion of an Incompressible Fluid," published in the *Transactions of the Cambridge Philosophical Society* for 1842. So far as those principles will be referred to in the present paper, they may be summed up as follows:—

When a liquid mass of indefinite extent flows past a solid body in such a manner that as the distance from the solid body in any direction increases without limit, the motion of the liquid particles approaches continually to uniformity in velocity and direction, the *condition of perfect fluidity* requires that the three components u, v, w of the velocity of a liquid particle should be the three differential coefficients of one function of the co-ordinates (ϕ); viz.—

$$u = \frac{d\phi}{dx};\ v = \frac{d\phi}{dy};\ w = \frac{d\phi}{dz}; \quad . \quad . \quad . \quad (1.)$$

* Read before the Royal Society of London on November 26, 1863, and published in the *Philosophical Transactions* for 1864.

and the *condition of constant density* requires that the said function should fulfil the following condition,

$$\frac{d^2 \phi}{d x^2} + \frac{d^2 \phi}{d y^2} + \frac{d^2 \phi}{d z^2} = 0. \quad . \quad . \quad . \quad . \quad (2.)$$

By giving to the function ϕ a series of different constant values, a series of surfaces are represented, to which each water-line curve is an orthogonal trajectory, so that if U = constant be the equation of a series of surfaces each containing a continuous series of water-line curves (and one of which surfaces must be that of the solid body), the function U must satisfy the following condition,

$$\frac{d\,U}{d x} \cdot \frac{d \phi}{d x} + \frac{d\,U}{d y} \cdot \frac{d \phi}{d y} + \frac{d\,U}{d z} \cdot \frac{d \phi}{d z} = 0; \quad . \quad . \quad (3.)$$

or if $d\,s'$ be an elementary arc of a water-line curve, and x', y', z' its co-ordinates, the following conditions must be satisfied,

$$\frac{d\,x'}{d\,s'} : \frac{d\,y'}{d\,s'} : \frac{d\,z'}{d\,s'} :: \frac{d \phi}{d x} : \frac{d \phi}{d y} : \frac{d \phi}{d z}; \quad . \quad . \quad . \quad (4.)$$

and these are the most general expressions of the geometrical properties of water-line curves in three dimensions.

When the inquiry is restricted to motion in two dimensions only, x and y, the terms containing $d\,z$ and $d\,z'$ disappear from the preceding equations; and it also becomes possible to express the same conditions by means of equations of a kind which are more convenient for the purposes of the present investigation, and which are as follows: Conceive the plane layer of liquid under consideration of thickness unity, to be divided into a series of elementary streams by a series of water-line curves, one of which must be the outline of the solid body; let U = constant be the equation of any one of those curves, U being a function of such a nature that d U is the volume of liquid which flows in a second along a given elementary stream; then the components of the velocity of a particle of liquid are

$$u = \frac{d\,U}{d y}; \; v = -\frac{d\,U}{d x}; \quad . \quad . \quad . \quad . \quad (5.)$$

the condition of continuity is satisfied; and the condition of perfect fluidity requires that the function U should fulfil the following equation,

$$\frac{d^2\,U}{d x^2} + \frac{d^2\,U}{d y^2} = 0. \quad . \quad . \quad . \quad . \quad (6.)$$

(When the motion of the liquid is not subject to the condition of being uniform in velocity and direction at an infinite distance in every direction from the solid, it is sufficient that

$$\frac{d^2 U}{d x^2} + \frac{d^2 U}{d y^2} = \text{function of } U;$$

but cases of that kind do not occur in the present paper.)*

3. *Notation.*—It is purely a question of convenience whether the infinitely distant particles of the fluid are to be regarded as fixed and the solid as moving uniformly, or the solid as fixed, and the infinitely distant particles of the fluid as moving uniformly with an equal speed in the contrary direction. Throughout the present paper the solid will be supposed to move along the axis of x; so that v will represent the transverse component of the velocity of a particle of liquid on either supposition. The longitudinal component of the velocity of a liquid particle *relatively to the solid* will be denoted by u; and when that particle is at an infinite distance from the solid, by c; so that when the infinitely distant part of the liquid is regarded as fixed, the solid is to be conceived as moving with the velocity $-c$; and the longitudinal component of the velocity of a liquid particle relatively to the indefinitely distant part of the liquid will be denoted by $u - c$.

It is convenient to regard the function U as equivalent to an expression of the following kind,

$$U = b\,c, \qquad (7.)$$

c being the uniform velocity of flow at an infinite distance, and b what the value of y would be for the water-line under consideration if the solid were removed; in which case that line would become a straight line parallel to the axis of x. This enables us to substitute for equations (5) and (6) the following, in which *proportionate* velocities only are considered:—

$$\frac{u}{c} = \frac{d\,b}{d\,y}; \qquad \frac{v}{c} = -\frac{d\,b}{d\,x}; \qquad (8.)$$

$$\frac{d^2 b}{d y^2} + \frac{d^2 b}{d x^2} = 0. \qquad (9.)$$

4. *General Characteristics of Water-Line Functions.*—Since at an infinite distance from the solid body we have $u = c$, $v = 0$, it follows that, if the

* Professor William Thomson, in 1858, completed an investigation of the motion of a solid through a perfect liquid, so as to obtain expressions for the motion of the solid itself, involving twenty-one constants depending on the figure and mass of the solid and the density of the liquid; but as that investigation, though on the eve of publication, has not yet been published, I shall not here refer to it further.

origin of co-ordinates be taken in or near the solid body, b must be a function of such a kind that, when either $x = \infty$, or $y = \infty$,

$$b = y.$$

Hence, in a great number of cases that function is of the form

$$b = y + F(x, y); \quad . \quad . \quad . \quad . \quad (10.)$$

where F is a function which either vanishes or becomes constant when x or y increases indefinitely.

It is plain that when the function b takes this form, the term F is the function for the motions of the liquid particles *relatively to still water;* that is to say,

$$\frac{u - c}{c} = \frac{d\,b}{d\,y} - 1 = \frac{d\,F}{d\,y}; \quad \frac{x}{c} = -\frac{d\,b}{d\,x} = -\frac{d\,F}{d\,x}; \quad . \quad (11.)$$

and also that the term F fulfils the equation

$$\frac{d^2\,F}{d\,y^2} + \frac{d^2\,F}{d\,x^2} = 0. \quad . \quad . \quad . \quad . \quad (12.)$$

When the solid is symmetrical at either side of the axis of x (as it is in all the cases that will be considered in this paper), the axis of x itself, so far as it lies beyond the outline of the solid, is a water-line. Hence it is necessary that the equation of that axis, viz.—

$$y = 0,$$

should be one of the solutions of the equation

$$b = y + F(x, y) = 0, \quad . \quad . \quad (13.)$$

and, consequently, that F should vanish with y.

The vanishing of F when $x = \infty$, indicates that every straight line given by the equation $y = b$ either forms part of, or is an asymptote to, a water-line curve.

The vanishing of F when $y = \infty$, indicates that the farther the water-lines are from the generating solid, the more nearly they approximate to parallel straight lines.

Every water-line curve is itself the outline of a solid capable of moving smoothly through a liquid.

5. *Water-Line Curves generated by a Circle, or Cyclogenous Neoïds.*—Conceive that a circular cylinder of indefinite height, and of the radius l, described about the axis of z, moves through the liquid along the axis of x. Then it is already known that the general equation of the water-line curves is the following,

$$b = y\left(1 - \frac{l^2}{x^2 + y^2}\right), \quad . \quad . \quad . \quad (14.)$$

giving a series of curves of the third order. When $b = 0$ this equation resolves itself into two, viz.—

$$y = 0;\ x^2 + y^2 = l^2;$$

the first of which represents the axis of x, and the second the circular outline of the cylinder. For each other value of b, equation (14) represents a curve having two branches: one of them is an oval, contained within the circle, and not relevant to the problem in question; the other, being the real water-line, is convex in the middle and concave towards the ends, and has for an asymptote in both directions the straight line $y = b$.

For brevity's sake, let $x^2 + y^2 = r^2$. Then the component velocities of a particle of water *relatively to the solid* are given by the equations

$$\left.\begin{array}{l} \dfrac{u}{c} = \dfrac{d\,b}{d\,y} = 1 - \dfrac{l^2}{r^2} + \dfrac{2\,l^2\,y^2}{r^4} = 1 + \dfrac{l^2\,(y^2 - x^2)}{r^4} \\ \dfrac{v}{c} = -\dfrac{d\,b}{d\,x} = -\dfrac{2\,l^2\,x\,y}{r^4}, \end{array}\right\} . \quad (15.)$$

and the square of their resultant by the equation

$$\frac{u^2 + v^2}{c^2} = \left(1 - \frac{l^2}{r^2}\right)^2 + \frac{4\,l^2\,y^2}{r^4}; \quad . \quad . \quad (16.)$$

while the component and resultant velocities *relatively to still water* are given by the following equations:—

$$\frac{u}{c} - 1 - \frac{l^2\,(y^2 - x^2)}{r^4};\ \frac{v}{c} = -\frac{2\,l^2\,x\,y}{r^4};\ \frac{\sqrt{\{(u - c)^2 + v^2\}}}{c} = \frac{l^2}{r^2}. \quad (17.)$$

As a convenient name for water-line curves of this sort, it is proposed to call them *Cyclogenous Neoïds,* that is, *ship-shape curves generated from a circle.*

The water-line surfaces generated by a sphere are known; but no use will be made of them in this paper.*

SECTION II.—PROPERTIES OF WATER-LINE CURVES GENERATED FROM OVALS, OR OÖGENOUS NEOÏDS.

6. *Derivation of other Water-Line Curves from Cyclogenous Neoids.*—When a form of the function F has been found which satisfies equation (12) of

* See Paper by Dr. Hoppe, *Quart. Journ. Math.*, March, 1856.

Art. 4 (that is to say, which fulfils the condition of liquidity), an endless variety of other forms of that function possessing the same property may be derived from the original form by differentiation and integration.

The original form, and also the derived forms, must possess the properties of vanishing for $x = \infty$ and for $y = \infty$, and of becoming $= 0$, or a constant for $y = 0$. The first of those properties excludes trigonometrical functions, and consequently exponential functions also, which are always accompanied by trigonometrical functions, and leaves available functions of the nature of potentials. The second property excludes derivation by means of differentiation and integration with respect to y, and leaves available differentiation and integration with respect to x.

The original form of the function F which will be used in this paper is that appropriate to cyclogenous neoïds, or water-line curves generated from a circle, as given in equation (14) of Art. 5, viz.:—

$$\mathrm{F} = \frac{y}{r^2} \times \text{constant}.$$

When one or more differentiations with respect to x are performed on this function, and the results substituted for F in equation (10), there are obtained curves which are real water-lines, but which are not suitable for the figures of ships, some of them being lemniscates, others shaped like an hour-glass, and others looped and foliated in various ways. It is otherwise as regards integration with respect to x; for that operation, being performed once, gives the expression for the ordinate in a class of curves all of which resemble possible forms of ships, and which are so various in their proportions, that every form of ships' water-lines which has been found to succeed in practice may be closely imitated by means of them. As that class of curves consists of certain ovals, and of other water-lines generated from those ovals, it is proposed to call them *Oögenous Neoïds* (from Ὠογενής).

7. *General Equation of Oögenous Neoïds.*—The integration with respect to x, already referred to, is performed as follows:—The co-ordinates of a particle of water being x and y, let x' denote the position of a movable point in the axis of x: then the function to be integrated is

$$\frac{y}{(x - x')^2 + y^2},$$

for all values of x' between two arbitrary limits. Let $2a$ denote the distance between those limits: the most convenient position for the origin of co-ordinates is midway between them, so as to make the limits

$$x' = + a, \; x' = - a \text{ respectively.}$$

Then the following is the integral sought:

Plate III.

FIG 1.

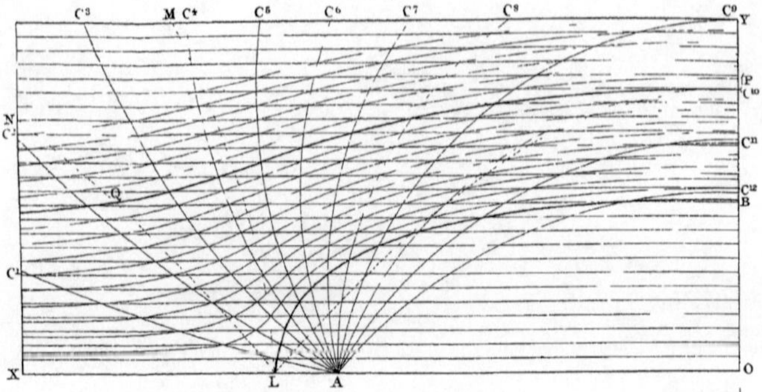

FIG. 2.

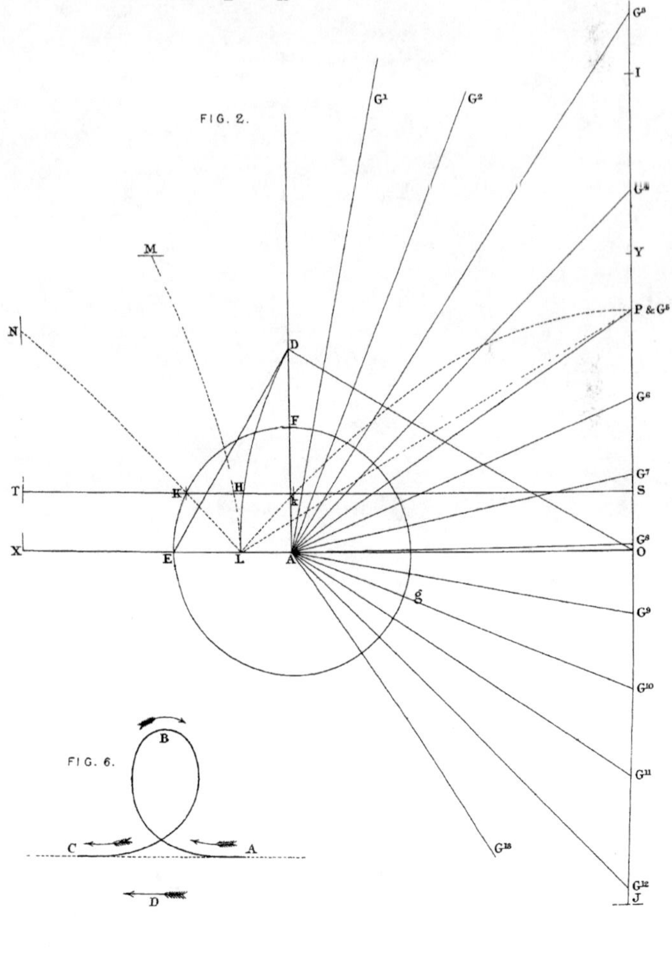

FIG. 6.

$$\int_{x'=-a}^{x'=+a} \frac{y\,dx'}{(x-x')^2+y^2} = \tan^{-1}\frac{a-x}{y} + \tan^{-1}\frac{a+x}{y}. \quad (18.)$$

This quantity evidently denotes *the angle contained between two lines drawn from the point* (x, y) *to the points* $(+a, 0)$ *and* $(-a, 0)$. For brevity's sake, in the sequel that angle will be occasionally denoted by θ; the points $(+a, 0)$ and $(-a, 0)$ will be called the *foci;* and their distance a from the centre will be called the *eccentricity.*

Substituting this integral in the general equation (10), we find, for the water-line curves now under consideration, the following equation, which is the general equation of *oögenous neoïds:*—

$$b = y - f\theta = y - f\left(\tan^{-1}\frac{a-x}{y} + \tan^{-1}\frac{a+x}{y}\right) \quad (19.)$$

The coefficient f denotes an arbitrary length, which will be called the *parameter.*

8. *Geometrical Meaning of that Equation.*—The equation (19) represents a curve at each point of which the excess $(y - b)$ of the ordinate (y) above a certain minimum value (b) is proportional to the angle (θ) contained at that point between two straight lines drawn to the two foci. Except when $b = 0$, the curve has an asymptote at the distance b from the axis of x, and parallel to that axis. Since the value of b is not altered by reversing the signs of x, and is only changed from positive to negative by reversing the sign of y, it follows that each curve consists of two halves, symmetrical about the axis of y; and that there are pairs of curves symmetrical about the axis of x.

In Plate III., Fig. 1, therefore, which represents a series of such curves, one quadrant only of the space round the origin or centre O is shown, the other three quadrants being symmetrical. A is one of the foci, at the distance O A $= a$ from the centre; the other focus, not shown in the figure, is at an equal distance from the centre in the opposite direction. B L is one quadrant of the primitive oval; and the wave-like curves outside of it are a series of water-lines generated from it, having for their respective asymptotes the series of straight lines parallel to O X, and whose distances from O X are a series of values of b.

The equation (19) embraces also a set of curves contained within the oval, and all traversing the two foci; but as these curves are not suited for the forms of ships' water-lines, no detailed description of them needs be given.

9. *Properties of Primitive Oval Neoïds.*—When in equation (19) b is made $= 0$, so that the equation becomes

$$y - f\theta = 0, \quad . \quad . \quad . \quad . \quad (20.)$$

there are two solutions; one of which, viz. $y = 0$, represents the axis of x, agreeably to the condition stated in Art. 4, equations (13). The other solution represents the oval L B.

The greater semi-axis of that oval, O L, will be called the *base* of the series of water-lines generated by the oval, and denoted by l; its value is found as follows:

$$\frac{d\,b}{d\,y} = 1 + f\frac{d}{d\,y}\left(\tan^{-1}\frac{y}{a-x} + \tan^{-1}\frac{y}{a+x}\right)$$

$$= 1 + f\left\{\frac{a-x}{(a-x)^2+y^2} + \frac{a+x}{(a+x)^2+y^2}\right\};$$

but at the point L we have

$$x = l;\ y = 0;\ \frac{d\,b}{d\,y} = 0;$$

and, therefore,

$$0 = 1 + f\left(\frac{1}{a-l} + \frac{1}{a+l}\right);$$

whence

$$l^2 = a^2 + 2\,a\,f. \quad . \quad . \quad . \quad . \quad (21.)$$

To find the parameter f when the base l and eccentricity a are given, we have the formula

$$f = \frac{l^2 - a^2}{2\,a}. \quad . \quad . \quad . \quad . \quad (22.)$$

The half-breadth, or minor semi-axis of the oval, O B $= y_0$, is the root of the following transcendental equation, found by making $x = 0$ in equation (19),

$$y_0 - 2\,f\tan^{-1}\frac{a}{y_0} = 0, \quad . \quad . \quad . \quad (23.)$$

which may be otherwise written as follows:—

$$\tan\frac{y_0}{2\,f} - \frac{a}{y_0} = 0. \quad . \quad . \quad . \quad (23\text{ A.})$$

When the minor semi-axis y_0 and eccentricity a are given, the parameter f is found by the equation

$$f = \frac{y_0}{2\tan^{-1}\frac{a}{y_0}}; \quad . \quad . \quad . \quad . \quad (24.)$$

and thence the base l can be computed by equation (21).

When the base l and half-breadth y_0 are given, the eccentricity a is found by solving the following transcendental equation:—

$$a y_0 - (l^2 - a^2) \tan^{-1} \frac{a}{y_0} = 0. \quad . \quad . \quad (24\text{ A.})$$

An oval neoid differs from an ellipse in being fuller towards the ends and flatter at the sides; and that difference is greater the more elongated the oval is.

10. *Varieties of Oval Neoids, and extreme cases.*—The eccentricity a may have any value, from nothing to infinity; and the base l may bear to the half-breadth y_0 any proportion, from equality to infinity. When the eccentricity $a = 0$, the two foci coalesce with the centre O; the base l becomes equal to the half-breadth b, the oval becomes a circle of the radius l; and the water-lines generated by it become cyclogenous neoïds, already described in Art. 5.

As the eccentricity increases, the oval becomes more elongated. In Plate IV., Fig. 3, P L is an oval whose length is to its breadth as $\sqrt{3} : 1$, its focus being at A_0. The oval BL in Plate III., Fig. 1, is more elongated, its length being to its breadth as 17 : 6 nearly. When the eccentricity is infinite, the centre O and the farther focus go off to infinity, leaving only one focus. The parameter f becomes equal to the focal distance L A. The oval is converted into a curve bearing the same sort of analogy to a parabola that an oval neoïd bears to an ellipse;* but instead of spreading to an infinite breadth like a parabola, it has a pair of asymptotes parallel to the axis of x, and at the distance $\pm \pi f$ to either side of it; and each generated water-line has two parallel asymptotes, at the respective distances b and $b + \pi f$ from the axis of x. The properties of these curves may be easily investigated by placing the origin of co-ordinates at the focus A, and substituting, in equation (19), $\tan^{-1}\frac{y}{x}$ for θ; but as their figure is not suitable for ships' water-lines, it is unnecessary here to discuss them in detail; and the same may be said of a class of curves analogous to hyperbolas, whose equation is formed by putting $-$ instead of $+$ between the two terms of the right-hand member of equation (18).

11. *Graphic Construction of Oval and Oögenous Neoïds.*—For the sake of distinctness, the processes of drawing these curves are represented in two figures,—Fig. 2 showing the preliminary, and Fig. 1 the final processes (see Plate III.)

The axis O Y is to be divided into equal parts of any convenient length (which will be denoted by δy in what follows), and through the divisions are to be drawn a series of straight lines parallel to O X. (It is convenient to print those lines from a copper-plate divided and ruled by machinery.) They are shown in Fig. 1 only, and not in Fig. 2, to avoid confusion.

* This curve is identical with the quadratrix of Tschirnhausen.

Suppose, now, that the problem is as follows:—*The base* O L *and eccentricity* O A *being given, it is required to construct the oval neoid and the water-lines generated by it.*

Through the focus A (Plate III., Fig. 2) draw A D perpendicular to O X; about O, with the radius O L, describe the circular arc L D, cutting A D in D; from D draw D E perpendicular to O D, cutting O X in E; then (as equation (22) shows) A E will be $= 2f$, the *double parameter.*

About A, with the radius $AE = 2f$ thus found, describe a circle cutting A D in F. Then commencing at F, lay off on that circle a series of arcs, each equal to $2\,\delta y$ (the double of the length of the equal divisions of the axis O Y). Through the points of division of the circle draw a series of radii, AG_1, AG_2, &c., cutting the axis O Y in a series of points (some of which, from G_3 to G_{12}, are marked in Fig. 2).* (These radii make, with the line A D, a series of angles, $\frac{\delta y}{f}$, $\frac{2\,\delta y}{f}$, $\frac{3\,\delta y}{f}$, &c.)

Then about each of the points in the axis O Y thus found, with the outer leg of the compasses starting from the focus A, describe a series of circles (shown in Plate III., Fig. 1), AC_1, AC_2, AC_3, &c.

Each of those circles traverses the two foci; and the equation of any one of them is

$$-F = f\theta = n\,\delta y, \quad . \quad . \quad . \quad (25.)$$

where θ denotes the angle made at any point of the circle by straight lines drawn to the two foci, and n has the series of values 1, 2, 3, &c. Since F, as explained in Art. 4, is the characteristic function for the motion of the liquid particles relatively to still water, it is plain that each of the circles for which $F =$ constant is a tangent to the directions of motion of all the particles that it traverses.

The paper is now covered, as in Fig. 1, with a network made by a series of straight lines, whose equations are of the form $y = n'\,\delta y$, crossed by a series of circles, whose equations are of the form $f\theta = n\,\delta y$.

Consequently, any curve drawn like those in Plate III., Fig. 1, diagonally through the corners of the quadrangles of that network, will have for its equation

$$y - f\theta = (n' - n)\,\delta y = b,$$

and will accordingly be an oögenous neoid, having for its asymptote the line $y = b$.

* When the parameter is small, it is sometimes advisable to use a circle (such as a protractor) with a radius which is a larger multiple of the parameter than double, the length of the divisions being increased in the same proportion; or the points on the axis O Y may be laid down by means of their distances from O, calculated by the formula $OG = a \cdot \cot \theta$.

The primitive oval is drawn by starting from the point L, and traversing the network diagonally. As many curves as are required can be drawn by the eye with great precision, and the whole process is very rapid and easy (see Appendix).

When the problem is, *with a given base and eccentricity to draw an oögenous neoïd through a given point in the axis* O Y, such as P, the process is modified as follows:—The axis O Y must be so divided that P shall be at a point of division. Then, up to the describing of the circle about A with the radius A E, the process is the same as before. Then, join A P (Plate III., Fig. 2), and draw A g, making the angle P A g = A P O, and cutting the axis O Y in a point (such as G_{10}), which will be the centre of the circle traversing A and P. Then, on the circumference of the circle about A, from g towards F, lay off a series of arcs each $= 2\,\delta y$; through the points of division draw radii cutting the axis O Y in the points G_9, G_8, &c., and complete the process as before.

12. *Graphic Construction of Cyclogenous and Parabologenous Neoïds.*—When the eccentricity vanishes and the oval becomes a circle, all the circles composing the network become tangents to O X at the point O. They pass through the points where the primitive circular water-line is cut by the equidistant parallel lines. Their radii are in harmonic progression; the equation of any one of them is of the form

$$- F = \frac{l^2 y}{x^2 + y^2} = n\,\delta y, \quad . \quad . \quad . \quad (26.)$$

n having the series of values 1, 2, 3, &c.; and its radius is given by the formula

$$\frac{l^2}{n\,\delta y}. \quad . \quad . \quad . \quad . \quad (26\text{ A.})$$

When there is but one focus, as in the infinitely long curve described in Art. 10, the network of circles is changed into a set of straight lines radiating from the focus, and making with A X the series of angles given by the formula

$$f\theta = n\,\delta y. \quad . \quad . \quad . \quad . \quad (27.)$$

13. *Component and Resultant Velocities of Gliding.*—The component and resultant velocities with which the liquid particles glide along the water-lines are given by the following equations, in terms of the eccentricity a, the parameter f, and the co-ordinates:—

$$\left.\begin{aligned} \frac{u}{c} &= \frac{d\,b}{d\,y} = 1 + \frac{f\,(a - x)}{(a - x)^2 + y^2} + \frac{f\,(a + x)}{(a + x)^2 + y^2}; \\ \frac{v}{c} &= -\frac{d\,b}{d\,x} = -\frac{f\,y}{(a - x)^2 + y^2} + \frac{f\,y}{(a + x)^2 + y^2}; \end{aligned}\right\} (28.)$$

$$\left.\begin{aligned}\frac{u^2+v^2}{c^2} = \frac{d\,b^2}{d\,y^2} + \frac{d\,b^2}{d\,x^2} = 1 + \frac{2\,f\,(a-x)}{(a-x)^2+y^2} + \frac{2\,f\,(a+x)}{(a+x)^2+y^2} \\ + \frac{4\,f^2\,a^2}{\{(a-x)^2+y^2\}\,.\,\{(a+x)^2+y^2\}}.\end{aligned}\right\} \quad (28.)$$

At the point of greatest breadth (that is, at the axis of y) these expressions take the following values:—

$$\frac{u_0}{c} = 1 + \frac{2\,f\,a}{a^2+y_0^2} = 1 + \frac{l^2-a^2}{a^2+y_0^2} = \frac{l^2+y_0^2}{a^2+y_0^2};\ \frac{v_0}{c} = 0. \quad (28\,\text{A.})$$

These equations are applicable to a *whole series of water-lines* (such as those shown in Fig. 1), including the generator oval, and are the best suited for solving questions relating to such a series.

But when *one particular water-line* is in question, it is sometimes more convenient to use another set of equations, formed from the equations (28) by the aid of the following substitutions, in which θ, as before, denotes $\frac{y-b}{f}$:—

$$\left.\begin{aligned}&\{(a-x)^2+y^2\}\,.\,\{(a+x)^2+y^2\} = \frac{4\,a^2\,y^2}{\sin^2\theta};\\ &\{(a-x)^2+y^2\} + \{(a+x)^2+y^2\} = 2\,a^2 + 2\,x^2\\ &\qquad + 2\,y^2 = 4\,a^2 + 4\,a\,y\,\text{cotan}\,\theta;\\ \therefore\ \ &x^2+y^2 = a^2 + 2\,a\,y\,\text{cotan}\,\theta;\\ &x = \sqrt{\{a^2-y^2+2\,a\,y\,\text{cotan}\,\theta\}}.\end{aligned}\right\} \quad (29.)$$

These substitutions being made in the equations (28), give the following results:—

$$\left.\begin{aligned}&\frac{u}{c} = 1 + \frac{f}{a}\sin^2\theta - \frac{f}{y}\cdot\cos\theta\sin\theta = 1 + \frac{f}{2\,a}\\ &\qquad\qquad \frac{f\cos 2\,\theta}{2\,a} - \frac{f\sin 2\,\theta}{2\,y};\\ &\frac{v}{c} = -\frac{f\,x}{a\,y}\sin^2\theta = -\frac{f}{a\,y}\sqrt{\{a^2-y^2+2\,a\,y\,\text{cotan}\,\theta\}}\,\sin^2\theta;\\ &\frac{u^2+v^2}{c^2} = 1 + \frac{2\,f}{a}\sin^2\theta - \frac{2\,f}{y}\cos\theta\sin\theta + \frac{f^2}{y^2}\sin^2\theta\\ &\qquad = 1 + \frac{f}{a} + \frac{f^2}{2\,y^2} - \left(\frac{f}{a}+\frac{f^2}{2y^2}\right)\cos 2\,\theta - \frac{f}{y}\sin 2\,\theta.\end{aligned}\right\} \quad (30.)$$

14. *Trajectories of Normal Displacement, and of Swiftest and Slowest Gliding.*—By the "trajectory of normal displacement" is meant a curve traversing all the points in a series of water-lines at which the directions of motion of the liquid particles relatively to still water are perpendicular to the water-lines; or, speaking geometrically, a curve traversing all the points at which the circles $A\,C_1$, $A\,C_2$, &c., of Fig. 1, Plate III., cut the water-lines at right angles. To find the form of that trajectory it is sufficient to make

$$\frac{u^2 + v^2}{c^2} - \frac{u}{c} = 0; \qquad (31.)$$

employing the values of those ratios given by the equations (28). This having been done, it appears, after some simple reductions, that the equation of the *trajectory of normal displacement* is the following,

$$x^2 - y^2 = l^2, \qquad (32.)$$

being that of a rectangular hyperbola L M, Fig. 1, having its vertex at L, and its centre at O. Hence, that curve is *similar for all oögenous and cyclogenous neoids whatsoever*, being independent of the eccentricity, and is identical for all oögenous and cyclogenous neoids having the same base l.

By the "trajectory of swiftest and slowest gliding" is meant a curve traversing every point in a series of water-lines at which the velocity of gliding, $\sqrt{u^2 + v^2}$, is a maximum or a minimum for the water-line on which that point is situated. To find the equation of that curve, it is necessary to solve the following equation,

$$\frac{d}{c\,d\,t}\left(\frac{u^2 + v^2}{c^2}\right) = \left(\frac{u}{c} \cdot \frac{d}{d\,x} + \frac{v}{c} \cdot \frac{d}{d\,y}\right)\left(\frac{u^2 + v^2}{c^2}\right) = 0, \qquad (33.)$$

the expression employed for $\frac{u^2 + v^2}{c^2}$ being that given by the third of the equations (28). After a tedious but not difficult process of differentiation and reduction, which it is unnecessary to give in detail, an equation is found which resolves itself into three factors, viz.—

$$x = 0, \qquad (34.)$$

being the equation of the axis O Y, and

$$\sqrt{x^2 + y^2} + y \pm \sqrt{l^2 + y^2} = 0, \qquad (35.)$$

being the equations of the two branches L N and L P of a curve of the fourth order. This curve, too, is independent of the eccentricity, and therefore *similar for all oögenous and cyclogenous neoïds whatsoever*, and identical for those having the same base l. It has also the following properties:—The straight line joining L with P makes an angle of 30° with the axis O X; there are a pair of straight asymptotes through O, making angles of 30° to either side of O X; and the two branches of the curve cut O X in the point L, at angles of 45°.

15. *Graphic Construction of those Trajectories.*—The curves described in the preceding article are easily and quickly constructed, with the aid of the series of equidistant lines parallel to O X, as follows:—In Fig. 2, Plate III., let S T be any one of those lines. With the distance S L in the compasses, lay off S H on that line; H will be a point in the hyperbola L M. Also from S lay off, on the axis of y, S I and S J, each equal to the same distance S L. About the centre O, with the radius O I, draw a circular arc cutting S T in K; this will be a point in the branch L N. About the centre O, with the radius O J, draw a circular arc cutting S T in k; this will be a point in the branch L P.

16. *Properties of the Trajectory of Swiftest and Slowest Gliding.*—The branch L N traverses a series of points of slowest gliding, where the water-lines are farthest apart; the branch L P traverses a set of points of swiftest gliding, where the water-lines are closest together; from O to P the axis of y traverses points of slowest gliding, and beyond P, points of swiftest gliding.

Hence every complete oögenous neoïd which cuts the axis of y between O and P, contains two points of swiftest and three of slowest gliding; and every complete oögenous or cyclogenous neoïd which cuts the axis of y at or beyond P contains only one point of swiftest and two of slowest gliding.

17. *Water-Lines of Smoothest Gliding, or Lissoneoïds.*—At the point P itself, situated at the distance

$$\text{O P} = \frac{l}{\sqrt{3}} \quad . \quad . \quad . \quad . \quad (36.)$$

from the centre, two maxima and a minimum of the velocity of gliding coalesce; and therefore not only the first, but the second and third differential coefficients of the velocity of gliding vanish; from which it follows that the velocity of gliding changes more gradually on those water-lines which pass through the point P, than on any other class of oögenous or cyclogenous neoïds.

It is proposed, therefore, to call this class of water-lines *Lissoneoïds* (from λισσός).

The oval neoïd, whose length is to its breadth as $\sqrt{3} : 1$, is itself a lissoneoïd; and every series of water-lines generated by an oval *more elongated* than this contains one lissoneoïd; for example, in the series of water-lines shown in Fig. 1, the lissoneoïd is marked P Q.

The eccentricity of the oval lissoneoïd is computed by solving equation

Plate IV.

FIG. 3.

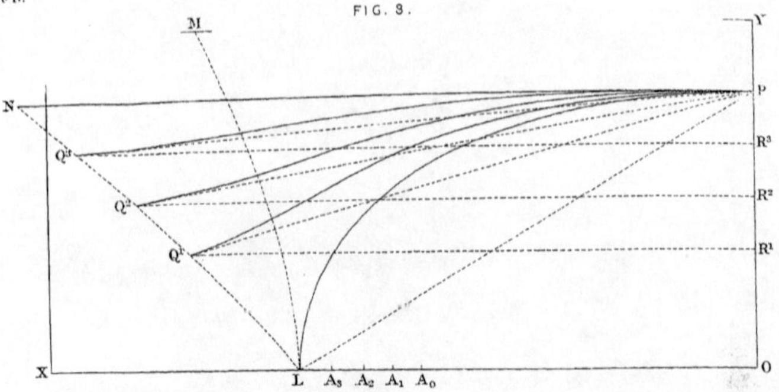

FIG. 4.

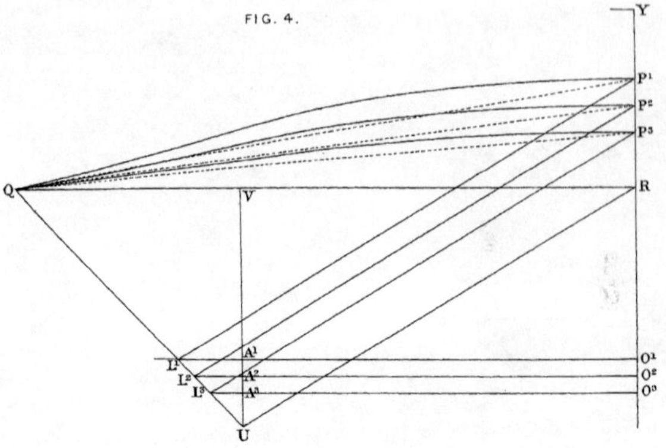

FIG. 5.

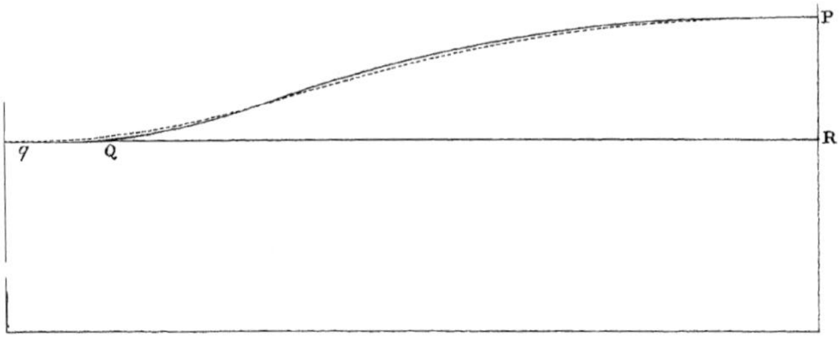

(24 A) of Art. 9, when $y_0 = \frac{l}{\sqrt{3}}$; and it is found to be

$$a = \cdot 732\, l, \text{ or nearly } (\sqrt{3} - 1)\, l. \quad . \quad . \quad (36\text{A.})$$

By giving the eccentricity values ranging from ·732l to l, there are produced a series of lissoneoids ranging from the oval P L, in Fig. 3, Plate IV., whose focus is at A_0, to the straight line P N, whose focus coalesces with L. P Q_1, P Q_2, and P Q_3 are specimens of the intermediate forms, having their foci respectively at A_1, A_2, and A_3. For a reason which will be explained in Section III., those curves are not shown beyond the trajectory of slowest gliding.

The *greatest speed of gliding*, for a lissoneoïd, is found by making $y_0^2 = \frac{l^2}{3}$ in equation (28A) of Art. 13; that is to say,

$$\frac{u_0}{c} = \frac{4\, l^2}{3\, a^2 + l^2}. \quad . \quad . \quad . \quad . \quad (37.)$$

18. *Orbits of the Particles of Water.*—The general expressions for the components of the velocity of a liquid particle relatively to still water have been given in equation (11) of Art. 4; and to apply those to the case of oögenous neoïds, it is only necessary to modify the equations (28) of Art. 13, by introducing the expression for $\frac{u - c}{c}$ instead of that for $\frac{u}{c}$, as follows:—

$$\left.\begin{aligned} \frac{u - c}{c} &= \frac{(l^2 - a^2)\,.\,(a^2 - x^2 + y^2)}{\{(a - x)^2 + y^2\}\,.\,\{(a + x)^2 + y^2\}}; \\ \frac{v}{c} &= \frac{-\,2\,(l^2 - a^2)\, x\, y}{\{(a - x)^2 + y^2\}\,.\,\{(a + x)^2 + y^2\}}; \\ \frac{(u - c)^2 + v^2}{c} &= \frac{(l^2 - a^2)^2}{\{(a - x)^2 + y^2\}\,.\,\{(a + x)^2 + y^2\}} \end{aligned}\right\} \quad . \quad (38.)$$

From the last of these equations it appears that *the velocity of a particle relatively to still water is inversely as the product of its distances from the two foci.*

The only other investigation which will here be made respecting the orbit of a particle of water, is that of the relation between its direction and curvature at a given point, and its ordinate y.

It has already been explained, in Art. 11, that the direction of motion of a particle is a tangent to a circle traversing it and the two foci. The radius of that circle is

$$\frac{a}{\sin \frac{y-b}{f}} = \frac{a}{\sin \theta};$$

and if ϕ be taken to denote the angle which the direction of the particle's motion relatively to still water makes with the axis of x, it is easily seen that

$$\cos \phi = \cos \theta - \frac{y}{a} \sin \theta. \quad . \quad . \quad . \quad (39.)$$

While that angle undergoes the increment $d\,\phi$, the particle moves through an arc of its orbit whose length is $\frac{d\,y}{\sin \phi}$; consequently the curvature of that orbit at the arc in question is

$$\frac{1}{\rho} = \frac{\sin \phi \, d\,\phi}{d\,y} = -\frac{d\,.\cos \phi}{d\,y} = \left(\frac{1}{f} + \frac{1}{a}\right) \sin \theta + \frac{y}{f\,a} \cos \theta$$

$$= \frac{2}{l^2 - a^2} \cdot \left\{ \frac{l^2 + a^2}{2\,a} \cdot \sin \theta + y \cos \theta \right\} \quad . \quad . \quad (40.)$$

For cyclogenous neoids, we obtain the value of this expression by making

$$\sin \theta = \frac{y-b}{f}, \cos \theta = 1,$$

substituting $l^2 - a^2$ for $2\,f\,a$, and then making $a = 0$; the result being as follows,

$$\frac{1}{\rho} = \frac{4}{l^2}\left(y - \frac{b}{2}\right); \quad . \quad . \quad . \quad . \quad (40\text{ A.})$$

that is to say, the *curvature of the orbit varies as the distance of the particle from a line parallel to the axis of* x, *and midway between that axis and the undisturbed position of the particle.* This is the property of the looped or coiled elastic curve; therefore, *when the water-lines are cyclogenous, the orbit of each particle of water forms one loop of an elastic curve.*

The general appearance of such an orbit is shown in Fig. 6, Plate III. The arrow D shows the direction of motion of the solid body. The dotted line A C is supposed to be at the distance b from the axis of x. The particle starts from A, is at first pushed forwards, then deviates outwards and turns backwards, moving directly against the motion of the solid body as it passes the point of greatest breadth, as shown at B. The particle then turns inwards, and ends by following the body, and coming to rest at C, in advance of its original position.

When the water-lines are oögenous, the equations (39) and (40) show that the orbit is of the same general character with the looped elastic curve in Fig. 6, but differs from it in detail to an extent which is greater the greater the eccentricity a; and the difference consists mainly in a flattening of the loop, so as to make it less sharply curved at B.

When the eccentricity increases without limit, the orbit approximates indefinitely to a "curve of pursuit," for which

$$\phi = \theta, \frac{1}{\rho} = \frac{\sin\theta}{f}. \quad . \quad . \quad . \quad (40\text{ B.})$$

19. *Trajectory of Transverse Displacement.—Of Speed of Gliding equal to Speed of Ship.—Orthogonal Trajectories.*—The trajectories described in this article differ from those described in Arts. 14, 15, and 16 by being dependent upon the eccentricity, and therefore not similar for all sets of oögenous neoids.

By the "trajectory of transverse displacement" is meant the curve traversing all the points at which the liquid particles are moving at right angles to the axis O X, relatively to still water. It is determined from the first of the equations (28), by making

$$\frac{u}{c} - 1 = 0;$$

from which is easily deduced the following equation,

$$x^2 - y^2 = a^2, \quad . \quad . \quad . \quad . \quad (41.)$$

being that of a rectangular hyperbola, with its centre at O and its vertex at the focus A.

The trajectory of the points where the speed of gliding is equal to the speed of the solid body, is found from the third of the equations (28) by making

$$\frac{u^2 + v^2}{c^2} - 1 = 0.$$

Its equation is

$$x^2 - y^2 = \frac{l^2 + a^2}{2}, \quad . \quad . \quad . \quad . \quad (42.)$$

being that of a rectangular hyperbola, with its centre at O and its vertex between A and L, at a distance from O equal to half the hypothenuse of a right-angled triangle whose other sides are equal to the base and the eccentricity respectively.

Let $q =$ constant be the equation of one out of an indefinite number of orthogonal trajectories to a set of oögenous neoids. The function q, as is well known, must satisfy the equation

$$\frac{d\,q}{d\,x} \cdot \frac{d\,b}{d\,x} + \frac{d\,q}{d\,y} \cdot \frac{d\,b}{d\,y} = 0.$$

Referring to equation (19) of Art. 7 for the value of b, it is easily seen that this condition is fulfilled by the following function,

$$q = x + \frac{f}{2} \text{hyp. log.} \frac{(a + x)^2 + y^2}{(a - x)^2 + y^2} \quad . \quad . \quad (43.)$$

which has also the following properties,

$$\frac{d\,q}{d\,x} = \frac{d\,b}{d\,y} = \frac{u}{c};\ \frac{d\,q}{d\,y} = -\frac{d\,b}{d\,x} = \frac{v}{c};\ \frac{d^2\,q}{d\,x^2} + \frac{d^2\,q}{d\,y^2} = 0. \quad (44.)$$

Every orthogonal trajectory has a straight asymptote parallel to the axis of y, and expressed by the equation $x = q$.

The perpendicular distance between two consecutive orthogonal trajectories, like that between two consecutive water-lines, is inversely proportional to the velocity of gliding; hence, if a complete set of orthogonal trajectories were drawn on Fig. 1, they would divide it into a network of small rectangles, the dimensions and area of any one of which would be expressed as follows:—

$$\frac{c\,d\,b}{\sqrt{u^2 + v^2}} \times \frac{c\,d\,q}{\sqrt{u^2 + v^2}} = \frac{c^2\,d\,b\,d\,q}{u^2 + v^2}. \quad . \quad . \quad (45.)$$

For a series of cyclogenous neoïds, the equation of the orthogonal trajectories takes the following form,

$$q = x\left(1 + \frac{l^2}{x^2 + y^2}\right). \quad . \quad . \quad . \quad (45\text{A}.)$$

20. *Disturbances of Pressure and Level.*—Let h denote the *head* at a given particle of liquid, being the sum of its elevation above a fixed level, and of its pressure expressed in units of height of the liquid itself. In a mass of liquid which is at rest, the head has a uniform value for every particle of the mass; let that value be denoted by h_0. Then, when the mass of liquid is in the state of motion produced by the passage of a solid through it, the head at each particle, according to well-known principles, undergoes the change expressed by the following equation,

$$h - h_0 = \frac{c^2 - u^2 - v^2}{2\,g}, \quad . \quad . \quad . \quad (46.)$$

being the height due to the difference between the squares of the speed of the solid body and of the speed of gliding; and in an open mass of water with a vessel floating in it, that change will take place by alterations in the level of surfaces of equal pressure. The trajectory of slowest gliding, L N (Plate III., Fig. 1), will mark the summit of a swell thus produced, and so also will the axis of y between O and P; while the trajectory of swiftest gliding O P, and the axis of y beyond P, will mark the bottom of a hollow. These are the principal vertical disturbances which, throughout this investigation, have been assumed to be so small, compared with the dimensions of the body, as not to produce any appreciable error in the consequences of the supposition of motion in plane layers.

21. *Integral on which the Friction depends.*—Suppose a portion of an oögenous neoid to be taken for the water-line of part of the side of a vessel, which part is of the depth δz, and that the resistance arising from friction between the water and the vessel is to be expressed—the law of that friction being, that it varies as the square of the velocity of gliding, and as the extent of rubbing surface.

That resistance is to be found (as already explained in a paper on Waves, published in the *Philosophical Transactions* for 1863) (*See p. 481*) by determining the *work performed* in a second in overcoming friction, and dividing by the speed of the vessel; for thus is taken into account not only the direct resistance caused by the longitudinal component of the friction, but the resistance caused indirectly through the increase of pressure at the bow, and diminution of pressure at the stern, assuming the vertical disturbance to be unimportant.

Then, for a part of the water-line which measures longitudinally $d x$, the extent of surface is

$$\delta z \, . \, \frac{\sqrt{u^2 + v^2}}{u} \, d x;$$

the friction on the unit of surface is

$$\frac{\mathrm{K\,W}\,(u^2 + v^2)}{2\,g},$$

where W is the weight of a unit of volume of water, and K a coefficient of friction; and that friction has to be overcome through the distance $\sqrt{u^2 + v^2}$, while the vessel advances through the distance c, giving as a factor

$$\frac{\sqrt{u^2 + v^2}}{c}.$$

Those three factors being multiplied together, and the result put under the sign of integration, give the following expression for the resistance,

$$R = \frac{K W c^2}{2 g} \delta z . \int \left(\frac{u^2 + v^2}{c^2} \right)^2 . \frac{c}{u} d x . . \quad . \quad (46 \text{ A.})$$

Another form of expression for the same integral is obtained by putting $\frac{c}{x} d y$ or $f \frac{c}{v} d \theta$ instead of $\frac{c}{u} d x$; and a third form by putting for the elementary area of the rubbing surface the following value,

$$d z . \frac{c}{\sqrt{u^2 + v^2}} d q ; \quad .$$

where $d q$ is the distance between the asymptotes of a pair of orthogonal trajectories, as explained in Art. 19. This gives for the resistance

$$R = \frac{K W c^2}{2 g} \delta z . \int \frac{u^2 + v^2}{c^2} d q . \quad . \quad . \quad (47.)$$

In preparing these formulæ for integration, it is necessary to express the function to be integrated in terms of constants and of the independent variable only, x, y, θ, or q, as the case may be; for example, if y or θ is the independent variable, the expression of the function to be integrated is to be taken from the equations (30) of Art. 13.

Owing to the great complexity of that function, its exact integration presents difficulties which have not yet been overcome, although a probable approximate formula for the resistance has been arrived at by methods partly theoretical and partly empirical, as to which some further remarks will be made in the third section of this paper.*

There is one particular case only in which the exact integration of equation (46 A) is easy, that of a complete circular water-line of the radius l; and the result is as follows:—

$$R = \frac{K W c^2}{2 g} \delta z \times 21\tfrac{1}{3} l . \quad . \quad . \quad (48.)$$

22. *Statement of the General Problem of the Water-Line of Least Friction.*— It is evident that, by introducing under the sign of integration in equation (18) of Art. 7 an arbitrary function of x', the integral may be made capable of representing an arbitrary function of x and y, and will still satisfy the condition of perfect liquidity; and thus the equation

$$b = y + \int_{-a}^{+a} \frac{y \phi (x') d x'}{(x - x')^2 + y^2} = 0, \quad . \quad (48 \text{ A.})$$

may be made to represent an arbitrary form of primitive water-line.

* See *The Civil Engineer and Architect's Journal* for October, 1861, *The Philosophical Transactions* for 1863, *The Transactions of the Institution of Naval Architects* for 1864, and a *Treatise on Shipbuilding*, published in 1864.

To find therefore, by the calculus of variations, the water-line enclosing a given area which shall have the least friction, will require the solution of the following problem:—To determine the function $\phi(x')$ so that, with a fixed value of the integral $\int x\,dy$, the integral in equation (46 A) shall be a minimum.

22 A. *Another Class of Plane Water-Line Equations.*—A mode of expressing the conditions of the flow of water in plane layers past a solid differing in form from that made use of in the preceding parts of this paper, consists in taking for independent variables, not the co-ordinates of the water-lines themselves, x and y, but the co-ordinates of their asymptotes (b), and of the asymptotes of their orthogonal trajectories (q). These new variables are connected with x and y, and with the velocity of gliding, by the following equations:—

$$\frac{u^2+v^2}{c^2} = \frac{dq}{dx}\cdot\frac{db}{dy} - \frac{dq}{dy}\cdot\frac{db}{dx} = \frac{1}{\frac{dx}{dq}\cdot\frac{dy}{db} - \frac{dy}{dq}\cdot\frac{dx}{db}}. \qquad (49.)$$

It can be shown that, in order to satisfy the condition of liquidity, we must have

$$x = \frac{d\psi}{db}, \quad y = \frac{d\psi}{dq}; \quad . \quad . \quad . \quad (50.)$$

where ψ denotes a function of b and q, such that

$$\frac{d^2\psi}{db^2} + \frac{d^2\psi}{dq^2} = 0; \quad . \quad . \quad . \quad (51.)$$

and, consequently, that

$$\frac{c^2}{u^2+v^2} = \left(\frac{d^2\psi}{db\,dq}\right)^2 + \left(\frac{d^2\psi}{db^2}\right)^2. \quad . \quad . \quad (52.)$$

The curves to which this method of investigation leads are inferior to oögenous neoïds as water-lines for ships, because they have comparatively sharp curvature amidships, which causes them to have small capacity for their length and breadth, and would give rise to comparatively sudden changes in the speed of gliding. They will therefore not be further discussed in the present paper, except to state that the simplest of them is the well-known cissoid.

SECTION III.—REMARKS ON THE PRACTICAL USE OF OÖGENOUS WATER-LINES.

23. *Previous Systems of Water-lines.*—Owing principally to the great antiquity of the art of shipbuilding, and the immense number of practical

experiments of which it has been the subject, that part of it which relates to the forms of water-lines has in many cases attained a high degree of excellence through purely empirical means. Excellence attained in that manner is of an uncertain and unstable kind; for as it does not spring from a knowledge of general principles, it can be perpetuated by mere imitation only.

The existing forms of water-lines, whose merits are known through their practical success, constitute one of the best tests of a mathematical theory of the subject; for if that theory is a sound one, it will reproduce known good forms of water-line; and if it is a comprehensive one, it will reproduce their numerous varieties, which differ very much from each other.

The geometrical system of Chapman for constructing water-lines is wholly empirical; it consists in the use of parabolas of various orders, chosen so as to approximate to figures that have been found to answer in practice, and it has no connection with any mechanical theory of the motion of the particles of water.

The first theory of ships' water-lines which was at once practically useful, and based on mechanical principles, was that of Mr. Scott Russell, explained in the first and second volumes of the *Transactions of the Institution of Naval Architects.* It consists of two parts; the first has reference to the dimensions of water-lines intended for a given maximum speed, and prescribes a certain relation between the length of those lines and the length of a natural wave which travels with that speed; the second part relates to the form of those lines, and prescribes for imitation the figures of certain natural waves, as being lines along which water is more easily displaced than along other lines. The figures thus obtained are known to be successful in practice; but it is also well known that there are other figures which answer well in practice, differing considerably from those wave-lines; and it is desirable that the mathematical theory of the subject should embrace those figures also. It may further be observed, that the figure of the solitary wave, as investigated experimentally by Mr. Scott Russell (*Reports of the British Association*, 1845), and mathematically by Mr. Earnshaw (*Camb. Trans.*, 1845), is that of a wave propagated in a canal of small breadth and depth as compared with the dimensions of the wave, and in which particles of water originally in a plane at right angles to the direction of motion, continue to be very nearly in a plane at right angles to the direction of motion, so as to have sensibly the same longitudinal velocity. This state of things is so different from the circumstances of the motions of the particles in the open sea, that it appears desirable to investigate the subject with special reference to a mass of water of unlimited breadth and depth, as has been done in the previous sections of this paper.

24. *Variety of Forms of Oögenous Neoids, and their Likeness to good known Forms of Water-line.*—The water-lines generated from ovals which have been described in the second section of this paper, are remarkable for the great varieties of form and proportions which they present, and for the resemblance of their figures to those of the water-lines of the different varieties of existing vessels. There is an endless series of ovals, having all proportions of length to breadth, from equality to infinity; and each of those ovals generates an endless series of water-lines, with all degrees of fulness or fineness, from the absolute bluffness of the oval itself to the sharpness of the knife-edge. Further variations may be made by taking a greater or a less length of the curve chosen.

The ovals are figures suitable for vessels of low speed, it being only necessary, in order to make them good water-lines, that the vertical disturbance (as explained in Art. 20) should be small compared with the vessel's draught of water. At higher speeds the sharper water-lines, more distant from the oval, become necessary. The water-lines generated by a circle, or "cyclogenous neoids," are the "leanest" for a given proportion of length to breadth; and as the eccentricity increases, the lines become "fuller." The lines generated from a very much elongated oval approximate to a straight middle body with more or less sharp ends. In short, there is no form of water-line that has been found to answer in practice which cannot be imitated by means of oögenous neoids.

25. *Discontinuity at the Bow and Stern.—Best limits of Water-Lines.*—Amongst the endless variety of forms presented by oögenous water-lines, it may be well to consider whether there are any which there are reasons for preferring to the others. One of the questions which thus arise is the following:—Inasmuch as all the water-line curves of a series, except the primitive oval, are infinitely long and have asymptotes, there must necessarily be an abrupt change of motion at either end of the limited portion of a curve which is used as a water-line in practice, and the question of the effect of such abrupt change or discontinuity of motion is one which at present can be decided by observation and experiment only. Now it appears from observation and experiment, that the effect of the discontinuity of motion at the bow and stern of a vessel, which has an entrance and run of ordinary sharpness and not convex, extends to a very thin layer of water only; and that beyond a short distance from the vessel's side the discontinuity ceases, through some slight modification of the water-lines, of which the mathematical theory is not yet adequate to give an exact account.*

* In confirmation of this, experiments made on the steamers "Admiral" and "Lancefield," by Mr. J. R. Napier and the author, may be specially referred to. The water-lines of the "Admiral" are complete trochoids, and tangents to the longitudinal axis at the bow and stern. The engine-power required to drive her at her intended speed

Still, although the effect of the discontinuity in increasing resistance may not yet have been reduced to a mathematical expression, and although it may be so small that our present methods of experimenting have not yet detected it, it must have some value; and it is desirable so to select the limits of the water-line as to make that value as small as possible. In order that the abrupt change of motion may take place in as small a mass of water as possible, it would seem that the limits of the water-line employed in practice should be at or near the point of *slowest gliding;* that is, where the water-line curve is cut by the trajectory of slowest gliding LN, in Plate III., Fig. 1, and Plate IV., Fig. 3, as explained in Arts. 14, 15, and 16; and that conclusion is borne out by the figures of many vessels remarkable for economy of power.

26. *Preferable Figures of Water-Lines.*—In forming a probable opinion as to which, out of all the water-lines generated by a given oval, is to be preferred to the others, regard is to be had to the fact, that every point of maximum disturbance of the level of the water, whether upwards or downwards, that is to say, every point of maximum or minimum speed of gliding (see Art. 20), forms the origin of a wave, which spreads out obliquely from the vessel (as may easily be observed in smooth water), and so transfers mechanical energy to distant particles of water, which energy is lost. Hence such points should be as few as possible; and the changes of motion at them should be as gradual as possible; and these conditions are fulfilled by the curves described in Art. 17, by the name of "lissoneoïds," being those which traverse the point P in the figures, and which may have any proportion of length to breadth, from $\sqrt{3}$ to infinity.

27. *Approximate Rules for Construction and Calculation.*—The description of those curves, already given in Art. 17, has been confined to those properties which are exactly true. The following rules are convenient approximations for practical purposes, *when the proportion of length to breadth is not less than* 4 : 1 (see Plate IV., Figs. 3 and 4).

I. A tangent to the curve at Q, the point of slowest gliding, passes very nearly through the point P of greatest breadth.

II. The area P Q R enclosed within the water-line is very nearly equal to the rectangle of the breadth P R and eccentricity a. (When the length is not less than six times the breadth, this rule is almost perfectly exact.)

was computed from the frictional resistance, according to principles explained in publications already referred to in the note to Art. 21; and the result of the calculation was closely verified by experiment. The water-lines of the "Lancefield" are only partly trochoidal, being straight from the point of contrary flexure to the bow, so that, instead of being tangents there to the longitudinal axis, they form with it angles of about 13½°. Yet the same formula which gave the resistance of the "Admiral" has been found to give also the resistance of the "Lancefield" without any addition on account of the discontinuity of motion at the bow.

III. For the trajectory of slowest gliding, L N, there may be substituted, without practical error, a straight line cutting the axis O X in L at an angle of 45°; and when this has been done, the eccentricity O A or a is almost exactly equal to the length

$$\times \cdot 634 \left(= \frac{3}{3 + \sqrt{3}} \right);$$

and this of course is also the ratio of the area to the circumscribed rectangle. The base O L or l also is very nearly equal to (the sum of the length and breadth) × ·634.

IV. Hence the following approximate construction: Given, the common length Q R of a set of water-lines of smoothest gliding, which are to have a common termination at Q, and their breadths R P_1, R P_2, R P_3, &c.: required, to find their areas, bases, and foci.

Through Q and R draw the straight lines Q U and R U, making the angles R Q U = 45°, Q R U = 30°. Through their intersection U draw U V perpendicular to R Q. All the required foci will be in U V; and R V will be the length of the rectangles equivalent to each of the water-line areas; so that

$$\text{area } P_1 Q R_1 = RV \times R P_1,$$
$$\text{area } P_2 Q R_2 = RV \times R P_2,$$
$$\&c. \qquad \&c.$$

Through P_1, P_2, P_3, &c., draw lines parallel to R U, cutting Q U in L_1, L_2, L_3, &c.: these points will be the ends of the bases required, through which draw the bases $L_1 O_1$, $L_2 O_2$, $L_3 O_3$, &c., parallel to Q R, and cutting V U in A_1, A_2, A_3, &c.: these will be the required foci.

The bases and foci and the points P_1, P_2, P_3, &c., being given, the water-lines are to be constructed by the rules given in Art. 11.

28. *Lissoneoids compared with Trochoids.*—In Fig. 5, Plate IV., the full line P Q is a lissoneoïd, and the dotted line Pq a trochoid of the same breadth and area. The curves lie very near together throughout their whole course—the only difference being, that the trochoid is slightly less full and more hollow than the lissoneoïd, but at the same time the trochoid is the longer, and has a greater frictional surface. Had the entrance of the trochoid consisted of a straight tangent from its point of contrary flexure (as in the bow of the "Lancefield," mentioned in the note to Art. 25), the two curves would have lain still closer together. The same likeness to a trochoid is found in all lissoneoïds whose length is more than about 3½ times the breadth.

29. *Combinations of Bow and Stern.*—Although there is reason to believe that water-lines of equal length and similar form at the bow and stern,

such as are produced by using one neoïd curve throughout, are the best on the whole, still the naval architect, should he think fit, can combine two different oögenous neoïds for the bow and stern; or, according to a frequent practice, he may adapt the figure of the stern to motion of the particles in vertical layers instead of horizontal layers; provided he takes care in every case that the midship velocity of gliding (u_0, as given by equation (28A) of Art. 13) is the same for each bow water-line and stern water-line at their point of junction.

30. *Provisional Formula for Resistance.*—Until the difficulty of integration, mentioned in Art. 30, shall have been overcome, or until more exact experimental data than we have at present shall have been obtained, the following provisional formula, analogous to that which has been found to agree with the results of experiment on trochoidal and nearly trochoidal lines, as well as some others, may be considered as a probable approximation for lissoneoïds,

$$R = \frac{K\,W\,c^2}{2\,g} \cdot \left(1 + 4\,\frac{(u_0 - c)^2}{c^2}\right) L\,G; \qquad (53.)$$

where G is the mean girth of the vessel under water; L her total length; u_0 the midship velocity of gliding, found, for a lissoneoïd, by equation (37) of Art. 17; c the speed of the ship; W the heaviness of water; and K a coefficient of friction (= about ·0036 for a clean surface of paint).

APPENDIX.

Note to Article 11.—The general process of constructing a series of curves whose equation is $\phi(x, y) + \psi(x, y) =$ constant, by drawing lines diagonally through a network consisting of two sets of curves whose equations are respectively $\phi(x, y) =$ constant and $\psi(x, y) =$ constant, is due to Professor Clerk Maxwell.

Summary of the Contents.

Section I.—Introduction, and Summary of known Principles.

Section II.—Properties of Water-Line Curves generated from Ovals, or Oögenous Neoïds.

Section III.—Remarks on the Practical Use of Oögenous Water-Lines.

XXXI.—ELEMENTARY DEMONSTRATIONS OF PRINCIPLES RELATING TO STREAM-LINES.*

1. *Object of this Communication.*—The object of this communication is to explain some very elementary demonstrations of certain propositions in hydrodynamics which bear upon important practical questions. The propositions themselves are not new; but all the previously published demonstrations of them with which I am acquainted involve the use of mathematical methods of some difficulty, and especially, of the solution of differential equations of the second order. The demonstrations now given (which have hitherto been made public in the form of lectures only), are intended to enable persons who have not mastered the higher mathematics to understand the propositions in question, and to satisfy themselves of their truth.

2. *Stream-Lines explained.*—A stream-line is the line, whether straight or curved, that is traced by a particle in a current of fluid. In what is termed a "steady current," each individual stream-line preserves its figure and position unchanged, and marks the path or track of a *filament*, or continuous series of particles that follow each other. The direction of the motions in different parts of a steady current may be represented to the eye by drawing the group of stream-lines traced by different particles in that current, and indicating by one or more arrows in which of two contrary directions the motion takes place. The wavy lines in Fig. 3 (*See p. 527*) represent an example of this.

3. *Relation between Velocity and Transverse Area.*—It is obvious that if the area of a transverse section of a current be multiplied by the mean velocity of the particles of fluid in the act of traversing that transverse section, the product will be the *flow;* that is, the volume of fluid which passes through that transverse section in an unit of time; and conversely, that if the flow be divided by the transverse area, the quotient will be the mean velocity of the particles that traverse that section. By a transverse section is to be understood a surface that cuts all the stream-lines at right angles. Moreover, in a *liquid of invariable density* (to which class of fluids alone this communication is restricted), the flow through each

* From *The Engineer* of Oct. 16, 1868.

transverse section of a steady current is of equal volume; therefore, in a steady current, *the mean velocity of the particles at a given transverse section is inversely proportional to the area of that section.*

4. *Elementary Streams.*—A current may be conceived mentally to be divided by insensibly thin partitions, following the course of the stream-lines, into a number of *elementary streams;* and the positions of those partitions may be conceived to be so adjusted that the volumes of flow in all the elementary streams shall be equal. The use of this conception is to represent to the mind the velocity, as well as the direction of motion, of the particles in different parts of the current; for it is obvious that in a set of elementary streams of equal flow, *the velocity of a particle at any point is inversely proportional to the area of the transverse section, through that point, of the elementary stream to which the particle belongs.* This is the principle which, when expressed in the symbols of the differential calculus, is called "the equation of continuity" of a liquid.*

5. *Component Velocities.*—The component velocity of a particle in a direction oblique to its actual direction of motion, may be found by the help of elementary streams; for it is only necessary to divide the elementary volume of flow by the area of an *oblique section* of the elementary stream, made by a plane perpendicular to the direction of the required component. For example, in Fig 1, let A A, A′ A′, be the boundaries of an elementary stream, and let C C′ be a plane cutting it obliquely; then, if the volume of flow be divided by the area of the oblique section made by the plane C C′, the quotient will be the component velocity of a particle in a direction perpendicular to that plane, being less than the total velocity along the stream in the same ratio in which the area of a transverse section of the stream is less than the area of the oblique section.

6. *Representation of the Elementary Streams in a Layer of a Current.*—In considering the motion of the particles of one layer of a current, we may conceive the paper of a diagram, such as Fig. 3, to represent either one of the actual surfaces of that layer, if it is plane, or if it is not plane, the same surface *developed*, that is, spread out flat. The layer may be conceived to be divided into elementary streams of equal flow by partitions perpendicular to the surfaces of the layer; and in the diagram those partitions will be represented by stream-lines, such as the wavy lines in Fig. 3. Such a diagram exhibits to the eye the velocity, as well as the direction, of the motions of the particles in every part of the layer; for, if the layer is uniformly thick, the velocity of any particle is inversely proportional simply to the perpendicular distance between the two adjacent

* Let δQ denote the volume of flow in each of the elementary streams of which a steady current of liquid consists; and at a given point let δS be the transverse area of an elementary stream, and v the velocity; then $v = \frac{\delta Q}{\delta S}$.

stream-lines; and if the thickness of the layer varies at different points, that velocity is inversely proportional to the same perpendicular distance multiplied by the thickness of the layer.

7. *Composition of Elementary Streams.*—If a layer of liquid is acted upon at the same time by two sets of forces, which, if acting separately, would produce currents consisting of two different sets of elementary streams, the combined action of those two sets of forces will produce currents consisting of a third set of elementary streams, which may be regarded as the *resultant* of the two former sets. The stream-lines marking the boundaries of the first two sets of elementary streams may be called the *component stream-lines,* and those marking the boundaries of the third set, the *resultant stream-lines.* Then the principle which connects the resultant current with the component currents is as follows:—

The resultant stream-lines pass diagonally through all the angles of the network formed by the component stream-lines.

For example, in Fig. 1, A A, A′ A′, are a pair of lines belonging to one set of component stream-lines, and B B, B′ B′, a pair belonging to another

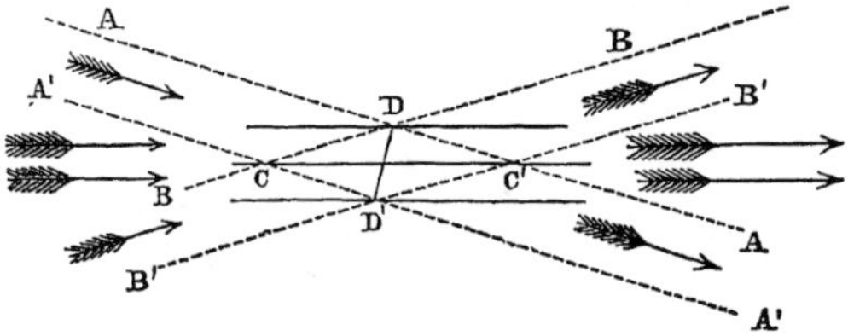

Fig. 1.

set. The line C C′, drawn through two of the intersections, is one of the set of resultant stream-lines; and the lines parallel to C C′, drawn through D and D′, are two more. Also, in Fig. 2, the straight dotted lines diverging from A, and the straight dotted lines converging towards B, are two sets of component stream-lines; and the curved lines which traverse the intersections of the straight lines are the resultant stream-lines. For a third example, in Fig. 3, the straight lines parallel to X O, and the arcs diverging from A, are two sets of component stream-lines; and the wavy lines drawn through the intersections of the first two sets are the resultant stream-lines.

To demonstrate this principle it is to be considered—First, as regards the direction of the resultant stream-lines: that in Fig. 1, C C′ represents a plane, which is an oblique section at once of the elementary stream A A, A′ A′, and of the elementary stream B B, B′ B′. The forces which produce the elementary stream A A, A′ A′, tend to send a certain volume of liquid per second through that oblique section, from the side next D to the side next D′. The forces which produce the elementary stream B B, B′ B′, tend to send an equal volume of liquid through the same oblique section in the contrary direction. Therefore, the effect of the combination of the forces is that there is *no flow* through the oblique section C C′; therefore C C′ is part of one of the resultant stream-lines. Secondly, as to the number and closeness of those stream-lines, it is to be considered that D D′ also represents a plane, which is an oblique section at once of both the elementary streams. The forces which produce the elementary stream A A, A′ A′, tend to send a certain volume of liquid per second through that section in a certain direction; and the forces which produce the elementary stream B B, B′ B′, tend to send an equal volume per second through in the same direction. Therefore, the effect of the combination of the forces is that a *double volume* per second passes through the section D D′; therefore, the space between D and D′ contains *two resultant elementary streams;* therefore, each of the points D and D′ is traversed by one of the resultant stream-lines. Thus it is proved that *all* the intersections of the component stream-lines are traversed by resultant stream-lines.

8. *Condition of Perfect Fluidity.*—The characteristic property of a perfect fluid—in other words, a fluid absolutely free from viscosity—is that the particles have no tendency to preserve any definite figure, and are incapable of exerting any force against a surface which they touch except normal pressure; that is to say, pressure in a direction at right angles to that surface. One consequence of this is that no particle of a perfect fluid can have rotation impressed upon it; for normal pressure can impress rotation only on a body which tends to preserve a definite figure. No existing fluid is absolutely free from viscosity; and therefore the mechanical consequences of the supposition of perfect fluidity are not realised exactly, but only approximately. Nevertheless, there are cases in which the errors caused by neglecting viscosity are unimportant; and hence the use of investigating the properties of stream-lines in a perfect fluid.

9. *Rectilinear Motions in a Perfect Fluid.*—Another way of stating the absence of rotation in the motion of a perfect fluid is to say that *any two particles which move side by side in straight lines must move with equal velocities;* for if their velocities are different the larger particle formed by uniting them is in a state of rotation, one side moving faster than the other.

There are three modes of rectilinear motion in a perfectly fluid liquid which fulfil this condition, and by combining which an immense number of modes of curvilinear motion may be generated; and all those curvilinear resultant motions fulfil the condition of perfect fluidity, because their components do so. Those three modes of rectilinear motion are the following:—

I. Motion in parallel straight lines, with an uniform velocity. Here the elementary streams are everywhere of equal transverse area.

II. Motion in straight lines converging towards or diverging from an axis, to which they are all perpendicular. Here we may consider the motion of the particles in a layer of uniform thickness perpendicular to the axis. The elementary streams in such a layer are of the form of wedges, separated from each other by planes radiating from the axis, and making equal angles with each other. The area of a transverse section of an elementary stream varies directly as the distance from the axis, and the velocity of a particle varies inversely as that distance.

III. Motion in straight lines converging towards or diverging from a central point. Here the elementary streams are of the forms of cones or of pyramids, having their summits at the central point, and of such shapes and sizes as to divide the surface of a sphere described about that point into equal areas. The area of a transverse section of an elementary stream varies directly as the square of the distance from the central point, and the velocity of a particle consequently varies inversely as the square of that distance.

It is easy to see that in the last two modes of motion the elementary streams cannot actually extend in a pointed form to the axis or to the central point, but must be deflected in its neighbourhood, so as to afford an inlet or an outlet for the liquid, as the case may be.

10. *Rectilinear Stream-Lines in an Uniformly Thick Layer.*—The stream-

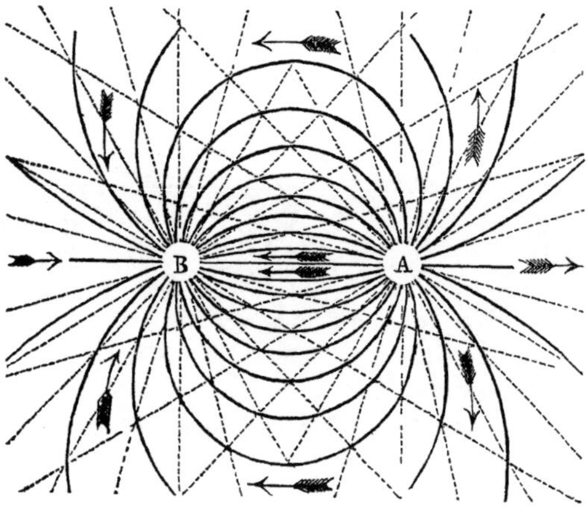

Fig. 2

lines which represent an uniform straight current in an uniformly thick layer of liquid are simply parallel equidistant straight lines, such as those shown in Fig 3. The stream-lines which, in a similar layer, represent a current diverging from or converging towards an axis, are straight lines radiating from a point, and making equal angles with each other, like either of the two sets of dotted straight lines in Fig 2.

11. *Circular Stream-Lines in an Uniformly Thick Layer.*—The simplest example of a set of resultant stream-lines is that obtained, as in Fig. 2, by combining together a pair of equal and similar sets of radiating stream-lines, one set diverging from a point at A, and the other converging towards a point at B. Those two points may be called *foci*. According to well-known geometrical principles, the resultant stream-lines, which traverse the intersections of the network formed by the two sets of radiating stream-lines, are a series of circles, each of which traverses the foci A and B, the only exception being the straight line through A and B. The radii of those circles are proportional to the secants of a series of angles, increasing by equal intervals from 0° to 90°. These resultant stream-lines represent the motion of a layer of liquid of uniform thickness, under the action of forces which urge the particles to move from an axis at A, and towards another axis at B.

In a paper published in the *Philosophical Transactions* for 1863 (*See p. 495*), the properties of those circular stream-lines traversing two foci were arrived at by the integration of a differential equation of the second order. They have now been demonstrated by a very elementary method; and to do so was one of the chief objects of the communication.

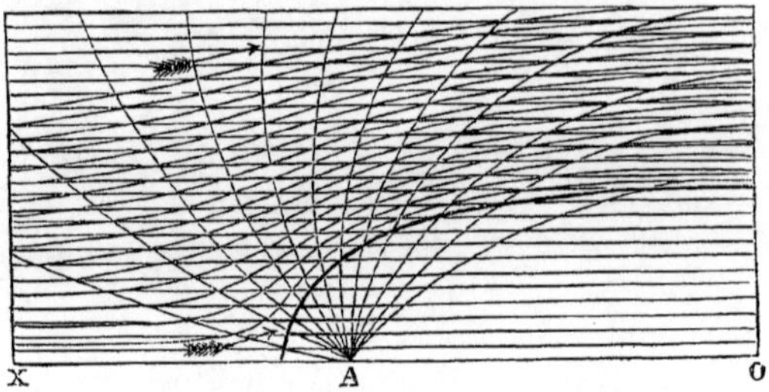

Fig. 3.

12. *Various Resultant Stream-Lines in a Layer of Uniform Thickness.*—By compounding the circular stream-lines of Art. 11 with the equidistant straight stream-lines of Art. 10, and drawing curves through the angles of the network, an endless variety of stream-lines is obtained of figures closely resembling the lines of ships of various degrees of fineness, and of various proportions of length to breadth. These, under the name of *Neoïds* (or ship-like curves), have been fully explained and illustrated in previous papers and publications; and especially in the paper already referred to as having been published in the *Philosophical Transactions* for 1863, from which Fig. 3 is copied; and it is unnecessary to enter into

details respecting them here. The figure shows only one quadrant of the complete set of stream-lines, the other three quadrants being symmetrical to the first. The curves diverging from the focus A are circular stream-lines, and they converge to another focus at an equal distance from O, in the opposite direction.

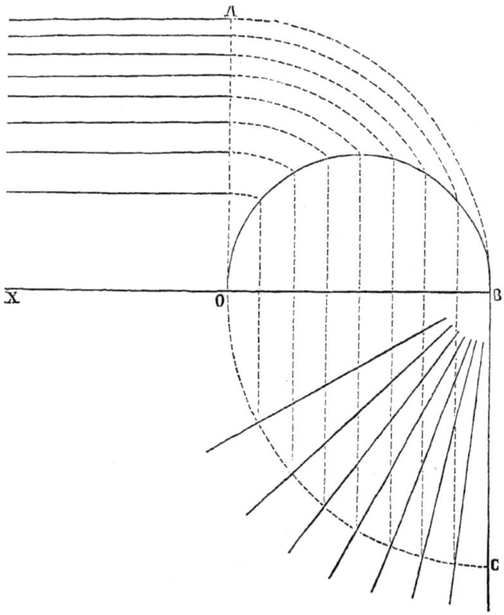

Fig. 4.

13. *Parallel Straight Stream-Lines in a Wedge-shaped Layer.*—When a current flows past a solid of revolution, the figure and arrangement of the stream-lines are to be determined by considering the motion in a wedge-shaped layer of indefinite length and breadth, having its edge at the axis of the solid. The thickness of such a layer varies as the distance from the edge. In Fig. 4, let O X represent the axis of the solid and edge of the wedge-shaped layer, and let the paper represent one of the plane surfaces of the layer. The stream-lines representing an uniform straight current must be so arranged as to divide the layer into elementary streams of uniform transverse area; and, in order that they may do so, the *squares* of their distances from the axis O X must increase by uniform differences. Let O A be the total breadth which it is desired to subdivide into elementary streams. Make O B = O A, and divide it into as many equal parts as there are to be elementary streams. On O B as a

diameter draw a semicircle, and from the points of division of O B draw ordinates perpendicular to it, and cutting the semicircle. Then lay off from O, along O A, a series of distances equal respectively to the chords measured from O to the points of division of the semicircle; the required stream-lines will be straight lines drawn parallel to O X through the points of division of O A.

14. *Radiating Straight Stream-Lines in a Wedge-shaped Layer.*—To divide such a wedge-shaped layer, as has been described in the preceding article, into equal elementary streams radiating from a point B in the axis X O B, lay off along the axis as many equal divisions as there are to be elementary streams in one quadrant of the space round B. Let B O be the distance containing all those divisions. About B, with the radius B C, draw the quarter circle O C; and from the points of division B O, and perpendicular to it, draw ordinates cutting the quarter circle. Then draw lines radiating from B to the points of division of the quarter circle. These will be the required stream-lines for one quadrant of the space round B. Those of the other quadrants are symmetrical to them. The reason for this construction is the well-known geometrical proposition, that if O C is the trace of a spherical surface, and if the dotted ordinates are the traces of a set of parallel planes perpendicular to the radius O B, and dividing it into equal parts, those planes divide the spherical surface into zones of equal area.

15. *Compound Stream-Lines in a Wedge-shaped Layer.*—By compounding two sets of straight stream-lines, like those shown in the lower part of Fig. 4, radiating from a pair of foci in the same axis (that is, in the edge of the wedge-shaped layer), and drawing curves diagonally through the network, there are obtained a set of oval stream-lines, representing the motion of a current which diverges in all directions from one of the foci, and converges towards the other. These ovals all pass through the foci, and are arranged like the circular stream-lines of Fig. 2. It may be mentioned that they are of the same figure with the lines of force of a two-poled magnet.

Then, by combining these oval stream-lines with the parallel straight stream-lines of the upper part of Fig. 4, there are obtained a great variety of curved lines, representing the stream-lines of a current flowing past a solid of revolution. Their figures resemble in a general way those of the stream-lines of an uniformly thick layer, exemplified in Fig. 3.

XXXII.—ON THE THERMODYNAMIC THEORY OF WAVES OF FINITE LONGITUDINAL DISTURBANCE.*

1. THE object of the present investigation is to determine the relations which must exist between the laws of the elasticity of any substance, whether gaseous, liquid, or solid, and those of the wave-like propagation of a finite longitudinal disturbance in that substance; in other words, of a disturbance consisting in displacements of particles along the direction of propagation, the velocity of displacement of the particles being so great that it is not to be neglected in comparison with the velocity of propagation. In particular, the investigation aims at ascertaining what conditions as to the transfer of heat from particle to particle must be fulfilled in order that a finite longitudinal disturbance may be propagated along a prismatic or cylindrical mass without loss of energy or change of type: the word *type* being used to denote the relation between the extent of disturbance at a given instant of a set of particles, and their respective undisturbed positions. The disturbed matter in these inquiries may be conceived to be contained in a straight tube of uniform cross-section and indefinite length.

2. *Mass-Velocity.*—A convenient quantity in the present investigation is what may be termed the *mass-velocity* or *somatic velocity*—that is to say, the mass of matter through which a disturbance is propagated in a unit of time while advancing along a prism of the sectional area unity. That mass-velocity will be denoted by m.

Let S denote the *bulkiness*, or the space filled by unity of mass, of the substance in the undisturbed state, and a the linear velocity of advance of the wave; then we have evidently

$$a = m\,\mathrm{S}. \qquad . \quad . \quad . \quad . \quad (1.)$$

3. *Cinematical Condition of Permanency of Type.*—If it be possible for a wave of disturbance to be propagated in an uniform tube without change of type, that possibility is expressed by the uniformity of the mass-velocity m for all parts of the wave.

Conceive a space in the supposed tube, of an invariable length Δx, to

* Read before the Royal Society of London on Dec. 16, 1869, and published in the *Philosophical Transactions* for 1870.

be contained between a pair of transverse planes, and let those planes advance with the linear velocity a in the direction of propagation. Let the values of the bulkiness of the matter at the foremost and aftermost planes respectively be denoted by s_1 and s_2, and those of the velocity of longitudinal disturbance by u_1 and u_2. Then the linear velocities with which the particles traverse the two planes respectively are as follows: for the foremost plane $u_1 - a$, for the aftermost plane $u_2 - a$. The uniformity of type of the disturbance involves, as a condition, that equal masses of matter traverse the two planes respectively in a given time, being each, in unity of time, expressed by the *mass-velocity;* hence we have, as the *cinematical condition* of uniformity of type, the following equation:

$$\frac{a - u_1}{s_1} = \frac{a - u_2}{s_2} = \frac{a}{\mathrm{S}} = m. \quad . \quad . \quad . \quad (2.)$$

Another way of expressing the same condition is as follows:

$$\Delta u = m \Delta s. \quad . \quad . \quad . \quad . \quad . \quad (3.)$$

4. *Dynamical Condition of Permanency of Type.*—Let p_1 and p_2 be the intensities of the longitudinal pressure at the foremost and aftermost advancing planes respectively. Then in each unit of time the difference of pressure, $p_2 - p_1$, impresses on the mass m the acceleration $u_2 - u_1$, and consequently, by the second law of motion, we have the following value for the difference of pressure:

$$p_2 - p_1 = m(u_2 - u_1). \quad . \quad . \quad . \quad (4.)$$

Then, substituting for the acceleration $u_2 - u_1$ its value in terms of the change of bulkiness as given by equation (3), we obtain, for the *dynamical condition* of permanency of type, the following equation,

$$p_2 - p_1 = m^2 (s_1 - s_2), \quad . \quad . \quad . \quad (5.)$$

which may also be put in the form of an expression giving the value of the square of the mass-velocity, viz.—

$$m^2 = -\frac{\Delta p}{\Delta s} = -\frac{dp}{ds}. \quad . \quad . \quad . \quad (6.)$$

The square of the linear velocity of advance is given by the following equation:

$$a^2 = m^2 \mathrm{S}^2 = -\mathrm{S}^2 \frac{dp}{ds}. \quad . \quad . \quad . \quad (7.)$$

The integral form of the preceding equations may be expressed as follows. Let S, as before, be the bulkiness in the undisturbed state, and P the longitudinal pressure; then in a wave of disturbance of permanent type we must have the following condition fulfilled:

$$p + m^2 s = \mathrm{P} + m^2 \mathrm{S}. \qquad (8.)$$

5. *Waves of Sudden Disturbance.*—The condition expressed by the equations of the preceding section holds for any type of disturbance, continuous or discontinuous, gradual or abrupt. To represent, in particular, the case of a single abrupt disturbance, we must conceive the foremost and aftermost advancing planes already mentioned to coalesce into one. Then P is the longitudinal pressure, and S the bulkiness, in front of the advancing plane; p is the longitudinal pressure, and s the bulkiness, behind the advancing plane; and the advancing plane is a wave-front of *sudden compression* or of *sudden rarefaction,** according as p is greater or less than P. The squares of the *mass-velocity* and of the linear velocity of advance are respectively as follows:

$$m^2 = \frac{p - \mathrm{P}}{\mathrm{S} - s}; \qquad (9.)$$

$$a^2 = m^2 \mathrm{S}^2 = \frac{p - \mathrm{P}}{\mathrm{S} - s} . \mathrm{S}^2. \qquad (10.)$$

The velocity of the disturbed particles is as follows:

$$u = m(\mathrm{S} - s) = \frac{p - \mathrm{P}}{m} = \sqrt{(p - \mathrm{P}).(\mathrm{S} - s)}; \qquad (11.)$$

and it is forward or backward according as the wave is one of compression or of rarefaction.

The energy expended in unity of time, in producing any such wave, is expressed by $p\,u$; for the wave may be conceived to be produced in a tube closed at one end by a movable piston of inappreciable mass, to which there is applied a pressure p different from the undisturbed pressure P, and which consequently moves with the velocity u. The way in which that energy is disposed of is as follows: actual energy of the disturbance, $\frac{m\,u^2}{2}$; work done in altering bulkiness, $\frac{m\,(p + \mathrm{P})\,(\mathrm{S} - s)}{2}$; and the equation of the conservation of energy is

* *Note, added 1st August, 1870.*—Sir William Thomson has pointed out to the author, that a wave of sudden rarefaction, though mathematically possible, is an unstable condition of motion; any deviation from absolute suddenness tending to make the disturbance become more and more gradual. Hence the only wave of sudden disturbance whose permanency of type is physically possible, is one of sudden compression; and this is to be taken into account in connection with all that is stated in the paper respecting such waves.

$$p\,u = \frac{m}{2}\,\{u^2 + (p + \mathrm{P})\,(\mathrm{S} - s)\}. \quad . \quad . \quad (11\ \text{A.})$$

6. *Thermodynamic Conditions.*—While the equations of the two preceding sections impose the constancy of the rate of variation of pressure with bulkiness during the disturbance $\left(\frac{d\,p}{d\,s} = -\,m^2\right)$ as an indispensable condition of permanency of type of the wave, they leave the limits of pressure and of bulkiness, being four quantities, connected by one equation only $\left(\frac{p_2 - p_1}{s_1 - s_2} = -\,\frac{d\,p}{d\,s} = m^2\right)$. Two only of those quantities can be arbitrary; therefore, one more equation is required, and that is to be determined by the aid of the laws of thermodynamics.

It is to be observed, in the first place, that no substance yet known fulfils the condition expressed by the equation $\frac{d\,p}{d\,s} = -\,m^2 = \text{constant}$, between finite limits of disturbance, at a constant temperature, nor in a state of non-conduction of heat (called the *adiabatic* state). In order, then, that permanency of type may be possible in a wave of longitudinal disturbance, there must be both change of temperature and conduction of heat during the disturbance.

The cylindrical or prismatic tube in which the disturbance is supposed to take place being ideal, is to be considered as non-conducting. Also, the foremost and aftermost transverse advancing planes, or front and back of the wave, which contain between them the particles whose pressure and bulkiness are in the act of varying, are to be considered as non-conducting, because of there being an indefinite length of matter before the foremost and behind the aftermost plane, to resist conduction.

The transfer of heat, therefore, takes place wholly amongst the particles undergoing variation of pressure and bulkiness; and therefore for any given particle, during its passage from the front to the back of the wave, the *integral amount of heat received must be nothing;* and this is the thermodynamic condition which gives the required equation. That equation is expressed as follows:

$$\int_{\phi_1}^{\phi_2} \tau\,d\,\phi = 0; \quad . \quad . \quad . \quad . \quad (12.)$$

in which τ denotes absolute temperature, and ϕ the "thermodynamic function." The value of that function, as explained in various papers and treatises on thermodynamics, is given by the following formula:

$$\phi = \mathrm{J}\,c\ \text{hyp. log.}\ \tau + \chi\,(\tau) + \frac{d\,\mathrm{U}}{d\,\tau}, \quad . \quad . \quad (12\ \text{A.})$$

in which J is the dynamical value of a unit of heat; c, the real specific heat of the substance; $\chi(\tau)$, a function of the temperature alone, which is $= 0$ for all temperatures at which the substance is capable of approximating indefinitely to the perfectly gaseous state, and is introduced into the formula solely to provide for the possible existence of substances which at some temperatures are incapable of approximating to the perfectly gaseous state; and U, the work which the elastic forces in unity of mass are capable of doing at the constant temperature τ. The substitution for the integral in equation (12) of its value in terms of p and s for any particular substance, gives a relation between the limits of pressure p_1 and p_2, and the limits of bulkiness s_1 and s_2, which being combined with equation (5), or with any one of the equivalent equations (6), (8), or (9), completes the expression of the laws of the propagation of waves of finite longitudinal disturbance and permanent type in that particular substance.

7. *Assumption as to Transfer of Heat.*—In applying the principles of the preceding section to the propagation of waves of longitudinal disturbance, it is obviously assumed that the transfer of heat takes place between the various particles which are undergoing disturbance at a given time, in such a manner as to ensure the fulfilment of the dynamical condition of permanency of type. It appears highly probable, that how great soever the resistance of the substance to the conduction of heat may be, that assumption as to the transfer is realised when the disturbance is *sudden*, as described in sec. 5; for then particles in all the successive stages of the change of pressure and bulkiness within the limits of the disturbance are at inappreciable distances from each other; so that the resistance to the transfer of heat between them is inappreciable.

But when the disturbance is not sudden, it is probable that the assumption as to the transfer of heat is fulfilled in an approximate manner only; and if such is the case, it follows that *the only longitudinal disturbance which can be propagated with absolute permanence of type is a sudden disturbance.*

8. *Combination of the Dynamic and Thermodynamic Equations.*—In every fluid, and probably in many solids, the quantity of heat received during an indefinitely small change of pressure $d\,p$ and of bulkiness $d\,s$ is capable of being expressed in either of the following forms:

$$\frac{\tau\, d\,\phi}{\mathrm{J}} = c_s \frac{d\,\tau}{d\,p}\, d\,p + c_p \frac{d\,\tau}{d\,s}\, d\,s;$$

in which c_s and c_p denote the specific heat at constant bulkiness and at constant pressure respectively; and the differential coefficients $\frac{d\,\tau}{d\,p}$ and $\frac{d\,\tau}{d\,s}$ of the absolute temperature are taken, the former on the supposition that the bulkiness is constant, and the latter on the supposition that the pressure is constant. Let it now be supposed that the bulkiness varies with the pressure according to some definite law; and let the actual rate of variation of the bulkiness with the pressure be denoted by $\frac{ds}{dp}$. Then equation (12) may be expressed in the following form:

$$\int_{p_1}^{p_2} dp \cdot \left\{ c_s \frac{d\tau}{dp} + c_p \frac{d\tau}{ds} \cdot \frac{ds}{dp} \right\} = 0.$$

Now, according to the dynamic condition of permanence of type, we have by equation (6),

$$\frac{ds}{dp} = -\frac{1}{m^2};$$

which, being substituted in the preceding integrals, gives the following equations from which to deduce *the square of the mass-velocity:*

$$\int_{p_1}^{p_2} dp \cdot \left\{ m^2 c_s \frac{d\tau}{dp} - c_p \frac{d\tau}{ds} \right\} = 0. \quad . \quad . \quad (13.)$$

It is sometimes convenient to substitute for $c_p \frac{d\tau}{ds}$ the following value, which is a known consequence of the laws of thermodynamics:

$$c_p \frac{d\tau}{ds} = c_s \frac{d\tau}{ds} + \frac{\tau\, dp}{J\, d\tau}, \quad . \quad . \quad . \quad (13\text{ A}.)$$

the differential coefficient $\frac{dp}{d\tau}$ being taken on the supposition that s is constant. The equations (13) and (13 A) are applicable to all fluids, and probably to many solids also, especially those which are isotropic.

The determination of the squared mass-velocity, m^2, enables the bulkiness s for any given pressure p, and the corresponding velocity of disturbance u, to be found by means of the following formulæ, which are substantially identical with equations (8) and (3) respectively:

$$s = S + \frac{P - p}{m^2}; \quad . \quad . \quad . \quad (14.)$$

$$u = m(S - s) = \frac{p - P}{m}. \quad . \quad . \quad . \quad (15.)$$

Equation (15) also serves to calculate the pressure p corresponding to a given velocity of disturbance u. It may here be repeated that the linear velocity of advance is $a = m\,S$ (equation 1).

9. *Application to a Perfect Gas.*—In a perfect gas, the specific heat at constant volume, c_s, and the specific heat at constant pressure, c_p, are both constant; and, consequently, bear to each other a constant ratio, $\frac{c_p}{c_s}$, whose value for air, oxygen, nitrogen, and hydrogen is nearly 1·41, and for steam-gas nearly 1·3. Let this ratio be denoted by γ. Also, the differential coefficients which appear in equations (13) and (13 A) have the following values:—

$$\left.\begin{aligned}
\frac{d\tau}{dp} &= \frac{\tau}{p} = \frac{s}{\mathrm{J}\,(c_p - c_s)} = \frac{s}{\mathrm{J}\,(\gamma - 1)\,c_s}; \\
\frac{d\tau}{ds} &= \frac{\tau}{s} = \frac{p}{\mathrm{J}\,(c_p - c_s)} = \frac{p}{\mathrm{J}\,(\gamma - 1)\,c_s}; \\
\frac{dp}{d\tau} &= \frac{p}{\tau} = \frac{\mathrm{J}\,(c_p - c_s)}{s} = \frac{\mathrm{J}\,(\gamma - 1)\,c_s}{s}.
\end{aligned}\right\} \quad . \quad . \quad (16.)$$

When these substitutions are made in equation (13), and constant common factors cancelled, it is reduced to the following:

$$\int_{p_1}^{p_2} dp \,.\, \{m^2 s - \gamma p\} = 0. \quad . \quad . \quad . \quad (17.)$$

But according to the dynamical condition of permanence of type, as expressed in equation (8), we have $m^2 s = m^2 \mathrm{S} + \mathrm{P} - p$; whence it follows that the value of the integral in equation (17) is

$$\int_{p_1}^{p_2} dp \,.\, \{m^2 \mathrm{S} + \mathrm{P} - (\gamma + 1) p\} = (m^2 \mathrm{S} + \mathrm{P})\,(p_2 - p_1)$$

$$- \frac{\gamma + 1}{2}\,(p_2^2 - p_1^2) = 0\,;$$

which, being divided by $p_2 - p_1$, gives for the square of the mass-velocity of advance the following value:

$$m^2 = \frac{1}{\mathrm{S}}\left\{(\gamma + 1)\,.\,\frac{p_2 + p_1}{2} - \mathrm{P}\right\}. \quad . \quad . \quad (18.)$$

The square of the linear velocity of advance is

$$a^2 = m^2 \mathrm{S}^2 = \mathrm{S}\left\{(\gamma + 1)\,.\,\frac{p_2 + p_1}{2} - \mathrm{P}\right\}. \quad . \quad (19.)$$

The velocity of disturbance u corresponding to a given pressure p, or, conversely, the pressure p corresponding to a given velocity of disturbance, may be found by means of equation (15).

Such are the general equations of the propagation of waves of longitudinal disturbance of permanent type along a cylindrical mass of a perfect gas whose undisturbed pressure and bulkiness are respectively P and S. In the next two sections particular cases will be treated of.

10. *Wave of Oscillation in a Perfect Gas.*—Let the mean between the two extreme pressures be equal to the undisturbed pressure; that is, let

$$\frac{p_2 + p_1}{2} = \mathrm{P}; \quad . \quad . \quad . \quad . \quad (20.)$$

then equations (18) and (19) become simply

$$m^2 = \frac{\gamma \mathrm{P}}{\mathrm{S}}, \quad . \quad . \quad . \quad . \quad (21.)$$

and

$$a^2 = \gamma \mathrm{P}\,\mathrm{S}; \quad . \quad . \quad . \quad . \quad (22.)$$

the last of which is Laplace's well-known law of the propagation of sound. The three equations of this section are applicable to an indefinitely long series of waves, in which equal disturbances of pressure take place alternately in opposite directions.

11. *Wave of Permanent Compression or Dilatation in a Tube of Perfect Gas.*—To adapt equation (18) to the case of a wave of permanent compression or dilatation in a tube of perfect gas, the pressure at the front of the wave is to be made equal to the undisturbed pressure; and the pressure at the back of the wave to the final or permanently altered pressure. Let the final pressure be denoted simply by p; then $p_1 = \mathrm{P}$, and $p_2 = p$; giving for the square of the mass-velocity

$$m^2 = \frac{1}{\mathrm{S}}\left\{(\gamma + 1)\frac{p}{2} + (\gamma - 1)\frac{\mathrm{P}}{2}\right\}, \quad . \quad . \quad (23.)$$

for the square of the linear velocity of advance

$$a^2 = m^2\,\mathrm{S}^2 = \mathrm{S}\left\{(\gamma + 1)\frac{p}{2} + (\gamma - 1)\frac{\mathrm{P}}{2}\right\}, \quad . \quad (24.)$$

and for the final velocity of disturbance

$$u = \frac{p - \mathrm{P}}{m} = (p - \mathrm{P})\sqrt{\left\{\frac{\mathrm{S}}{(\gamma + 1)\frac{p}{2} + (\gamma - 1)\frac{\mathrm{P}}{2}}\right\}}. \quad (25.)$$

Equations (23) and (24) show that a wave of condensation is propagated faster, and a wave of rarefaction slower, than a series of waves of oscillation. They further show that there is no upper limit to the velocity of propagation of a wave of condensation; and, also, that to the velocity of propagation of a wave of rarefaction there is a lower limit, found by making $p = 0$ in equations (23) and (24). The values of that lower limit, for the squares of the mass-velocity and linear velocity respectively, are as follows :—

$$m^2 (p = 0) = \frac{(\gamma - 1) \mathrm{P}}{2 \mathrm{S}}; \quad . \quad . \quad . \quad (26.)$$

$$a^2 (p = 0) = \frac{(\gamma - 1) \mathrm{P S}}{2}; \quad . \quad . \quad . \quad (27.)$$

and the corresponding value of the velocity of disturbance, being its negative limit, is

$$u (p = 0) = - \sqrt{\left\{ \frac{2 \mathrm{P S}}{\gamma - 1} \right\}}. \quad . \quad . \quad (28.)$$

It is to be borne in mind that the last three equations represent a state of matters which may be approximated to, but not absolutely realised.

Equation (25) gives the velocity with which a piston in a tube is to be moved inwards or outwards, as the case may be, in order to produce a change of pressure from P to p, travelling along the tube from the piston towards the farther end. Equation (25) may be converted into a quadratic equation, for finding p in terms of u; in other words, for finding what pressure must be applied to a piston in order to make it move at a given speed along a tube filled with a perfect gas, whose undisturbed pressure and bulkiness are P and S. The quadratic equation is as follows:

$$p^2 - \left(2 \mathrm{P} + \frac{\gamma + 1}{2 \mathrm{S}} u^2 \right) p - \frac{\gamma - 1}{2} \cdot \frac{\mathrm{P} u^2}{\mathrm{S}} + \mathrm{P}^2 = 0;$$

and its alternative roots are given by the following formula:

$$p = \mathrm{P} + \frac{\gamma + 1}{4 \mathrm{S}} u^2 \pm \sqrt{\left\{ \frac{\gamma \mathrm{P} u^2}{\mathrm{S}} + \frac{(\gamma + 1)^2 u^4}{16 \mathrm{S}^2} \right\}} \quad . \quad (29.)$$

The sign + or − is to be used, according as the piston moves inwards, so as to produce condensation, or outwards so as to produce rarefaction. Suppose, now, that in a tube of unit area, filled with a perfect gas whose undisturbed pressure and volume are P and S, there is a piston dividing the space within that tube into two parts, and moving at the uniform velocity u: condensation will be propagated from one side of the piston, and rarefaction from the other; the pressures on the two sides of the piston will be expressed by the two values of p in equation (29); and the force required in order to keep the piston in motion will be the difference of these values; that is to say,

$$\Delta p = 2u \cdot \sqrt{\left\{\frac{\gamma P}{S} + \frac{(\gamma+1)^2 u^2}{16 S^2}\right\}} \quad . \quad . \quad . \quad (30.)$$

Two limiting cases of the last equation may be noted: first, if the velocity of the piston is very small compared with the velocity of sound, that is, if $\frac{S u^2}{\gamma P}$ is very small, we have

$$\Delta p \text{ nearly} = 2u \cdot \sqrt{\left(\frac{\gamma P}{S}\right)}; \quad . \quad . \quad (30\text{ A.})$$

secondly, if the velocity of the piston is very great compared with the velocity of sound, that is, if $\frac{\gamma P}{S u^2}$ is very small, we have

$$\Delta p \text{ nearly} = \frac{(\gamma+1) u^2}{2 S}. \quad . \quad . \quad . \quad (30\text{ B.})$$

12. *Absolute Temperature.*—The absolute temperature of a given particle of a given substance, being a function of the pressure p and bulkiness s, can be calculated for a point in a wave of disturbance for which p and s are given. In particular, the absolute temperature in a perfect gas is given by the following well-known thermodynamic formula:

$$\tau = \frac{p s}{(J c_p - c_s)}; \quad . \quad . \quad . \quad . \quad (31.)$$

and if, in that formula, there be substituted the value of s in terms of p, given by equations (8) and (18) combined, we find, for the absolute temperature of a particle at which the pressure is p, in a wave of permanent type, the following value:

$$\tau = \frac{P S}{J(c_p - c_s)} \cdot \frac{(\gamma+1)(p_1+p_2)p - 2p^2}{(\gamma+1)(p_1+p_2)P - 2P^2}; \quad . \quad (32.)$$

in which the first factor $\frac{P S}{J(c_p - c_s)}$ is obviously the *undisturbed* value of the absolute temperature. For brevity's sake let this be denoted by T.

The following particular cases may be noted. In a wave of oscillation, as defined in sec. 10, we have $p_1 + p_2 = 2P$; and, consequently,

$$\tau = T \cdot \frac{(\gamma+1)Pp - p^2}{\gamma P^2}. \quad . \quad . \quad . \quad (32\text{ A.})$$

In a wave of permanent condensation or rarefaction, as described in sec. 11, let $p_1 = P$, $p_2 = P$; then the final temperature is

$$\tau = T \cdot \frac{(\gamma + 1) P p + (\gamma - 1) p^2}{(\gamma + 1) P p + (\gamma - 1) P^2}.$$

13. *Types of Disturbance capable of Permanence.*—In order that a particular type of disturbance may be capable of permanence during its propagation, a relation must exist between the temperatures of the particles and their relative positions, such that the conduction of heat between the particles may effect the transfers of heat required by the thermodynamic conditions of permanence of type stated in sec. 6.

During the time occupied by a given phase of the disturbance in traversing a unit of mass of the cylindrical body of area unity in which the wave is travelling, the quantity of heat received by that mass, as determined by the thermodynamic conditions, is expressed in dynamical units by

$$\tau \, d \, \phi.$$

The time during which that transfer of heat takes place is the reciprocal $\frac{1}{m}$ of the mass-velocity of the wave. Let $\frac{d \tau}{d x}$ be the rate at which temperature varies with longitudinal distance, and k the conductivity of the substance, in dynamical units; then the same quantity of heat, as determined by the laws of conduction, is expressed by

$$\frac{1}{m} \cdot d \left(k \frac{d \tau}{d x} \right).$$

The equality of these two expressions gives the following general differential equation for the determination of the types of disturbance that are capable of permanence:

$$m \tau d \cdot \phi = d \cdot \left(k \frac{d \tau}{d x} \right). \quad . \quad . \quad . \quad (33.)$$

The following are the results of two successive integrations of that differential equation:

$$\frac{d x}{d \tau} = \frac{k}{A + m \int \tau \, d \phi}, \quad . \quad . \quad . \quad (33\text{ A.})$$

$$x = B + \int \frac{k \, d \tau}{A + m \int \tau \, d \phi}; \quad . \quad . \quad (33\text{ B.})$$

in which A and B are arbitrary constants. The value of A depends on the magnitude of the disturbance, and that of B upon the position of the point from which x is reckoned. In applying these general equations to particular substances, the values of τ and ϕ are to be expressed in terms of the pressure p, by the aid of the formulæ of the preceding section, when equation (33 B) will give the value of x in terms of p, and thus will show the type of disturbance required.

Our knowledge of the laws of the conduction of heat is not yet sufficient to enable us to solve such problems as these for actual substances with certainty. As a hypothetical example, however, of a simple kind, we may suppose the substance to be perfectly gaseous and of constant conductivity. The assumption of the perfectly gaseous condition gives, according to the formulæ of the preceding sections,

$$\tau = \frac{\mathrm{P\,S}}{(\gamma - 1)\,\mathrm{J}\,c_s} \cdot \frac{(\gamma + 1)\,(p_1 + p_2)\,p - 2\,p^2}{(\gamma + 1)\,(p_1 + p_2)\,\mathrm{P} - 2\,\mathrm{P}^2}$$

and

$$\tau\,d\,\phi = \frac{\gamma + 1}{m^2\,(\gamma - 1)} \left\{ \frac{p_2 + p_1}{2} - p \right\} d\,p.$$

It is unnecessary to occupy space by giving the whole details of the calculation; and it may be sufficient to state that the following are the results. Let

$$p - \frac{p_1 + p_2}{2} = q,$$

$$\frac{p_2 - p_1}{2} = q_1;$$

then

$$\frac{d\,x}{d\,p} = \frac{d\,x}{d\,q} = \frac{k}{(\gamma + 1)\,m\,\mathrm{J}\,c_s} \cdot \frac{(\gamma - 1)\,(p_1 + p_2) - 4\,q}{q_1^2 - q^2} \qquad (34.)$$

$$x = \frac{k}{(\gamma + 1)\,m\,\mathrm{J}\,c_s} \left\{ \frac{(\gamma - 1)\,(p_1 + p_2)}{2\,q_1} \,.\, \text{hyp. log.}\, \frac{q_1 + q}{q_1 - q} \right.$$

$$\left. + \, 2 \text{ hyp. log.} \left(1 - \frac{q^2}{q_1^2}\right) \right\}. \qquad . \quad . \quad . \quad (34\,\text{A.})$$

In equation (34 A) it is obvious that x is reckoned from the point where $q = 0$: that is, where the pressure $p = \frac{p_2 + p_1}{2}$; a mean between the greatest and least pressures. The direction in which x is positive may be either the same with or contrary to that of the advance of the wave; the former case represents the type of a wave of rarefaction, the latter that of a wave of compression. For the two limiting pressures when $q = \pm q_1$, $\frac{dx}{dq}$ becomes infinite, and x becomes positively or negatively infinite; so that the wave is infinitely long. The only exception to this is the limiting case, when the conductivity k is indefinitely small; and then we have the following results: when $p = p_1$, or $p = p_2$, $\frac{dx}{dp}$ is infinite, and x is indefinite; and for all values of p between p_1 and p_2, $\frac{dx}{dp}$ and x are each indefinitely small. These conditions evidently represent the case of a wave of abrupt rarefaction or compression, already referred to in secs. 6 and 7.

SUPPLEMENT (Dec., 1869).

Note as to previous investigations.—Four previous investigations on the subject of the transmission of waves of finite longitudinal disturbance may be referred to, in order to show in what respects the present investigation was anticipated by them, and in what respects its results are new.

The first is that of Poisson, in the *Journal de l'Ecole Polytechnique*, Vol. VII., Cahier 14, p. 319. The author arrives at the following general equations for a gas fulfilling Mariotte's law:—

$$\frac{d\phi}{dx} = f\left\{x - at - \frac{d\phi}{dx}t\right\},$$

$$\frac{d\phi}{dt} + a\frac{d\phi}{dx} + \frac{1}{2}\cdot\frac{d\phi^2}{dx^2} = 0;$$

in which ϕ is the velocity-function; $\frac{d\phi}{dx}$ the velocity of disturbance, at the time t, of a particle whose distance from the origin is x; a is the limit to which the velocity of propagation of the wave approximates when $\frac{d\phi}{dx}$ becomes indefinitely small, viz. $\sqrt{\frac{dp_0}{d\rho_0}}$, p_0 being the undisturbed pressure and ρ_0 the undisturbed density; and f denotes an arbitrary function. This equation obviously indicates the quicker propagation of the parts of the wave where the disturbance is forward (that is, the compressed parts), and the slower propagation of the parts where the disturbance is backward (that is, the dilated parts).

The second is that of Mr. Stokes, in the *Philosophical Magazine* for November, 1848, 3rd series, Vol. XXXIII., p 349, in which that author shows how the type of a series of waves of finite longitudinal disturbance in a perfect gas alters as it advances, and tends ultimately to become a series of sudden compressions followed by gradual dilatations.

The third is that of Mr. Airy, Astronomer-Royal, in the *Philosophical Magazine* for June, 1849, 3rd series, Vol. XXXIV., p. 401, in which is pointed out the analogy between the above-mentioned change of type in waves of sound, and that which takes place in sea-waves when they roll into shallow water.

The fourth and most complete, is that of the Rev. Samuel Earnshaw, received by the Royal Society in November, 1858, read in January, 1859, and published in the *Philosophical Transactions* for 1860, page 133. That author obtains exact equations for the propagation of waves of finite longitudinal disturbance in a medium in which the pressure is any function of the density; he shows what changes of type, of the kind already mentioned, must go on in such waves; and he points out, finally, that in order that the type may be permanent $\rho^2 \frac{dp}{d\rho}$ ($= -\frac{dp}{ds}$ in the notation of the present paper) must be a constant quantity; being the proposition which is demonstrated in an elementary way near the beginning of the present paper. Mr. Earnshaw regards that condition as one which cannot be realised.

The *new results*, then, obtained in the present paper may be considered to be the following:—The conditions as to transformation and transfer of heat which must be fulfilled, in order that permanence of type may be realised, exactly or approximately; the types of wave which enable such conditions to be fulfilled, with a given law of the conduction of heat; and the velocity of advance of such waves.

The *method of investigation* in the present paper, by the aid of *mass-velocity* to express the speed of advance of a wave is new, so far as I know; and it seems to me to have great advantages in point of simplicity, enabling results to be demonstrated in a very elementary manner, which otherwise would have required comparatively long and elaborate processes of investigation.

XXXIII.—ON THE THEORETICAL LIMIT OF THE EFFICIENCY OF PROPELLERS.*

1. THE following statement of a certain theoretical limit towards which the efficiency of propellers may be made to approximate by mechanical improvements, and of certain causes which make the actual efficiency fall short of that limit, although it involves no new principle, may be useful in the present state of the question of propulsion.

To avoid complexity, let the water be still when the action of the propelling apparatus begins; so that its velocity relatively to the vessel (which may be called the *velocity of feed* of the propelling apparatus), is simply equal and opposite to the speed of the vessel. Let that velocity be denoted by v.

2. Let s be the *true slip*, or *acceleration*, or additional velocity, impressed on the water by the propelling apparatus; so that $v + s$ is what may be called the *velocity of discharge* from the propelling apparatus, relatively to the vessel.

3. If W is the weight of the mass of water acted upon in each second, and g the acceleration produced by gravity in one second, the *reaction* of the water, equal and opposite to the resistance of the ship, is well known to be given by the following formula,

$$\mathrm{R} = \frac{\mathrm{W}\,s}{g}; \quad . \quad . \quad . \quad . \quad (1.)$$

and the *effective power*, or *useful work per second*, done in driving the ship, by the formula,

$$\mathrm{R}\,v = \frac{\mathrm{W}\,v\,s}{g}. \quad . \quad . \quad . \quad . \quad (2.)$$

4. When the apparatus first takes up a supply of water, then carries it for a time along with the vessel, and then discharges it, the reaction R may be the resultant of a forward reaction $\frac{\mathrm{W}}{g}(v + s)$ exerted by the

* From *The Engineer* of Jan. 11, 1867.

water when discharged, and a backward reaction— $\frac{W v}{g}$, exerted by the water when received; but in this, as in other cases, the resultant reaction is $\frac{W s}{g}$.

5. In order that the loss of work may be the least possible, the propelling instrument should be so contrived as to act on each particle of water with a velocity at first simply equal to the velocity of feed v, and gradually increasing at an uniform rate up to the velocity of discharge $v + s$. If this condition were fulfilled, the *mean velocity* with which the propelling apparatus would have to work against the reaction R would be $v + \frac{s}{2}$; and the *total work per second* would be

$$R\left(v + \frac{s}{2}\right) = \frac{W v s}{g} + \frac{W s^2}{2 g}; \quad . \quad . \quad (3.)$$

in which equation the first term is the useful work per second, as already given in equation (2), and the second term is the *lost work, reduced to a minimum;* for it is easy to see that this lost work is simply the actual energy of the discharged water, moving astern with the velocity s relatively to still water; and that quantity of energy must necessarily be lost under all circumstances.

6. The corresponding value of the efficiency, or ratio of the useful to the total work, is

$$\frac{v}{v + \frac{s}{2}}; \quad . \quad . \quad . \quad . \quad (4.)$$

and this is the *theoretical limit to the efficiency* of a propeller.

7. It is certain that no actual propelling instrument has ever attained the limit of efficiency stated above. It is probable that the nearest approach to the theoretical limit of efficiency is made by the oar; for the skilful rower pulls with a nearly uniform force, and thus produces a gradual acceleration of the water laid hold of by the blade.

In the following articles are described some causes of additional loss of work, irrespective of friction. Those causes may be briefly enumerated thus:—Suddenness of change from the velocity of feed to the velocity of discharge; transverse motions impressed on the water; and waste of the energy of the feed water; and the effect of each of them is to waste work in the production of eddies.

8. Suddenness of the change from the velocity of feed to the velocity of discharge operates to the full extent in every case in which the propeller, instead of beginning its action with the velocity of feed v, and

gradually increasing its speed to the velocity of discharge $v + s$, acts throughout with the velocity of discharge $v + s$. Thus the total work per second becomes

$$\mathrm{R}(v + s) = \frac{\mathrm{W}\, v\, s}{g} + \frac{\mathrm{W}\, s^2}{g}; \qquad (5.)$$

so that the lost work, instead of being simply equal to the actual energy of the water discharged per second, is increased to double that quantity of energy; and thus besides the unavoidable loss of work, there is a *waste*, or unnecessary loss of work per second, expressed by $\frac{\mathrm{W}\, s^2}{2\, g}$. The corresponding value of the efficiency is

$$\frac{v}{v + s} \qquad (6.)$$

The object of such inventions as Woodcroft's gaining pitch screw, and Mangin's screw, is to diminish waste of the kind that has now been described; and in Ruthven's form of centrifugal pump the same principle appears to be kept in view. The same is also the object of making paddles feather, so as to enter the water edgewise. It is probable that the object is partly attained by all those inventions, but by none of them wholly; and such being the case, the loss of work may be expressed by $\frac{c\, \mathrm{W}\, s^2}{2\, g}$; c being a multiplier not exceeding unity, depending on the mode of action of the particular propeller employed. It is probable that in a well-designed centrifugal pump c may be very small; while for ordinary paddles and screws it is $= 1$.

9. Transverse motions are impressed on the discharged water by all forms of the screw and paddle.

Let u denote the transverse component (whether vertical, horizontal, or inclined) of the velocity of the discharged water. Then, if that motion is impressed gradually, the work wasted per second in producing it is $\frac{\mathrm{W}\, u^2}{2\, g}$, and if more or less suddenly $(1 + c_1)\, \frac{\mathrm{W}\, u^2}{g}$; c_1 being a multiplier not exceeding unity; and the latter is the more common case. The jet propeller is free from this cause of waste of work.

10. Waste of the energy of the feed water may occur in those cases in which the water acted upon by the propelling apparatus is received into the vessel, and carried along with her before being discharged; that is to say, in certain forms of jet propeller. The feed water has, relatively to the ship, the velocity v; and in order that the energy due to that relative velocity may not be wasted, it is necessary either that each particle of

water should begin to be acted upon by the propelling apparatus without losing any part of that relative velocity (as in the case of the screw and the paddle), or that any loss of velocity should be compensated by a corresponding increase of pressure, to co-operate with the propelling apparatus in producing the velocity of discharge $v + s$. For example, if the feed water is taken into a space in which it is sensibly at rest relatively to the ship, it should produce by its impulse on the water previously in that space the whole head of pressure due to its relative velocity $\frac{v^2}{2g}$, otherwise energy will be wasted in producing eddies in the confined water, to an amount per second which may be expressed by $\frac{f W v^2}{2g}$; f being a multiplier whose value may range from an insensibly small fraction to unity, according to the degree of suddenness with which the velocity of feed is checked.

11. The multiplier f may even take values greater than unity, if the feed water is "throttled:" that is, if it is drawn through openings so narrow that the velocity becomes for a time greater than v, and then falls suddenly by the water entering a large receiver.

12. The following is a summary of the results arrived at in the preceding articles:—

		Ratio to the Useful Work.
A. Useful work per second, . .	$R v = \frac{W v s}{g}$	1
B. Work unavoidably lost, being the energy of the discharged water, .	$\frac{R s}{2} = \frac{W s^2}{2 g}$	$\frac{s}{2 v}$
C. Additional work wasted, through suddenness of action of the propeller on the water, . . .	$\frac{c R s}{2} = \frac{c W s^2}{2 g}$	$\frac{c s}{2 v}$
D. Work wasted through transverse motion of the water if produced gradually, . .	$\frac{W u^2}{2 g}$	$\frac{u^2}{2 v s}$
E. Additional work wasted if transverse motion is produced suddenly, . . .	$\frac{c_1 W u^2}{2 g}$	$\frac{c_1 u}{2 v s}$
F. Work wasted, through loss of energy of feed water, ...	$\frac{f W v^2}{2 g}$	$\frac{f v}{2 s}$

13. The following particular case may be specially mentioned. Suppose that the velocity of discharge is impressed gradually (so that $c = 0$),

that there is no transverse motion of the discharged water (so that $u = 0$), and that all the energy due to the velocity of feed is lost (so that $f = 1$). Then the total work per second is—

$$R x\left(1 + \frac{s}{2v} + \frac{v}{2s}\right) = \frac{W (x + s)^2}{2g} \quad . \quad . \quad (7.)$$

being the actual energy corresponding to the velocity of discharge; and the corresponding efficiency is—

$$\frac{2 v s}{(v + s)^2} \quad . \quad . \quad . \quad . \quad . \quad (8.)$$

In this case the lost work becomes a minimum, and the efficiency a maximum* when $s = v$; and such is very nearly the case in the "Nautilus" and the "Waterwitch."

14. In the following example the data assumed are—

W = 5 tons, or 11,200 lbs. per second:

x = 15 ft. per second; s = 15 ft. per second;

so that $x + s = 30$ ft. per second, and $R = 5217$ lbs.; the velocity is supposed to be impressed gradually; and $u = 0$. Then—

	Ratio to Useful Work.	Foot-pounds per second.	H.P.
A. Useful work, ...	1	78,255	142
B. Necessary loss of work, ...	½	39,127½	71
F. Additional loss if energy of feed water is wholly wasted, ...	½	39,127½	71
Total work, including the above losses, but exclusive of friction (the efficiency being 0·5), ...	2	156,510	284
C. Additional loss if the velocity s is impressed suddenly, ...	½	39,127½	71
Total work with that addition, but still exclusive of friction (the efficiency being 0·4), ...	2½	195,637½	355

* This case of maximum efficiency has been pointed out by Mr. R. D. Napier. See *Engineer*, November 30, 1866, page 424.

15. As another example, let—

$$W = 10 \text{ tons, or } 22{,}400 \text{ lbs. per second;}$$

$$x = 15 \text{ ft. per second; } s = 7\tfrac{1}{2} \text{ ft. per second;}$$

(so that $v + s = 22\frac{1}{2}$ ft. per second, and $R = 5217$ lbs., as before); let $u = 0$; and let the velocity s be impressed suddenly. Then—

	Ratio to Useful Work.	Foot-pounds per second.	H.P.
A. Useful work, ...	1	78,255	142
B. Necessary loss of work, . . .	$\frac{1}{4}$	19,563$\frac{3}{4}$	35$\frac{3}{4}$
C. Additional loss through suddenness of action, ...	$\frac{1}{4}$	19,563$\frac{3}{4}$	35$\frac{3}{4}$
Total work, exclusive of friction (the efficiency being 0 07), . . .	1$\frac{1}{2}$	117,382$\frac{1}{2}$	213$\frac{1}{2}$

XXXIV.—REPORT ON THE DESIGN AND CONSTRUCTION OF MASONRY DAMS.*

1. *Subjects of Report.*—I have carefully considered the letter of Captain Tulloch, R.E., Executive Engineer of the Municipality of Bombay, dated the 10th December, 1870, on the subject of masonry dams or reservoir walls of great height, and also the papers on the same subject by M. Graeff and by M. Delocre, which appeared in the *Annales des Ponts et Chaussées*. These last I have studied both in the original and in the very faithful translation by Mr. J. G. Fife. I have also made mathematical investigations as to the proper figure and dimensions of such dams, which are given in an appendix to this report.

2. *Material.*—As regards the material best suited for a reservoir wall or embankment, I consider that it must be determined by the nature of the foundation. That foundation should be sound rock, if practicable; and should a rock foundation be unattainable, firm impervious earth. It may be doubted whether any earthen foundation is thoroughly to be relied on where the depth of water exceeds 100 or 120 feet. It is not advisable to build a high masonry dam on an earthen foundation; for the base of the dam must be spread to a width sufficient to distribute the pressure, so that it shall not be more intense than the earthen foundation can bear; and this involves the use of a quantity of material which would lead to immoderate expense, if the material used were masonry.

3. *Mode of Building.*—In the case of a rock foundation, the proper material is unquestionably rubble masonry, laid in hydraulic mortar; and the opinion of M. Graeff that continuous courses in building that masonry are to be avoided, is fully corroborated by experience; for the bed-joints of such courses tend to become channels for the leakage of the water.

4. *Precaution.*—The very fact, however, of the irregular structure of that masonry renders necessary unusual care and vigilance in superintending its erection, in order to insure that every stone shall be thoroughly and firmly bedded, and that there shall be no empty hollows in the interior of the

* From *The Engineer* for Jan. 5, 1872.

wall, nor spaces filled with mortar alone where stones ought to be placed. The practice of "grouting," or filling hollows by pouring in liquid mortar, should be strictly prohibited. Should it be resolved to insert in the face of the wall headers, or long bond-stones, with or without projecting ends to form corbels, as in the dam of the river Furens, those stones ought to be laid with their lengths *not horizontal*, but *normal to the face of the wall.*

5. *Principles determining Profile.*—With respect to the profile of the wall, its figure is in the main to be determined by principles nearly the same with those laid down by the French engineers already referred to, and put in practice in the dams of the rivers Furens and Ban; that is to say, the intensity of the vertical pressure at the inner face of the wall should at no point exceed a certain limit when the reservoir is empty, and the intensity of the vertical pressure at the outer face of the wall should at no point exceed a certain limit when the reservoir is full.

6. *Limits of Vertical Pressure.*—In the theoretical investigations of M. Delocre, and the practical examples given by M. Graeff, the same limit is assigned to the intensity of the vertical pressure at both faces of the wall. But it appears to me that there are the following reasons for adopting a lower limit at the outer than at the inner face. The direction in which the pressure is exerted amongst the particles close to either face of the masonry, is necessarily that of a tangent to that face; and, unless the face is vertical, the vertical pressure found by means of the ordinary formula is not the whole pressure, but only its vertical component; and the whole pressure exceeds the vertical pressure in a ratio which becomes the greater the greater the "batter," or deviation of the face from the vertical. The outer face of the wall has a much greater batter than the inner face; therefore, in order that the masonry of the outer face may not be more severely strained when the reservoir is full, than that of the inner face when the reservoir is empty, a lower limit must be taken for the intensity of the vertical pressure at the outer face than at the inner face.

7. *Weight of Wall to be Thrown Inwards.*—The proposal of the executive engineer to throw the weight of the wall farther inwards than in the French designs, tends to realise the principles just stated, and so far I fully approve of it, and have carried it out in the profile which accompanies this report.

8. *Wall not to Overhang Inwards.*—I do not, however, concur with the executive engineer in the proposal to throw the weight of the wall so far inwards as to make it overhang, for the following reason—the additional stability against the horizontal thrust of the water gained by giving the wall an overhanging batter inwards, is not that due to the whole weight of the overhanging masonry, but only to the excess of that weight above the weight of water which it displaces; in other words, about half the effect of the weight of the overhanging mass of masonry in giving stability is

lost through its buoyancy, and hence the additional stability gained by making the wall overhang inwards is not proportionate to the additional load thrown upon the lower parts of the inner face; and more stability would be gained by placing a given mass of masonry, so as to form an uniform addition to the thickness of the wall, than by making it overhang inwards.

9. *Limits of Vertical Pressure, how Fixed.*—In choosing limits for the intensity of the vertical pressure at the inner and outer faces of the wall represented by the accompanying profile, I have not attempted to deduce the ratio which those quantities ought to bear to each other from the theory of the distribution of stress in a solid body; for the data on which any such theoretical determination would have to be based are too uncertain. The limits which I have chosen are as follows, and they are given, in the first place, in feet of a vertical column of masonry whose weight would be equivalent to the pressure, and are then reduced to various other measures:—

Limits of vertical pressure at	Inner face.	Outer face.
Feet of masonry, ...	160	125
Feet of water, ...	320	250
Pounds on the square foot (nearly), .	20,000	15,625
Metres of masonry (nearly), . .	49	38
Metres of water (nearly), . . .	98	76
Kilog. on the square centimetre (nearly),	9·8	7·6

In choosing these two limits I have been guided by the consideration of the following facts. As regards the inner face, where the deviation of the direction of the stress from the vertical is unimportant, it is certain, from practical experience, that rubble masonry laid in strong hydraulic mortar, and on good rock foundations, will safely bear a vertical pressure equivalent to the weight of a column of masonry 160 feet high, if not higher. As regards the outer face, the practical data given by M. Graeff show that masonry of the same quality in the sloping outer face of a dam will safely bear a pressure whose vertical component, as found by the ordinary rules, is equivalent to the weight of a column 125 feet high.

10. *Diminution of Vertical Pressure towards Foot of Slope.*—The same reasons which show that the intensity of the vertical component of the pressure ought to be less for a battering than for a vertical face, show also that this intensity ought gradually to diminish at the lower part of the outer face, where the batter gradually increases. In the present state of our knowledge we should not be warranted in forming any definite theory as to the law which this diminution ought to follow; and, there-

fore, in preparing the accompanying design, I have thought it best to be guided in this, as in the previous case, by practical examples, and to consider it sufficient to make the law of diminution such, that at the depth of 150 feet below the surface, the intensity of the vertical component of the pressure at the outer face becomes nearly equal to what it is

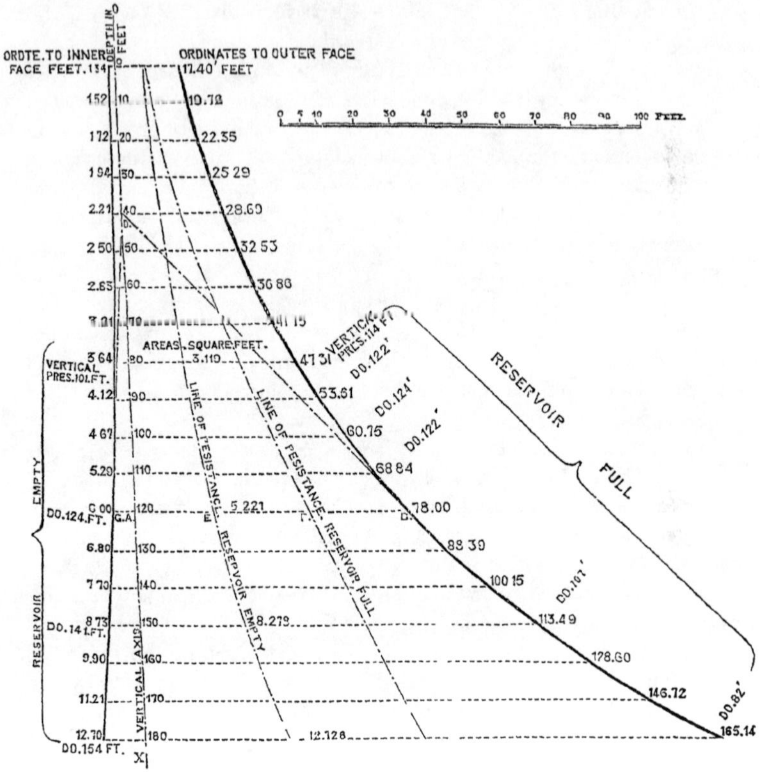

at the same depth in the outer face of the dam across the Furens—viz., 107 feet of masonry, or about 6½ kilogrammes on the square centimetre.

11. *Tension to be avoided.*—I have kept in view another principle, not referred to by the French authors—viz., that there ought to be no practically appreciable tension at any point of the masonry, whether at the outer face when the reservoir is empty, or at the inner face when the reservoir is full. Experience has shown that in structures of brickwork and masonry that are exposed to the overturning action of forces which fluctuate in amount and direction (as when a factory chimney is exposed to the pressure of the wind), the tendency to give way first shows itself

at that point at which the tension is greatest. In order that this principle may be fulfilled, the line of resistance should not deviate from the middle of the thickness of the wall to an extent materially exceeding one-sixth of the thickness. In other words, the lines of resistance when the reservoir is empty and full respectively, should both lie within, or but a small distance beyond, the middle third of the thickness of the wall.

12. *Horizontal Curvature of Wall.*—As regards the effect of giving the wall a curvature in plan convex towards the reservoir, I look upon this as a desirable, and in many cases an essential precaution, in order to prevent the wall from being bent by the pressure of the water into a curved shape concave towards the water, and thus having its outer face brought into a state of tension horizontally, which would probably cause the formation of vertical fissures, and perhaps lead to the destruction of the dam. I consider, however, that calculations of stability which treat the dam as a horizontal arch are so uncertain as to be of very doubtful utility; and I would not rely upon them in designing the profile. In fixing the radius of horizontal curvature, I consider that the engineer should be guided by the form of the gorge in which the dam is to be built, making that radius as short as may be consistent with convenience in execution, and with making the ends of the dam abut normally against the sound rock at the sides of the gorge.

13. *Summary of Conditions to be fulfilled by Profile; Logarithmic Curves chosen.*—The conditions which have been observed in designing the accompanying profile may be summed up as follows:—A. The vertical pressure at the inner face not to exceed 160 feet of masonry. B. The vertical pressure at the outer face not to exceed 125 feet of masonry at the point where it is most intense, and to diminish in going down from that point. C. The lines of resistance when the reservoir is full and empty respectively, to lie within or near to the middle third of the thickness of the wall. These are limiting conditions, and do not prescribe exactly any definite form. In choosing a form in order to fulfil them without any practically important excess in the expenditure of material beyond what is necessary, I have been guided by the consideration that a form whose dimensions, sectional area, and centre of gravity under different circumstances, are found by short and simple calculations, is to be preferred to one of a more complex kind, when their merits in other respects are equal; and I have chosen logarithmic curves for both the inner and the outer faces.

14. *Rule as to Thicknesses.*—The constant subtangent common to both curves (marked A D in the figure) is 80 feet; this bears relations to the vertical pressures which are stated in the appendix. The thickness C B at 120 feet below the top is 84 feet; and of this one-fourteenth, A C = 6 feet, lies inside the vertical axis O X, and thirteen-fourteenths,

AB = 78 feet, outside that axis. The formula for the thickness t at any depth x below the top is as follows:

$$t = t_1 e^{\frac{x - x_1}{a}}; \quad . \quad . \quad . \quad . \quad (1.)$$

or, in *common* logarithms,

$$\log. t = \log. t_1 + 0{\cdot}4343 \frac{x - x_1}{a} \quad . \quad . \quad (1\ \text{A}.)$$

in which a denotes the subtangent (80 feet), and t_1 the given thickness (84 feet) at the given depth ($x_1 = 120$ feet) below the top. The thickness at the top is 18·74 feet.

15. *Horizontal Ordinates.*—In the profile, horizontal ordinates are drawn at every 10 feet of depth, from the top down to 180 feet, and their lengths, from the vertical axis OX to the inner and outer faces respectively, are marked in feet and decimals. In each case those ordinates are respectively one-fourteenth and thirteen-fourteenths of the thickness. Intermediate ordinates, at intervals of 5 feet, can easily be calculated, if required, by taking mean proportionals between the adjacent pairs of ordinates at the intervals of 10 feet.

16. *Sectional Areas.*—The sectional area of the wall, from the top down to any given depth, is found by multiplying the constant subtangent ($a = 80$ feet) by the difference ($t - t_0$) between the thicknesses at the top and at the given depth; that is to say,

$$\int_0^x t\, dx = a\,(t - t_0). \quad . \quad . \quad . \quad . \quad (2.)$$

17. *Line of Resistance when Reservoir is Empty.*—The vertical line through the centre of gravity of the part of the wall above a given horizontal plane, stands midway between the middle of the thickness at the given horizontal plane and the middle of the thickness at the top of the wall; and thus have been found points in the curve marked "Line of resistance, reservoir empty."

18. *Moment of Pressure of Water.*—Supposing the reservoir filled to the level of the top of the wall, the moment of the pressure exerted horizontally by the water against each unit of length of wall, from the top down to a given depth (x), is found by multiplying the weight of a cubic unit of water by one-sixth of the cube of the depth; and if we take, for convenience, the weight of a cubic unit of masonry as the unit of weight, and suppose the masonry to have twice the heaviness of water, this gives us, for the moment of horizontal pressure

$$M = \frac{x^3}{12}. \quad . \quad . \quad . \quad . \quad (3.)$$

19. *Line of Resistance when Reservoir is Full.*—The moment of horizontal pressure, expressed as above stated, being divided by the area of cross-section above the given depth, gives the horizontal distance at the given depth between the lines of resistance with the reservoir empty and full respectively; that is to say,

$$\frac{M}{\int t\,dx} = \frac{x^3}{12\,a\,(t - t_0)}, \quad \ldots \quad (4.)$$

and thus have been found points in the curve marked "Line of resistance, reservoir full."

20. *Vertical Component of Water-Pressure neglected.*—In the preceding formulæ the pressure of the water against the inner face of the wall is treated as if it were wholly horizontal (as in the investigations of M. Graeff and M. Delocre). In fact, however, that pressure, being normal to the inner face of the wall, has a small inclination downwards, and, therefore, contains a small vertical component, which adds to the stability of the wall. The neglect of that vertical component is an error on the safe side.

21. *Intensity of Vertical Pressure in Masonry.*—To find the mean intensity of the vertical pressure on a given horizontal plane in the masonry, expressed in feet of masonry, divide the sectional area by the thickness at the given plane; that is to say,

$$\int \frac{t\,dx}{t} = a\left(1 - \frac{t_0}{t}\right). \quad \ldots \quad (5.)$$

To find the greatest intensity of that vertical pressure, according to the ordinary assumption that it is an *uniformly varying stress*—in other words, that it increases at an uniform rate from the face farthest from the line of resistance to the face nearest to that line, the mean intensity is to be increased by a fraction of itself expressed by the ratio which the deviation of the line of resistance from the middle of the thickness bears to one-sixth of the thickness; that is to say, let p denote that greatest intensity, expressed in feet of masonry, and x the deviation of the line of resistance from the middle of the thickness; then,

$$p = a\left(1 - \frac{t_0}{t}\right)\left(1 + \frac{6\,r}{t}\right). \quad \ldots \quad (6.)$$

When that deviation is appreciably greater than one-sixth of the thickness, the preceding rule is no longer applicable; but this case, as already explained, ought not to occur in a reservoir wall. The assumption on which this rule is based, of an uniform rate of variation of that component of the pressure which is normal to the pressed surface, is known to be sensibly correct in the case of beams, and is probably very near the truth in walls of uniform or nearly uniform thickness. Whether, or to what extent, it deviates from exactness in walls of varying thickness is uncertain in the present state of our experimental knowledge.

22. *Profiles for Different Depths.*—The range of different depths to which the same profile is applicable without any waste of material extends from the greatest depth shown on the figure 180 feet, up to 110 feet or thereabouts. For depths between 110 feet and 80 or 90 feet, or thereabouts, the waste of material is unimportant. For depths to any considerable extent less than 90 feet, the use of a part of the same profile gives a surplus of stability. For example, if the depth be 50 feet, the quantity of material is greater than that which is necessary in the ratio of 1·4 to 1 nearly. For the shallow parts, however, at the ends of a dam that is deep in the centre, I think it preferable to use the same profile as in the deep parts, notwithstanding this expenditure of material, in order that the full advantage of the abutment against the sides of the ravine may be obtained. In the case of a dam that is less deep in the centre than 120 feet, the following rule may be employed: construct a profile similar to that suited to a depth of 120 feet, with all the thicknesses and ordinates diminished in the same proportion with the depth. The intensity of the vertical pressure at each point will be diminished in the same proportion also, but this does not imply waste of material, the whole strength of the material being required in order that there will be no appreciable tension in any part of the wall.

APPENDIX.

Mathematical Principles of the Profile Curves.

I. *Principles Relating to all Forms of Profile.*—Let t, as before, be the thickness of the wall in a horizontal plane at the depth x below the top; then, taking the weight of a cubic unit of masonry as the unit of weight, the weight of each unit of length of the wall above that plane is expressed by

$$\int_0^x t\,dx.$$

In order that there may be no appreciable tension at the outer edge of the given plane when the reservoir is empty, nor at the inner edge when it is full, the centre of resistance of that plane ought not to deviate from the middle of the thickness by more than about one-sixth of the thickness inwards when the reservoir is empty, outwards when it is full.

Let y denote the deviation of the centre line of the thickness of the wall outwards from a vertical axis O X; so that $y - \frac{t}{2}$ and $y + \frac{t}{2}$ are the ordinates of the inner and outer faces of the wall respectively; and when $x = 0$, let $y = y_0$. The line of resistance when the reservoir is empty cuts the horizontal plane at the depth x, in a point vertically below the centre of gravity of the part of the wall above that plane; and in order that the weight of the wall may be thrown as far inwards as is consistent with there being no appreciable tension at the outer face when the reservoir is empty, the deviation of that line of resistance from the middle of the thickness of the wall ought not materially to exceed one-sixth of the thickness; hence, if x_1 be taken to denote the inward deviation in question,

$$x_1 = y - \frac{\int_0^x y\,t\,d\,x}{\int_0^x t\,d\,x} = \text{or} < \frac{t}{6} \text{ nearly.} \qquad \text{(A.)}$$

Let w be the ratio in which the masonry is heavier than water. Then the moment of the horizontal pressure of the water above the same plane on each unit of the length of wall is,

$$\text{M} = \frac{x^3}{6\,w}.$$

The vertical component of that pressure is neglected, as explained in the body of the report. The extent to which the centre of resistance at the given horizontal plane is shifted outwards by the pressure of the water is

$$r_1 + r = \frac{\text{M}}{\int_0^x t\,d\,x} = \frac{x^3}{6\,w\int_0^x t\,d\,x}, \qquad \text{(B.)}$$

in which r denotes the outward deviation of the line of resistance from the middle of the thickness when the reservoir is full; and the condition that the centre of resistance, when the reservoir is full, is not to deviate from the middle of the thickness by more than about one-sixth of the thickness, is expressed by the following formula :—

$$r = \frac{\frac{x^3}{6\,w} + \int_0^x y\,t\,d\,x}{\int_0^x t\,d\,x} - y = \text{or} < \frac{t}{6} \text{ nearly.} \qquad \text{(C.)}$$

The formulæ (A) and (C) express the condition that there shall be no practically important tension in the masonry at any horizontal plane. Let p_1 and p be the vertical pressures at the inner and outer faces respectively at the depth x; and P_1 and P the limits which those pressures are not to exceed. Then we have, as another pair of equations to be satisfied,

$$p_1 = \left(1 + \frac{6\,r_1}{t}\right) \frac{\int_0^x t\,d\,x}{t} = \text{or} < P_1. \qquad \text{(D.)}$$

$$p = \left(1 + \frac{6\,r}{t}\right) \frac{\int_0^x t\,d\,x}{t} = \text{or} < P. \qquad \text{(E.)}$$

II. *Principles Relating to the Logarithmic-Curve Profile.*—As a means of satisfying the equations of condition to a degree of approximation sufficient for practical purposes, let the inner and outer boundaries and the centre line of the profile be all three logarithmic curves, with the vertical axis O X for their common asymptote, and having one common constant subtangent a. It may be remarked that one reason for adopting the logarithmic curve is its giving a thickness at the top of the wall sufficient for the formation of a roadway; and that another reason is, its giving values to the intensity of the pressure at the outer face below the point of maximum pressure, which diminish as the batter increases. Let the ratio borne by the deviation y of the centre line of the thickness from the vertical axis to the thickness t be expressed by $c = \frac{y}{t}$.

Then we have the following equations:—

$$t = t_0\, e^{\frac{x}{a}}. \qquad \text{(F.)}$$

$$y = c\,t = c\,t_0\, e^{\frac{x}{a}}. \qquad \text{(G.)}$$

$$\int_0^x t\,d\,x = a\,t_0\left(e^{\frac{x}{a}} - 1\right) = a\,(t - t_0). \qquad \text{(H.)}$$

$$r_1 = \frac{c\,t_0}{2}\left(e^{\frac{x}{a}} - 1\right) = c\,\frac{t - t_0}{2} = \frac{y - y_0}{2}. \qquad \text{(K.)}$$

$$r = \frac{x^3}{6\,w\,a\,(t - t_0)} - \frac{c\,(t - t_0)}{2}$$

$$= \frac{x^3}{6\,w\,a\,t_0^2\left(e^{\frac{x}{a}} - 1\right)} - \frac{c}{2}\left(e^{\frac{x}{a}} - 1\right) \quad . \quad . \quad \text{(L.)}$$

$$p_1 = a\left(1 - e^{-\frac{x}{a}}\right)\left\{1 + 3\,c\left(1 - e^{-\frac{x}{a}}\right)\right\} \quad . \quad \text{(M.)}$$

$$p = a\left\{1 - e^{-\frac{x}{a}} - 3\,c\left(1 - e^{-\frac{x}{a}}\right)^2\right\} . \quad . \quad \text{(N.)}$$

When the values given above are substituted in the expressions of conditions, A, C, D, and E, the formulæ obtained are of a kind incapable of solution by any direct process. They can, however, be solved approximately without much difficulty by the process of trial and error; and such is the method by which the dimensions of the profile sent with the report have been obtained; the constants employed being

$$w = 2;\ P_1 = 160 \text{ feet};\ P = 125 \text{ feet}.$$

The general nature of the process of approximation followed may be summed up as follows:—By making $\frac{d\,p}{d\,x} = 0$, an equation is obtained involving the value of $\frac{x}{a}$, which makes p a maximum. That equation shows that as a first approximation to that value we may take $\frac{3}{2}$. This first approximation is inserted in equation (K); and by making $r_1 = \frac{t}{6}$, there is deduced from that equation an approximate value of c. Then, in equation (M), by inserting the approximate values of $\frac{x}{a}$ and of c, and making $p_1 = P_1$ (the limit of p_1), there is obtained an approximate value of a; and by making $r = \frac{t}{6}$ in equation (L), an approximate value of t_0. The several first approximate values being then inserted in $\frac{d\,p}{d\,x} = 0$, there is obtained a corrected value of $\frac{x}{a}$, which is found to be about $\frac{11}{8}$; and thence by means of equation (N), the actual maximum value of p is computed, and found to fall slightly within the prescribed limit. Finally, as a test of the approximations, equations (K), (L), (M), and (N) are applied to a series of values of x, extending from the top to the bottom of the wall. As to the degree of approximation obtained, the greatest values p_1 and p are respectively 154 feet and 124 feet, instead of 160 feet and 125 feet; and there are, as the drawing shows, some small deviations of the lines of resistance beyond the middle third of the thickness, but not sufficient to be of practical importance.

XXXV.—ON THE APPLICATION OF BARYCENTRIC PERSPECTIVE TO THE TRANSFORMATION OF STRUCTURES.*

1. THIS paper contains the substance of some remarks which I made at the recent meeting of the British Association, on the elegant investigation by Professor Sylvester of the principles of Barycentric Perspective and Homalographic Projection.

2. In the *Proceedings of the Royal Society* for the 6th March, 1856, I published a theorem called that of "The Transformation of Structures," which may be briefly expressed as follows:—

If a structure of a given figure be balanced and stable under forces represented by given lines, then will any structure whose figure is a parallel projection of the original figure be balanced and stable under forces represented by the corresponding projections of the lines representing the original forces.

3. By a parallel projection of a figure is meant a figure derived from the original figure by altering the co-ordinates in uniform proportions, or by substituting oblique for rectangular co-ordinates; and it is called *parallel* because to every pair of equal and parallel lines in the original figure there correspond a pair of equal and parallel lines in the transformed figure. For example, every orthographic projection of a plane figure is a parallel projection; all ellipsoids are parallel projections of each other and of a sphere, &c.

4. That theorem was applied in *A Manual of Applied Mechanics* to the deduction of the figures of a skew arch and of a ramping arch from that of a common arch, of an equilibrated rib from a common catenary, of arches for supporting earth from arches for supporting the pressure of a liquid, &c.

5. Its applications, however, were limited by the condition of *parallel projection;* and there were, consequently, many conceivable transformations of structures to which it could not be applied.

6. The theorems discovered by Mr. Sylvester now afford the means of

* From the *Philosophical Magazine* for Nov., 1863.

greatly extending the art of designing structures by transformation from structures of more simple figures; for they obviously give at once the solution of the question—*given the figure of a structure which is balanced and stable under a load distributed in a given way; given also any perspective or homalographic projection of that figure; to find how the load must be distributed on the transformed structure, in order that it also may be balanced and stable.*

7. This is not the first instance in which theorems of pure science have proved to be capable of practical applications unexpected, perhaps, by their discoverers.

XXXVI.—PRINCIPLE OF THE EQUILIBRIUM OF POLYHEDRAL FRAMES.*

THE following theorem is the extension to polyhedral frames of a principle which is proved for polygonal frames in *A Manual of Applied Mechanics*, Art. 150.

THEOREM.—If planes diverging from a point or line be drawn normal to the lines of resistance of the bars of a polyhedral frame, then the faces of a polyhedron whose edges lie in those diverging planes (in such a manner that those faces, together with the diverging planes which contain their edges, form a set of contiguous diverging pyramids or wedges) will represent, and be normal to, a system of forces which, being applied to the summits of the polyhedral frame, will balance each other—each such force being applied to the summit of meeting of the bars whose lines of resistance are normal to the set of diverging planes that enclose that face of the polyhedron of forces which represents and is normal to the force in question. Also, the areas of the diverging planes will represent the stresses along the bars to whose lines of resistance they are respectively normal.

It is obvious that the polyhedron of forces and the polyhedral frame are reciprocally related as follows:—their numbers of edges are equal, and their corresponding pairs of edges perpendicular to each other; and the number of faces in each polyhedron is equal to the number of summits in the other.

* From the *Philosophical Magazine* for Feb., 1864.

XXXVII.—ON A PROPERTY OF CURVES FULFILLING THE CONDITION $\frac{d^2\phi}{dx^2} + \frac{d^2\phi}{dy^2} = 0$.*

1. In a paper "On Stream-Lines," published in the *Philosophical Magazine* for October, 1864, I stated, and, in a Supplement to the same paper, published in the *Philosophical Magazine* for January, 1865, I proved the proposition that "all waves in which molecular rotation is null, begin to break when the two slopes of the crest meet at right angles."

2. I have now to state the purely geometrical proposition of which that mechanical proposition is a consequence. *If a plane curve which fulfils the condition*

$$\frac{d^2\phi}{dx^2} + \frac{d^2\phi}{dy^2} = 0$$

cuts itself in a double point, it does so at right angles.

3. The following is the demonstration. It is well-known that the inclination of any plane curve to the axes at an ordinary point is given by the equation

$$\frac{d\phi}{dx}\,dx + \frac{d\phi}{dy}\,dy = 0;$$

also, that at a double point $\frac{d\phi}{dx}$ and $\frac{d\phi}{dy}$ both vanish, so that the inclinations of the two branches to the axes are given by the two roots of the quadratic equation

$$\frac{d^2\phi}{dx^2}\cdot dx^2 + 2\frac{d^2\phi}{dx\,dy}\cdot dx\,dy + \frac{d^2\phi}{dy^2}\cdot dy^2 = 0;$$

whence it follows that the product of the two values of $\frac{dy}{dx}$, which are the two values of the tangent of the inclination to the axis of x, is

* From the *Proceedings of the Royal Society* for 1867.

$$= \frac{\frac{d^2 \phi}{d x^2}}{\frac{d^2 \phi}{d y^2}}.$$

In a curve which fulfils the before-mentioned condition, the value of that product is -1; and when such is the case with the product of the tangents of two angles, the difference of those angles is a right angle; therefore, the two branches cut each other at right angles. Q.E.D.

4. The proposition just demonstrated is so simple and so obvious, that I was at first disposed to think it must have been known and published previously; and had I not been assured by several eminent mathematicians that it had not been previously published to their knowledge, I should not have ventured to put it forth as new.

Supplement to the preceding Paper.

Professor Stokes, D.C.L., has pointed out to me an extension of the preceding theorem—viz., *that at every multiple point in a plane curve which fulfils the condition*

$$\frac{d^2 \phi}{d x^2} + \frac{d^2 \phi}{d y^2} = 0,$$

the branches make equal angles with each other; so that, for example, if n branches cut each other at a multiple point, they make with each other $2n$ equal angles of $\frac{\pi}{n}$.

The following appears to me to be the simplest demonstration of the extended theorem: At a point where n branches cut each other the following equation is fulfilled by all curves:

$$\left(d x \frac{d}{d x} + d y \frac{d}{d y}\right)^n \phi = 0.$$

Let θ be the angle made by any branch with the axis of x; then

$$\left(\cos \theta \frac{d}{d x} + \sin \theta \frac{d}{d y}\right)^n \phi = 0.$$

But in a curve which fulfils the equation

$$\frac{d^2 \phi}{d x^2} + \frac{d^2 \phi}{d y^2} = 0,$$

we have

$$\frac{d}{d y} = \sqrt{-1} \cdot \frac{d}{d x};$$

whence it follows that in such a curve the equation of a multiple point of n branches is

$$\left\{ (\cos \theta + \sqrt{-1} \cdot \sin \theta) \frac{d}{d x} \right\}^n \phi = 0.$$

Choose for the axis of x a tangent to one of the branches at the multiple point. Then it is evident that the preceding equation is satisfied by the $2n$ values of θ corresponding to the $2n$th roots of unity; that is to say, by

$$\theta = 0, \frac{\pi}{n}, \frac{2\pi}{n}, \text{\&c.}, \quad . \quad . \quad . \quad \frac{(2n-1)\pi}{n};$$

therefore, the n branches make with each other $2n$ equal angles of $\frac{\pi}{n}$. Q.E.D.

BELL AND BAIN, PRINTERS, GLASGOW.

A SELECTION

FROM

CHARLES GRIFFIN AND COMPANY'S

CATALOGUE.

*** General Catalogue of Standard Publications, in all Departments of Literature, forwarded on Application.

10 STATIONERS' HALL COURT, LONDON.

Published with the Approval of the Director-General of Telegraphs in India.

Crown 8vo, cloth, 15s. Second Edition.

A MANUAL OF

TELEGRAPH CONSTRUCTION:

THE MECHANICAL ELEMENTS OF ELECTRIC TELEGRAPH ENGINEERING.

BY

JOHN CHRISTIE DOUGLAS,

MEM. S. C. TELEGRAPH ENGINEERS; EAST INDIA GOVT. TELEGRAPH DEPARTMENT.

THIS Work has been prepared to supply the existing want of a Text-Book on the Mechanical Principles and Practice of Telegraph Engineering, and is the only treatise in the language specially devoted to the subject. It gives the result of many years' practical experience on the part of the Author, and is intended for use by the Student as well as by the Engineer actually engaged in Design and Construction.

GENERAL CONTENTS.

PART I.—GENERAL PRINCIPLES OF STRENGTH AND STABILITY, comprising the Strength of Materials; the Distribution of Load and Stress in Telegraph Structures, such as Poles—simple, strutted, tied, stayed, coupled, and trussed; the Catenary, with application of its Formulæ to the cases of Wires and Cables; Theory of the Submersion of Cables, &c.

PART II.—PROPERTIES AND APPLICATIONS OF MATERIALS, OPERATIONS, AND MANIPULATION, including the Principles and Practice of, and Numerical Data for, designing Simple Structures, such as Poles of Iron and Wood; Iron and Wooden Masts—simple and compound; Specifications for Wire, &c.; Soldering; Surveying; the Raising of heavy Masts; Insulating Materials and their Applications, &c.

PART III.—TELEGRAPH CONSTRUCTION, MAINTENANCE, AND ORGANISATION, treating of the Application of the Information conveyed in Parts I. and II. to the case of Combined Structures, including the Construction of Overground, Subterranean, and Subaqueous Lines; Office Fittings; Estimating; Organisation, &c.

"Mr. Douglas deserves the thanks of Telegraph Engineers for the excellent 'Manual' now before us . . . he has ably supplied an existing want . . . the subject is treated with great clearness and judgment . . . good practical information given in a clear, terse style."—*Engineering.*

"Mr. Douglas's work is, we believe, the first of its kind. . . . The author is evidently a practical Telegraphic Engineer. . . . The amount of information given is such as to render this volume a most useful guide to any one who may be engaged in any branch of Electric Telegraph Engineering."—*Athenæum.*

"The book is calculated to be of great service to Telegraphic Engineers . . . the arrangement is so judicious that with the aid of the full Table of Contents, reference to any special point should be easy."—*Iron.*

LONDON: CHARLES GRIFFIN & CO., 10 STATIONERS' HALL COURT.

Published with Concurrence of the Surveyors-General of New South Wales and Victoria.

Half-Bound, Folio, Price 30s.

TRAVERSE TABLES:

Computed to 4 Places Decimals for every Minute of Angle up to 100 of Distance,

FOR THE USE OF SURVEYORS AND ENGINEERS.

BY

R. LLOYD GURDEN,

AUTHORISED SURVEYOR FOR THE GOVERNMENTS OF NEW SOUTH WALES AND VICTORIA.

—

DISTINCTIVE FEATURES:

1. The Tables are calculated to Single Minutes, and to 100 of Distance.
2. The Traverses are given to four places of decimals, so that the Sines and Cosines for a distance of twelve miles can be ascertained correctly to within half an inch.
3. Simplicity and economy of labour in calculation. ONE OPENING OF THE TABLES gives the information which, when sought by the usual method, involves four references to the book of logarithms, two additions, and the writing out of forty-eight more figures than are required in the use of Traverse Tables.

"Those who have experience in exact SURVEY-WORK will best know how to appreciate the enormous amount of labour represented by this valuable book. The computations enable the user to ascertain the sines and cosines for a distance of twelve miles to within half an inch, and this BY REFERENCE TO BUT ONE TABLE, in place of the usual fifteen minute computations required. This alone is evidence of the assistance which the Tables ensure to every user, and as every Surveyor in active practice has felt the want of such assistance, few knowing of their publication will remain without them."—*Engineer*.

"We cannot sufficiently admire the heroic patience of the author, who, in order to prevent error, calculated each result by two different modes, and, before the work was finally placed in the Printer's hands, repeated the operation for a third time on revising the Proofs."—*Engineering*.

"Mr. GURDEN is to be thanked for the extraordinary labour which he has bestowed on facilitating the work of the Surveyor. . . . An almost unexampled instance of professional and literary industry . . . When the anxious and laborious work of one man affords the means of such a saving of toil for all those who avail themselves of his work, the patient and careful tabulator deserves the name of a benefactor of his profession, and of a good servant of his fellows."—*Athenæum*.

"These Tables are characterised by ABSOLUTE SIMPLICITY, and the saving of time effected by their use is most material. . . . The Author has done much to reduce the cost and burden of the Surveyor's work. Every one connected with Engineering or Survey should be made aware of the existence of this elaborate and useful set of Tables."—*Builder*.

"From the enormous amount of time and labour which the Tables will save, they may safely be recommended to every Surveyor and Engineer."—*Mining Journal*.

"Up to the present time, no Tables for the use of Surveyors have been prepared which in minuteness of detail can be compared with those compiled by Mr. R. L. GURDEN. . . . With the aid of this book the TOIL OF CALCULATION IS REDUCED TO A MINIMUM; and not only is time saved, but the risk of error is avoided. . . . The profession is under an obligation to Mr. GURDEN for ensuring that in the calculation of triangles and traverses inaccuracies are for the future impossible. . . . Mr. GURDEN'S BOOK HAS BUT TO BE KNOWN, AND NO ENGINEER'S OR SURVEYOR'S OFFICE WILL BE WITHOUT A COPY."—*Architect*.

LONDON: CHARLES GRIFFIN & CO., 10 STATIONERS' HALL COURT.

TECHNICAL AND SCIENTIFIC PUBLICATIONS.

ELECTRO-METALLURGY.

ELECTRO-METALLURGY (A Manual of). By JAMES NAPIER, F.R.S.E., F.C.S. Comprising—

ELECTROTYPE PROCESSES.	BRONZING.
ELECTRO-PLATING.	COATING WITH COPPER.
ELECTRO-GILDING.	DEPOSITION OF OTHER METALS.

A HISTORY OF THE ART, &c., &c.

With Illustrations, crown 8vo, cloth, 7s. 6d. *Fifth Edition.*

"The FIFTH EDITION has all the advantages of a NEW WORK, and of a PROVED AND TRIED FRIEND. A work calculated to inspire invention."—*Jeweller and Watchmaker.*

DYEING.

DYEING AND DYEING RECEIPTS (A Manual of). By JAMES NAPIER, F.R.S.E., F.C.S. Comprising—

1. CHEMISTRY OF DYEING.	4. ANIMAL DYES.
2. MORDANTS AND ALTERANTS.	5. ANILINE COLOURS.
3. VEGETABLE DYES.	6. PRACTICAL MANIPULATION.

With Diagrams and 57 Specimens of Dyed Cotton, Silk, and Woollen Fabrics.

Demy 8vo, cloth bevelled, 21s. *Third Edition.*

"EXCEEDINGLY VALUABLE to the Practical Dyer. . . . A Manual of necessary reference to all who wish to keep pace with the scientific discoveries of the time."—*Journal of Applied Science.*

CHEMISTRY.

CHEMICAL RECREATIONS: A Popular Manual of Experimental Chemistry. By JOHN JOSEPH GRIFFIN, F.C.S. With 540 Engravings of Apparatus. Crown 4to, cloth. *Tenth Edition.*

Part I. ELEMENTARY CHEMISTRY. Price 2s.

Part II. THE CHEMISTRY OF THE NON-METALLIC ELEMENTS, including a Comprehensive Course of Class Experiments. Price 10s. 6d.

Or, complete in one volume, cloth, gilt top, 12*s.* 6*d.*

PRACTICAL CHEMISTRY (A Manual of): The ANALYSIS OF FOODS and the DETECTION OF POISONS. By A. WYNTER BLYTH, F.C.S., M.R.C.S, Public Analyst for the County of Devon.

Part I. FOODS: SUGAR, STARCHES; FLOUR, BREAD; MILK, BUTTER, CHEESE; TEA, COFFEE, COCOA; ALCOHOL AND ALCOHOLIC LIQUIDS; CONDIMENTS.

Part II. POISONS: ORGANIC AND INORGANIC, THEIR DETECTION AND ESTIMATION.

Crown 8vo, cloth, with Numerous Tables and Diagrams, price 12s. 6d.

"Will be used by every Analyst."—*Lancet.*

"STANDS UNRIVALLED for completeness of information."—*Sanitary Record.*

GENERAL SCIENTIFIC PUBLICATIONS.

GEOLOGY.

A MANUAL OF GEOLOGY:

PRACTICAL AND THEORETICAL.

BY

JOHN PHILLIPS, M.A., F.R.S., F.G.S.,

LATE PROFESSOR OF GEOLOGY IN THE UNIVERSITY OF OXFORD.

REVISED AND EDITED BY

ROBERT ETHERIDGE, F.R.S., F.G.S.,

OF THE MUSEUM OF PRACTICAL GEOLOGY,

AND

H. GOVIER SEELEY, F.R.S,

OF KING'S COLLEGE, LONDON.

WITH NUMEROUS ILLUSTRATIONS.

(In Preparation.)

NATURAL HISTORY.

THE STUDENT'S NATURAL HISTORY; a Dictionary of the Natural Sciences: Botany, Conchology, Entomology, Geology, Mineralogy, Palæontology, and Zoology. By W. Baird, M.D., F.L.S., late of the British Museum. With a Zoological Chart, showing the Distribution and Range of Animal Life, and over Two hundred and fifty Illustrations. Demy 8vo. Cloth gilt, 10s. 6d.

"The work is a very useful one, and will contribute, by its cheapness and comprehensiveness, to foster the extending taste for Natural Science."—*Westminster Review.*

THE NATURAL HISTORY OF THE INANIMATE CREATION, recorded in the Structure of the Earth, the Plants of the Field, and the Atmospheric Phenomena. By Professor Ansted, M.A., F.R.S. With numerous Illustrations. Large post 8vo. Cloth, 8s. 6d.

A POPULAR HISTORY OF THE ANIMAL CREATION: being a Systematic and Popular Description of the Habits, Structure, and Classification of Animals. By W. S. Dallas, F.L.S. With coloured Frontispiece and many hundred Illustrations. Crown 8vo. Cloth, 8s. 6d. *New Edition.*

GENERAL SCIENTIFIC PUBLICATIONS.

HYGIÈNE AND PUBLIC HEALTH.

A DICTIONARY OF HYGIÈNE AND PUBLIC HEALTH.

BY

ALEXANDER WYNTER BLYTH, M.R.C.S., F.C.S.,

ANALYST FOR THE COUNTY OF DEVON.

COMPRISING—

1. Sanitary Chemistry.
2. Sanitary Engineering.
3. Sanitary Legislation.
4. Epidemic and Epizootic Diseases.
5. Hygiène—Military, Naval, Private, Public, School.

Royal 8vo, cloth bevelled, with Illustrations, price 28s.

"A work of extreme value to all specially interested in Sanitation."—*Medical Times and Gazette.*

Fifteenth Edition.

DOMESTIC MEDICINE AND HOUSEHOLD SURGERY (A DICTIONARY OF). By Spencer Thomson, M.D., Edin., L.R.C.S. With Appendix on the Management of the Sick-room, and many Hints for the Diet and Comfort of Invalids.

Large 8vo, cloth, with numerous Illustrations, price 8s. 6d.

"Dr. Thomson has fully succeeded in conveying to the public a vast amount of useful professional knowledge."—*Dublin Journal of Medical Science.*

MILITARY SURGERY.

INTERNATIONAL PRIZE ESSAY.

THE SURGEON'S POCKET-BOOK: an Essay on the Best Treatment of the Wounded in War; for which a prize was awarded by Her Majesty the Empress of Germany. Specially adapted to the PUBLIC MEDICAL SERVICES. By Surgeon-Major J. H. Porter, Hon. Assoc. of the Order of St. John of Jerusalem, late Assistant-Professor of Military Surgery in the Army Medical School.

16*mo, roan, with* 152 *Illustrations, price* 7*s.* 6*d.* *Second Edition, revised and enlarged.*

"Every medical officer is recommended to have the 'Surgeon's Pocket-Book,' by Surgeon-Major Porter, accessible, to refresh his memory and fortify his judgment."—*Precis of Field-Service Medical Arrangements for Afghan War.*

"This capital little book . . . of the greatest practical value. . . . A surgeon with this Manual in his pocket becomes a man of resource at once."—*Westminster Review.*

Published under the Sanction of the National Society for Aid to the Sick and Wounded in War.

A MANUAL OF INSTRUCTION FOR ATTENDANTS ON THE SICK AND WOUNDED IN WAR. By Staff-Assistant-Surgeon A. Moffitt, of the Royal Victoria Hospital, Netley.

With numerous Illustrations, post 8vo, cloth, 5s.

GENERAL SCIENTIFIC PUBLICATIONS.

MEDICINE.

WORKS BY PROFESSOR AITKEN, M.D., F.R.S.,

Professor of Pathology in the Army Medical School;
Examiner in Medicine for the Military Medical Services of the Queen.

THE SCIENCE AND PRACTICE OF MEDICINE. In Two Volumes, Royal 8vo, cloth, with Map, and Numerous Illustrations. *Seventh Edition*, thoroughly revised, in part re-written, and greatly enlarged. Price, 42s.

"The STANDARD TEXT-BOOK in the English language. . . . There is no work more indispensable for the Practitioner and Student."—*Edin. Medical Journal.*

OUTLINES OF THE SCIENCE AND PRACTICE OF MEDICINE. A Text-Book for Students. In Crown 8vo. *Second Edition in preparation.*

"Well digested, clear, and well written, the work of a man conversant with every detail of his subject, and a thorough master of the art of teaching."—*British Medical Journal.*

THE GROWTH OF THE RECRUIT, and the Young Soldier, with a view to the Selection of "Growing Lads" and their Training. 2s. 6d.

"This little work should be in the hands of all instructors of youth, and all employers of youthful labour."—*Lancet.*

OUTLINE FIGURES OF THE TRUNK OF THE HUMAN BODY, on which to indicate the areas of physical signs in the Clinical Diagnosis of Disease. For the use of Students and Practitioners of Medicine. 1s. 6d.

IMPERFECT DIGESTION: Its Causes and Treatment. By ARTHUR LEARED, M.D., F.R.C.P., late Senior Physician to the Great Northern Hospital. Post 8vo, Cloth, 4s. 6d. *Sixth Edition.*

"It now constitutes about the best work on the subject."—*Lancet.*

"Dr. Leared has treated a most important subject in a practical spirit and popular manner."—*Medical Times and Gazette.*

"A useful manual of the subject upon which it treats, and we welcome it as an addition to our Medical Literature."—*Dublin Quarterly Journal of Medical Science.*

UNIVERSITY TEXT-BOOKS.

MR. CRUTTWELL'S CLASSICAL WORKS.

Second Edition, crown 8vo, cloth, 8s. 6d.

1. **A HISTORY OF ROMAN LITERATURE,** from the Earliest Period to the Times of the Antonines. By C. T. CRUTTWELL, M.A., Fellow of Merton College, Oxford, Head Master of Malvern College.

"Nothing at all equal to it has hitherto been published in England."—*British Quarterly Review.*

"A most serviceable—indeed indispensable—guide for the student. . . . The 'general reader' will be both charmed and instructed."—*Saturday Review.*

COMPANION VOLUME.

2. **SPECIMENS OF ROMAN LITERATURE,** from the Earliest Period to the Times of the Antonines.

PART I.—ROMAN THOUGHT—RELIGION, PHILOSOPHY AND SCIENCE, ART AND LETTERS.

PART II.—ROMAN STYLE—DESCRIPTIVE, RHETORICAL, AND HUMOROUS PASSAGES.

With SYNOPSIS and INDICES complete.

Edited by C. T. CRUTTWELL, M.A., Merton College, Oxford; and PEAKE BANTON, M.A., some time Scholar of Jesus College, Cambridge. Crown 8vo, cloth, 10s. 6d.

"A work with a standing-ground of its own. . . . Not only useful, but necessary, for many classes of readers. . . . The sound judgment exercised in arranging the plan and in the selection of the Passages calls for hearty commendation."—*Saturday Review.*

PROFESSOR RAMSAY'S CLASSICAL WORKS.

In crown 8vo, cloth.

1. **A MANUAL OF ROMAN ANTIQUITIES:** Law, Constitution, Army and Navy, Public and Social Life, Agriculture, Religion, &c. For the Use of Advanced Students. By WILLIAM RAMSAY, M.A., Trinity College, Cambridge, late Professor of Humanity in the University of Glasgow. With Map, numerous Engravings, and very copious Index. *Eleventh Edition.* 8s. 6d.

2. **AN ELEMENTARY MANUAL OF ROMAN ANTIQUITIES.** Adapted for Junior Classes. With numerous Illustrations. *Seventh Edition.* 4s.

3. **A MANUAL OF LATIN PROSODY.** Illustrated by Copious Examples and Critical Remarks. *Sixth Edition.* 5s.

DR. BRYCE'S VIRGIL.

VIRGILII OPERA. Edited by A. HAMILTON BRYCE, D.C.L., LL.D., Senior Classical Moderator in the University of Dublin. Text from HEYNE and WAGNER. English Notes, Original and Selected, from the leading German and English Commentators. Illustrations from the Antique. *Thirteenth Edition.* In 1 Vol., fcap. 8vo, cloth, 6s.; or in Three Parts:—

PART I.—BUCOLICS AND GEORGICS, 2s. 6d.
PART II.—THE ÆNEID, BOOKS I.—VI., 2s. 6d.
PART III.—THE ÆNEID, BOOKS VII.—XII., 2s. 6d.

"Contains the pith of what has been written by the best scholars on the subject."—*Athenæum.*

UNIFORM WITH THE ABOVE.

HORATII OPERA. Edited by JOSEPH CURRIE, formerly Head Classical Master in Glasgow Academy. Text from ORELLIUS. English Notes, Original and Selected, from the best Commentators. Illustrations from the Antique. In 1 Vol., fcap. 8vo, cloth, 5s.; or in Two Parts:—

PART I.—CARMINA, 3s.
PART II.—SATIRES AND EPISTLES, 3s.

"The notes are excellent and exhaustive."—*Quarterly Journal of Education.*

THE VOCABULARY OF PHILOSOPHY: Mental, Moral, and Metaphysical. With Quotations and References for the Use of Students. By WILLIAM FLEMING, D.D., late Professor of Moral Philosophy in the University of Glasgow. Revised and Edited by HENRY CALDERWOOD, LL.D., Professor of Moral Philosophy in the University of Edinburgh. Crown 8vo, cloth, bevelled, 10s. 6d. *Third Edition.*